Fundamentals of Electrical Power Systems Analysis

Md. Abdus Salam

Fundamentals of Electrical Power Systems Analysis

Springer

Md. Abdus Salam
Department of Electrical
and Computer Engineering
Faculty of Engineering
Western University
London, ON, Canada

ISBN 978-981-15-3211-5 ISBN 978-981-15-3212-2 (eBook)
https://doi.org/10.1007/978-981-15-3212-2

This Springer imprint is published by the registered company Springer Nature Singapore Pte Ltd.
The registered company address is: 152 Beach Road, #21-01/04 Gateway East, Singapore 189721, Singapore

To my parents, wife Asma Ara Bagum, children Syeed Hasan, Yusra Salam, and Sundus Salam for their constant support and patience during the writing of this book. As well as to all of my teachers and well-wishers who have helped me grow professionally over the years.

Md. Abdus Salam

Preface

Power generation, transmission, and distribution are the core topics usually taught in power system courses for electrical and electronics engineering students. In addition, the demand for renewable energy and its integration with the existing grid system is increasing and will be increasing in the future due to the depletion of conventional fuel. Therefore, undergraduate and graduate students need basic concepts to advance their knowledge in power systems. At least six credit hours are allocated in a standard electrical and electronic engineering course curriculum for the power system courses in all universities and colleges that provide this knowledge. I have taught several courses, including power transmission, power distribution, and power systems at different universities around the world. The main purpose of this book *Fundamentals of Electrical Power Systems Analysis* is to cover various aspects of power systems. This book introduces straightforward and step-by-step methods to discuss different theories along with their mathematical representation aiming at solving the problems. This book also explores a different power system software to ease calculations by exposing the underlying principles of complex power system networks.

I have targeted two main audiences for this book, namely students and researchers who are learning power systems engineering in their academic institutions and who need to enhance their understanding. Students with different backgrounds will also get fundamental knowledge of power systems through this book.

The second major audience consists of electrical engineers and other professionals working in the power industry. Other professionals may be those whose educational background is not in electrical engineering, but intend to learn the fundamental knowledge of power systems. These groups of the audience could also gain more knowledge of power systems.

I believe that this book can make a meaningful contribution to the students, researchers, and power system engineers in their practical field.

Acknowledgements

I would like to acknowledge Karin Koonings and Maureen Mallonee, Fluke Corporation, USA, and Elang Kumaran Suppiah, Jeremy Ang, Eric Ng, Jauharie Azhary, and Gracelyn Soon, Fluke South East Asia Pte Ltd, Singapore, for providing relevant information and photographs of some of their equipment and giving permission to include in this book. I would also like to acknowledge Mr. Paul Bernat, Principal Software Engineer, TNEI, 2nd Floor Bainbridge House, 86–90 London Road, Manchester, M1 2PW, UK, for giving permission to use IPSA software (latest version) in some of the worked-out example simulations. I like to acknowledge my former student Mr. Mustafa Nasir Hassan (P. Eng.), Project Management Engineer, Electric Generation Utilities Division, Medicine Hat, The Gas City, Alberta, for providing the photograph of transmission lines covered with ice. I also like to acknowledge Dr. Loyola D'Silva, Springer Nature Singapore Pte Ltd, and Mr. Boopalan Renu, Springer Nature Scientific Publishing Services (P) Ltd, India.

Last but not least, I would like to acknowledge my son Syeed Hasan, Associate Manufacturing Equipment Engineer, Tesla Inc., for revising the whole manuscript.

Features

The key features of this book are as follows.

- Basics of AC circuit, three-phase circuit, and different types of electrical power.
- Each topic is presented clearly and represented with their mathematical illustrations.
- Worked-out examples and practice problems are included after each topic.
- Worked-out examples are solved by step-by-step methods for easy understanding.
- Power system simulation softwares such as CYME, IPSA, and PowerWorld are used for easy calculation and comparison.
- Large numbers of exercise problems are included at the end of each chapter.
- Answers to practice and exercise problems are included.

Organization of Book

Basics of instantaneous power, apparent power, average power, and reactive power analysis of electrical power have been discussed in Chap. 1.

In Chap. 2, fundamentals of a transformer, equivalent circuit, efficiency, three-phase transformer connection, vector diagrams, and per unit systems have been discussed.

Power generation, turbines, different types of power system-related factors, and tariff have been included in Chap. 3.

In Chap. 4, transmission line parameters and their analysis are discussed. Modeling and performance of transmission lines are discussed in Chap. 5.

In Chap. 6, the analysis of symmetrical and unsymmetrical faults is included. The load flow analysis is mentioned in Chap. 7.

In Chap. 8, underground cables and their detailed analysis have been discussed. Power system stability has been discussed in Chap. 9.

Power system harmonics are discussed in Chap. 10, and insulators, sag, and their analysis are discussed in Chap. 11.

Aids for Instructors

Instructors, who will adopt this publication as a textbook, may obtain the solution manual as a supplement copy by contacting the publishers.

Finally, I am confident that this book is free of factual errors and omissions. However, I encourage all readers to send their comments to my email msalam7@uwo.ca, if any. I would like to thank all production staff of Springers who tirelessly work to publish this book successfully.

London, Canada Md. Abdus Salam

Contents

Chapter 1
Analysis of Electrical Power

1.1 Introduction

AC generator, transformer, and other related high-voltage devices are required to generate, transmit, and distribute alternating voltage and current. These types of equipment are rated by megawatt (MW) and mega voltage-ampere (MVA). Electrical appliances, such as DVD player, television, microwave, light bulb, refrigerator, ceiling fan, and many more are rated in a Watt (W). Careful attention is required in designing these kinds of equipment to address their power ratings, which in turn addresses the varying electricity bill. In this case, the analysis of AC power plays an important role. This chapter will discuss instantaneous power, average power, complex power, power factor, maximum power transfer theorem, and power factor correction.

1.2 Instantaneous Power

Figure 1.1 shows a circuit with an AC voltage source $v(t)$ with an impedance connected in series. In this case, the resultant current $i(t)$ varies with time. The instantaneous power also varies with time, and it is defined as the product of the instantaneous voltage and the instantaneous current. Let us consider that the expression of the instantaneous voltage is

$$v(t) = V_m \sin \omega t \tag{1.1}$$

The expression of the current can be derived as [1, 2]

$$i(t) = \frac{v(t)}{Z \angle \theta} \tag{1.2}$$

Md. A. Salam, *Fundamentals of Electrical Power Systems Analysis*,
https://doi.org/10.1007/978-981-15-3212-2_1

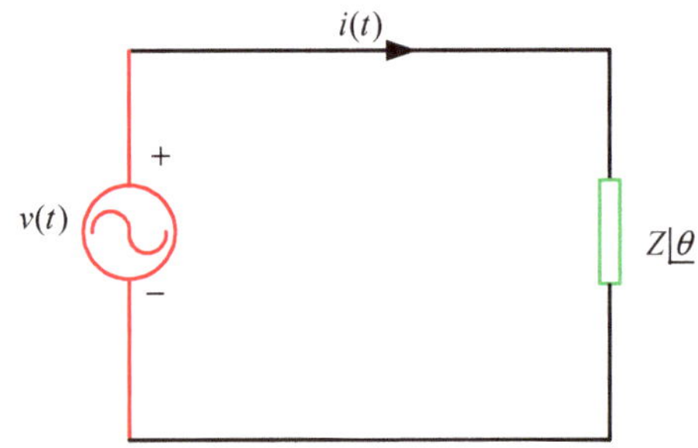

Fig. 1.1 Circuit with an impedance

Substituting Eq. (1.1) into Eq. (1.2) yields

$$i(t) = \frac{V_m \sin \omega t}{Z \angle \theta} \tag{1.3}$$

$$i(t) = \frac{V_m \angle 0^\circ}{Z \angle \theta} = \frac{V_m}{Z} \angle -\theta \tag{1.4}$$

$$i(t) = I_m \angle -\theta = I_m \sin(\omega t - \theta) \tag{1.5}$$

The expression of instantaneous power is written as

$$p(t) = v(t) \times i(t) \tag{1.6}$$

Substituting Eqs. (1.1) and (1.5) into Eq. (1.6) yields

$$p(t) = V_m \sin \omega t \times I_m \sin(\omega t - \theta) \tag{1.7}$$

$$p(t) = \frac{V_m I_m}{2} 2 \sin \omega t \times \sin(\omega t - \theta) \tag{1.8}$$

$$p(t) = \frac{V_m I_m}{2} [\cos(\omega t - \omega t + \theta) - \cos(\omega t + \omega t - \theta)] \tag{1.9}$$

$$p(t) = \frac{V_m I_m}{2} \cos \theta - \frac{V_m I_m}{2} \cos(2\omega t - \theta) \tag{1.10}$$

Equation (1.10) provides the expression for the instantaneous power for a series AC circuit. The first part of Eq. (1.10) is the average power.

Example 1.1
The excitation voltage and impedance of a series circuit are given by $v(t) = 15 \sin \omega t$ V and $Z = 5 \angle 10^\circ\ \Omega$, respectively. Calculate the instantaneous power.

Solution

The value of the current is calculated as

$$i(t) = \frac{15\,\angle 0^\circ}{5\,\angle 10^\circ} = 3\,\angle -10^\circ\ \mathrm{A} \tag{1.11}$$

The instantaneous power is calculated as

$$p(t) = \frac{15 \times 3}{2} 2 \sin \omega t \times \sin(\omega t - 10^\circ) \tag{1.12}$$

$$p(t) = \frac{45}{2} \cos 10^\circ - \frac{45}{2} \cos(2\omega t - 10^\circ) \tag{1.13}$$

$$p(t) = 22.16 - 22.5 \cos(2\omega t - 10^\circ) \tag{1.14}$$

Practice Problem 1.1

The current and impedance of a series circuit are given by $i(t) = 10 \sin(\omega t + 15^\circ)$ A and $Z = 2\,\angle 20^\circ\ \Omega$, respectively. Calculate the instantaneous voltage and power.

1.3 Average Power and Reactive Power

The average power is related to the sinusoidal voltage and current, which are shown in Eqs. (1.1) and (1.5), respectively. The average power for a periodic waveform over one cycle can be derived as [3, 4]

$$P = \frac{1}{T} \int_0^T p(t) \tag{1.15}$$

Substituting Eq. (1.10) into the Eq. (1.15) yields

$$P = \frac{V_m I_m}{2T} \int_0^T \cos\theta \,\mathrm{d}t - \frac{V_m I_m}{2T} \int_0^T \cos(2\omega t - \theta)\,\mathrm{d}t \tag{1.16}$$

$$P = \frac{V_m I_m}{2T} \cos\theta [T] - \frac{V_m I_m}{2T} \int_0^T \cos(2\omega t - \theta)\,\mathrm{d}t \tag{1.17}$$

The second term of Eq. (1.17) is a cosine waveform. The average value of any cosine waveform over one cycle is zero. Therefore, from Eq. (1.17), the final expression of the average power can be represented as

$$P = \frac{V_m I_m}{2} \cos\theta \tag{1.18}$$

Similarly, the expression of the average reactive power can be written as

$$Q = \frac{V_m I_m}{2} \sin\theta \tag{1.19}$$

The term $\cos\theta$ in Eq. (1.18) is the power factor of the circuit, and it is determined by the phase angle θ of the circuit impedance, where θ is the phase difference between the voltage and current phases, i.e., $\theta = \theta_v - \theta_i$. The average power is often known as the true power or real power. The units of average power and reactive power are watts (W) and volt-ampere reactive (Var), respectively. The average power from Eq. (1.18) can be represented in terms of rms values of the voltage and current as

$$P = \frac{V_m}{\sqrt{2}} \frac{I_m}{\sqrt{2}} \cos\theta = \frac{V_m}{\sqrt{2}} \frac{I_m}{\sqrt{2}} \cos(\theta_v - \theta_i) \tag{1.20}$$

Substituting $V_{\mathrm{rms}} = \frac{V_m}{\sqrt{2}}$ and $I_{\mathrm{rms}} = \frac{I_m}{\sqrt{2}}$ into Eq. (1.20) yields

$$P = V_{\mathrm{rms}} I_{\mathrm{rms}} \cos\theta = V_{\mathrm{rms}} I_{\mathrm{rms}} \cos(\theta_v - \theta_i) \tag{1.21}$$

Similarly, from Eq. (1.19), reactive power can be represented as

$$Q = V_{\mathrm{rms}} I_{\mathrm{rms}} \sin\theta = V_{\mathrm{rms}} I_{\mathrm{rms}} \sin(\theta_v - \theta_i) \tag{1.22}$$

Due to a sufficient magnitude of the reactive power, the current flows back and forth between the source and the network. The reactive power does not dissipate any energy in the load. However, in practice, it produces energy losses in the line. Therefore, extra care needs to be taken in designing a power system network.

For a purely resistive circuit, the voltage ($V = V_m \underline{|\theta_v}$) and the current ($I = I_m \underline{|\theta_i}$) are in phase. It means that the phase angle between them is zero, and it can be expressed as

$$\theta = \theta_v - \theta_i = 0 \tag{1.23}$$

Substituting Eq. (1.23) into Eq. (1.18) yields

$$P_R = \frac{V_m I_m}{2} \cos 0^\circ \tag{1.24}$$

$$P_R = \frac{V_m I_m}{2} \tag{1.25}$$

Equation (1.25) can be rearranged by substituting $V_m = I_m R$ as

$$P_R = \frac{I_m^2 R}{2} = \frac{V_m^2}{2R} \quad (1.26)$$

The phase difference between the voltage and current due to inductance and capacitance is

$$\theta = \theta_v - \theta_i = \pm 90^\circ \quad (1.27)$$

Substituting Eq. (1.27) into Eq. (1.18) yields the average power for either an inductance or a capacitance

$$P_L = P_C = \frac{V_m I_m}{2} \cos 90^\circ = 0 \quad (1.28)$$

The reactive power is usually stored in a circuit, and it can be expressed for the inductor and capacitor as

$$Q_L = I_L^2 X_L = \frac{V_L^2}{X_L} \quad (1.29)$$

$$Q_C = I_C^2 X_C = \frac{V_C^2}{X_C} \quad (1.30)$$

From Eqs. (1.26) and (1.28), it can be concluded that the resistive load absorbs power, whereas inductive or capacitive loads do not absorb any power.

Example 1.2
An electrical series circuit with resistance, inductive, and capacitive reactance is shown in Fig. 1.2. Calculate the average power supplied by the source and the power absorbed by the resistor.

Solution
The net impedance is calculated as

$$Z_t = 2 + j5 - j9 = 4.47 \angle{-63.43^\circ}\ \Omega \quad (1.31)$$

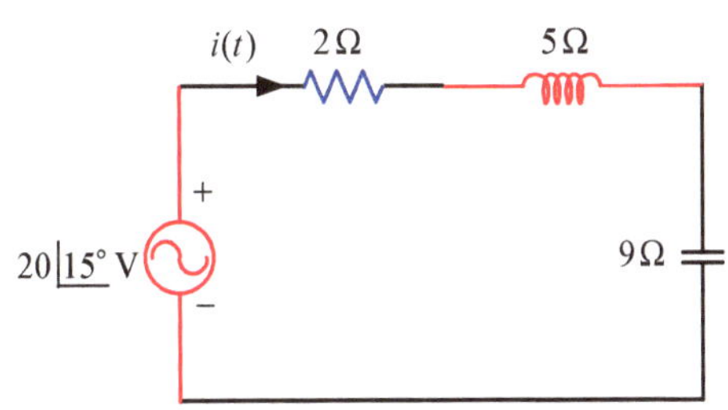

Fig. 1.2 Circuit for Example 1.2

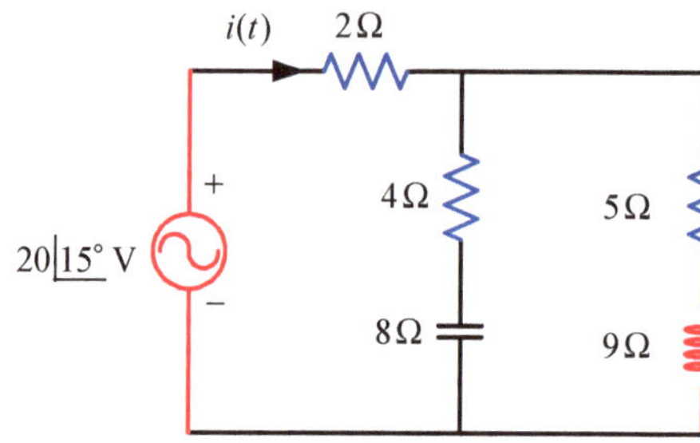

Fig. 1.3 Circuit for Practice Problem 1.2

The source current in phasor form is calculated as

$$I = \frac{20\ \angle 15^\circ}{4.47\ \angle -63.43^\circ} = 4.47\ \angle 78.43^\circ\ \text{A} \tag{1.32}$$

The average power supplied by the source is calculated as

$$P_s = \frac{20 \times 4.47}{2} \cos(15^\circ - 78.43^\circ) = 20\,\text{W} \tag{1.33}$$

The average power absorbed by the resistor is calculated as

$$P_R = \frac{4.47^2 \times 2}{2} = 20\,\text{W} \tag{1.34}$$

Practice Problem 1.2

A series–parallel circuit with resistance, inductive, and capacitive reactance is shown in Fig. 1.3. Determine the average power supplied by the source and the power absorbed by the resistors.

1.4 Apparent Power

The apparent power can be derived from the average power. But, the apparent power is related to the sinusoidal voltage and current. Let us consider that the expressions for sinusoidal voltage and current are [5, 6]

$$v(t) = V_m \sin(\omega t + \theta_v) \tag{1.35}$$

$$i(t) = I_m \sin(\omega t + \theta_i) \tag{1.36}$$

The phasor forms of these voltage and current components are

$$V = V_m \angle \theta_v \tag{1.37}$$

$$I = I_m \angle \theta_i \tag{1.38}$$

According to Eqs. (1.20) and (1.21), the average power is

$$P = \frac{V_m I_m}{2} \cos(\theta_v - \theta_i) = V_{\mathrm{rms}} I_{\mathrm{rms}} \cos(\theta_v - \theta_i) \tag{1.39}$$

The apparent power is the product of the rms voltage and rms current. The unit of apparent power is volt-amps (VA) and is denoted by the letter S. The apparent power can be expressed as

$$S = \frac{V_m I_m}{2} = V_{\mathrm{rms}} I_{\mathrm{rms}} \tag{1.40}$$

Substituting Eq. (1.40) into Eq. (1.39) yields

$$P = S \cos(\theta_v - \theta_i) \tag{1.41}$$

In addition, the apparent power can be determined by the vector sum of the real power (P) and the reactive power (Q). In this case, the expression of reactive power becomes

$$S = P + jQ \tag{1.42}$$

The power triangle (Power Triangle: A right-angle triangle that shows the vector relationship between active power, reactive power, and apparent power) with a lagging and leading power factor is shown in Fig. 1.4. The power triangles with an inductance and the capacitance loads will be lagging and leading, respectively.

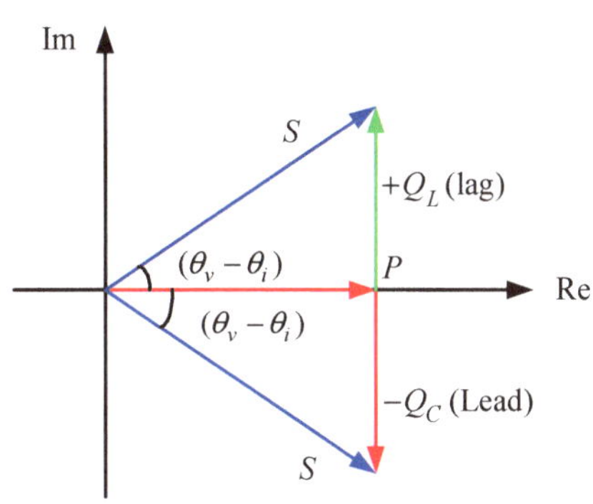

Fig. 1.4 Power triangles

Example 1.3
An industrial load draws a current of $i(t) = 16\sin(314t + 25°)$ A from an alternating voltage source of $v(t) = 220\sin(314t + 60°)$ V. Determine the apparent power, circuit resistance, and inductance.

Solution
The apparent power is calculated as

$$S = \frac{V_m I_m}{2} = \frac{220 \times 16}{2} = 1.76\,\text{kVA} \tag{1.43}$$

The circuit impedance is calculated as

$$Z = \frac{220\,\angle 60°}{16\,\angle 25°} = 11.26 + j7.89\,\Omega \tag{1.44}$$

The value of the circuit resistance is

$$R = 11.26\,\Omega \tag{1.45}$$

The circuit inductance is calculated as

$$L = \frac{7.89}{314} = 0.025\,\text{H} \tag{1.46}$$

Practice Problem 1.3
An industrial load draws a current of $i(t) = 10\sin(100t + 55°)$ A from an alternating voltage source of $v(t) = 120\sin(100t + 10°)$ V. Calculate the apparent power, circuit resistance, and capacitance.

1.5 Complex Power

The complex power is the combination of real power and reactive power. The reactive power creates an adverse effect on power generation, which can be studied by analyzing the complex power. Mathematically, the product of half of the phasor voltage and the conjugate of the phasor current is known as complex power. The complex power is represented by the letter S_c and is expressed as

$$S_c = \frac{1}{2} V I^* \tag{1.47}$$

Substituting Eqs. (1.37) and (1.38) into Eq. (1.47) yields

$$S_c = \frac{1}{2} V_m \underline{|\theta_v} \times I_m \underline{|-\theta_i} \tag{1.48}$$

$$S_c = V_{\text{rms}} \underline{|\theta_v} \times I_{\text{rms}} \underline{|-\theta_i} \tag{1.49}$$

The complex power in terms of phasor form of rms voltage and current can be written as

$$S_c = \mathbf{V}_{\text{rms}} \times \mathbf{I}^*_{\text{rms}} \tag{1.50}$$

From Eq. (1.50), the complex power is defined as the product of rms voltage and the conjugate of the rms current.

Equation (1.50) can be rearranged as

$$S_c = V_{\text{rms}} I_{\text{rms}} \underline{|\theta_v - \theta_i} \tag{1.51}$$

$$S_c = V_{\text{rms}} I_{\text{rms}} \cos(\theta_v - \theta_i) + jV_{\text{rms}} I_{\text{rms}} \sin(\theta_v - \theta_i) \tag{1.52}$$

Consider the circuit as shown in Fig. 1.5 to explain complex power. The impedance of this circuit is

$$Z = R + jX \tag{1.53}$$

Equation (1.54) can be represented by the impedance triangle as shown in Fig. 1.6.

The rms value of the current is

$$I_{\text{rms}} = \frac{V_{\text{rms}}}{Z} \tag{1.54}$$

Substituting Eq. (1.54) into Eq. (1.50) yields

$$S = V_{\text{rms}} \frac{V^*_{\text{rms}}}{Z^*} = \frac{V^2_{\text{rms}}}{Z} \tag{1.55}$$

Fig. 1.5 Simple AC circuit

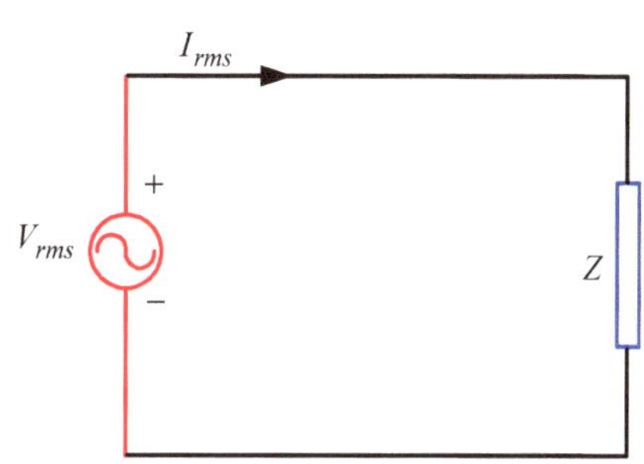

Fig. 1.6 Impedance triangle

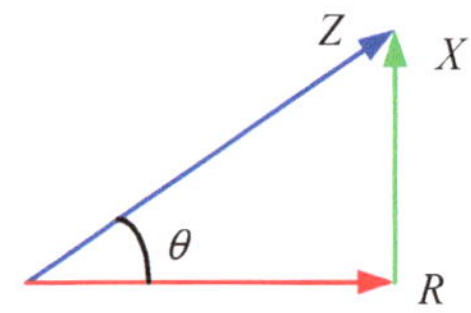

Equation (1.50) again can be represented as

$$S_c = I_{\text{rms}} Z I_{\text{rms}}^* = I_{\text{rms}}^2 Z \tag{1.56}$$

Substituting Eq. (1.54) into Eq. (1.56) yields

$$S_c = I_{\text{rms}}^2 (R + jX) = I_{\text{rms}}^2 R + jI_{\text{rms}}^2 X = P + jQ \tag{1.57}$$

where P and Q are the real and the imaginary parts of the complex power, and in this case, the expressions of P and Q can be written as

$$P = \text{Re}(S_c) = I_{\text{rms}}^2 R \tag{1.58}$$

$$Q = \text{Im}(S_c) = I_{\text{rms}}^2 X \tag{1.59}$$

The real power, reactive power, and apparent power of Eq. (1.57) are shown in Fig. 1.7. The complex power for a resistive branch can be written as

$$S_{cR} = P_R + jQ_R = I_{\text{rms}}^2 R \tag{1.60}$$

From Eq. (1.60), the real power and the reactive power for the resistive branch can be expressed as

$$P_R = I_{\text{rms}}^2 R \tag{1.61}$$

$$Q = 0 \tag{1.62}$$

The complex power for an inductive branch is

$$S_{cL} = P_L + jQ_L = jI_{\text{rms}}^2 X_L \tag{1.63}$$

Fig. 1.7 Power triangle

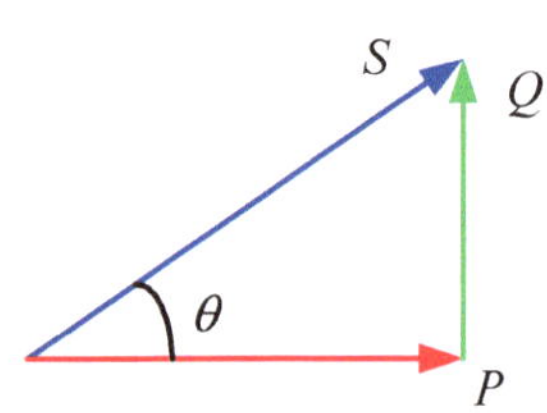

From Eq. (1.63), the real power and the reactive power for an inductive branch can be separated as

$$P_L = 0 \tag{1.64}$$

$$Q_L = I_{\text{rms}}^2 X_L \tag{1.65}$$

The complex power for a capacitive branch is

$$S_{cC} = P_C + jQ_C = -jI_{\text{rms}}^2 X_C \tag{1.66}$$

From Eq. (1.66), the real power and the reactive power for a capacitive branch can be separated as

$$P_c = 0 \tag{1.67}$$

$$Q_C = I_{\text{rms}}^2 X_C \tag{1.68}$$

In the case of reactive power, the following points are summarized:

$Q = 0$ for resistive load, i.e., unity power factor,
$Q > 0$ for inductive load, i.e., lagging power factor,
$Q < 0$ for capacitive load, i.e., leading power factor.

Example 1.4
A series AC circuit is shown in Fig. 1.8. Calculate the source current, apparent, real, and reactive powers.

The expression of the alternating voltage source is $v(t) = 16\sin(10t + 25°)$ V.

Solution
The rms value of the source voltage is calculated as

$$V_{\text{rms}} = \sqrt{2} \times 16 \angle 25° = 22.63 \angle 25° \text{ V} \tag{1.69}$$

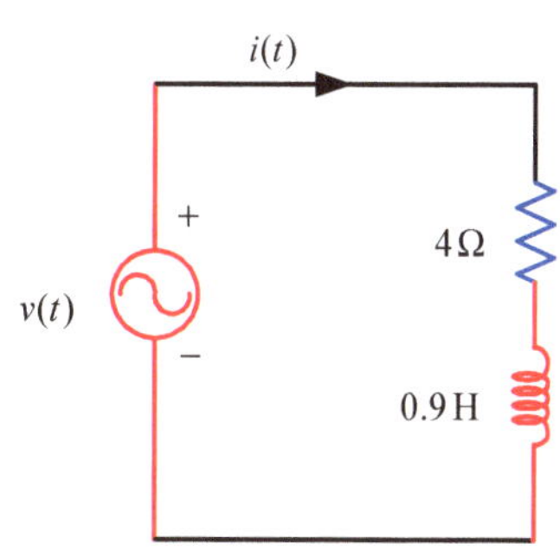

Fig. 1.8 Circuit for Example 1.4

The inductive reactance is

$$X_L = 10 \times 0.09 = 9\,\Omega \tag{1.70}$$

The circuit impedance is calculated as

$$Z = 4 + j9 = 9.85\ \underline{|66.04^\circ}\ \Omega \tag{1.71}$$

The source current is calculated as

$$I_{\text{rms}} = \frac{22.63\ \underline{|25^\circ}}{9.85\ \underline{|66.04^\circ}} = 2.30\ \underline{|-41.04^\circ}\ \text{A} \tag{1.72}$$

The complex power is calculated as

$$S = V_{\text{rms}} I_{\text{rms}}^* = 22.63\ \underline{|25^\circ} \times 2.30\ \underline{|-41.04^\circ} = 50.02\,\text{W} - j14.38\ \text{Var} \tag{1.73}$$

The apparent power is determined as

$$S_{\text{ap}} = |S| = 52.05\,\text{VA} \tag{1.74}$$

The real power is calculated as

$$P = \text{Re}(S) = 52.02\,\text{W} \tag{1.75}$$

The reactive power is determined as

$$Q = \text{Im}(S) = 14.38\,\text{Var} \tag{1.76}$$

Example 1.5
A 220 V rms delivers power to a load. The load absorbs an average power of 10 kW at a leading power factor of 0.9. Determine the complex power and the impedance of the load.

Solution
The power factor is

$$\cos\theta = 0.9 \tag{1.77}$$

$$\theta = 25.84^\circ \tag{1.78}$$

The reactive component is calculated as

$$\sin\theta = \sin 25.84^\circ = 0.44 \tag{1.79}$$

The magnitude of the complex power is calculated as

$$|S| = \frac{P}{\cos\theta} = \frac{10}{0.9} = 11.11\,\text{kVA} \tag{1.80}$$

The reactive power is determined as

$$Q = |S|\sin\theta = 11.11 \times 0.44 = 4.89\,\text{kVar} \tag{1.81}$$

The complex power is calculated as

$$S = P + jQ = 10\,\text{kW} - j4.89\,\text{kVar} \tag{1.82}$$

The rms voltage can be determined as

$$P = V_{\text{rms}} I_{\text{rms}} \cos\theta = 10000 \tag{1.83}$$

$$I_{\text{rms}} = \frac{10000}{220 \times 0.9} = 50.51\,\text{A} \tag{1.84}$$

The value of the impedance is calculated as

$$|Z| = \frac{|V_{\text{rms}}|}{|I_{\text{rms}}|} = \frac{220}{50.51} = 4.36\,\Omega \tag{1.85}$$

$$Z = 4.36\,\underline{|-25.84^\circ}\,\Omega \tag{1.86}$$

Practice Problem 1.4

A series RC circuit is shown in Fig. 1.9. Find the source current, apparent, real, and reactive powers. The expression of the alternating voltage source is $v(t) = 10\sin(2t + 12^\circ)$ V.

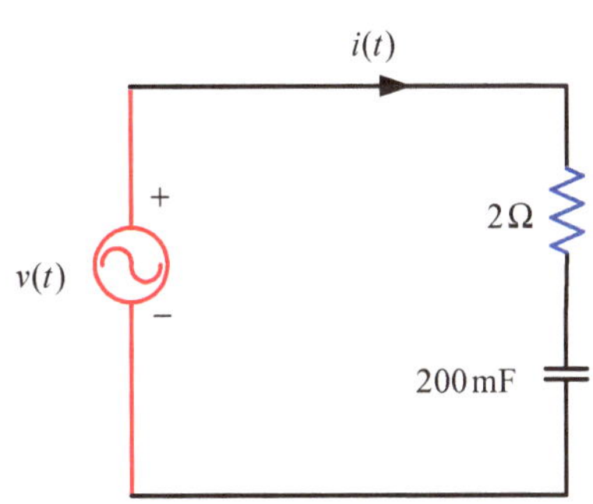

Fig. 1.9 Circuit for Practice Problem 1.4

Practice Problem 1.5
An electrical load absorbs an average power of 12 kW from a source of 230 V rms at a lagging power factor of 0.95. Calculate the complex power and the impedance of the load.

1.6 Complex Power Balance

Two electrical loads are connected in parallel with a voltage source as shown in Fig. 1.10. According to the conservation of energy, the real power delivered by the source will be equal to the total real power absorbed by the loads [7, 8]. Similarly, the complex power delivered by the source will be equal to the total complex power absorbed by the loads. According to KCL, the rms value of the source current is equal to the sum of the rms values of the branch currents I_1 and I_2, i.e.,

$$I = I_1 + I_2 \tag{1.87}$$

The total complex power is defined as the product of the rms value of the source voltage and the conjugate of the current supplied by the source, and it is expressed as

$$S = V_{\text{rms}} I^* \tag{1.88}$$

Substituting Eq. (1.87) into Eq. (1.88) yields the complex power of the parallel circuit as

$$S_p = V_{\text{rms}} [I_1 + I_2]^* \tag{1.89}$$

$$S_p = V_{\text{rms}} I_1^* + V_{\text{rms}} I_2^* \tag{1.90}$$

$$S_p = S_1 + S_2 \tag{1.91}$$

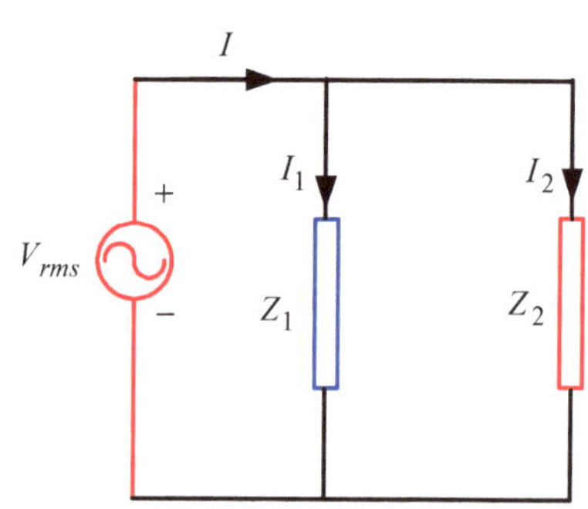

Fig. 1.10 Circuit with two parallel impedances

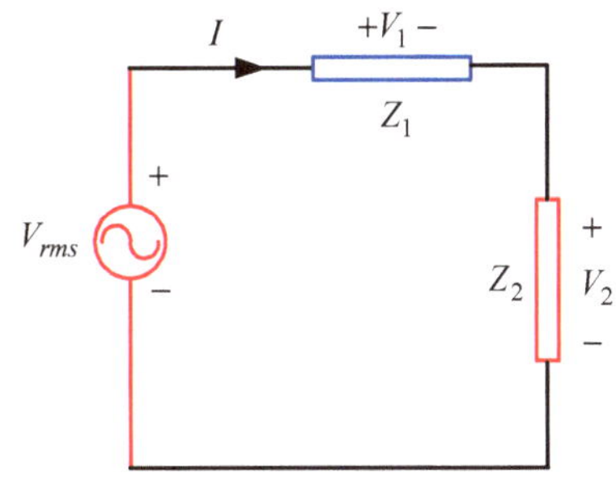

Fig. 1.11 Circuit with series impedances

The electrical loads are again connected in series with a voltage source as shown in Fig. 1.11. According to KVL, the rms value of the source voltage is equal to the sum of the rms values of the load voltages, and it is written as

$$V_{\mathrm{rms}} = V_1 + V_2 \tag{1.92}$$

Substituting Eq. (1.92) into Eq. (1.88) yields the complex power of the series circuit as

$$S_s = (V_1 + V_2)I^* \tag{1.93}$$

$$S_s = V_1 I^* + V_2 I^* \tag{1.94}$$

$$S_s = S_1 + S_2 \tag{1.95}$$

From Eqs. (1.91) and (1.95), it is observed that the total complex power delivered by the source is equal to the sum of the individual complex power absorbed by the loads.

Example 1.6
A series–parallel circuit is supplied by a source of 60 V rms as shown in Fig. 1.12. Find the complex power for each branch and the total complex power.

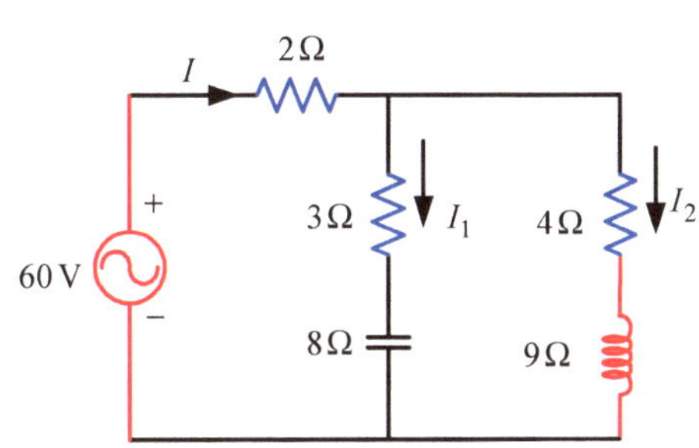

Fig. 1.12 Circuit for Example 1.6

Solution

The circuit impedance is calculated as

$$Z_t = 2 + \frac{(4 + j9)(3 - j8)}{4 + j9 + 3 - j8} = 13.87 \angle -9.88^\circ \ \Omega \tag{1.96}$$

The source current is determined as

$$I = \frac{60}{13.87 \angle -9.88^\circ} = 4.33 \angle 9.88^\circ \ \text{A} \tag{1.97}$$

The branch currents can be calculated as

$$I_1 = 4.33 \angle 9.88^\circ \times \frac{4 + j9}{7 + j1} = 6.03 \angle 67.79^\circ \ \text{A} \tag{1.98}$$

$$I_2 = 4.33 \angle 9.88^\circ \times \frac{3 - j8}{7 + j1} = 5.23 \angle -67.69^\circ \ \text{A} \tag{1.99}$$

The voltage across the parallel branches is

$$V_p = 60 - 2 \times 4.33 \angle 9.88^\circ = 51.49 \angle -1.65^\circ \ \text{V} \tag{1.100}$$

The complex power in the branches is calculated as

$$S_1 = 51.49 \angle -1.65^\circ \times 6.03 \angle -67.79^\circ = 310.48 \angle -69.44^\circ \ \text{VA} \tag{1.101}$$

$$S_2 = 51.49 \angle -1.65^\circ \times 5.23 \angle 67.69^\circ = 269.29 \angle 66.04^\circ \ \text{VA} \tag{1.102}$$

The voltage drops across the 2Ω resistor are calculated as

$$V_{2\Omega} = 2 \times 4.33 \angle 9.88^\circ = 8.66 \angle 9.88^\circ \ \text{V} \tag{1.103}$$

The complex power for 2Ω resistor is calculated as

$$S_{2\Omega} = 8.66 \angle 9.88^\circ \times 4.33 \angle -9.88^\circ = 37.50 \, \text{VA} \tag{1.104}$$

The total complex power is calculated as

$$S_t = 310.48 \angle -69.44^\circ + 269.29 \angle 66.04^\circ + 37.50 = 259.76 \angle -9.89^\circ \ \text{VA} \tag{1.105}$$

Alternatively, the total complex power can be calculated as

$$S = 60 \times 4.33 \angle -9.88^\circ = 259.8 \angle -9.88^\circ \ \text{VA} \tag{1.106}$$

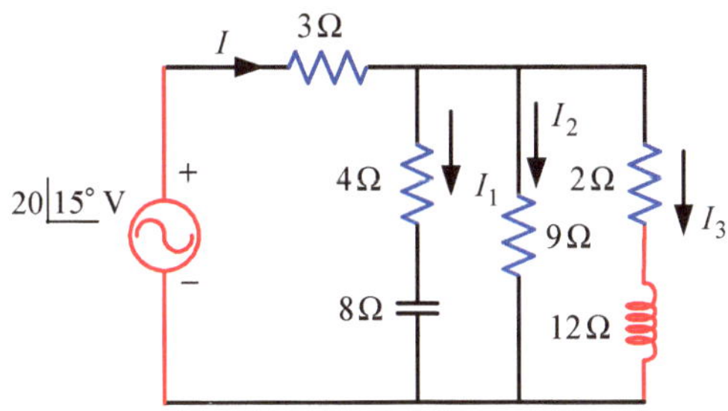

Fig. 1.13 Circuit for Practice Problem 1.6

Practice Problem 1.6

A series–parallel circuit is supplied by a rms voltage source as shown in Fig. 1.13. Determine the complex power for each branch and the total complex power.

1.7 Power Factor and Reactive Power

In an AC circuit, power is calculated by multiplying a factor with the rms values of current and voltage. This factor is known as power factor. The power factor is defined as the cosine of the difference in phase angles between the voltage and the current, whereas the reactive factor is defined as the sine of the difference in the phase angles between the voltage and the current. The power factor is also defined as the cosine of the phase angle of the load impedance. A close to unity power factor represents an efficient power transfer from the source to a load, whereas a low power factor identifies an inefficient transmission of power. Low power factor usually affects power generation devices. Mathematically, the power factor is written as [9]

$$\text{pf} = \cos(\theta_v - \theta_i) \tag{1.107}$$

The reactive factor is written as

$$\text{rf} = \sin(\theta_v - \theta_i) \tag{1.108}$$

From impedance and power triangles as shown in Figs. 1.6 and 1.7, the power factor can be written as

$$\text{pf} = \cos\theta = \frac{R}{Z} = \frac{\text{kW}}{\text{kVA}} \tag{1.109}$$

The angle θ is positive if the current lags the voltage, and in this case, the power factor is considered as lagging. Whereas, the angle θ is negative if the current leads the voltage, and in this case, the power factor is considered as leading. The leading power factor is usually considered for capacitive loads. The industrial loads are inductive and have a low lagging power factor.

A low power factor has many disadvantages, which are outlined below:

- kVA rating of electrical machines is increased,
- larger conductor size is required to transmit or distribute electric power at a constant voltage,
- copper losses are increased, and
- voltage regulation is small.

1.8 Power Factor Correction

In the electrical domain, heavy and medium-sized industry applications contain inductive loads which draw a lagging current from the source. As a result, the reactive power for these applications is increased. In this scenario, the transformer rating and the conductor size need to be increased to carry out the additional reactive power.

In order to cancel this reactive component of power, an opposite type of reactance needs to be included in the circuit. Let us consider that a single-phase inductive load is connected across a voltage source as shown in Fig. 1.14, and this load draws a current with a lagging power factor of $\cos\theta_1$.

Figure 1.15 shows a circuit where the capacitor is connected in parallel with the load to improve the power factor. The capacitor will draw current from the source that leads the source voltage by 90°. The line current is the vector sum of the currents in the inductive load and the capacitor. The current in the inductive load circuit lags the supply voltage by θ_1, and the current in the capacitor leads the voltage by 90° as shown in the vector diagram in Fig. 1.16.

The exact value of the capacitor needs to be identified to improve the power factor from $\cos\phi_1$ to $\cos\phi_2$ without changing the real power. A power triangle is drawn using the inductive load and the capacitor as shown in Fig. 1.17.

Fig. 1.14 Single-phase inductive circuit

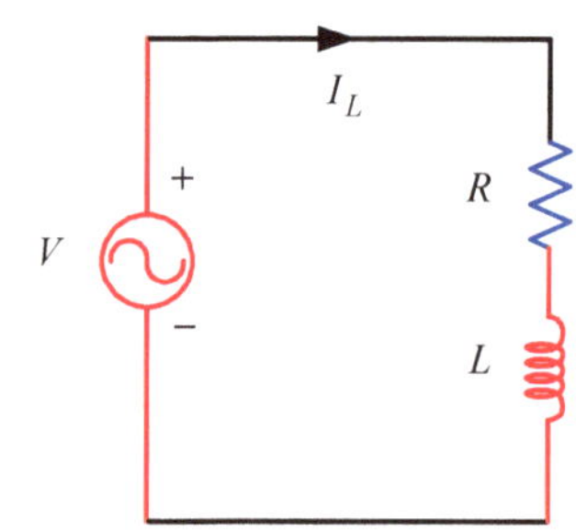

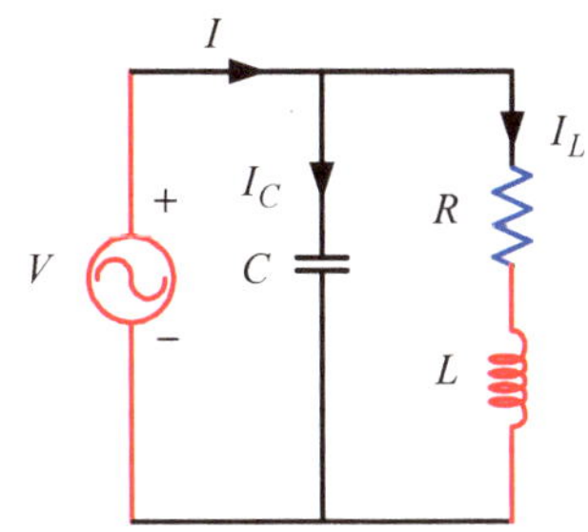

Fig. 1.15 Capacitor is in parallel with inductive load

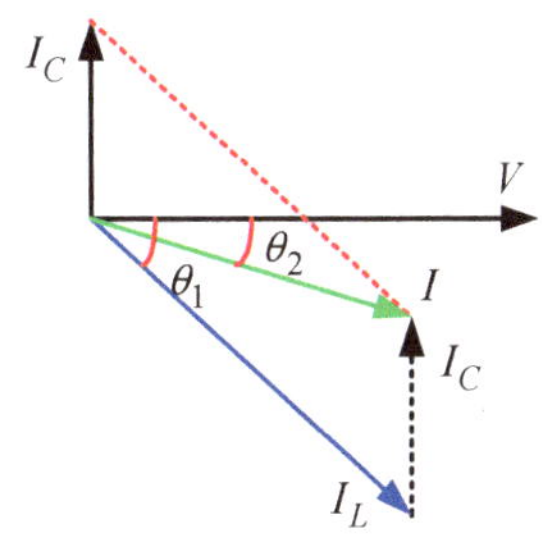

Fig. 1.16 Vector diagram with different currents

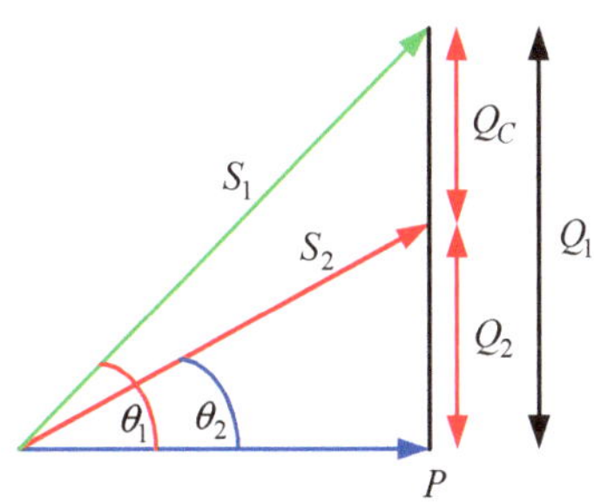

Fig. 1.17 Power triangles for inductive load and capacitor

The reactive power of the original inductive load is written as

$$Q_1 = P \tan \phi_1 \tag{1.110}$$

The expression of new reactive power is written as

$$Q_2 = P \tan \phi_2 \tag{1.111}$$

The reduction in reactive power due to parallel capacitor is expressed as

$$Q_C = Q_1 - Q_2 \tag{1.112}$$

Substituting Eqs. (1.110) and (1.111) into Eq. (1.112) yields

$$Q_C = P(\tan\phi_1 - \tan\phi_2) \tag{1.113}$$

The reactive power due to the capacitor can be calculated as

$$Q_C = \frac{V_{\text{rms}}^2}{X_C} = \omega C V_{\text{rms}}^2 \tag{1.114}$$

Substituting Eq. (1.114) into Eq. (1.115) yields the expression for the capacitor as

$$\omega C V_{\text{rms}}^2 = P(\tan\phi_1 - \tan\phi_2) \tag{1.115}$$

$$C = \frac{P(\tan\phi_1 - \tan\phi_2)}{\omega V_{\text{rms}}^2} \tag{1.116}$$

Example 1.7
A load of 6 kVA, 50 Hz, 0.75 lagging power factor is connected across a voltage source of 120 V rms as shown in Fig. 1.18. A capacitor is connected across the load to improve the power factor to 0.95 lagging. Determine the capacitance of the connected capacitor.

Solution
The initial power factor is

$$\cos\phi_1 = 0.75 \tag{1.117}$$

$$\phi_1 = 41.41^\circ \tag{1.118}$$

The final power factor is

$$\cos\phi_2 = 0.95 \tag{1.119}$$

$$\phi_2 = 18.19^\circ \tag{1.120}$$

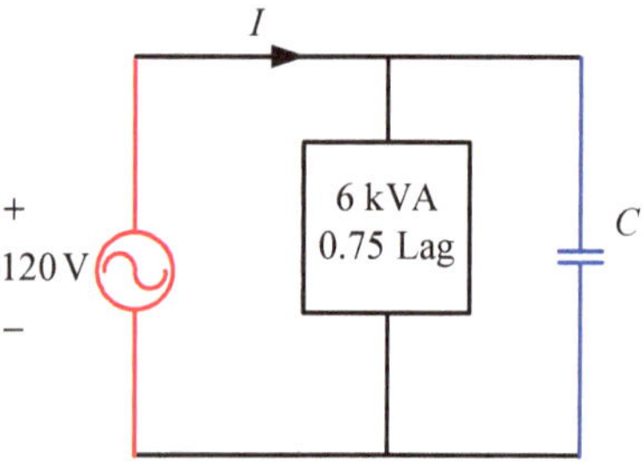

Fig. 1.18 Circuit for Example 1.7

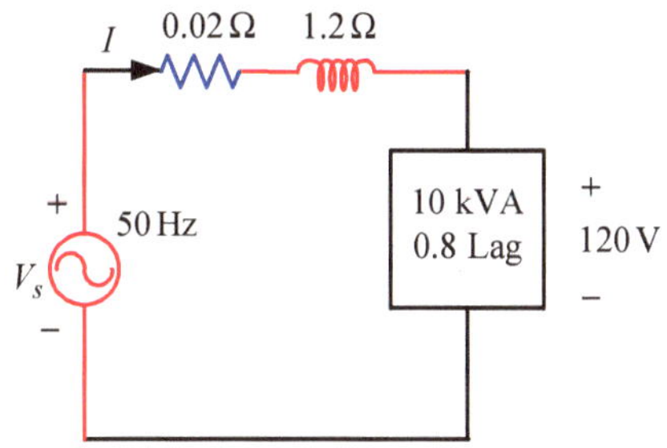

Fig. 1.19 Circuit for Example 1.8

The real power is calculated as

$$P = 6 \times 0.75 = 4.5\,\text{kW} \tag{1.121}$$

The value of the parallel capacitance can be calculated as

$$C = \frac{4.5 \times 1000(\tan 41.41^\circ - \tan 18.19^\circ)}{2\pi \times 50 \times 120^2} = 0.55\,\text{mF} \tag{1.122}$$

Example 1.8
A load of 10 kVA, 50 Hz, 0.8 lagging power factor is connected across the voltage source shown in Fig. 1.19. A capacitor is connected across the load to improve the power factor to 0.90 lagging. Calculate the value of the capacitance and line loss with and without the capacitor.

Solution
The initial power factor is

$$\cos\phi_1 = 0.8 \tag{1.123}$$

$$\phi_1 = 36.87^\circ \tag{1.124}$$

The final power factor is

$$\cos\phi_2 = 0.9 \tag{1.125}$$

$$\phi_2 = 25.84^\circ \tag{1.126}$$

The power of the load is calculated as

$$P = 10 \times 0.8 = 8\,\text{kW} \tag{1.127}$$

The capacitor is calculated as

$$C = \frac{8 \times 1000(\tan 36.87^\circ - \tan 25.84^\circ)}{2\pi \times 50 \times 120^2} = 0.47\,\text{mF} \tag{1.128}$$

The line current before adding the capacitor is calculated as

$$I_1 = \frac{8000}{0.8 \times 120} = 83.33\,\text{A} \tag{1.131}$$

The power loss in the line before the adding capacitor is calculated as

$$P_1 = 83.33^2 \times 0.02 = 138.88\,\text{W} \tag{1.130}$$

The apparent power with a power factor of 0.9 lagging is calculated as

$$S = \frac{8000}{0.9} = 8888.89\,\text{VA} \tag{1.131}$$

The line current after adding the capacitor is calculated as

$$I_2 = \frac{8888.89}{120} = 74.07\,\text{A} \tag{1.132}$$

The power loss in the line after adding the capacitor is calculated as

$$P_2 = 74.07^2 \times 0.02 = 109.73\,\text{W} \tag{1.133}$$

Practice Problem 1.7

A load of 0.85 lagging power factor is connected across a voltage source of 220 V rms as shown in Fig. 1.20. A 0.56 mF capacitor is connected across the load to improve the power factor to 0.95 lagging. Find the value of the load, *P*.

Practice Problem 1.8

Two loads are connected to a source through a line as shown in Fig. 1.21. Determine the value of the voltage source.

Fig. 1.20 Circuit for Practice Problem 1.7

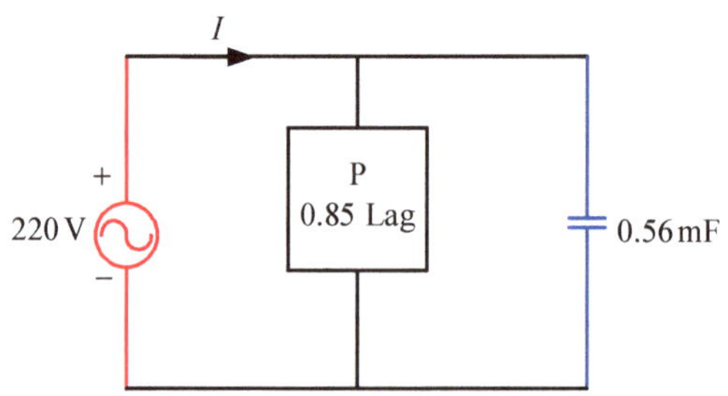

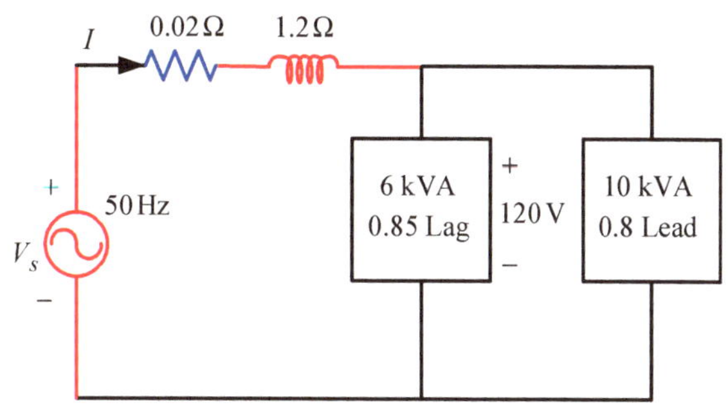

Fig. 1.21 Circuit for Practice Problem 1.8

1.9 Three-Phase Voltage Generation

Figure 1.22 shows a two-pole three-phase AC generator for three-phase voltage generation. Coils aa', bb', and cc' represent the whole coils into a three-phase system as shown in Fig. 1.22a. The rotor of the AC machine is energized by the DC source, which creates the magnetic field. This rotor is attached to the turbine through a soft coupling, and this turbine rotates the rotor. According to Faraday's law of electromagnetic induction, three-phase voltages V_{an}, V_{bn}, and V_{cn} will be generated across the generator terminals.

The magnitudes of these voltages are constant and are displaced from each other by 120 electrical degrees as shown in Fig. 1.22b. The waveforms of the generated voltages are shown in Fig. 1.23. The expression of the generated voltages can be represented as [1, 2]

$$v_{an} = V_{an} \sin \omega t \tag{1.134}$$

$$v_{bn} = V_{bn} \sin(\omega t - 120^\circ) \tag{1.135}$$

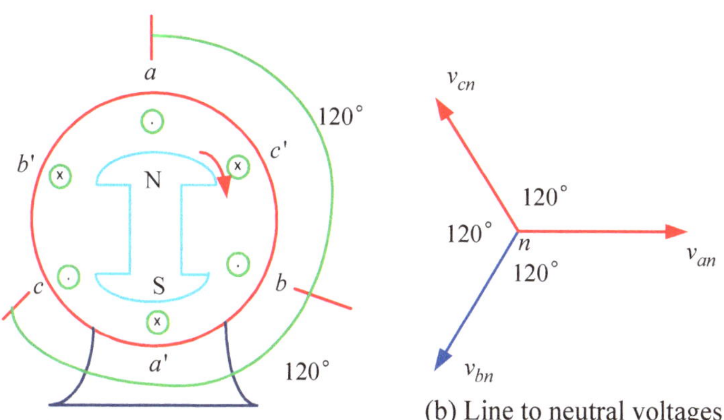

Fig. 1.22 Schematic of AC generator and phase voltages

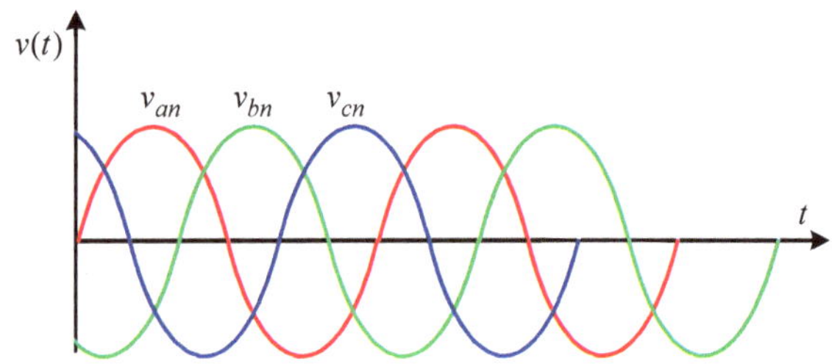

Fig. 1.23 Three-phase voltage waveforms

$$v_{cn} = V_{cn} \sin(\omega t - 240^\circ) \tag{1.136}$$

where V_{an}, V_{bn}, and V_{cn} are the magnitudes of the line to neutral or phase voltages. These voltages are constant in magnitude, and it can be expressed as

$$|V_{an}| = |V_{bn}| = |V_{cn}| = V_p \tag{1.137}$$

The phasor of the generated voltages can be written as

$$V_{an} = V_P \underline{|0^\circ} \tag{1.138}$$

$$V_{bn} = V_P \underline{|-120^\circ} \tag{1.139}$$

$$V_{cn} = V_P \underline{|-240^\circ} = V_P \underline{|120^\circ} \tag{1.140}$$

The sum of phasor voltages and the sum of sinusoidal voltages are zero, and these can be expressed as

$$V = V_{an} + V_{bn} + V_{cn} = V_P \underline{|0^\circ} + V_P \underline{|-120^\circ} + V_P \underline{|-240^\circ} = 0 \tag{1.141}$$

$$v = v_{an} + v_{bn} + v_{cn} = V_{an} \sin \omega t + V_{bn} \sin(\omega t - 120^\circ) + V_{cn} \sin(\omega t - 240^\circ) = 0 \tag{1.142}$$

1.10 Phase Sequence

The phase sequence is very important for the interconnection of the three-phase transformer, motor, and other high-voltage equipment. The three-phase systems are numbered either by the numbers 1, 2, and 3 or by the letters *a*, *b*, and *c*. Sometimes, these are labeled by the colors red, yellow, and blue or *RYB* in short. The generator is said to have a positive phase sequence when the generated voltages reach their maximum or peak values in the sequential order of *abc*, whereas the generator is said to have a negative phase sequence when the generated voltages reach their maximum or peak values in the sequential order of *acb*. Figure 1.24 shows the

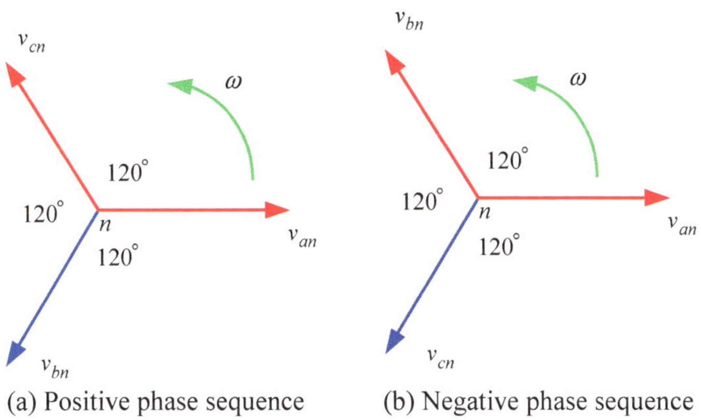

Fig. 1.24 Phase sequence identification

positive and negative phase sequences. Here, the voltage V_{an} is considered to be the reference voltage while the direction of rotation is considered to be anticlockwise. In the positive phase sequence, the crossing sequence of voltage rotation is identified by $V_{an} - V_{bn} - V_{cn}$, whereas for negative phase sequence, it is identified as $V_{an} - V_{cn} - V_{bn}$.

1.11 Wye Connection

Three-phase transformer, AC generator, and induction motor are connected either in wye or in delta connection. In wye connection, one terminal of each coil is connected to form a common or neutral point, and other terminals are to the three-phase supply. The voltage between any line and neutral is known as the phase voltage, and the voltage between any two lines is called the line voltage. The line voltage and phase voltage are usually represented by V_L and V_P, respectively. The important points of this connection are the line voltage is equal to $\sqrt{3}$ times the phase voltage, the line current is equal to the phase current, and the current (I_n) in the neutral wire is equal to the phasor sum of the three-line currents. For a balanced three-phase load, the neutral current is zero i.e., $I_n = 0$. The wye-connected generator and load are shown in Fig. 1.25.

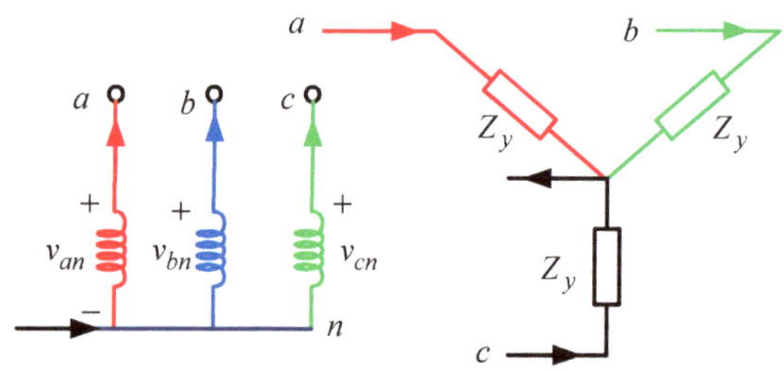

Fig. 1.25 Wye-connected generator and load

1.12 Analysis for Wye Connection

A three-phase wye-connected AC generator shown in Fig. 1.26 is considered for analysis. Here, V_{an}, V_{bn}, and V_{cn} are the phase voltages, whereas V_{ab}, V_{bc}, and V_{ca} are the line voltages. Applying KVL to the circuit to find the line voltage between lines a and b yields

$$V_{an} - V_{bn} - V_{ab} = 0 \tag{1.143}$$

$$V_{ab} = V_{an} - V_{bn} \tag{1.144}$$

Substituting Eqs. (1.138) and (1.139) into Eq. (1.144) yields

$$V_{ab} = V_P \underline{|0^\circ} - V_P \underline{|-120^\circ} \tag{1.145}$$

$$V_{ab} = \sqrt{3}\, V_P \underline{|30^\circ} \tag{1.146}$$

Applying KVL between lines b and c yields the expression of line voltage as

$$V_{bc} = V_{bn} - V_{cn} \tag{1.147}$$

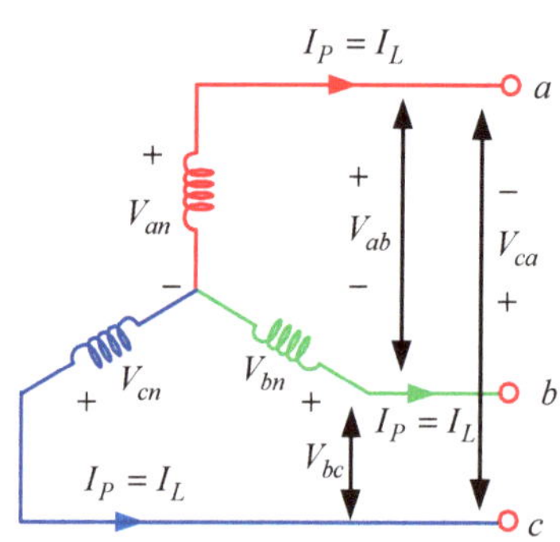

Fig. 1.26 Wye-connected generator

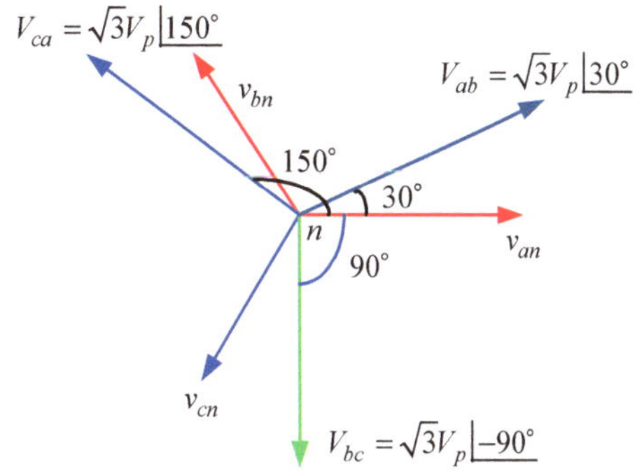

Fig. 1.27 Phasor diagram with line and phase voltages

Substituting Eqs. (1.139) and (1.140) into Eq. (1.147) yields

$$V_{bc} = V_P \underline{\lfloor -120^\circ} - V_P \underline{\lfloor -240^\circ} = \sqrt{3}\, V_P \underline{\lfloor -90^\circ} \tag{1.148}$$

Applying KVL between lines c and a yields the expression of the line voltage as

$$V_{ca} = V_{cn} - V_{an} \tag{1.149}$$

Substituting Eqs. (1.138) and (1.140) into Eq. (1.149) yields

$$V_{ca} = V_P \underline{\lfloor -240^\circ} - V_P \underline{\lfloor 0^\circ} = \sqrt{3}\, V_P \underline{\lfloor 150^\circ} \tag{1.150}$$

Line voltages with angles are drawn as shown in Fig. 1.27. From Eqs. (1.146), (1.148), and (1.150), it is seen that the magnitude of the line voltage is equal to $\sqrt{3}$ times the magnitude of the phase voltage. The general relationship between the line voltage and the phase voltage can be written as

$$V_L = \sqrt{3}\, V_P \tag{1.151}$$

From Fig. 1.26, it is also observed that the phase current is equal to the line current, and it is written as

$$I_L = I_P \tag{1.152}$$

Alternative approach: A vector diagram with phase voltages is drawn using the lines a and c as shown in Fig. 1.28. A perpendicular line is drawn from point A, which divides the line BD equally. From the triangle ABC, the following expression relation can be written as

$$\cos 30^\circ = \frac{BC}{AB} \tag{1.153}$$

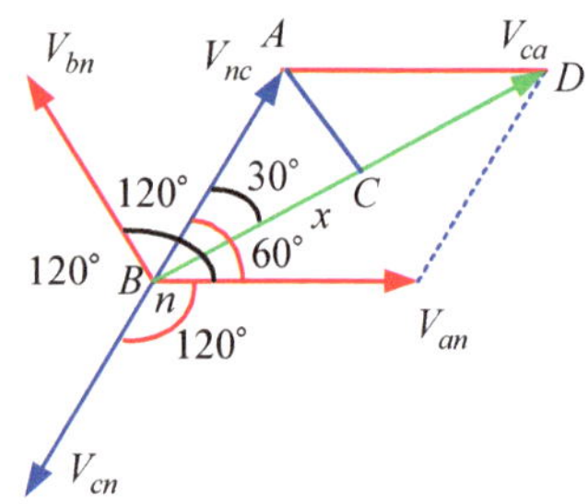

Fig. 1.28 Vector diagram

$$\frac{\sqrt{3}}{2} = \frac{x}{|Vnc|} \tag{1.154}$$

According to Fig. 1.28, the following expression can be written as

$$BD = 2AC \tag{1.155}$$

$$V_{ca} = 2x \tag{1.156}$$

Substituting Eq. (1.154) into Eq. (1.156) yields

$$V_{ca} = 2 \times \frac{\sqrt{3}}{2} |V_{nc}| \tag{1.157}$$

$$V_{ca} = \sqrt{3} V_{nc} \tag{1.158}$$

In general, the following equation can be written as

$$V_L = \sqrt{3} V_p$$

Example 1.9
The phase voltage is given by $V_{an} = 230 \underline{|10^\circ}$ V. For *abc* phase sequence, determine V_{bn} and V_{cn}.

Solution
The phase voltage for line *a* is calculated as

$$V_{an} = 230 \underline{|10^\circ} \text{ V} \tag{1.159}$$

The phase voltage for line *b* is calculated as

$$V_{bn} = 230 \underline{|10^\circ - 120^\circ} = 230 \underline{|-110^\circ} \text{ V} \tag{1.160}$$

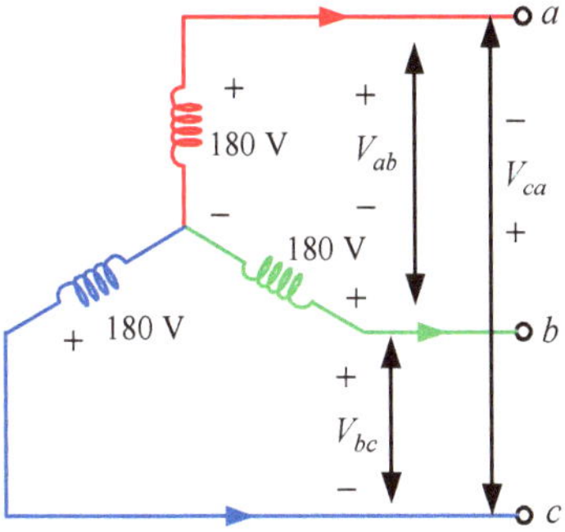

Fig. 1.29 Circuit for Example 1.10

The phase voltage for line c is calculated as

$$V_{cn} = 230\,\angle 10^\circ - 240^\circ = 230\,\angle -230^\circ \text{ V} \quad (1.161)$$

Practice Problem 1.9
The phase voltage is given by $V_{bn} = 200\,\angle 10^\circ$ V. For *abc* phase sequence, calculate V_{an} and V_{cn}.

Example 1.10
A wye-connected generator generates a voltage of 180 V rms as shown in Fig. 1.29. For *abc* phase sequence, write down the phase and line voltages.

Solution
The phase voltages are

$$V_{an} = 180\,\angle 0^\circ \text{ V} \quad (1.162)$$

$$V_{bn} = 180\,\angle -120^\circ \text{ V} \quad (1.163)$$

$$V_{cn} = 180\,\angle -240^\circ \text{ V} \quad (1.164)$$

The line voltages are calculated as

$$V_{ab} = \sqrt{3} \times 180\,\angle 30^\circ = 311.77\,\angle 30^\circ \text{ V} \quad (1.165)$$

$$V_{bc} = \sqrt{3} \times 180\,\angle -120^\circ + 30^\circ = 311.77\,\angle -90^\circ \text{ V} \quad (1.166)$$

$$V_{ca} = \sqrt{3} \times 180\,\angle -240^\circ + 30^\circ = 311.77\,\angle -210^\circ \text{ V} \quad (1.167)$$

Practice Problem 1.10
A wye-connected generator generates the line-to-line voltage of 200 V rms as shown in Fig. 1.30. For *abc* phase sequence, write down the phase voltages.

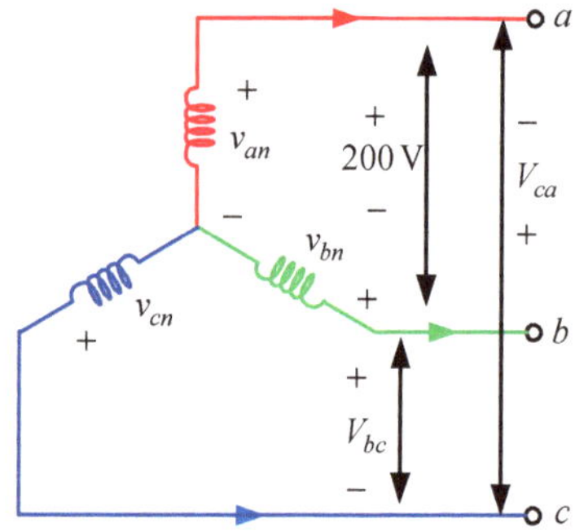

Fig. 1.30 Circuit for Practice Problem 1.10

1.13 Delta Connection

The coils in a delta-connected circuit are arranged in such a way that a looking structure is formed. The delta connection is formed by the connecting point a_2 of a_1a_2 coil to the point b_1 of b_1b_2 coil, the point b_2 of b_1b_2 coil to the point c_1 of c_1c_2 coil, and the point c_2 of c_1c_2 coil to the point a_1 of a_1a_2 coil. In this connection, the phase voltage is equal to the line voltage, and line current is equal to $\sqrt{3}$ times the phase current. Figure 1.31 shows delta-connected generator and load.

1.14 Analysis for Delta Connection

A three-phase delta-connected load is shown in Fig. 1.32. In this connection, I_{ab}, I_{bc}, and I_{ca} are the phase currents and I_a, I_b, and I_c are the line currents. For *abc* phase sequence, the phase currents can be written as

$$I_{ab} = I_P \angle 0^\circ \tag{1.168}$$

$$I_{bc} = I_P \angle -120^\circ \tag{1.169}$$

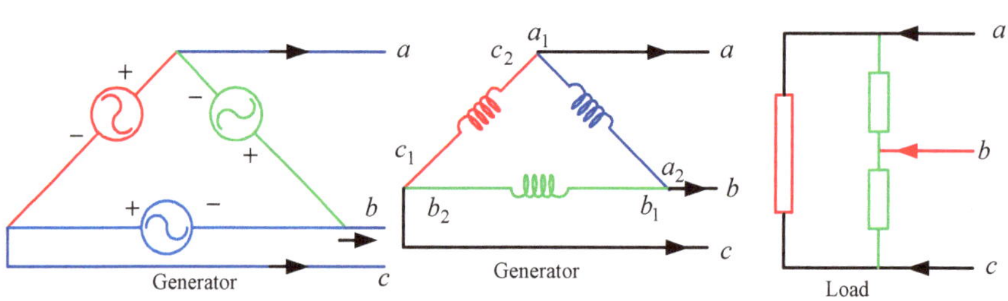

Fig. 1.31 Delta-connected generator and load

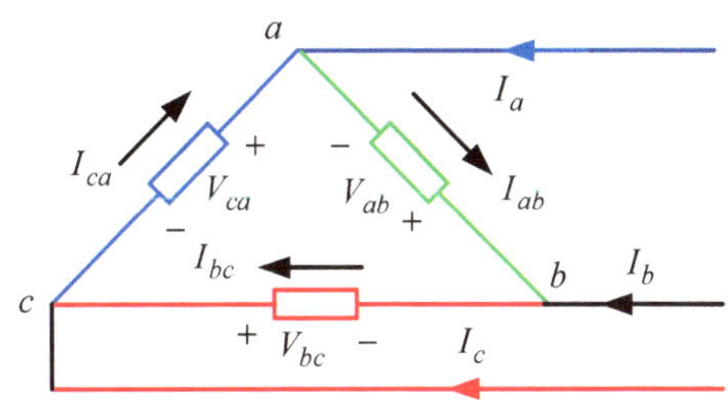

Fig. 1.32 Delta-connected load

$$I_{ca} = I_P \angle{-240^\circ} \tag{1.170}$$

Applying KCL at the node a of the circuit in Fig. 1.32 yields

$$I_a = I_{ab} - I_{ca} \tag{1.171}$$

Substituting Eqs. (1.168) and (1.170) into Eq. (1.171) yields

$$I_a = I_P \angle{0^\circ} - I_P \angle{-120^\circ} = \sqrt{3}\, I_P \angle{-30^\circ} \tag{1.172}$$

Applying KCL at the node b of the circuit in Fig. 1.32 yields

$$I_b = I_{bc} - I_{ab} \tag{1.173}$$

Substituting Eqs. (1.168) and (1.169) into Eq. (1.173) yields

$$I_b = I_P \angle{-120^\circ} - I_P \angle{0^\circ} = \sqrt{3}\, I_P \angle{-150^\circ} \tag{1.174}$$

Applying KCL at the node c of the circuit in Fig. 1.32 yields

$$I_c = I_{ca} - I_{bc} \tag{1.175}$$

Substituting Eqs. (1.169) and (1.170) into Eq. (1.175) yields

$$I_c = I_P \angle{-240^\circ} - I_P \angle{-120^\circ} = \sqrt{3}\, I_P \angle{90^\circ} \tag{1.176}$$

From Eqs. (1.172), (1.174), and (1.167), it is found that the magnitude of the line current is equal to $\sqrt{3}$ times the phase current. The general relationship between the line current and the phase current is

$$I_L = \sqrt{3}\, I_P \tag{1.177}$$

Fig. 1.33 Phasor diagram using line and phase currents

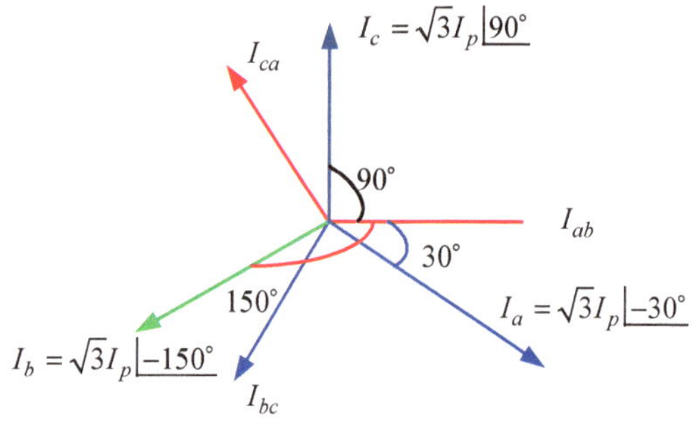

According to Fig. 1.32, it is observed that the phase voltage is equal to the line voltage i.e.,

$$V_L = V_P \tag{1.178}$$

The line and phase currents with their phase angles are drawn as shown in Fig. 1.33, where the phase current I_{ab} is arbitrarily chosen as reference.

1.15 Analysis for Three-Phase Power

Consider a balanced three-phase wye-connected generator that delivers power to the balanced three-phase wye-connected load as shown in Fig. 1.34. The total power of the three-phase system is calculated by considering the instantaneous voltages and currents.

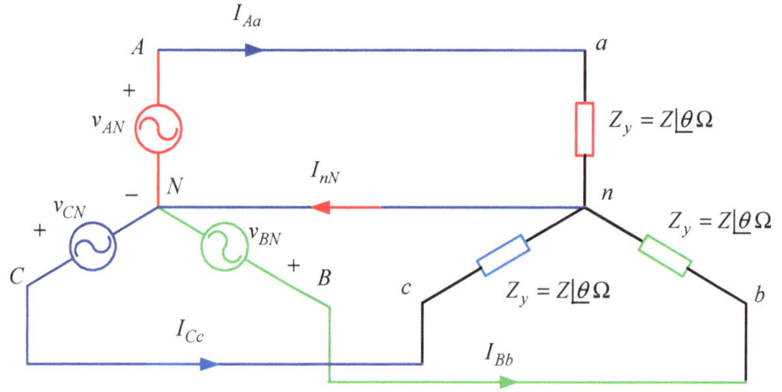

Fig. 1.34 Wye-wye system for power calculation

The instantaneous voltages are

$$v_{AN} = V_m \sin \omega t \tag{1.179}$$

$$v_{BN} = V_m \sin(\omega t - 120^\circ) \tag{1.180}$$

$$v_{CN} = V_m \sin(\omega t - 240^\circ) \tag{1.181}$$

The phase currents of the three-phase wye-connected load can be expressed as

$$i_{Aa} = \frac{v_{AN}}{Z_y} = \frac{V_m \ \sin \omega t}{Z \underline{|\theta}} = I_m \sin(\omega t - \theta) \tag{1.182}$$

$$i_{Bb} = \frac{v_{BN}}{Z_y} = \frac{V_m \ \sin(\omega t - 120^\circ)}{Z \underline{|\theta}} = I_m \sin(\omega t - \theta - 120^\circ) \tag{1.183}$$

$$i_{Cc} = \frac{v_{CN}}{Z_y} = \frac{V_m \ \sin(\omega t - 240^\circ)}{Z \underline{|\theta}} = I_m \sin(\omega t - \theta - 240^\circ) \tag{1.184}$$

The instantaneous power for phase a can be expressed as [8, 9]

$$p_a(t) = \frac{1}{T} \int_0^T v_{AN} i_{Aa} \, \mathrm{d}t \tag{1.185}$$

Substituting Eqs. (1.179) and (1.182) into Eq. (1.185) yields

$$p_a(t) = \frac{V_m I_m}{T} \int_0^T \sin \omega t \times \ \sin(\omega t - \theta) \, \mathrm{d}t \tag{1.186}$$

$$p_a(t) = \frac{V_m I_m}{2T} \int_0^T 2 \sin \omega t \times \ \sin(\omega t - \theta) \, \mathrm{d}t \tag{1.187}$$

$$p_a(t) = \frac{V_m I_m}{2T} \int_0^T [\cos \theta - \cos(2\omega t - \theta)] \, \mathrm{d}t \tag{1.188}$$

$$p_a(t) = \frac{V_m I_m}{2T} \times \ \cos \theta \times T - 0 \tag{1.189}$$

$$p_a(t) = \frac{V_m I_m}{\sqrt{2} \times \sqrt{2}} \cos \theta = V_P I_P \cos \theta \tag{1.190}$$

where V_p and I_p are the rms values of phase voltage and phase current.

Similarly, the expressions of the instantaneous power for the phase b and phase c can be written as

$$p_b(t) = V_P I_P \cos\theta \tag{1.191}$$

$$p_c(t) = V_P I_P \cos\theta \tag{1.192}$$

Therefore, the average three-phase power P can be calculated as

$$P = p_a(t) + p_b(t) + p_c(t) \tag{1.193}$$

Substituting Eqs. (1.190), (1.191) and (1.192) into Eq. (1.193) yields

$$P_t = 3V_p I_p \cos\theta \tag{1.194}$$

Similarly, the expression of three-phase reactive power can be expressed as

$$Q_t = 3\,V_p I_p \sin\theta \tag{1.195}$$

Therefore, the per-phase average (P_{pp}) and reactive (Q_{pp}) power can be written as

$$P_{pp} = V_P I_P \cos\theta \tag{1.196}$$

$$Q_{pp} = V_P I_P \sin\theta \tag{1.197}$$

The complex power per phase S_{pp} is represented as

$$S_{pp} = P_{pp} + jQ_{pp} \tag{1.198}$$

Substituting Eqs. (1.196) and (1.197) into Eq. (1.198) yields

$$S_{pp} = V_P I_P \cos\theta + jV_P I_P \sin\theta \tag{1.199}$$

Equation (1.199) can be expressed as

$$S_{pp} = V_P I_P \,\underline{|\theta} \tag{1.200}$$

From Eq. (1.200), it is seen that the per-phase complex power is equal to the product of the voltage per phase and the phase current with an angle.

Y-connection: Substituting Eqs. (1.151) and (1.152) into Eq. (1.194) yields

$$P_{tY} = 3 \times \frac{V_L}{\sqrt{3}} \times I_L \times \cos\theta \tag{1.201}$$

$$P_{tY} = \sqrt{3}\, V_L I_L \cos\theta \tag{1.202}$$

Substituting Eqs. (1.151) and (1.152) into Eq. (1.195) yields

$$Q_{tY} = 3 \times \frac{V_L}{\sqrt{3}} \times I_L \times \sin\theta \tag{1.203}$$

$$Q_{tY} = \sqrt{3}\, V_L I_L \sin\theta \tag{1.204}$$

Delta connection: Again, substituting Eqs. (1.177) and (1.178) into Eq. (1.196) yields

$$P_{t\Delta} = 3 \times \frac{I_L}{\sqrt{3}} \times V_L \times \cos\theta \tag{1.205}$$

$$P_{t\Delta} = \sqrt{3}\, V_L I_L \cos\theta \tag{1.206}$$

Substituting Eqs. (1.177) and (1.178) into Eq. (1.197) yields

$$Q_{t\Delta} = 3 \times \frac{V_L}{\sqrt{3}} \times I_L \times \sin\theta \tag{1.207}$$

$$Q_{t\Delta} = \sqrt{3}\, V_L I_L \sin\theta \tag{1.208}$$

In general, the total real and reactive power can be expressed as

$$P_t = \sqrt{3}\, V_L I_L \cos\theta \tag{1.209}$$

$$Q_t = \sqrt{3}\, V_L I_L \sin\theta \tag{1.210}$$

The total complex power can be written as

$$S_t = P_t + jQ_t \tag{1.211}$$

Substituting Eqs. (1.209) and (1.210) into Eq. (1.211) yields

$$S_t = \sqrt{3}\, V_L I_L \cos\theta + j\sqrt{3}\, V_L I_L \sin\theta \tag{1.212}$$

$$S_t = \sqrt{3}\, V_L I_L \underline{\left|\theta\right.} \tag{1.213}$$

Three-phase system uses less amount of copper wire than the single-phase system for the same line voltage and same power factor to transmit the same

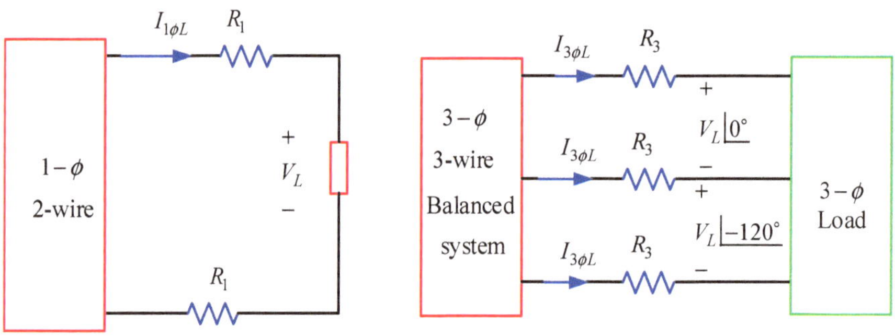

Fig. 1.35 Single-phase and three-phase systems with loads

amount of power over a fixed distance. From Fig. 1.35, the real power for a single-phase two wire system is

$$P_{1\phi 2w} = V_L I_{1\phi L} \cos\theta \tag{1.214}$$

From Fig. 1.35, the real power for a three-phase three wire system is

$$P_{3\phi 3w} = \sqrt{3}\, V_L I_{3\phi L} \cos\theta \tag{1.215}$$

Equations (1.214) and (1.215) will be equal for transmitting or delivering the same amount of power over a fixed distance. It can be expressed as

$$V_L I_{1\phi L} \cos\theta = \sqrt{3}\, V_L I_{3\phi L} \cos\theta \tag{1.216}$$

$$I_{1\phi L} = \sqrt{3}\, I_{3\phi L} \tag{1.217}$$

The power loss in the single-phase wire is

$$P_{1\phi 2\text{wloss}} = 2I_{1\phi L}^2 R_1 \tag{1.218}$$

$$P_{3\phi 3\text{wloss}} = 3I_{3\phi L}^2 R_3 \tag{1.219}$$

From Eqs. (1.218) and (1.219), the ratio of power loss of a single-phase system to a three-phase system can be derived as

$$\frac{P_{1\phi 2\text{wloss}}}{P_{3\phi 3\text{wloss}}} = \frac{2I_{1\phi L}^2 R_1}{3I_{3\phi L}^2 R_3} \tag{1.220}$$

For equal losses ($P_{1\phi 2\text{wloss}} = P_{3\phi 3\text{wloss}}$), Eq. (1.220) can be modified as

$$1 = \frac{2I_{1\phi L}^2 R_1}{3I_{3\phi L}^2 R_3} \tag{1.221}$$

$$\frac{3I_{3\phi L}^2}{2I_{1\phi L}^2} = \frac{R_1}{R_3} \tag{1.222}$$

Substituting Eq. (1.217) into Eq. (1.222) yields

$$\frac{3I_{3\phi L}^2}{2 \times 3I_{3\phi L}^2} = \frac{R_1}{R_3} \tag{1.223}$$

$$\frac{R_1}{R_3} = \frac{1}{2} \tag{1.224}$$

The following ratio can be written as

$$\frac{\text{Copper for } 3\phi \text{ system}}{\text{Copper for } 1\phi \text{ system}} = \frac{\text{number of wires in } 3\phi \text{ system}}{\text{number of wires in } 1\phi \text{ system}} \times \frac{R_1}{R_3} \tag{1.225}$$

Substituting Eq. (1.224) and the number of wires for both systems in Eq. (1.225) yields

$$\frac{\text{Copper for } 3\phi \text{ system}}{\text{Copper for } 1\phi \text{ system}} = \frac{3}{2} \times \frac{1}{2} \tag{1.226}$$

$$\text{Copper for } 3\phi \text{ system} = \frac{3}{4} \times \text{Copper for } 1\phi \text{ system} \tag{1.227}$$

From Eq. (1.227), it is seen that the copper required for three-phase system is equal to the three-fourths of the copper required for a single-phase system.

Example 1.11
A balanced three-phase wye-wye system is shown in Fig. 1.36. For *ABC* phase sequence, calculate the line current, power supplied to each phase, power absorbed by each phase, and the total complex power supplied by the source.

Solution
The line currents are calculated as

$$I_{Aa} = \frac{230\,\underline{|15^\circ}}{2 + j8} = 27.89\,\underline{|-60.96^\circ}\text{ A} \tag{1.228}$$

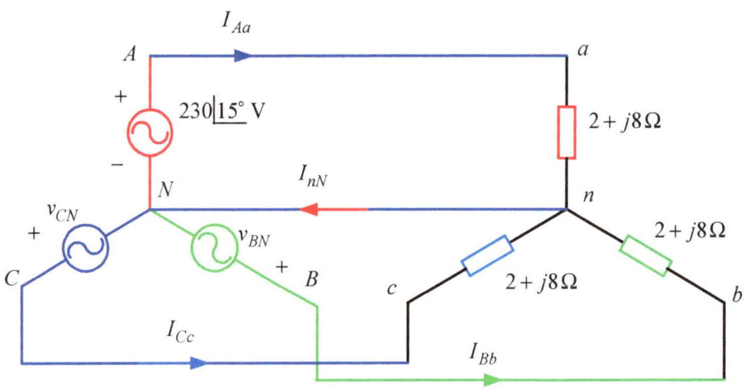

Fig. 1.36 Circuit for Example 1.11

$$I_{Bb} = I_{Aa}\angle{-120^\circ} = 27.89\angle{-180.96^\circ}\ \text{A} \tag{1.229}$$

$$I_{Cc} = I_{Aa}\angle{+120^\circ} = 27.89\angle{59.04^\circ}\ \text{A} \tag{1.230}$$

Power supplied to each phase is calculated as

$$P_A = V_p I_p \cos\theta = V_{AN} I_{Aa} \cos(\theta_v - \theta_i) = 230 \times 27.89 \times \cos(15 + 60.96) \\ = 1556.20\ \text{W} \tag{1.231}$$

Per-phase power absorbed by the load is calculated as

$$P_{L1\phi} = 27.89^2 \times 2 = 1555.70\ \text{W} \tag{1.232}$$

The total complex power supplied by the source is calculated as

$$S_t = 3V_{An} I_{Aa}^* = 3 \times 230\angle{15^\circ} \times 27.89\angle{60.96^\circ} = 4668.60 + j18669.21\ \text{VA} \tag{1.233}$$

Practice Problem 1.11

A balanced three-phase wye-delta system is shown in Fig. 1.37. For *ABC* phase sequence, find the line current, power supplied to each phase, power absorbed by each phase, and the total complex power supplied by the source.

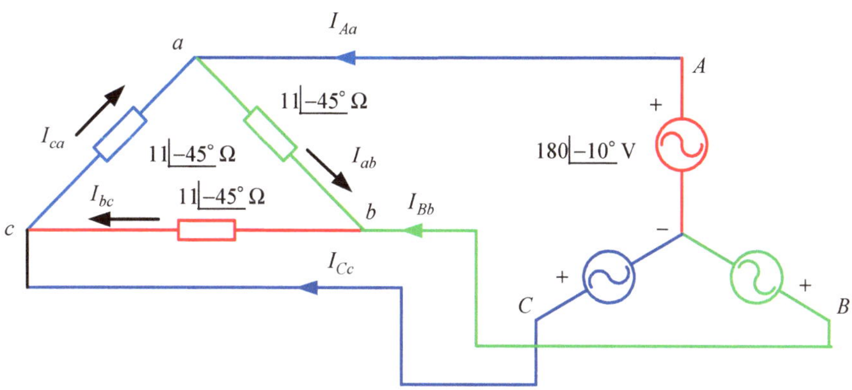

Fig. 1.37 Circuit for Practice Problem 1.11

1.16 Basic Measuring Equipment

In laboratory experiments, students usually verify fundamental electrical theories through the measurement of voltage, current, resistance, and power. The voltmeter, ammeter, ohmmeter, and wattmeter are used to measure those parameters. Nowadays, almost every educational laboratory is equipped with digital meters. An ohmmeter is used to measure resistance within a circuit, check the circuit continuity, identify short circuit and open circuit, and to identify the specific lead of a multilead cable. The voltmeter is used to measure the voltage of a circuit, and it always connects in parallel with a specific element. The voltmeter has high input impedance. An ammeter is an instrument that is used to measure the current in the circuit. An ammeter has low input impedance, and it connects in series with an element of the circuit. Under the energized condition, the ammeter cannot be disconnected from the circuit, whereas the voltmeter can be disconnected from the circuit. A wattmeter has two coils, namely a voltage coil and a current coil. The voltage coil is connected across the element, and the current coil is connected in series. The symbols of voltmeter, ammeter, wattmeter, and ohmmeter are shown in Fig. 1.38.

There are many advanced electrical meters available for use in the practical field and can be obtained at a low price from hardware stores. Fluke Corporation introduced different types of digital multimeter as shown in Fig. 1.39, which

Fig. 1.38 Symbols of basic electrical meters

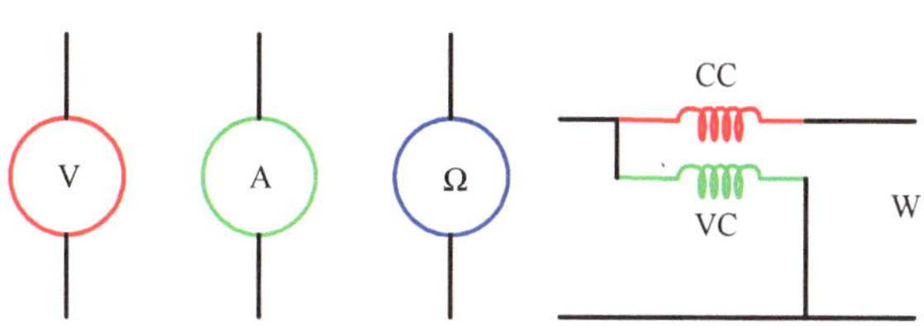

Fig. 1.39 Digital multimeters courtesy by Fluke Corporation

includes advanced safety features during practical measurements. These meters have proprietary functions that prevent accidents resulting from breakers that suddenly trip due to incorrect connections. These meters also have high accuracy, reliability, extensive additional functionality, and a broad range of measurement activities.

References

1. Alexander CK, Sadiku MNO (2016) Fundamentals of electric circuits, 6th edn. McGraw-Hill Higher Education
2. Boylestad RL (2016) Introductory circuit analysis, 13th edn. Pearson
3. Nilsson JW, Riedel SA (2015) Electric circuits, 10th edn. Prentice-Hall International Edition
4. Jackson HW, Temple D, Kelly BE (2015) Introduction to electric circuits, 9th edn. Oxford University Press
5. Bell D (2007) Fundamentals of electric circuits, 7th edn. Oxford University Press
6. Rizzoni G, Kearns J (2014) Principles and applications of electrical engineering, 6th edn. McGraw-Hill Education

7. Irwin JD, Nelms RM (2015) Basic engineering circuit analysis, 11th edn. Wiley, USA
8. Hayt W, Kemmerly J (2012) Engineering circuit analysis, 8th edn. McGraw-Hill Education
9. Salam MA, Rahman QM (2018) Fundamentals of electrical circuits analysis, 1st edn. Springer

Exercise Problems

1.1 The excitation voltage and the impedance of a series circuit are given by $v(t) = 8\sin 10t\,\text{V}$ and $Z = 5\,\underline{|10^\circ}\,\Omega$, respectively. Calculate the instantaneous power.

1.2 The excitation current and the impedance of a series circuit are given by $i(t) = 4\sin(100t - 20^\circ)\text{A}$ and $Z = 5\,\underline{|10^\circ}\,\Omega$, respectively. Determine the instantaneous power.

1.3 Calculate the average power supplied by the source and the power absorbed by the resistors as shown in Fig. P1.1.

1.4 Determine the average power supplied by the source and the power absorbed by the 8Ω resistor shown in Fig. P1.2.

1.5 Calculate the average power supplied by the source and the power absorbed by the 3Ω resistor as shown in Fig. P1.3.

1.6 Determine the average power supplied by the source and the power absorbed by the 2Ω resistor shown in Fig. P1.4.

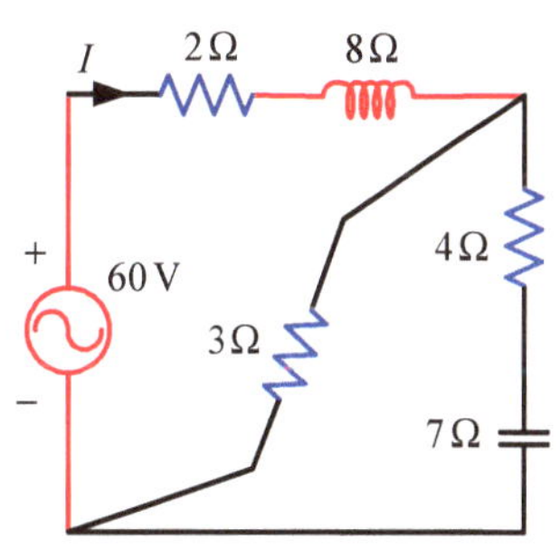

Fig. P1.1 Circuit for Problem 1.3

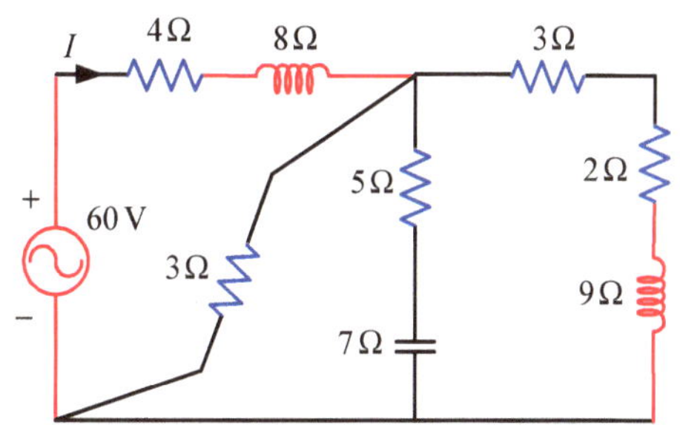

Fig. P1.2 Circuit for Problem 1.4

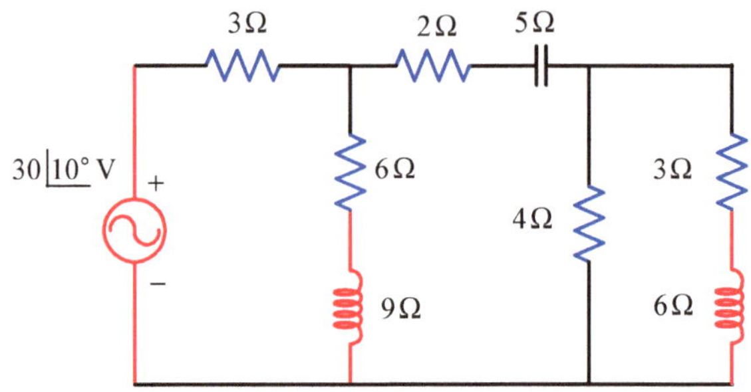

Fig. P1.3 Circuit for Problem 1.5

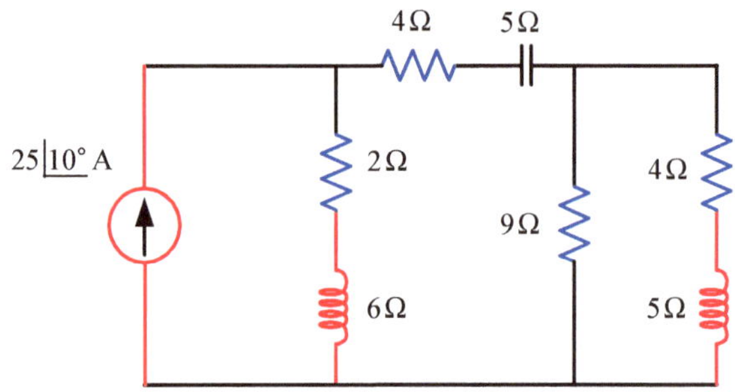

Fig. P1.4 Circuit for Problem 1.6

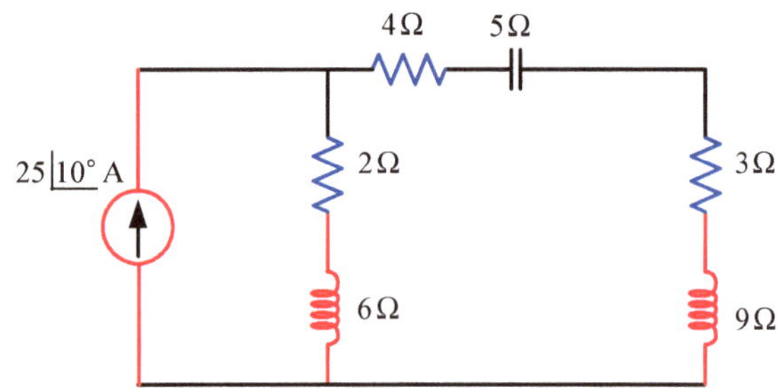

Fig. P1.5 Circuit for Problem 1.7

1.7 Find the total average power absorbed by all the resistors in the circuit shown in Fig. P1.5.

1.8 An industrial load is connected across an alternating voltage source $v(t) = 230\sin(314t + 20^\circ)$ V that draws a current of $i(t) = 15\sin(314t + 45^\circ)$ A. Determine the apparent power, circuit resistance and capacitance.

1.9 The rms values of voltage and current are given by $V = 20\,\angle -15^\circ$ V, and $I = 3\,\angle 25^\circ$ A. Calculate the complex power, real power and reactive power.

1.10 The rms values of voltage is given by $V = 34\,\angle 25^\circ$ V, and the impedance is $Z = 6\,\angle -15^\circ\ \Omega$. Determine the complex power, real power and reactive power.

Fig. P1.6 Circuit for Problem 1.11

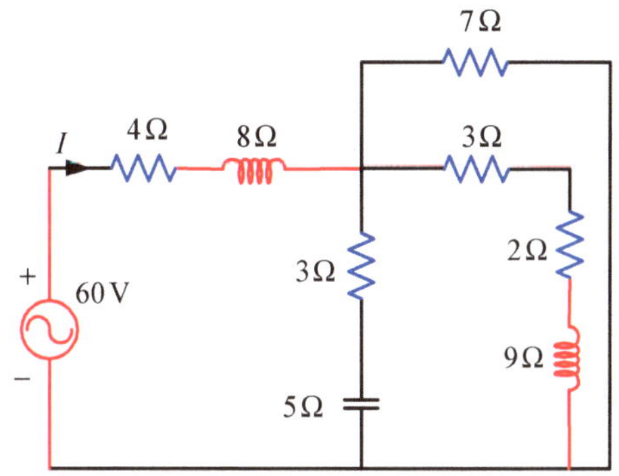

Fig. P1.7 Circuit for Problem 1.12

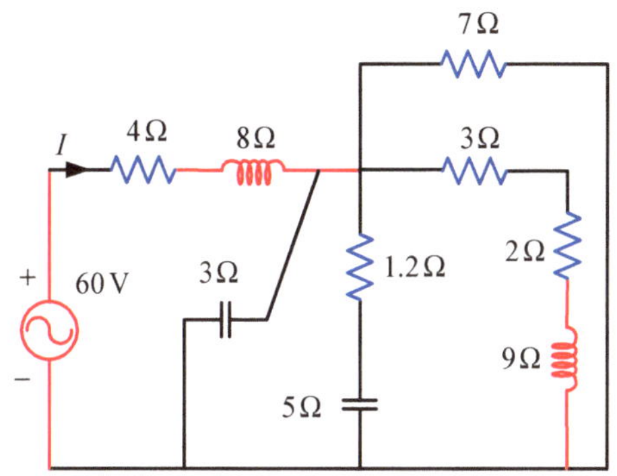

Fig. P1.8 Circuit for Problem 1.13

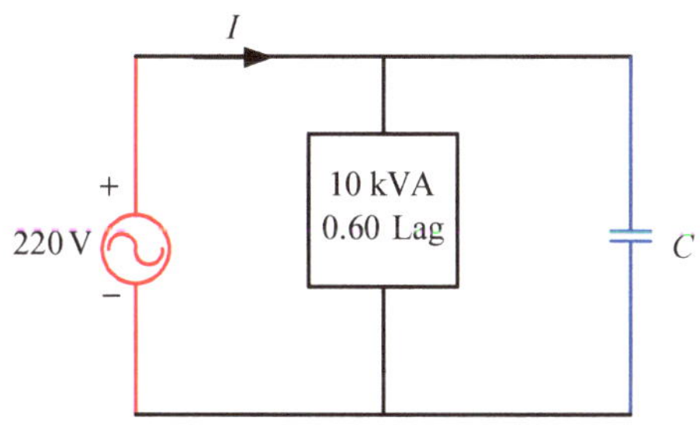

1.11 A series-parallel circuit is supplied by a rms source of 60 V as shown in Fig. P1.6. Find the total complex power.

1.12 Calculate the total complex power of the circuit shown in Fig. P1.7.

1.13 A 10 kVA, 50 Hz, 0.6 lagging power factor load is connected across an rms voltage source of 220 V as shown in Fig. P1.8. A capacitor is connected across the load to improve the power factor to 0.85 lagging. Find the value of the capacitor.

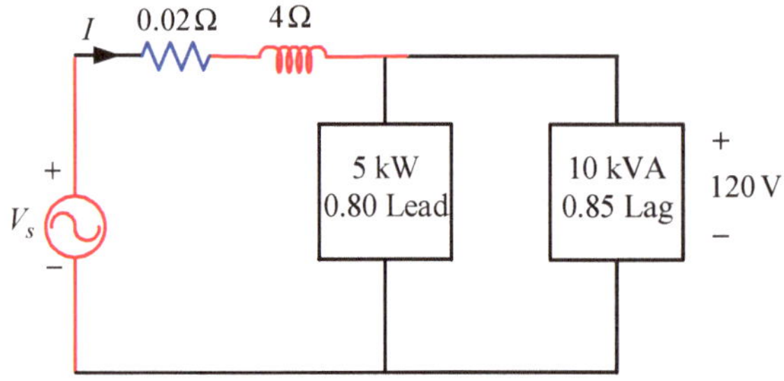

Fig. P1.9 Circuit for Problem 1.14

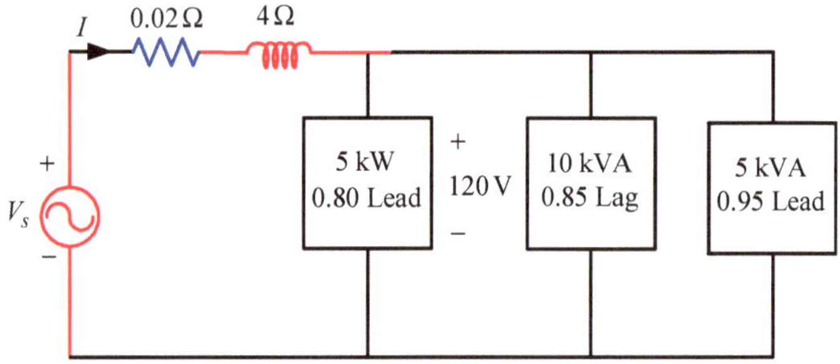

Fig. P1.10 Circuit for Problem 1.15

1.14 Two loads with different power factor are connected with the source through a transmission line as shown in Fig. P1.9. Determine the source current and the source voltage.

1.15 A voltage source delivers power to the three loads shown in Fig. P1.10. Find the source current and the source voltage.

1.16 The line voltage of a three-phase wye-connected generator is found to be 440 V. For *abc* phase sequence, calculate the phase voltages.

1.17 The phase voltage of a three-phase wye-connected generator is given by $V_{an} = 100 \angle -10^\circ$ V. Determine the voltages V_{bn} and V_{cn} for *abc* phase sequence.

Chapter 2
Transformer: Principles and Practices

2.1 Introduction

There are many devices such as three-phase AC generators, transformers, etc., which are used in a power station to generate and supply electrical power to a power system network. In the power station, the three-phase AC generator generates a three-phase alternating voltage in the range between 11 and 20 kV. The magnitude of the generated voltage is increased to 120 kV or more using a power transformer. This higher magnitude of voltage is then transmitted to the grid substation by three-phase transmission lines. A lower line voltage of 415 V is achieved by stepping down either from the 11 kV or 33 kV lines by a distribution transformer. In these cases, a three-phase transformer is used either to step up or step down the voltage. Since a transformer plays a vital role in feeding an electrical network with the required voltage, it becomes an important requirement of a power system engineer to understand the fundamental details about a transformer along with its analytical behavior in the circuit domain. This chapter is dedicated to this goal. On the onset of this discussion, it is worth mentioning that a transformer, irrespective of its type, contains the following characteristics: (i) it has no moving parts, (ii) no electrical connection between the primary and secondary windings, (iii) windings are magnetically coupled, (iv) rugged and durable in construction, (v) efficiency is very high, i.e., more than 95%, and (vi) frequency is unchanged.

2.2 Working Principle of Transformer

Figure 2.1 shows a schematic diagram of a single-phase transformer. There are two types of windings in a single-phase transformer. These are called primary and secondary windings or coils. The primary winding is connected to the alternating voltage source, and the secondary winding is connected to the load.

Md. A. Salam, *Fundamentals of Electrical Power Systems Analysis*,
https://doi.org/10.1007/978-981-15-3212-2_2

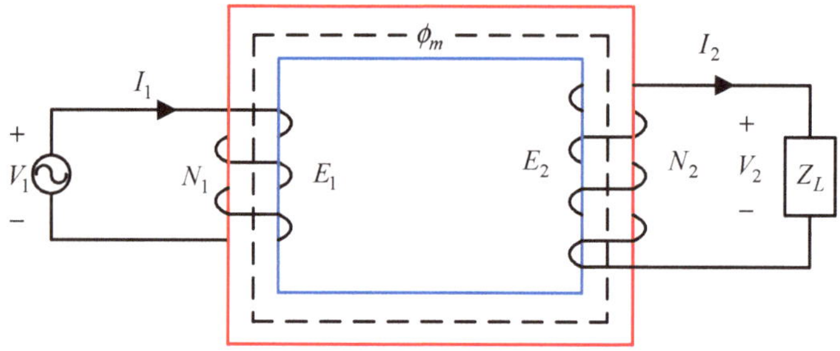

Fig. 2.1 Single-phase transformer

The primary and secondary winding parameters are represented by the suffix *p* or 1 and *s* or 2, respectively. A sinusoidal current flows in the primary winding when it is connected to an alternating voltage source. This current establishes a flux ϕ that moves from the primary winding to the secondary winding through low reluctance magnetic core [1].

About 95% of this flux moves from the primary to the secondary through the low reluctance path of the magnetic core, and this flux is linked by both windings and a small percent of this flux links to the primary winding. According to Faraday's laws of electromagnetic induction, a voltage will be induced across the secondary winding as well as in the primary winding. Due to this voltage, a current will flow through the load if it is connected with the secondary winding. Hence, the primary voltage is transferred to the secondary winding without a change in frequency.

2.3 Flux in a Transformer

The current in the primary winding establishes a flux. The flux that moves from primary to secondary and links both the windings is called the mutual flux, and its maximum value is represented by ϕ_m. Flux which links only the primary winding and completes the magnetic path through the surrounding air is known as primary leakage flux. The primary leakage flux is denoted by ϕ_{1l}. Similarly, secondary leakage flux is that flux which links only the secondary winding and completes the magnetic path through the surrounding air. The secondary leakage flux is denoted by ϕ_{2l}. Mutual and leakage fluxes are shown in Fig. 2.2.

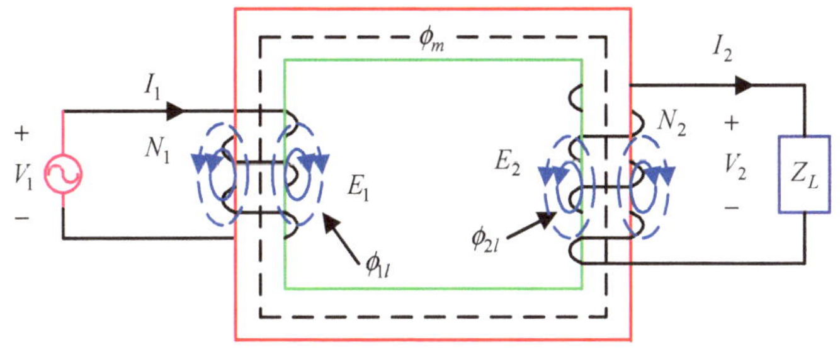

Fig. 2.2 Mutual and leakage fluxes

2.4 Ideal Transformer

An ideal transformer is one that does not supply any energy to the load, i.e., the secondary winding is open circuited. The main points of an ideal transformer are (i) no winding resistance, (ii) no leakage flux and leakage inductance, (iii) self-inductance and mutual inductance are zero, (iv) no losses due to resistance, inductance, hysteresis, or eddy current, and (v) coefficient of coupling is unity [2].

Figure 2.3a shows an ideal transformer where the secondary winding is left open. A small magnetizing current I_m will flow in the primary winding when it is connected to the alternating voltage source, V_1. This magnetizing current lags behind the supply voltage, V_1 by 90° and produces the flux ϕ, which induces the primary and secondary emfs. These emfs lag behind the flux, ϕ by 90°. The magnitude of primary induced emf E_1 and the supply voltage V_1 is the same, but are 180° out of phase as shown in Fig. 2.3b.

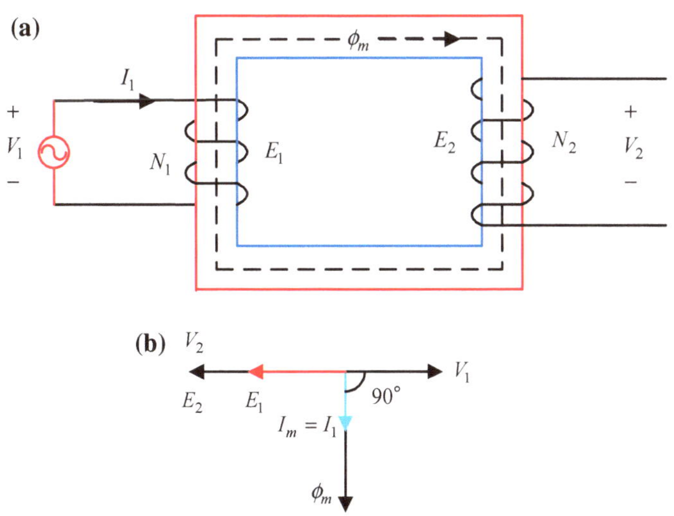

Fig. 2.3 Ideal transformer and phasor diagram

2.5 EMF Equation of Transformer

The primary winding draws a current when it is connected to an alternating voltage source. This primary sinusoidal current produces a sinusoidal flux ϕ that can be expressed as [2],

$$\phi = \phi_m \sin \omega t \tag{2.1}$$

Instantaneous emf induced in the primary winding is,

$$e_1 = -N_1 \frac{d\phi}{dt} \tag{2.2}$$

Similarly, instantaneous emf induced in the secondary winding is,

$$e_2 = -N_2 \frac{d\phi}{dt} \tag{2.3}$$

Substituting Eq. (2.1) into the Eq. (2.2) yields,

$$e_1 = -N_1 \frac{d}{dt}(\phi_m \sin \omega t) \tag{2.4}$$

$$e_1 = -N_1 \omega \phi_m \cos \omega t \tag{2.5}$$

$$e_1 = N_1 \omega \phi_m \sin(\omega t - 90^\circ) \tag{2.6}$$

The maximum value of e_1 is,

$$E_{m1} = N_1 \omega \phi_m \tag{2.7}$$

The rms value of the primary emf is,

$$E_1 = \frac{E_{m1}}{\sqrt{2}} \tag{2.8}$$

Substituting Eq. (2.7) into Eq. (2.8) yields,

$$E_1 = \frac{N_1 2\pi f \phi_m}{\sqrt{2}} \tag{2.9}$$

$$E_1 = 4.44 f \phi_m N_1 \tag{2.10}$$

Similarly, the expression of the secondary emf is,

$$E_2 = 4.44 f \phi_m N_2 \tag{2.11}$$

The primary and secondary voltages can be determined from equations Eqs. (2.10) and (2.11) if other parameters are known.

2.6 Turns Ratio of Transformer

Turns ratio is an important parameter for drawing an equivalent circuit of a transformer. The turns ratio is used to identify step-up and step-down transformers. According to Faraday's laws, the induced emfs in the primary (e_1) and the secondary (e_2) windings are,

$$e_1 = -N_1 \frac{d\phi}{dt} \tag{2.12}$$

$$e_2 = -N_2 \frac{d\phi}{dt} \tag{2.13}$$

Dividing Eq. (2.12) by Eq. (2.13) yields,

$$\frac{e_1}{e_2} = \frac{N_1}{N_2} \tag{2.14}$$

$$\frac{e_1}{e_2} = \frac{N_1}{N_2} = a \tag{2.15}$$

Similarly, dividing Eq. (2.10) by (2.11) yields,

$$\frac{E_1}{E_2} - \frac{N_1}{N_2} - a \tag{2.16}$$

where a is the turns ratio of a transformer. In case of $N_2 > N_1$, the transformer is called a step-up transformer. Whereas for $N_1 > N_2$, the transformer is called a step-down transformer. The losses are zero in an ideal transformer. In this case, the input power of the transformer is equal to its output power and this yield,

$$V_1 I_1 = V_2 I_2 \tag{2.17}$$

Equation (2.17) can be rearranged as,

$$\frac{V_1}{V_2} = \frac{I_2}{I_1} = a \tag{2.18}$$

The ratio of primary current to the secondary current is,

$$\frac{I_1}{I_2} = \frac{1}{a} \tag{2.19}$$

Again, the magnetomotive force produced by the primary current will be equal to the magnetomotive force produced by the secondary current and it can be expressed as,

$$\Im = \Im_1 - \Im_2 = 0 \tag{2.20}$$

$$N_1 I_1 = N_2 I_2 \tag{2.21}$$

$$\frac{I_1}{I_2} = \frac{N_2}{N_1} = \frac{1}{a} \tag{2.22}$$

From Eq. (2.22), it is concluded that the ratio of primary to secondary current is inversely proportional to the turns ratio of the transformer.

The input and output powers of an ideal transformer are,

$$P_{\text{in}} = V_1 I_1 \cos\phi_1 \tag{2.23}$$

$$P_{\text{out}} = V_2 I_2 \cos\phi_2 \tag{2.24}$$

For an ideal condition, the angle ϕ_1 is equal to the angle ϕ_2 and the output power can be rearranged as,

$$P_{\text{out}} = \frac{V_1}{a} a I_1 \cos\phi_1 \tag{2.25}$$

$$P_{\text{out}} = V_1 I_1 \cos\phi_1 = P_{\text{in}} \tag{2.26}$$

From Eq. (2.26), it is seen that the input and the output powers are the same in case of an ideal transformer. Similarly, the input and output reactive powers are,

$$Q_{\text{out}} = V_2 I_2 \sin\phi_2 = V_1 I_1 \sin\phi_1 = Q_{\text{in}} \tag{2.27}$$

From Eqs. (2.26) and (2.27), the input and output power and reactive power can be calculated if other parameters are given.

Example 2.1

The number of turns in the secondary coil of a 22 kVA, 2200 V/220 V single-phase transformer is 50. Find the number of primary turns, primary full load current, and secondary full load current. Neglect all kinds of losses in the transformer.

Solution

The value of the turns ratio is,

$$a = \frac{V_1}{V_2} = \frac{2200}{220} = 10 \tag{2.28}$$

The value of the primary turns can be determined as,

$$N_1 = aN_2 = 10 \times 50 = 500 \tag{2.29}$$

The value of the primary full load current is,

$$I_1 = \frac{22 \times 10^3}{2200} = 10\,\mathrm{A} \tag{2.30}$$

The value of the secondary full load current is,

$$I_2 = \frac{22 \times 10^3}{220} = 100\,\mathrm{A} \tag{2.31}$$

Example 2.2

A 25 kVA single-phase transformer has the primary and secondary number of turns of 200 and 400, respectively. The transformer is connected to a 220 V, 50 Hz source. Calculate the turns ratio and mutual flux in the core.

Solution

The turns ratio is,

$$a = \frac{N_1}{N_2} = \frac{200}{400} = 0.5 \tag{2.32}$$

The value of the mutual flux can be calculated as,

$$\phi_m = \frac{V_1}{4.44 f N_1} = \frac{220}{4.44 \times 50 \times 200} = 4.95\,\mathrm{mWb} \tag{2.33}$$

Practice Problem 2.1

The primary voltage of an iron core single-phase transformer is 220 V. The number of primary and secondary turns of the transformer is 200 and 50, respectively. Calculate the voltage of the secondary coil.

Practice Problem 2.2

The number of primary turns of a 30 kVA, 2200 V/220 V, 50 Hz single-phase transformer is 100. Find the turns ratio, the mutual flux in the core, and full load primary, and secondary currents.

2.7 Rules for Referring Impedance

For developing an equivalent circuit of a transformer, it is necessary to refer the parameters from the primary to the secondary or the secondary to the primary. These parameters are resistance, reactance, impedance, current, and voltage. The ratio of primary voltage to secondary voltage is,

$$\frac{V_1}{V_2} = a \tag{2.34}$$

The ratio of primary current to secondary current is,

$$\frac{I_1}{I_2} = \frac{1}{a} \tag{2.35}$$

Dividing Eq. (2.34) by Eq. (2.35) yields,

$$\frac{\frac{V_1}{V_2}}{\frac{I_1}{I_2}} = \frac{a}{\frac{1}{a}} \tag{2.36}$$

$$\frac{\frac{V_1}{I_1}}{\frac{V_2}{I_2}} = a^2 \tag{2.37}$$

$$\frac{Z_1}{Z_2} = a^2 \tag{2.38}$$

Alternative approach: The impedances in the primary and secondary windings are,

$$Z_1 = \frac{V_1}{I_1} \tag{2.39}$$

$$Z_2 = \frac{V_2}{I_2} \tag{2.40}$$

Dividing Eq. (2.39) by Eq. (2.40) yields,

$$\frac{Z_1}{Z_2} = \frac{\frac{V_1}{I_1}}{\frac{V_2}{I_2}} \tag{2.41}$$

$$\frac{Z_1}{Z_2} = \frac{V_1}{V_2} \times \frac{I_2}{I_1} \tag{2.42}$$

$$\frac{Z_1}{Z_2} = a \times a \tag{2.43}$$

$$\frac{Z_1}{Z_2} = a^2 \tag{2.44}$$

From Eq. (2.44), it can be concluded that the impedance ratio is equal to the square of the turns ratio. The important points for transferring parameters are (i) R_1 in the primary becomes $\frac{R_1}{a^2}$ when referred to the secondary, (ii) R_2 in the secondary becomes a^2R_2 when referred to the primary, (iii) X_1 in the primary becomes $\frac{X_1}{a^2}$ when referred to the secondary, and (iv) X_2 in the secondary becomes a^2X_2 when referred to the primary.

Example 2.3
The number of primary and secondary turns of a single-phase transformer is 300 and 30, respectively. The secondary coil is connected with a load impedance of 4 Ω. Calculate the turns ratio, load impedance referred to the primary, and primary current if the primary coil voltage is 220 V.

Solution
The value of the turns ratio is,

$$a = \frac{N_1}{N_2} = \frac{300}{30} = 10 \tag{2.45}$$

The value of the load impedance referred to the primary is,

$$Z_L' = a^2 Z_L = 10^2 \times 4 = 400\,\Omega \tag{2.46}$$

The value of the primary current is,

$$I_1 = \frac{V_1}{Z_L'} = \frac{220}{400} = 0.55\,\text{A} \tag{2.47}$$

Practice Problem 2.3
A load impedance of 8 Ω is connected to the secondary coil of 400/200 turns single-phase transformer. Determine the turns ratio, load impedance referred to primary and primary current if the primary coil voltage is 120 V.

2.8 Equivalent Circuit of a Transformer

Windings of a transformer are not connected electrically and are magnetically coupled with each other. In this case, it is tedious to do a proper analysis. Therefore, for easy computation and visualization, the practical transformer needs to be converted into an equivalent circuit by maintaining the same properties of the main transformer [3]. In the equivalent circuit, the related parameters need to be transferred either from the primary to the secondary or vice versa. A two windings ideal transformer is shown in Fig. 2.4.

2.8.1 *Exact Equivalent Circuit*

Figure 2.5 shows an exact equivalent circuit referred to the primary where all the parameters are transferred from the secondary to the primary and these parameters are,

$$R_2' = a^2 R_2 \tag{2.48}$$

$$X_2' = a^2 X_2 \tag{2.49}$$

$$Z_L' = a^2 Z_L \tag{2.50}$$

$$I_2' = \frac{I_2}{a} \tag{2.51}$$

$$V_2' = aV_2 \tag{2.52}$$

Figure 2.6 shows the exact equivalent circuit referred to the secondary where all the parameters are transferred from the primary to the secondary.

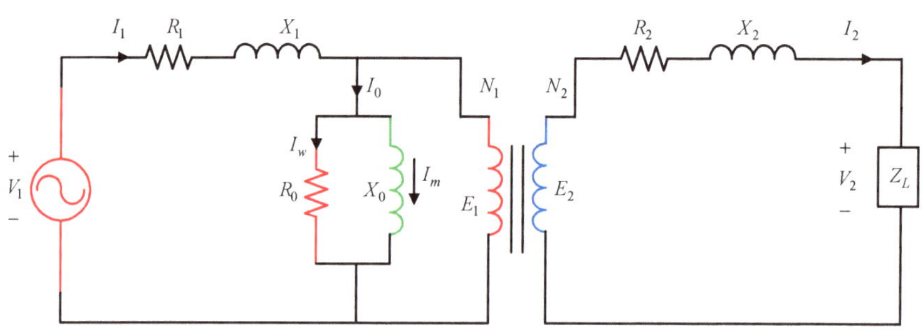

Fig. 2.4 Two windings transformer

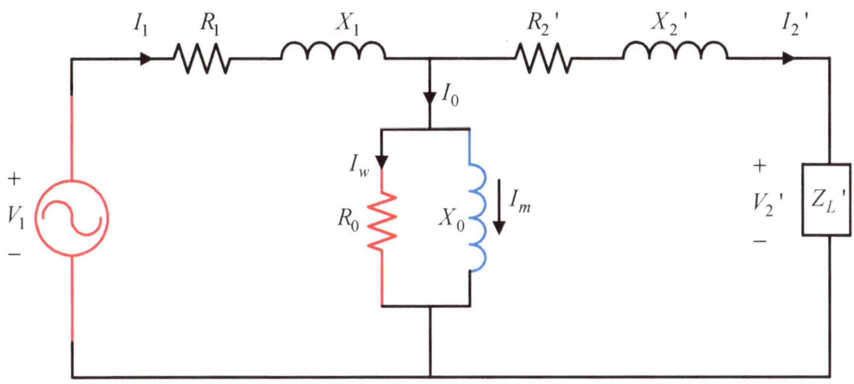

Fig. 2.5 Exact equivalent circuit referred to the primary

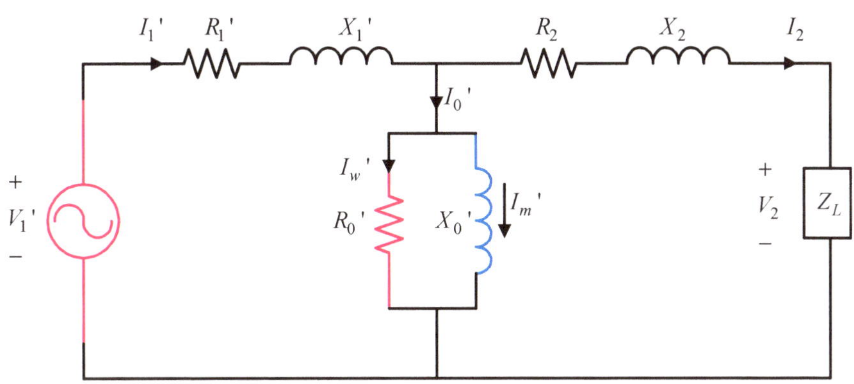

Fig. 2.6 Exact equivalent circuit referred to as the secondary

These parameters are,

$$R_1' = \frac{R_1}{a^2} \tag{2.53}$$

$$X_1' = \frac{X_1}{a^2} \tag{2.54}$$

$$I_1' = aI_1 \tag{2.55}$$

$$V_1' = \frac{V_1}{a} \tag{2.56}$$

$$I_w' = aI_w \tag{2.57}$$

$$I'_m = aI_m \quad (2.58)$$

$$I'_0 = aI_0 \quad (2.59)$$

2.8.2 Approximate Equivalent Circuit

The no-load current is very small as compared to the rated primary current. Therefore, there is a negligible voltage drop due to R_1 and X_1. In this condition, it can be assumed that the voltage drop across the no-load circuit is the same as the applied voltage without any significant error. The approximate equivalent circuit can be drawn by shifting the no-load circuit across the supply voltage, V_1. Figure 2.7 shows an approximate equivalent circuit referred to the primary. The total resistance, reactance, and impedance referred to the primary are,

$$R_{01} = R_1 + R'_2 = R_1 + a^2 R_2 \quad (2.60)$$

$$X_{01} = X_1 + X'_2 = X_1 + a^2 X_2 \quad (2.61)$$

$$Z_{01} = R_{01} + jX_{01} \quad (2.62)$$

The no-load circuit resistance and reactance are,

$$R_0 = \frac{V_1}{I_w} \quad (2.63)$$

$$X_0 = \frac{V_1}{I_m} \quad (2.64)$$

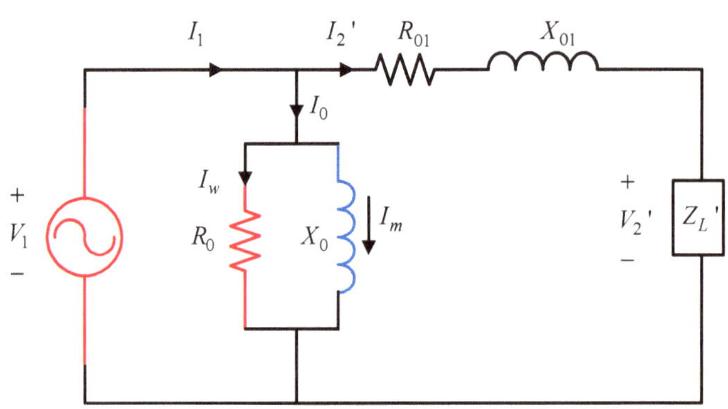

Fig. 2.7 Approximate equivalent circuit referred to primary

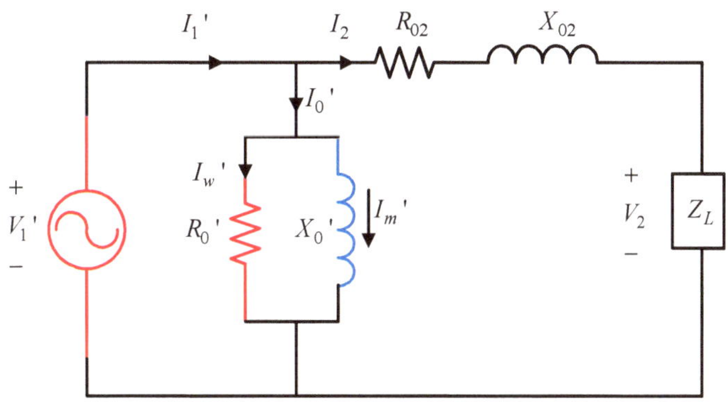

Fig. 2.8 Approximate equivalent circuit referred to secondary

Figure 2.8 shows an approximate equivalent circuit referred to the secondary. The total resistance, reactance, and impedance referred to the secondary are,

$$R_{02} = R_2 + R_1' = R_2 + \frac{R_1}{a^2} \tag{2.65}$$

$$X_{02} = X_2 + X_1' = X_2 + \frac{X_1}{a^2} \tag{2.66}$$

$$Z_{02} = R_{02} + jX_{02} \tag{2.67}$$

The no-load circuit resistance and reactance referred to the secondary are,

$$R_0' = \frac{V_1'}{I_w'} \tag{2.68}$$

$$X_0' = \frac{V_1'}{I_m'} \tag{2.69}$$

Example 2.4
A 2.5 kVA, 200 V/40 V single-phase transformer has the primary resistance and reactance of 3 Ω and 12 Ω, respectively. On the secondary side, these values are 0.3 Ω and 0.1 Ω, respectively. Find the equivalent impedance referred to the primary and the secondary.

Solution
The value of the turns ratio is,

$$a = \frac{V_1}{V_2} = \frac{200}{40} = 5 \tag{2.70}$$

The total resistance, reactance, and impedance referred to the primary can be determined as,

$$R_{01} = R_1 + a^2 R_2 = 3 + 25 \times 0.3 = 10.5\,\Omega \tag{2.71}$$

$$X_{01} = X_1 + a^2 X_2 = 12 + 25 \times 0.1 = 14.5\,\Omega \tag{2.72}$$

$$Z_{01} = \sqrt{10.5^2 + 14.5^2} = 17.9\,\Omega \tag{2.73}$$

The total resistance, reactance, and impedance referred to the secondary are calculated as,

$$R_{02} = R_2 + \frac{R_1}{a^2} = 0.3 + \frac{3}{25} = 0.42\,\Omega \tag{2.74}$$

$$X_{02} = X_2 + \frac{X_1}{a^2} = 0.1 + \frac{12}{25} = 0.58\,\Omega \tag{2.75}$$

$$Z_{01} = \sqrt{0.42^2 + 0.58^2} = 0.72\,\Omega \tag{2.76}$$

Practice Problem 2.4
A 1.5 kVA, 220 V/110 V single-phase transformer has the primary resistance and reactance of 6 Ω and 18 Ω, respectively. The resistance and reactance at the secondary side are 0.6 Ω and 0.5 Ω, respectively. Calculate the equivalent impedance referred to as the primary and the secondary.

2.9 Polarity of a Transformer

The relative directions of induced voltages between the high voltage and low voltage terminals are known as the polarity of a transformer. The polarity of a transformer is very important to construct a three-phase transformer bank, the parallel connection of transformer, the connection of the current transformer (CT), and potential transformer (PT) power with the metering device. Two polarities, namely additive and subtractive, are used in the transformer [4].

The polarity of a transformer is said to be an additive if the measured voltage between the high voltage and the low voltage terminals is greater than the supply voltage at the high voltage terminals. The additive polarity of a transformer is marked by the orientation of dots as shown in Fig. 2.9. Whereas, a polarity is said

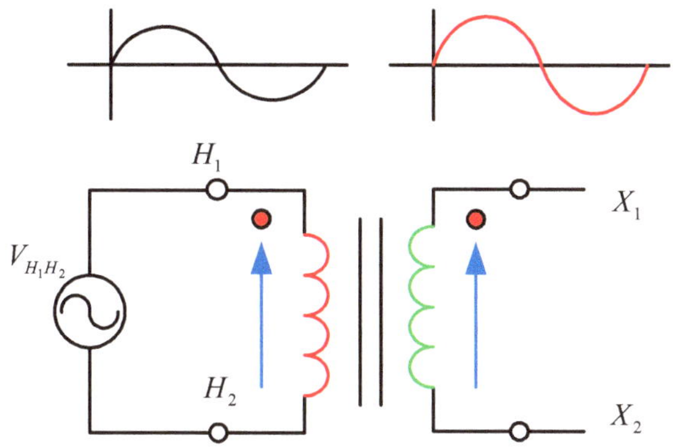

Fig. 2.9 Additive polarity

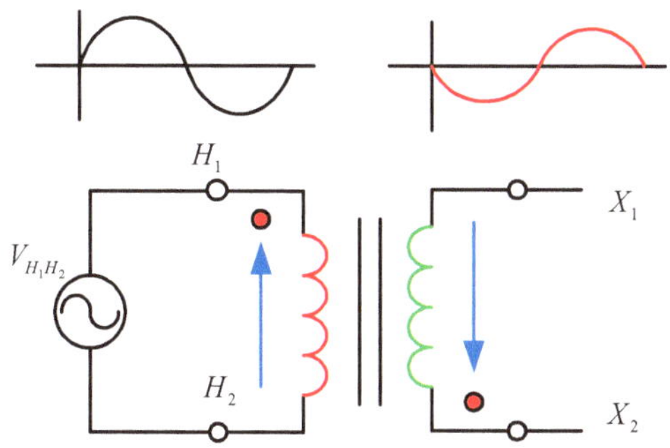

Fig. 2.10 Subtractive polarity

to be subtractive if the measured voltage between the high voltage and the low voltage terminals is lower than the supply voltage at the high voltage terminals. The subtractive polarity of a transformer is marked by the orientation of dots as shown in Fig. 2.10. Consider a 220 V/110 V single-phase transformer with the high voltage and the low voltage terminals for testing polarities. The high voltage terminal H_1 is connected to the low voltage terminal X_1 by a cable. The voltmeter is connected between H_2 and X_2. In this case, the turns ratio of the transformer is,

$$a = \frac{V_1}{V_2} = \frac{220}{110} = 2 \tag{2.77}$$

For safety, a lower voltage needs to be applied to the primary side, i.e., high voltage terminals. Suppose, a voltage of 110 V is applied to the primary side.

In this case, a voltage of 55 V (110/2) will appear at the secondary terminals. If the meter reads out the voltage of 165 V (110 + 55), then the transformer is said to be in additive polarity. This connection is shown in Fig. 2.11.

Whereas, if the voltmeter reads the voltage of 55 V (110 – 55), then the transformer is said to be in subtractive polarity as shown in Fig. 2.12.

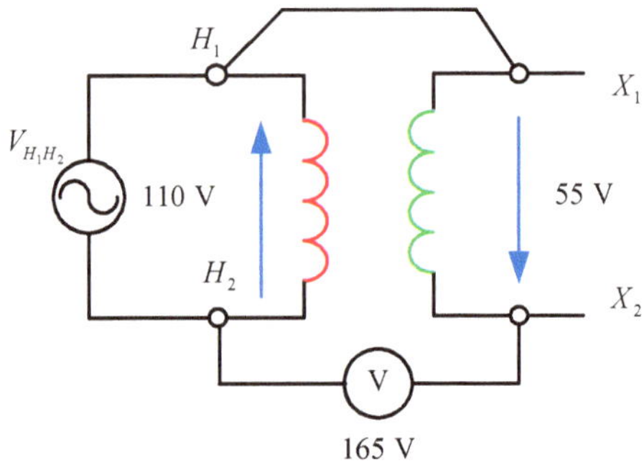

Fig. 2.11 Testing of additive polarity

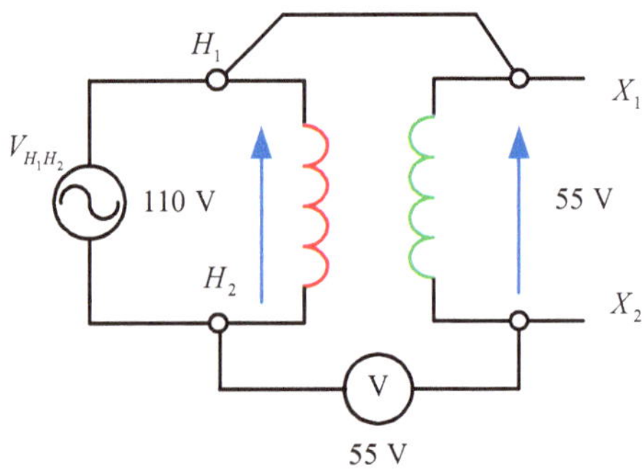

Fig. 2.12 Testing of subtractive polarity

2.10 Three-Phase Transformer

A three-phase power transformer is used at the power generating station to step up the voltage from 11 to 120 kV. Whereas in the power distribution substation, the three-phase voltage is again stepped down to 11 kV voltage through a three-phase distribution transformer.

Therefore, a three-phase transformer can be made either by three windings wound on a common core or by three single-phase transformers connected together in a three-phase bank. The first approach is a cheaper one that results in a transformer with a smaller size and less weight. The main disadvantage of the first approach is that if one phase becomes defective, then the whole transformer needs to be replaced. Whereas in the second approach, if one of the transformers becomes defective, then the system can be given power by an open delta at a reduced capacity. In this case, the defective transformer is normally replaced by a new one. A three-phase transformer with a wye-delta connection is shown in Fig. 2.13 and the three single-phase transformers are shown in Fig. 2.14.

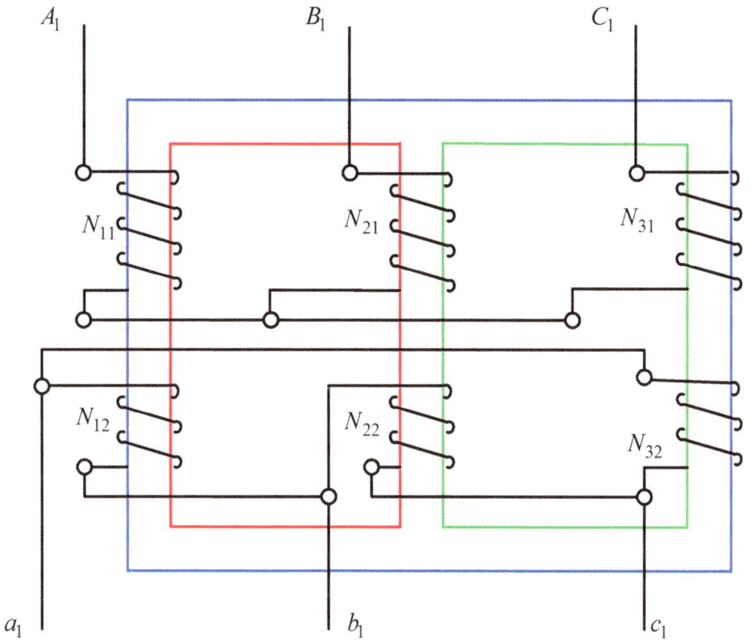

Fig. 2.13 Three windings on a common core and wye-delta connection

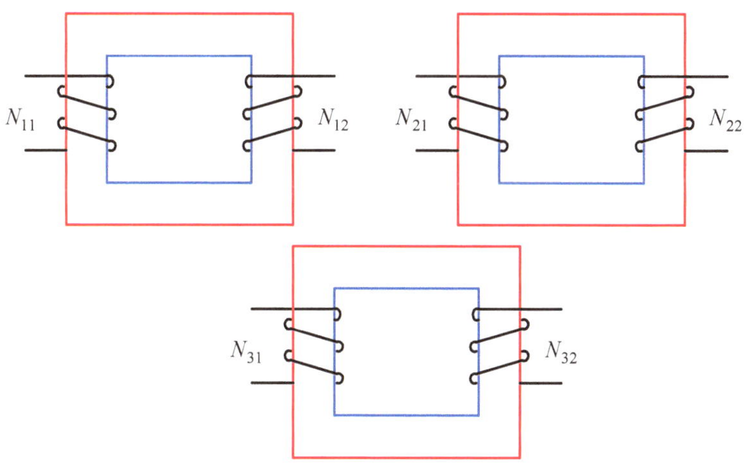

Fig. 2.14 Three single-phase transformers

2.11 Transformer Vector Group

The most common configurations of three-phase transformers are the delta and star configurations used in the power utility companies. The primary and secondary windings of a three-phase transformer are connected either in the same (delta-delta or star–star) or different (delta–star or star–delta) configuration pair. The secondary voltage waveforms of a three-phase transformer are in phase with the primary waveforms when the primary and secondary windings are connected in the same configuration. This condition is known as "no phase shift" condition. If the primary and secondary windings are connected in different configuration pairs, then the secondary voltage waveforms will differ from the corresponding primary voltage waveforms by 30 electrical degrees. This condition is called a "30-degree phase shift" condition. The windings and their position to each other are usually marked by a vector group. The vector group is used to identify the phase shift between the primary and secondary windings. In the vector group, the secondary voltage may have the phase shift of 30° lagging or leading, 0° (no phase shift) or 180° reversal with respect to the primary voltage. The transformer vector group is labeled by capital and small letters plus numbers from 1 to 12 in a typical clock-like diagram. The capital letter indicates the primary winding and the small letter represents secondary winding.

In the clock diagram, the minute hand represents the primary line to neutral line voltage, and its place is always in the 12. The hour hand represents the secondary line to neutral voltage and its position in the clock changes based on the phase shift as shown in Fig. 2.15. There are four vector groups used in the three-phase transformer connection. These vector groups are (i) Group I: 0 o'clock, zero phase displacement (Yy0, Dd0, Dz0), (ii) Group II: 6 o'clock, 180° phase displacement (Yy6, Dd6, Dz6), (iii) Group III: 1 o'clock, −30° lag phase displacement (Dy1, Yd1, Yz1), and (iv) Group IV: 11 o'clock, 30° lead phase displacement (Dy11, Yd11, Yz11). Here, Y represents wye connection, D represents delta connection, and z represents the zigzag connection. The connection diagrams for different combinations are shown in Figs. 2.16, 2.17, 2.18, 2.19, 2.20, 2.21, 2.22, 2.23, 2.24, 2.25, 2.26, 2.27.

2.12 Voltage Regulation of a Transformer

Different types of loads, such as domestic, commercial, and industrial, are usually connected with the secondary winding of a transformer. All these loads are operated with a constant magnitude of the voltage. The secondary voltage of a transformer changes under operation due to the voltage drop across the internal impedance and the load. The voltage regulation of a transformer is used to identify the characteristics of the secondary side voltage changes under different loading conditions. The voltage regulation of a transformer is defined as the difference between the

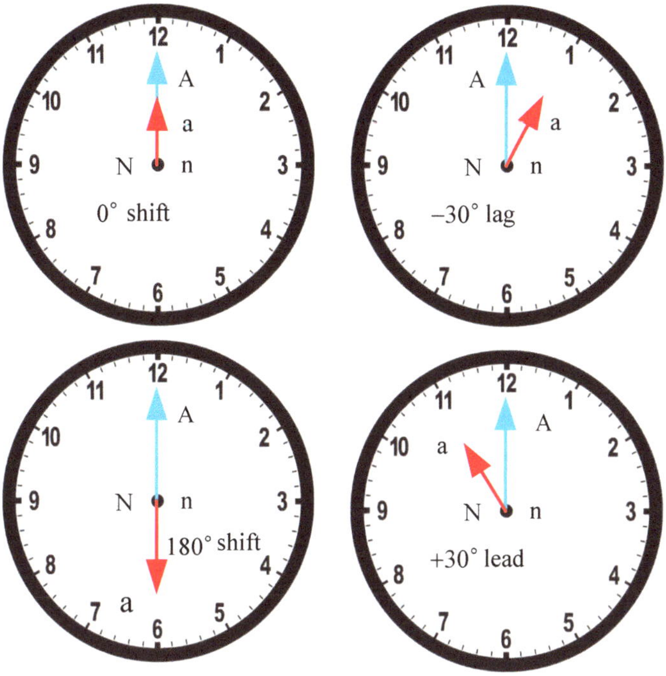

Fig. 2.15 Representation of vector groups with the clock

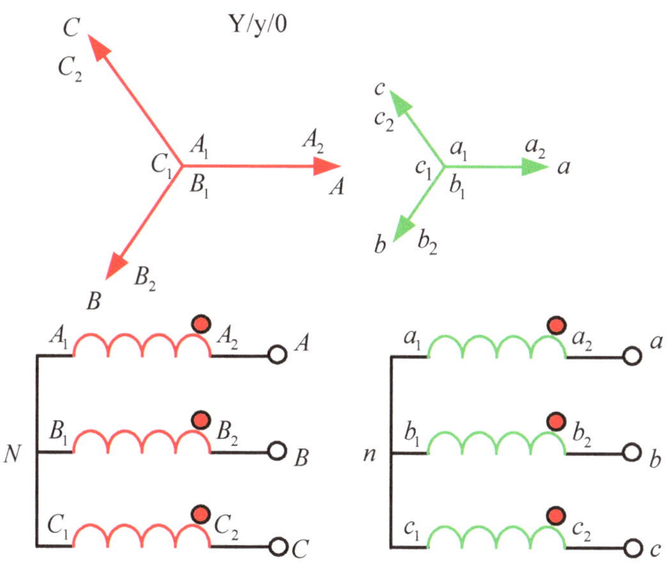

Fig. 2.16 Connection of Y-y-0

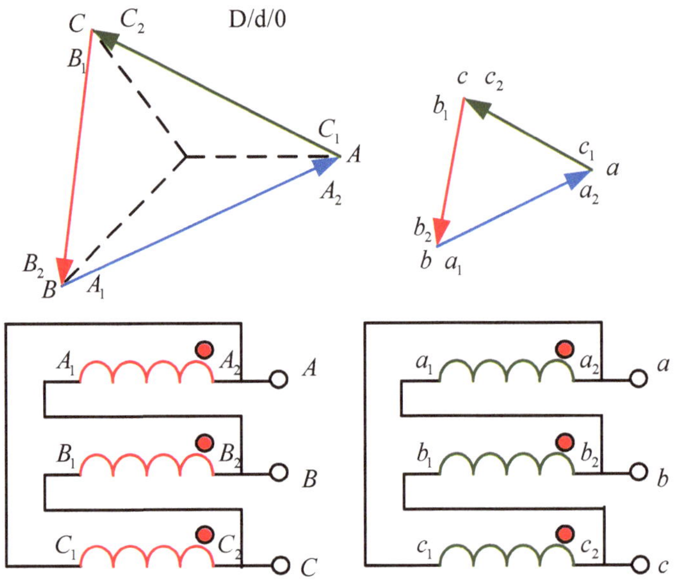

Fig. 2.17 Connection of D-d-0

no-load terminal voltage (V_{2NL}) to full load terminal voltage (V_{2FL}) and is expressed as a percentage of full load terminal voltage. It is, therefore, can be expressed as,

$$\text{Voltage regulation} = \frac{V_{2NL} - V_{2FL}}{V_{2FL}} \times 100\% = \frac{E_2 - V_2}{V_2} \times 100\% \tag{2.78}$$

Figure 2.28 shows an approximate equivalent circuit referred to as the secondary without a no-load circuit to find the voltage regulation for different power factors. Phasor diagrams for different power factors are shown in Fig. 2.29 where the secondary voltage V_2 is considered as the reference phasor.

The phasor diagram with a unity power factor is shown in Fig. 2.29a and the phasor form of the secondary induced voltage for a unity power factor can be written as,

$$E_2 = V_2 + I_2(R_{02} + jX_{02}) \tag{2.79}$$

Figure 2.29b shows the phasor diagram for a lagging power factor, and from this diagram, the following expressions can be written as,

$$AC = V_2 \cos \phi_2 \tag{2.80}$$

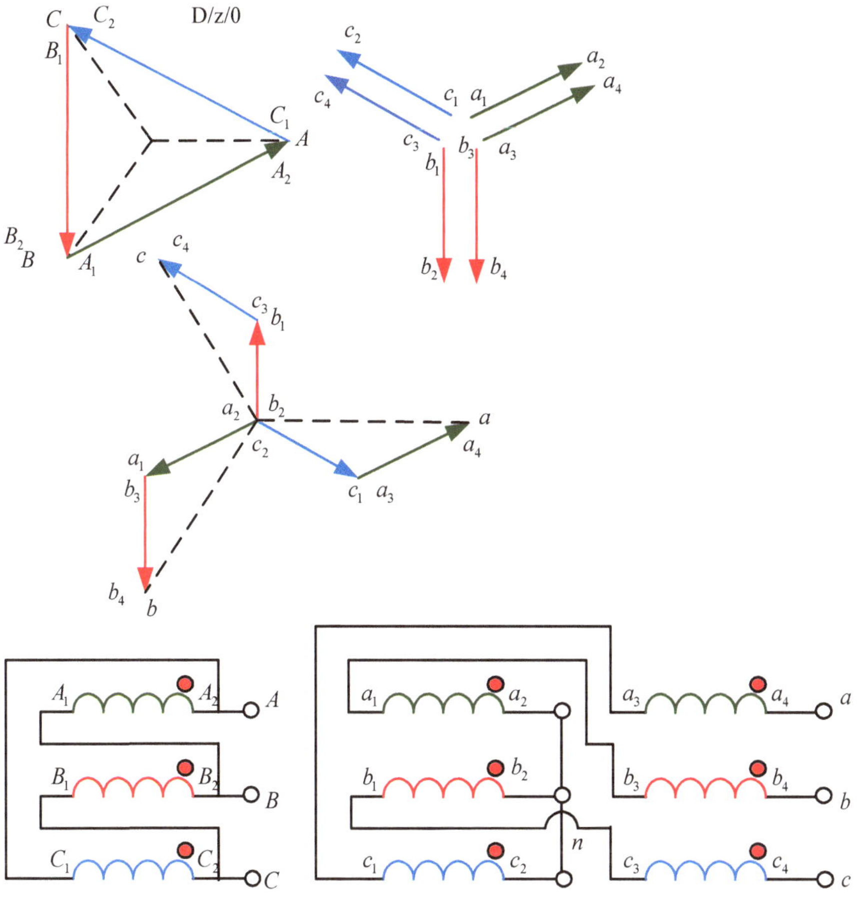

Fig. 2.18 Connection of D-z-0

$$BC = DE = V_2 \sin \phi_2 \tag{2.81}$$

$$CD = BE = I_2 R_{02} \tag{2.82}$$

$$EF = I_2 X_{02} \tag{2.83}$$

From the right-angle triangle-ADF, the expression of E_2 can be derived as,

$$AF^2 = AD^2 + DF^2 \tag{2.84}$$

$$AF^2 = (AC + CD)^2 + (DE + EF)^2 \tag{2.86}$$

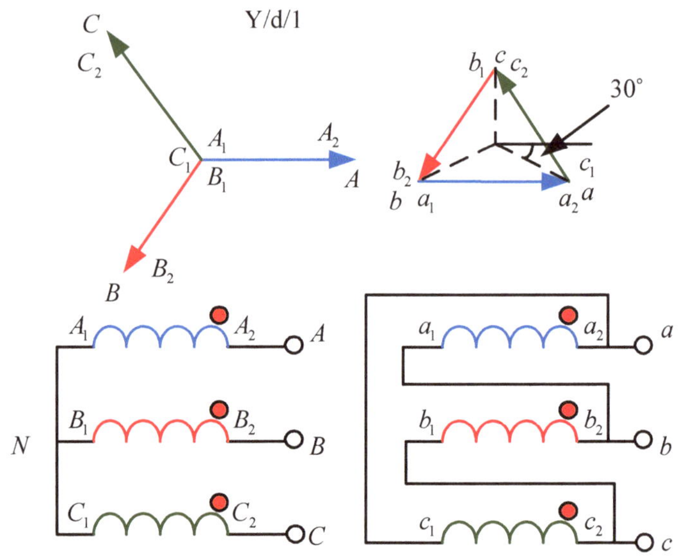

Fig. 2.19 Connection of Y-d-1

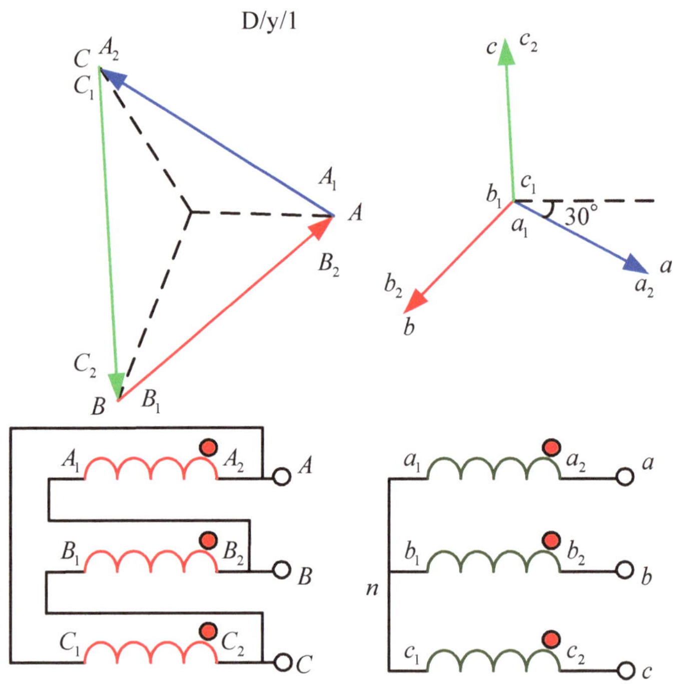

Fig. 2.20 Connection of D-y-1

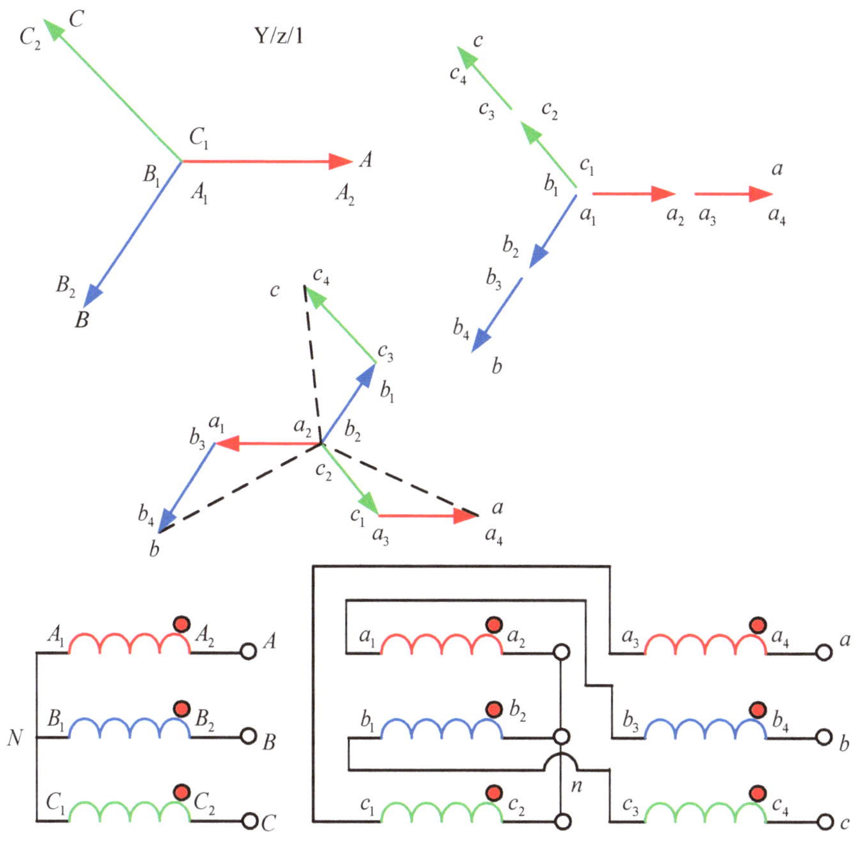

Fig. 2.21 Connection of Y-z-1

Substituting equations from (2.80) to (2.83) into Eq. (2.86) yields,

$$E_2 = \sqrt{(V_2 \cos \phi_2 + I_2 R_{02})^2 + (V_2 \sin \phi_2 + I_2 X_{02})^2} \tag{2.87}$$

In phasor form, Eq. (2.87) can be written as,

$$E_2 = (V_2 \cos \phi_2 + I_2 R_{02}) + j(V_2 \sin \phi_2 + I_2 X_{02}) \tag{2.88}$$

Figure 2.29c shows the phasor diagram for a leading power factor and from this diagram, the following expressions can be written,

$$MR = V_2 \cos \phi_2 \tag{2.89}$$

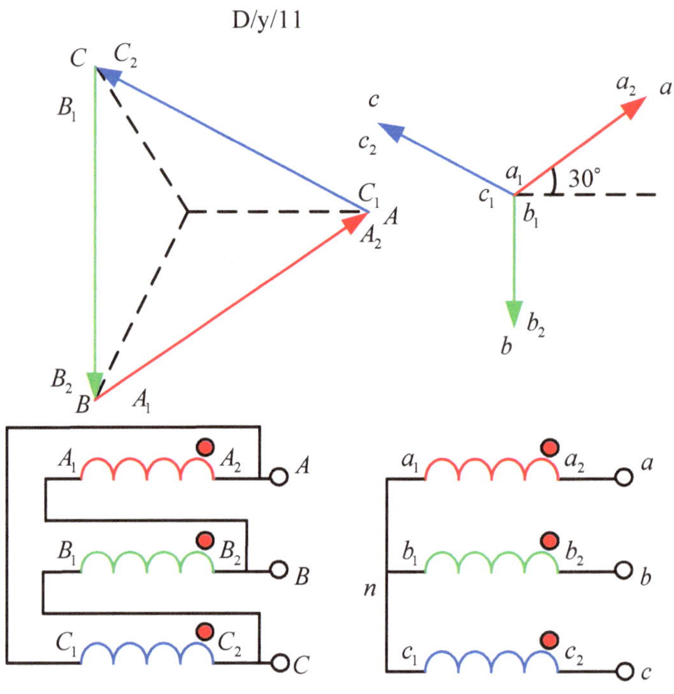

Fig. 2.22 Connection of D-y-11

$$RN = OQ = V_2 \sin \phi_2 \tag{2.90}$$

$$RQ = NO = I_2 R_{02} \tag{2.91}$$

$$PO = I_2 X_{02} \tag{2.92}$$

The expression of E_2 from the right-angle triangle MQP can be derived as,

$$MP^2 = MQ^2 + QP^2 \tag{2.93}$$

$$MP^2 = (MR + RQ)^2 + (OQ - PO)^2 \tag{2.94}$$

Substituting equations from (2.89) to (2.92) into Eq. (2.94) yields,

$$E_2 = \sqrt{(V_2 \cos \phi_2 + I_2 R_{02})^2 + (V_2 \sin \phi_2 - I_2 X_{02})^2} \tag{2.95}$$

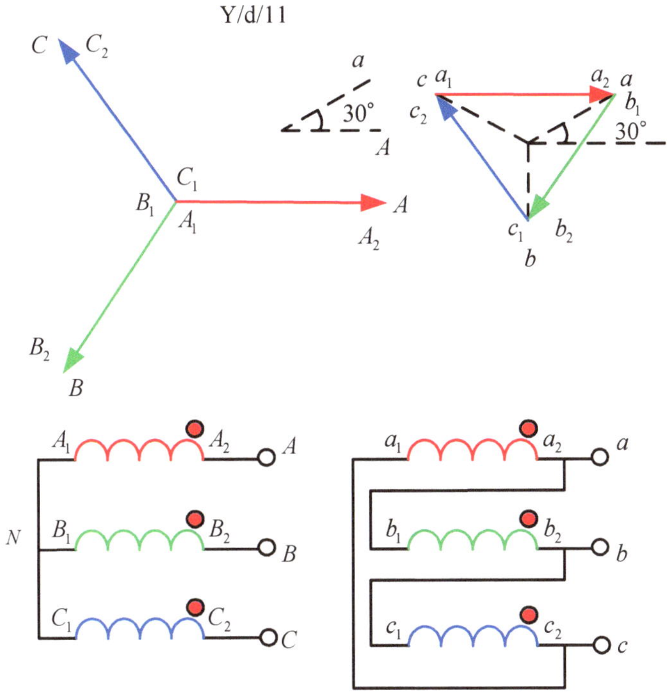

Fig. 2.23 Connection of Y-d-11

In phasor form, Eq. (2.95) can be written as,

$$E_2 = (V_2 \cos \phi_2 + I_2 R_{02}) + j(V_2 \sin \phi_2 - I_2 X_{02}) \tag{2.96}$$

Example 2.5

The primary coil resistance and reactance of a 200 V/400 V single-phase transformer are 0.3 Ω and 0.6 Ω, respectively. The secondary coil resistance and reactance are 0.8 Ω and 1.6 Ω, respectively. Calculate the voltage regulation if the secondary current of the transformer is 10 A at a 0.8 pf lagging.

Solution

The value of the turns ratio is,

$$a = \frac{V_1}{V_2} = \frac{200}{400} = 0.5 \tag{2.97}$$

The value of the total resistance referred to the secondary is,

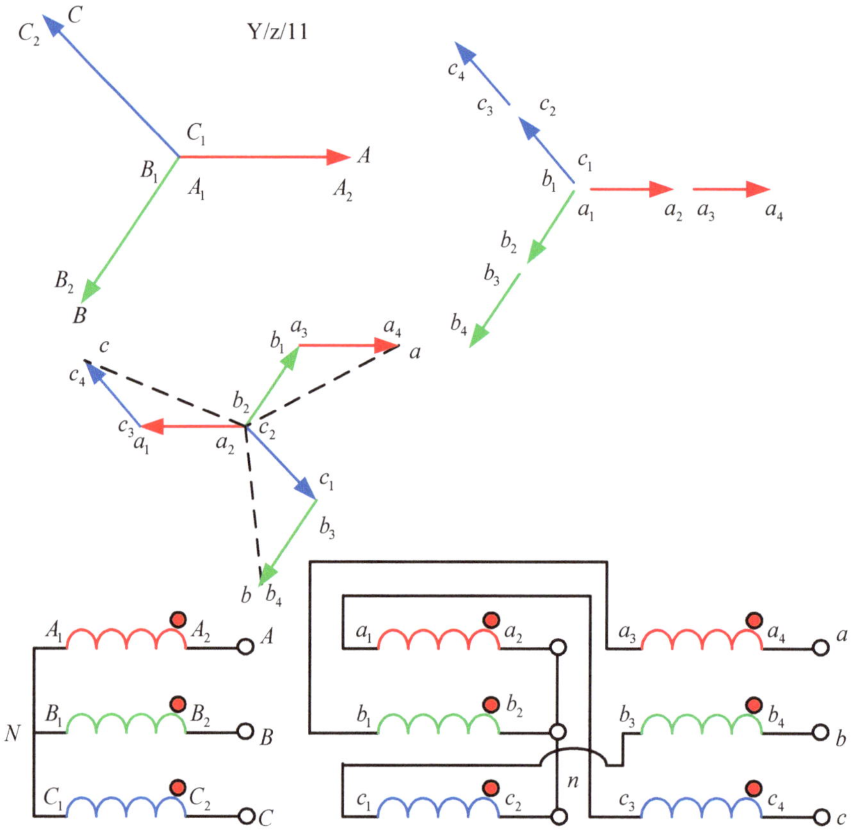

Fig. 2.24 Connection of Y-z-11

$$R_{02} = R_2 + \frac{R_1}{a^2} = 0.8 + \frac{0.3}{0.25} = 2\,\Omega \tag{2.98}$$

The total reactance referred to the primary is,

$$X_{02} = X_2 + \frac{X_1}{a^2} = 1.6 + \frac{0.6}{0.25} = 4\,\Omega \tag{2.99}$$

The no-load voltage is,

$$E_2 = (V_2 \cos\phi_2 + I_2 R_{02}) + j(V_2 \sin\phi_2 + I_2 X_{02}) = (400 \times 0.8 + 10 \times 2) + j(400 \times 0.6 + 40) = 440.5 \angle 39.5^\circ \text{ V} \tag{2.100}$$

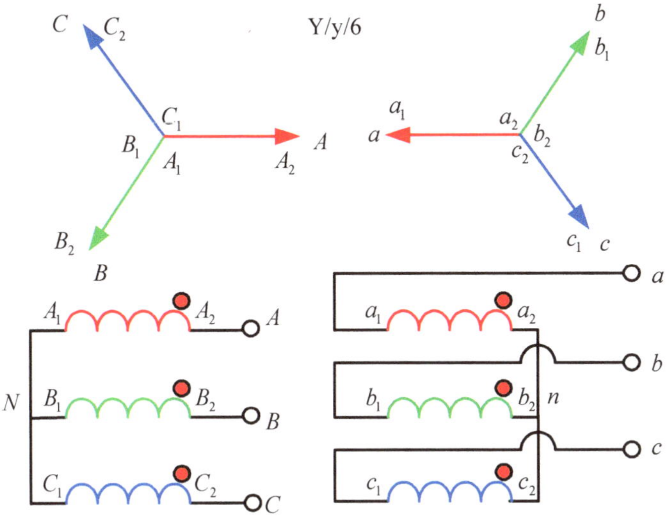

Fig. 2.25 Connection of Y-y-6

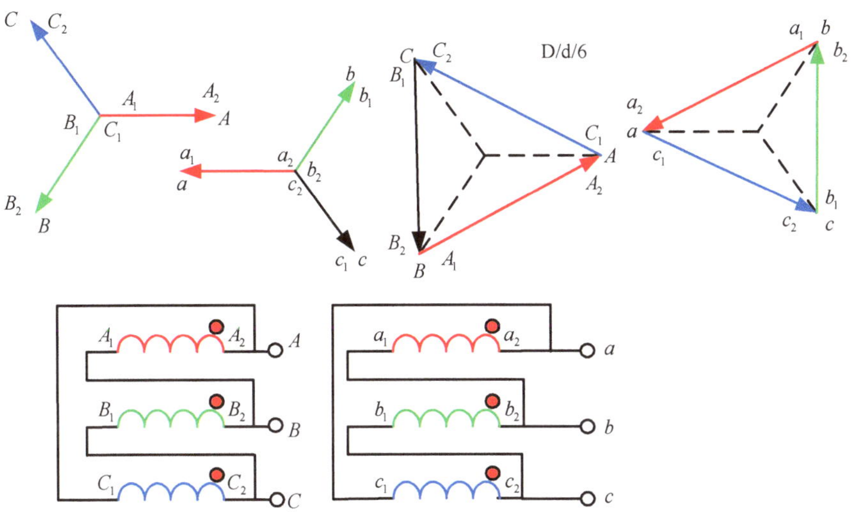

Fig. 2.26 Connection of D-d-6

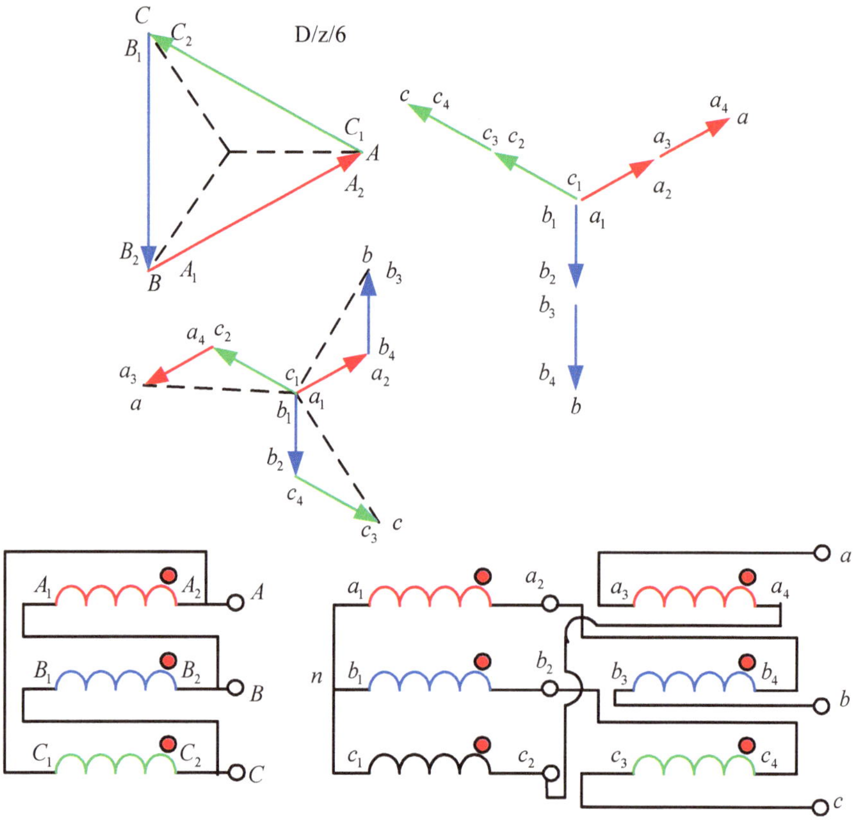

Fig. 2.27 Connection of D-y-6

Fig. 2.28 Approximate equivalent circuit referred to secondary

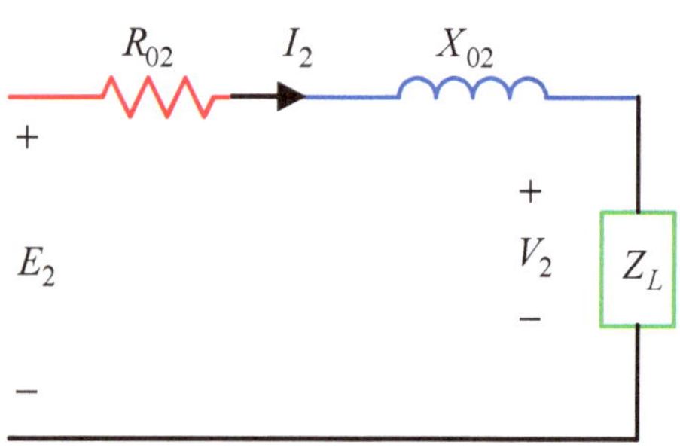

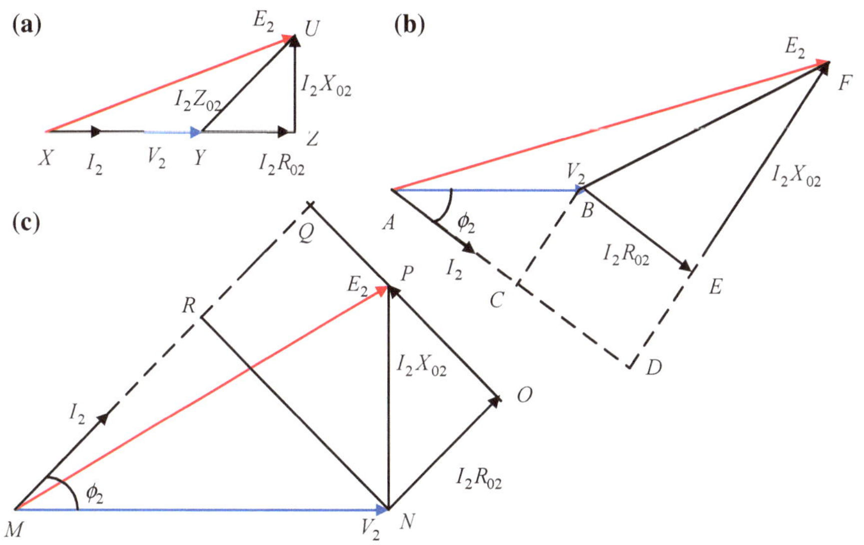

Fig. 2.29 Phasor diagram for different power factors

The voltage regulation can be determined as,

$$\text{Voltage regulation} = \frac{E_2 - V_2}{V_2} \times 100\% = \frac{440.5 - 400}{400} \times 100\% = 10\% \quad (2.101)$$

Practice Problem 2.5

A 110 V/220 V single-phase transformer has the resistance of 0.2 Ω and reactance of 0.8 Ω in the primary winding. The resistance and reactance in the secondary winding are 0.9 Ω and 1.8 Ω, respectively. Calculate the voltage regulation, when the secondary current is 6 A at a 0.85 power factor leading.

2.13 Efficiency of a Transformer

Efficiency is an important parameter to identify the characteristics of any machine. The efficiency η, of any machine, can be defined as the ratio of its output to the input. Mathematically, it can be expressed as,

$$\eta = \frac{\text{output}}{\text{input}} = \frac{\text{input} - \text{losses}}{\text{input}} = 1 - \frac{\text{losses}}{\text{input}} \quad (2.102)$$

Let us consider the following:

P_{out} is the output power, in W,
P_{in} is the input power, in W,
P_{losses} is the total losses, in W.

Equation (2.102) can be modified as,

$$\eta = \frac{P_{\text{out}}}{P_{\text{in}}} = 1 - \frac{P_{\text{losses}}}{P_{\text{in}}} \tag{2.103}$$

It is worth noting that the efficiency of a transformer is generally higher than other electrical machines because it has no moving parts.

2.14 Iron and Copper Losses

The iron loss of a transformer is often called a core loss, which is a result of an alternating flux in the core of the transformer. The iron loss consists of the eddy current loss and the hysteresis loss. In the transformer, most of the flux is transferred from the primary coil to the secondary coil through a low reluctance iron path. A few portions of that alternating flux links with the iron parts of the transformer and as a result, an emf is induced in the transformer core. A current will flow in that parts of the transformer. This current does not contribute to the output of the transformer but dissipated as heat. This current is known as eddy current and the power loss due to this current is known as eddy current loss. The eddy current loss (P_e) is directly proportional to the square of the frequency (f) times the maximum magnetic flux density (B_m) and the eddy current loss can be expressed as,

$$P_e = k_{ef}^2 B_m^2 \tag{2.104}$$

where

k_e is the proportionality constant

Steel is a very good ferromagnetic material that is used for the core of a transformer. This ferromagnetic material contains a number of domains in the structure and magnetized easily. These domains are like small magnets located randomly in the structure. When an mmf is applied to the core, then those domains change their position. After removing mmf, most of the domains come back to their original position and the remaining will be as it is. As a result, the substance is slightly permanently magnetized. An additional mmf is required to change the position of the remaining domains. Therefore, hysteresis loss is defined as the additional energy that is required to realign the domains in the ferromagnetic material. The hysteresis loss (P_h) is directly proportional to the frequency (f) and

2.6th power of the maximum magnetic flux density (B_m) and the expression of hysteresis loss is,

$$P_h = k_h f B_m^{2.6} \quad (2.105)$$

where k_e is the proportionality constant.

Now that the magnetic flux density is usually constant, Eqs. (2.104) and (2.105) can be modified as,

$$P_e \propto f^2 \quad (2.106)$$

$$P_h \propto f \quad (2.107)$$

Practically, hysteresis loss depends on the voltage and the eddy current loss depends on the current. Therefore, the total losses of the transformer depending on the voltage and the current but not on the power factor. That is why the transformer rating is always represented in kVA instead of kW.

In the transformer, copper losses occur due to the primary and the secondary resistances. The full load copper losses can be determined as,

$$P_{\text{culoss}} = I_1^2 R_1 + I_2^2 R_2 \quad (2.108)$$

$$P_{\text{culoss}} = I_1^2 R_{01} = I_2^2 R_{02} \quad (2.109)$$

2.15 Condition for Maximum Efficiency

The expression of the output power of a transformer is written as,

$$P_{\text{out}} = V_2 I_2 \cos \phi_2 \quad (2.110)$$

The expression of the copper loss is written as,

$$P_{\text{cu}} = I_2^2 R_{02} \quad (2.111)$$

The expression of an iron loss is written as,

$$P_{\text{iron}} = P_{\text{eddy}} + P_{\text{hys}} \quad (2.112)$$

According to Eq. (2.103), the efficiency can be expressed as,

$$\eta = \frac{\text{output}}{\text{input}} = \frac{\text{output}}{\text{output} + \text{losses}} \quad (2.113)$$

Substituting Eqs. (2.110), (2.111) and (2.112) into Eq. (2.113) yields,

$$\eta = \frac{V_2 I_2 \cos\phi_2}{V_2 I_2 \cos\phi_2 + P_{\text{iron}} + P_{cu}} \tag{2.114}$$

$$\eta = \frac{V_2 \cos\phi_2}{V_2 \cos\phi_2 + \frac{P_{\text{iron}}}{I_2} + \frac{I_2^2 R_{02}}{I_2}} \tag{2.115}$$

The terminal voltage at the secondary side is considered to be constant in case of a normal transformer. For a given power factor, the secondary current is varied with the variation of load. Therefore, the transformer efficiency will be maximum, if the denominator of Eq. (2.115) is minimum. The denominator of Eq. (2.115) will be minimum for the following condition.

$$\frac{d}{dI_2}\left(V_2 \cos\phi_2 + \frac{P_{\text{iron}}}{I_2} + \frac{I_2^2 R_{02}}{I_2}\right) = 0 \tag{2.116}$$

$$-\frac{P_{\text{iron}}}{I_2^2} + R_{02} = 0 \tag{2.117}$$

$$P_{\text{iron}} = I_2^2 R_{02} \tag{2.118}$$

From Eq. (2.118), it is concluded that the efficiency of a transformer will be maximum, when the iron loss is equal to the copper loss. From Eq. (2.118), the expression of secondary current can be written as,

$$I_2 = \sqrt{\frac{P_{\text{iron}}}{R_{02}}} \tag{2.119}$$

For maximum efficiency, the load current can be expressed as,

$$I_{2\eta} = \sqrt{\frac{P_{\text{iron}}}{R_{02}}} \tag{2.120}$$

Equation (2.120) can be rearranged as,

$$I_{2\eta} = I_2 \sqrt{\frac{P_{\text{iron}}}{I_2^2 R_{02}}} \tag{2.121}$$

Multiplying both sides of Eq. (2.121) by the secondary rated voltage V_2 yields,

$$V_2 I_{2\eta} = V_2 I_2 \sqrt{\frac{P_{\text{iron}}}{I_2^2 R_{02}}} \tag{2.122}$$

$$VI_{\text{max efficiency}} = VI_{\text{rated}} \times \sqrt{\frac{\text{iron loss}}{\text{full load copper loss}}} \tag{2.123}$$

Example 2.6
A 30 kVA transformer has the iron loss and full load copper loss of 350 W and 650 W, respectively. Determine the full load efficiency, output kVA corresponding to maximum efficiency, and maximum efficiency. Consider that the power factor of the load is 0.6 lagging.

Solution
The total value of full load loss is calculated as,

$$P_{\text{tloss}} = 350 + 650 = 1000\,\text{W} = 1\,\text{kW} \tag{2.124}$$

The output power at full load is calculated as,

$$P_{\text{out}} = 30 \times 0.6 = 18\,\text{kW} \tag{2.125}$$

The input power at full load is calculated as,

$$P_{\text{in}} = 18 + 1 = 19\,\text{kW} \tag{2.126}$$

The efficiency at full load is calculated as,

$$\eta = \frac{18}{19} \times 100 = 94.74\% \tag{2.127}$$

The output kVA corresponding to maximum efficiency is calculated as,

$$= kVA_{\text{rated}} \times \sqrt{\frac{\text{Iron loss}}{\text{Cu loss at full load}}} = 30 \times \sqrt{\frac{350}{650}} = 22\,\text{kVA} \tag{2.128}$$

The output power is calculated as,

$$P_{\text{o1}} = 22 \times 0.6 = 13.2\,\text{kW} \tag{2.129}$$

For maximum efficiency, iron loss is equal to copper loss.
The total value of the loss is calculated as,

$$P_{\text{tloss}-1} = 2 \times 350 = 700\,\text{W} \tag{2.130}$$

The value of the input power is calculated as,

$$P_{\text{in1}} = 13.2 + 0.7 = 13.9\,\text{kW} \tag{2.131}$$

The efficiency is calculated as,

$$\eta = \frac{13.2}{13.9} \times 100 = 94.96\% \tag{2.132}$$

Practice Problem 2.6

A 200 kVA transformer is having an iron loss of 1.5 kW and a copper loss of 5 kW at full load condition. Find the kVA rating at which the efficiency is maximum. Also, find the efficiency at a unity power factor.

2.16 Transformer Tests

Equivalent circuit parameters are very important to characterize the performance of the transformer. The parameters of a transformer equivalent circuit can be determined by the open circuit and the short circuit tests.

2.16.1 Open Circuit Test

The main objectives of open circuit tests are to determine the no-load current and iron loss. The components of the no-load current are used to determine the no-load circuit resistance and reactance.

In an open circuit test, the secondary side is considered to be an open circuit, and the primary coil is connected to the source as shown in Fig. 2.30a, where all measuring instruments are connected on the primary side. A specific alternating voltage is applied to the primary winding. Then the wattmeter will measure the iron loss and small amount of copper loss. The ammeter and voltmeter will measure the no-load current and the voltage, respectively. Since the no-load current is very

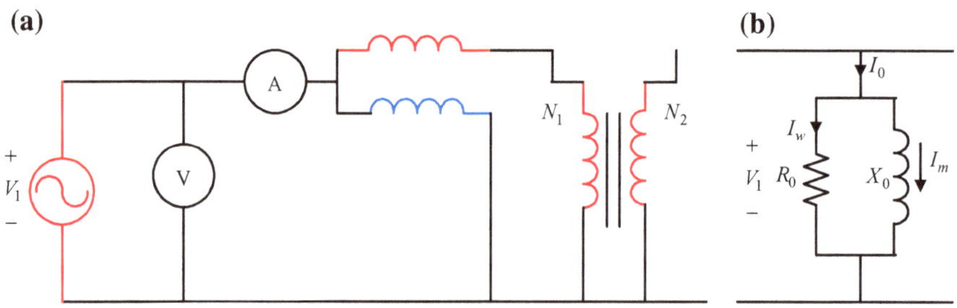

Fig. 2.30 Connection diagrams for open circuit test and no-load circuit

small, the copper losses can be neglected. Then the wattmeter reading can be expressed as,

$$P_0 = V_1 I_0 \cos \phi_0 \tag{2.132}$$

From Eq. (2.132), the no-load power factor can be determined as,

$$\cos \phi_0 = \frac{P_0}{V_1 I_0} \tag{2.133}$$

The working and magnetizing components of the current can be determined as,

$$I_w = I_0 \cos \phi_0 \tag{2.134}$$

$$I_m = I_0 \sin \phi_0 \tag{2.135}$$

Then the no-load circuit resistance and reactance can be determined as,

$$R_0 = \frac{V_1}{I_w} \tag{2.136}$$

$$X_0 = \frac{V_1}{I_m} \tag{2.137}$$

Example 2.7

A 200 V/400 V, 50 Hz single-phase transformer has the no-load test data of 200 V, 0.6 A, 80 W. Calculate the no-load circuit resistance and reactance.

Solution

The power factor can be determined as,

$$\cos \phi_0 = \frac{P_0}{V_1 I_0} = \frac{80}{200 \times 0.6} = 0.67 \tag{2.138}$$

$$\sin \phi_0 = 0.74 \tag{2.139}$$

The values of the working and magnetizing components of the no-load current are,

$$I_w = I_0 \cos \phi_0 = 0.6 \times 0.67 = 0.4\,\text{A} \tag{2.140}$$

$$I_m = I_0 \sin \phi_0 = 0.6 \times 0.74 = 0.44\,\text{A} \tag{2.141}$$

The no-load circuit parameters can be determined as,

$$R_0 = \frac{V_1}{I_w} = \frac{200}{0.4} = 500\,\Omega \tag{2.142}$$

$$X_0 = \frac{V_1}{I_m} = \frac{200}{0.44} = 454.5\,\Omega \tag{2.143}$$

2.16.2 Short Circuit Test

The main objectives of short circuit tests are to determine the equivalent resistance, reactance, impedance, and full load copper loss. In this test, the supply voltage and the measuring instruments (e.g., wattmeter, ammeter) are connected to the primary side, and the secondary winding is shorted out by wire or connected with an ammeter as shown in Fig. 2.31.

The primary voltage is adjusted until the current in the short-circuited winding is equal to the rated primary current. Under this condition, the wattmeter will measure the full load copper loss and it can be written as,

$$P_{\mathrm{sc}} = I_{\mathrm{sc}} V_{\mathrm{sc}} \cos\phi_{\mathrm{sc}} \tag{2.144}$$

From Eq. (2.144), the short circuit power factor can be calculated as,

$$\cos\phi_{\mathrm{sc}} = \frac{P_{\mathrm{sc}}}{I_{\mathrm{sc}} V_{\mathrm{sc}}} \tag{2.145}$$

The equivalent impedance can be calculated as,

$$Z_{01} = Z_{\mathrm{eq}} = \frac{V_{\mathrm{sc}}}{I_{\mathrm{sc}}} \tag{2.146}$$

In addition, the equivalent resistance and reactance can be calculated as,

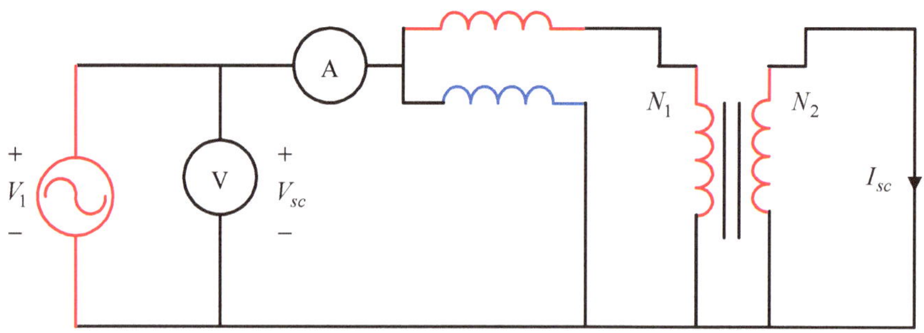

Fig. 2.31 Connection diagram for short circuit test

$$R_{01} = Z_{01} \cos \phi_{sc} \tag{2.147}$$

$$X_{01} = Z_{01} \sin \phi_{sc} \tag{2.148}$$

Example 2.8

A 25 kVA, 2200 V/220 V, 50 Hz single-phase transformer's low voltage side is short-circuited and the test data recorded from the high voltage side are $P = 150\,\text{W}$, $I_1 = 5\,\text{A}$, and $V_1 = 40\,\text{V}$. Determine the equivalent resistance, reactance, and impedance referred to primary, equivalent resistance, reactance, and impedance referred to secondary, and voltage regulation at unity power factor.

Solution

The parameters referred to primary are,

$$Z_{01} = \frac{V_1}{I_1} = \frac{40}{5} = 8\,\Omega \tag{2.149}$$

$$R_{01} = \frac{P}{I_1^2} = \frac{150}{25} = 6\,\Omega \tag{2.150}$$

$$\phi = \cos^{-1}\left(\frac{6}{8}\right) = 41.4^\circ \tag{2.151}$$

$$X_{01} = Z_{01} \sin \phi = 8 \sin 41.4^\circ = 5.2\,\Omega \tag{2.152}$$

The turns ratio is calculated as,

$$a = \frac{V_1}{V_2} = \frac{2200}{220} = 10 \tag{2.153}$$

The parameters referred to secondary are calculated as,

$$Z_{02} = \frac{Z_{01}}{a^2} = \frac{8}{100} = 0.08\,\Omega \tag{2.154}$$

$$R_{02} = \frac{R_{01}}{a^2} = \frac{6}{100} = 0.06\,\Omega \tag{2.155}$$

$$X_{02} = \frac{X_{01}}{a^2} = \frac{5.2}{100} = 0.052\,\Omega \tag{2.156}$$

The secondary side current is calculated as,

$$I_2 = \frac{25000}{220} = 113.6\,\text{A} \tag{2.157}$$

The secondary induced voltage is calculated as,

$$E_2 = V_2 + I_2 Z_{02} = 220 + 113.6 \times 0.08 = 229\,\text{V} \tag{2.158}$$

The voltage regulation is calculated as,

$$VR = \frac{E_2 - V_2}{V_2} = \frac{229 - 220}{220} = 4\% \tag{2.159}$$

Practice Problem 2.7

The no-load test data of a 220 V/2200 V, 50 Hz single-phase transformer are 220 V, 0.4 A, 75 W. Calculate the no-load circuit resistance and reactance.

Practice Problem 2.8

A 30 kVA, 2200 V/220 V, 50 Hz single-phase transformer's low voltage side is short-circuited by a thick wire. The test data recorded from the high voltage side are $P = 400\,\text{W}$, $I_1 = 8.5\,\text{A}$, and $V_1 = 65\,\text{V}$. Find the equivalent resistance, reactance, and impedance referred to a primary and equivalent resistance, reactance, and impedance referred to secondary, and voltage regulation at a 0.6 lagging power factor.

2.17 Autotransformer

A small rating transformer with a variable voltage output is usually used in the academic laboratory as well as in testing laboratories. This type of small rating transformer with variable output is known as an autotransformer. An autotransformer has one continuous winding that is common to both the primary and the secondary. Therefore, in an autotransformer, the primary and secondary windings are connected electrically. The advantages of an autotransformer over a two-winding transformer include lower initial investment, lower leakage reactance, lower losses compared to the conventional transformer, and lower excitation current.

An autotransformer with primary and secondary windings is shown in Fig. 2.32. In this connection, the suffix c indicates the common winding and the suffix s indicates the series winding. From Fig. 2.32, the following equations can be written as,

$$\frac{N_s}{N_c} = \frac{I_c}{I_s} = a \tag{2.160}$$

$$\frac{V_s}{V_c} = \frac{N_s}{N_c} = a \tag{2.161}$$

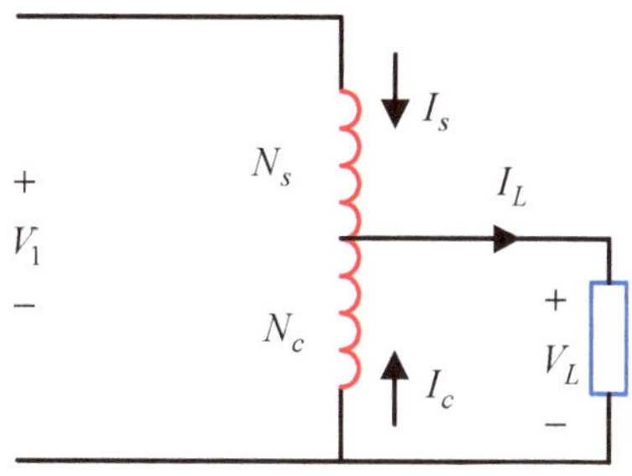

Fig. 2.32 Connection diagram for an autotransformer

The total voltage in the primary side is,

$$V_1 = V_s + V_c = V_c\left(1 + \frac{V_s}{V_c}\right) \tag{2.162}$$

Substituting Eq. (2.161) into Eq. (2.162) yields,

$$\frac{V_1}{V_c} = (1 + a) \tag{2.163}$$

$$\frac{V_L}{V_1} = \frac{1}{1 + a} \tag{2.164}$$

where the voltage at the load is equal to the voltage at the common terminals, i.e., $V_L = V_c$.

The expression of the load current can be written as,

$$I_L = I_s + I_c = aI_s + I_s \tag{2.165}$$

$$I_L = (1 + a)I_s \tag{2.166}$$

$$\frac{I_L}{I_s} = (1 + a) \tag{2.167}$$

Example 2.9

Figure 2.33 shows a single-phase 120 kVA, 2200 V/220 V, 50 Hz transformer which is connected as an autotransformer. The voltages of the upper and lower parts of the coil are 220 V and 2200 V, respectively. Calculate the kVA rating of the autotransformer.

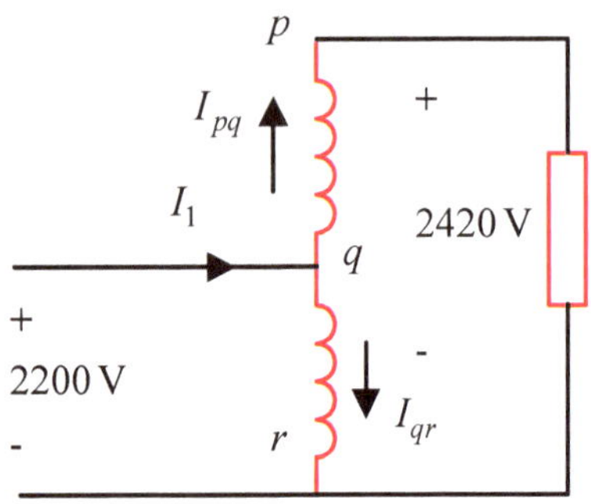

Fig. 2.33 An autotransformer for Example 2.9

Solution
The current ratings of the respective windings are,

$$I_{pq} = \frac{120,000}{220} = 545.5\,\text{A} \tag{2.168}$$

$$I_{qr} = \frac{120,000}{2200} = 54.5\,\text{A} \tag{2.169}$$

The current in the primary side is calculated as,

$$I_1 = 545.5 + 54.5 = 600\,\text{A} \tag{2.170}$$

The voltage across the secondary side is calculated as,

$$V_2 = 2200 + 220 = 2420\,\text{V} \tag{2.171}$$

Therefore, the kVA ratings of an autotransformer are calculated as,

$$\text{kVA}_1 = \frac{600 \times 2200}{1000} = 1320\,\text{kVA} \tag{2.172}$$

$$\text{kVA}_2 = \frac{545.5 \times 2420}{1000} = 1320\,\text{kVA} \tag{2.173}$$

Practice Problem 2.9
A single-phase 100 kVA, 1100 V/220 V, 50 Hz transformer is connected as an autotransformer as shown in Fig. 2.34. The voltages of the upper and lower parts of the coil are 220 V and 1100 V, respectively. Determine the kVA rating of the autotransformer.

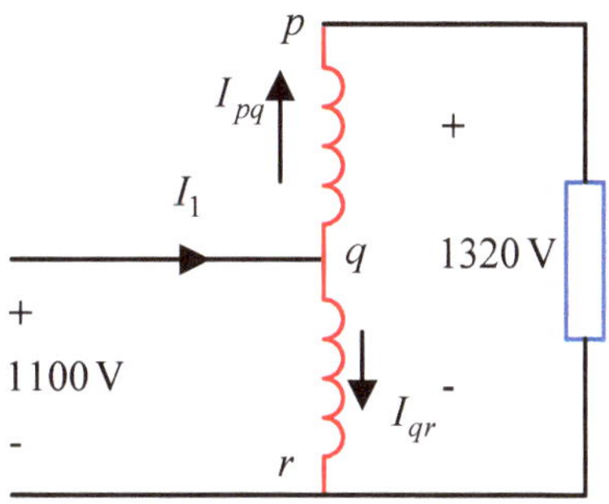

Fig. 2.34 An autotransformer for Practice Problem 2.9

2.18 Parallel Operation of a Single-Phase Transformer

Nowadays, the demand of load is increasing with the increase in population and industrial sector. Sometimes, it is difficult to meet the excess demand for power by the existing single unit transformer. Therefore, an additional transformer is required to connect in parallel with the existing one. The following points should be considered for making parallel connections of transformers:

- The terminal voltage of both transformers must be the same.
- Polarity must be the same for both transformers.
- For both transformers, the percentage of impedances should be equal in magnitude.
- The ratio of *R* to *X* must be the same for both transformers.
- Phase sequences and phase shifts must be the same (for a three-phase transformer).

2.19 Three-Phase Transformer Connections

The primary and secondary windings of the transformer may be connected in either wye (Y) or delta (Δ). Three-phase transformer connections are classified into four possible types, namely Y-Y (wye–wye), Y-Δ (wye-delta), Δ-Y (delta-wye), and Δ-Δ (delta-delta).

2.19.1 *Wye-Wye Connection*

Figure 2.35 shows the Y-Y connection diagram. This type of connection of a three-phase transformer is rarely used for a large amount of power transmission.

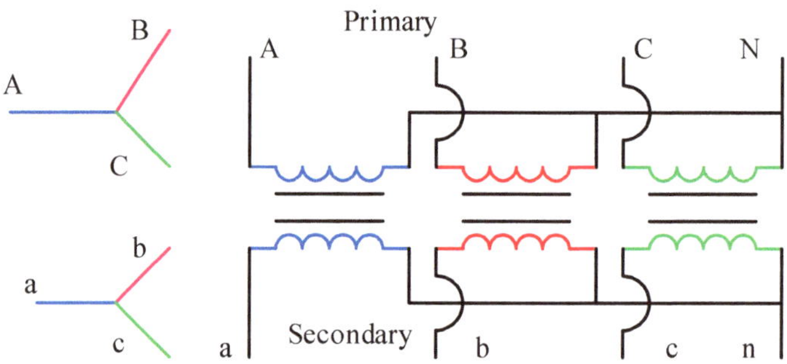

Fig. 2.35 Y–Y connection diagram

Neutral point is necessary for both primary and secondary sides in some cases. In balanced loads, this type of connection works satisfactorily and provides neutral to each side for grounding. At the primary side, the phase voltage can be written as,

$$V_{P1} = \frac{V_{L1}}{\sqrt{3}} \tag{2.174}$$

The secondary phase voltage can be written as,

$$V_{P2} = \frac{V_{L2}}{\sqrt{3}} \tag{2.175}$$

The ratio of the primary line voltage to the secondary line voltage of this connection is,

$$a = \frac{V_{L1}}{V_{L2}} \tag{2.176}$$

$$a = \frac{\sqrt{3}\,V_{P1}}{\sqrt{3}\,V_{P2}} = \frac{V_{P1}}{V_{P2}} \tag{2.177}$$

2.19.2 Wye-Delta Connection

The wye-delta connection is mainly used at the substation where the voltage is stepped down. In this connection, the current in the secondary coil is 57.7% of the load current. At the primary side of this connection, a copper wire is used to ground the neutral point. Figure 2.36 shows the connection diagram of the wye-delta transformer. In this connection, the expression of the primary line voltage is,

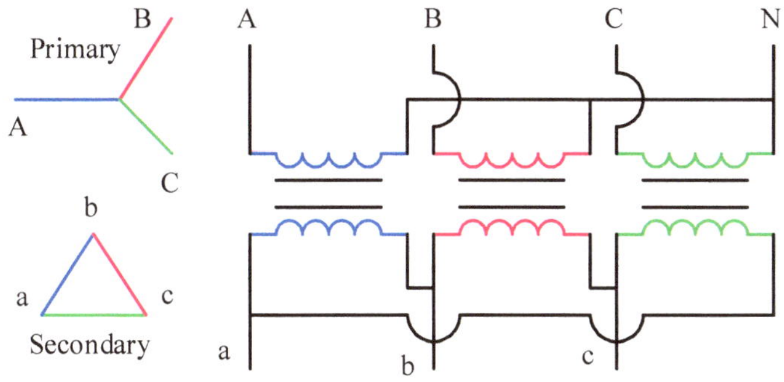

Fig. 2.36 Wye-delta connection diagram

$$V_{L1} = \sqrt{3}\, V_{P1} \tag{2.178}$$

At the secondary side, the line voltage is,

$$V_{L2} = V_{P2} \tag{2.179}$$

The ratio of primary phase voltage to secondary phase voltage is,

$$a = \frac{V_{P1}}{V_{P2}} \tag{2.180}$$

The ratio of primary line voltage to secondary line voltage is,

$$\frac{V_{L1}}{V_{L2}} = \frac{\sqrt{3}\, V_{P1}}{V_{P2}} = \sqrt{3}\, a \tag{2.181}$$

The primary phase current is,

$$I_{P1} = I_{L1} \tag{2.182}$$

In this case, the turns ratio is,

$$a = \frac{I_{P2}}{I_{P1}} \tag{2.183}$$

The expression of the secondary phase current is,

$$I_{P2} = aI_{P1} \tag{2.184}$$

The secondary line current is,

$$I_{L2} = \sqrt{3}\, I_{P2} = \sqrt{3}\, a I_{P1} \tag{2.185}$$

2.19.3 Delta-Wye Connection

The delta-wye connection is generally used at the power generating station to step up the voltage. Figure 2.37 shows the connection diagram of a delta-wye transformer. In this connection, the expression of the primary line voltage is,

$$V_{L1} = V_{P1} \tag{2.186}$$

The line voltage at the secondary side is,

$$V_{L2} = \sqrt{3}\, V_{P2} \tag{2.187}$$

The ratio of primary line voltage to secondary line voltage is,

$$\frac{V_{L1}}{V_{L2}} = \frac{V_{P1}}{\sqrt{3}\, V_{P2}} = \frac{a}{\sqrt{3}} \tag{2.188}$$

The phase current at the primary side is,

$$I_{P1} = \frac{1}{\sqrt{3}} I_{L1} \tag{2.189}$$

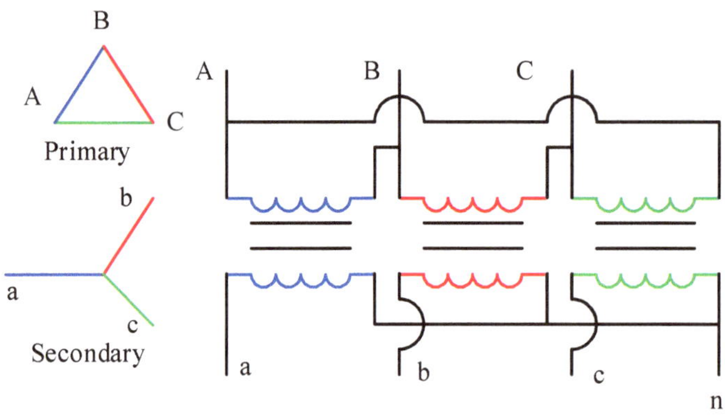

Fig. 2.37 Delta-wye connection diagram

In this connection, the turns ratio is expressed as,

$$a = \frac{I_{P2}}{I_{P1}} \tag{2.190}$$

In this case, the secondary phase current can be written as,

$$I_{P2} = aI_{P1} \tag{2.191}$$

The secondary line current is expressed as,

$$I_{L2} = I_{P2} = aI_{P1} \tag{2.192}$$

2.19.4 Delta-Delta Connection

The delta-delta connection is generally used for both high voltage and low voltage rating transformers where insulation is not an important issue. The connection diagram for delta-delta transformer is shown in Fig. 2.38.

In this case, the primary line voltage can be written as,

$$V_{L1} = V_{P1} \tag{2.193}$$

The secondary line voltage is expressed as,

$$V_{L2} = V_{P2} \tag{2.194}$$

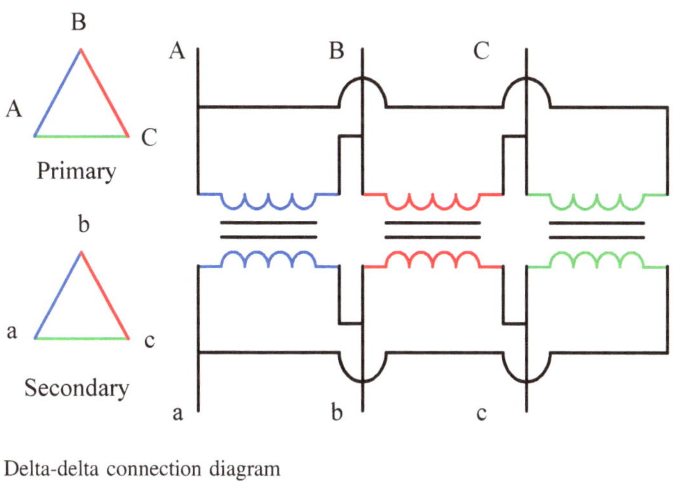

Fig. 2.38 Delta-delta connection diagram

The ratio of line-to-line voltage of this connection is,

$$\frac{V_{L1}}{V_{L2}} = \frac{V_{P1}}{V_{P2}} = a \tag{2.195}$$

The secondary line current is expressed as,

$$I_{L2} = \sqrt{3}\, I_{P2} \tag{2.196}$$

The output capacity in delta-delta connection can be expressed as,

$$P_{0\Delta-\Delta} = \sqrt{3}\, V_{L2} I_{L2} \tag{2.197}$$

Substituting Eq. (2.241) into Eq. (2.242) yields,

$$P_{0\Delta-\Delta} = \sqrt{3}\, V_{L2} \times \sqrt{3}\, I_{P2} = 3 V_{L2} I_{P2} \tag{2.198}$$

The remaining two transformers are able to supply three-phase power to the load terminal if one of the transformers is removed from the connection. This type of three-phase power supply by two transformers is known as open delta or V–V connection. This open delta connection is able to supply three-phase power at a reduced rate of 57.7%. The connection diagram for an open delta connection is shown in Fig. 2.39.

In the V–V connection, the secondary line current is,

$$I_{L2} = I_{P2} \tag{2.199}$$

The output capacity in V–V connection is expressed as,

$$P_{0\text{V}-\text{V}} = \sqrt{3}\, V_{L2} I_{P2} \tag{2.200}$$

Fig. 2.39 Open delta or V–V connection diagram

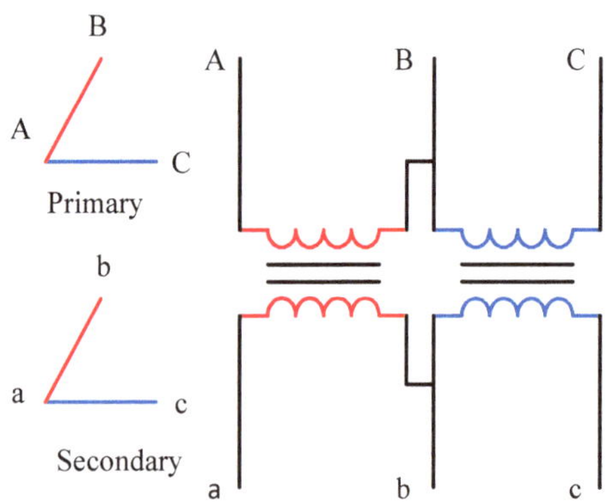

The ratio of delta-delta output capacity to the V–V output capacity is,

$$\frac{P_{0\text{V}-\text{V}}}{P_{0\Delta-\Delta}} = \frac{\sqrt{3}\,V_{L2} I_{P2}}{3 V_{L2} I_{P2}} = 0.577 \tag{2.201}$$

Example 2.10
A three-phase transformer is connected to an 11 kV supply and draws 6 A current. Determine line voltage at the secondary side and the line current in the secondary coil. Consider the turns ratio of the transformer is 11. Consider delta-wye and wye-delta connections.

Solution
Delta-wye connection:
The phase voltage at the primary side is,

$$V_{P1} = V_{L1} = 11\,\text{kV} \tag{2.202}$$

The phase voltage at the secondary side is calculated as,

$$V_{P2} = \frac{V_{P1}}{a} = \frac{11000}{11} = 1000\,\text{V} \tag{2.203}$$

The line voltage at the secondary is determined as,

$$V_{L2} = \sqrt{3}\,V_{P2} = \sqrt{3} \times 1000 = 1732\,\text{V} \tag{2.204}$$

The phase current in the primary is calculated as,

$$I_{P1} = \frac{I_{L1}}{\sqrt{3}} = \frac{11}{\sqrt{3}} = 6.35\,\text{A} \tag{2.205}$$

The line current in the secondary is calculated as,

$$I_{L2} = I_{P2} = aI_{P1} = 11 \times 6.35 = 69.85\,\text{A} \tag{2.206}$$

Wye-delta connection:
The line voltage at the primary side is,

$$V_{L1} = 11\,\text{kV} \tag{2.207}$$

The phase voltage at the primary side is calculated as,

$$V_{P1} = \frac{V_{L1}}{\sqrt{3}} = \frac{11000}{\sqrt{3}} = 6350.85\,\text{V} \tag{2.208}$$

The phase voltage at the secondary side is determined as,

$$V_{P2} = \frac{V_{P1}}{a} = \frac{6350.85}{11} = 577.35\,\text{V} \tag{2.209}$$

The line voltage at the secondary is,

$$V_{L2} = V_{P2} = 577.35\,\text{V} \tag{2.210}$$

The phase current in the secondary is calculated as,

$$I_{P2} = aI_{P1} = 11 \times 6 = 66\,\text{A} \tag{2.211}$$

The line current in the secondary is calculated as,

$$I_{L2} = \sqrt{3}\,I_{P2} = \sqrt{3} \times 66 = 114.32\,\text{A} \tag{2.212}$$

Practice Problem 2.10

A three-phase transformer is connected to the 33 kV supply and draws 15 A current. Calculate the line voltage at the secondary side and the line current in the secondary winding. Consider the turns ratio of the transformer is 8 and wye-delta connection.

2.20 Instrument Transformers

The magnitude of voltage and the current is normally high in the power system networks. To reduce the magnitude of voltage and current, instrument transformers are used. There are two types of instrument transformers, a current transformer (CT) and potential transformer (PT). If a power system carries an alternating current greater than 100 A, then it is difficult to connect measuring instruments like low range ammeter, wattmeter, energy meter, and relay.

The current transformer is then connected in series with the line to step down the high magnitude of the current to a rated value of about 5 A for the ammeter and the current coil of the wattmeter. The diagram of a current transformer is shown in Fig. 2.40. If the system voltage exceeds 500 V, then it is difficult to connect

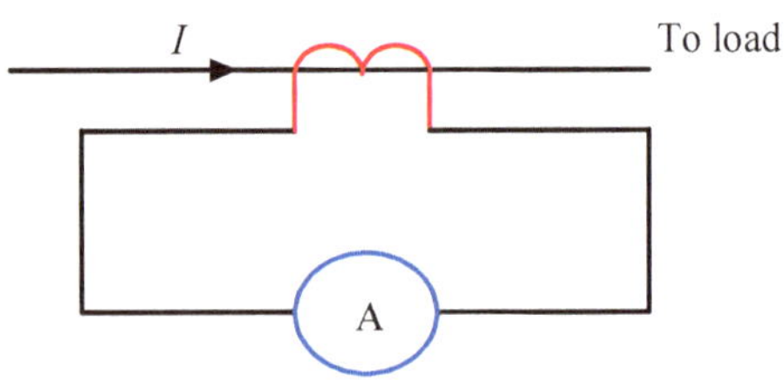

Fig. 2.40 Connection diagram of the current transformer

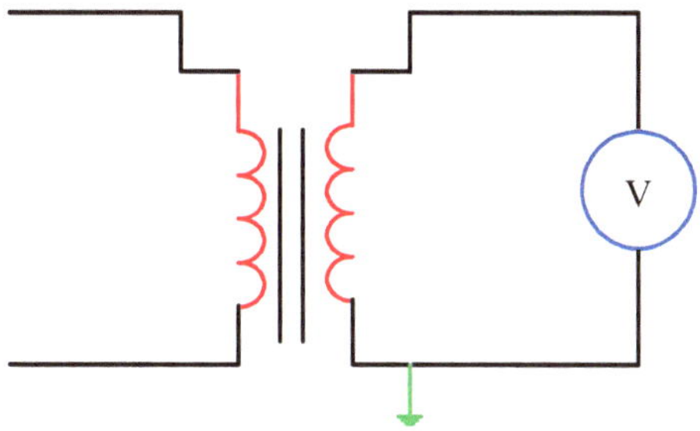

Fig. 2.41 Connection diagram of the potential transformer

measuring instruments directly to the high voltage terminals. The potential transformer is then used to step down to a suitable value of the voltage at the secondary for supplying to the voltmeter and the voltage coil of the wattmeter. The secondary of the instrument transformer is normally grounded for safety. The connection diagram of a potential transformer is shown in Fig. 2.41.

2.21 Transformer Oil Testing

Power transformer and distribution transformer are identified as the most important and critical component in the power utility companies in terms of price and reliability of service. Hydrotreated light naphthenic distillate chemical oil in different names such as "TRANSOL, POWER OIL, SAVITHA, APAR." is used as an insulating for inter-winding of a three-phase transformer. Due to age, transformer oil needs to be tested according to different standards such as IEEE Std 62-1995, IEC 60156, ASTM D1816, etc. The percentage saturation of water in oil is shown in Table 2.1. There are several apparatus available in the market to test the dielectric strength of insulating oils. This equipment automatically tests the dielectric strength of the insulating oil. In addition to that, it has a high precision oil breakdown voltage test (0–75 kV ± 1 kV) with a very accurate measurement principle performed directly on the HV unit as well as monitor the voltage slew (0.5 kV–10 kV/s) or real breakdown monitoring (RBM) as shown in Fig. 2.42. The equipment provides excellent short switch-off time (<10 μs) after a breakdown counteracts the contamination of the oil sample, thus ensuring the reliable reproducibility of the measurement results. Most of the available equipment complies with all types of standards, and its internal temperature recording of the oil sample varies from 0 to 99 °C.

Table 2.1 Representation of insulation levels

Insulation condition	Saturation of water in oil (%)
Dry insulation	0–5
Moderate-wet	6–20
Wet insulation	21–30
Extremely wet insulation	>30

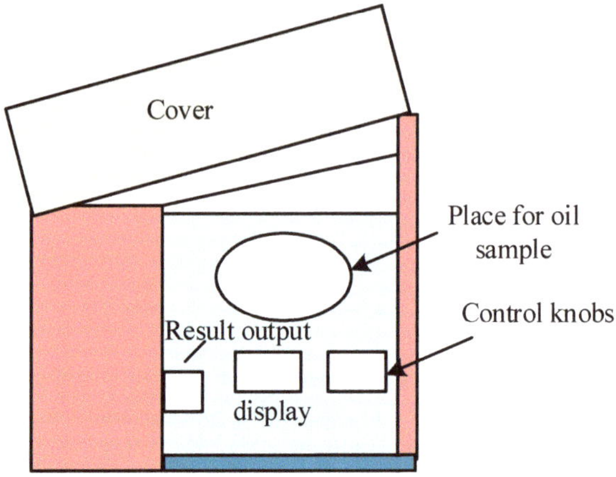

Fig. 2.42 A sample of oil tester equipment

2.22 Standard Symbols and Reactance Diagram

Standard symbols are used to represent the components of a typical power system as shown in Fig. 2.43. Generator, transformer, transmission line, and load are the basic components of a power system. These basic components or symbols are used to form the one-line or single-line diagram of a typical power system. A single-line diagram of a typical power system is shown in Fig. 2.44. This typical power system consists of a synchronous machine, load, two transformers, four buses, and a transmission line. Circuit breakers are generally used here to protect the power system components. Initially, a single-line diagram is considered to derive the impedance and reactance diagrams. Additional information such as ratings of synchronous machines and transformers can be obtained from the single-line diagram. Power consumed or delivered in the system and impedances of the components can be calculated from the single-line diagram. The details of a single-line diagram along with the simulation of a power system will be discussed in the other chapters. The following points are considered to derive the impedance diagram:

- The generator is represented by a voltage source in series with reactance,
- The motor is represented by a voltage source in series with resistance and reactance,
- The transformer is represented by resistance and reactance,
- The load is represented by resistance and reactance,
- The transmission line is represented by resistance and reactance.

The impedance diagram is shown in Fig. 2.45. The reactance diagram is normally drawn by neglecting the resistances and line capacitance from the impedance diagram shown in Fig. 2.46.

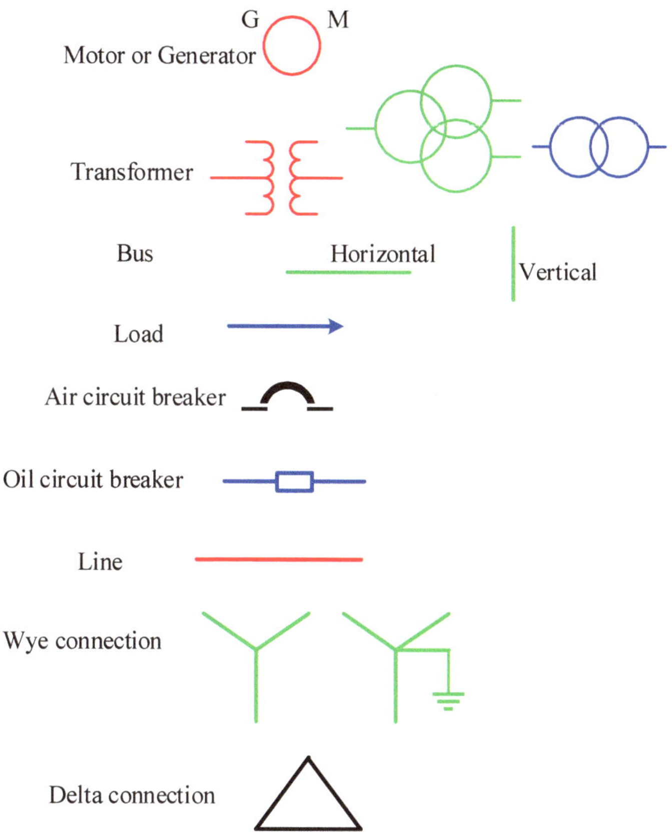

Fig. 2.43 Standard symbols of a power system

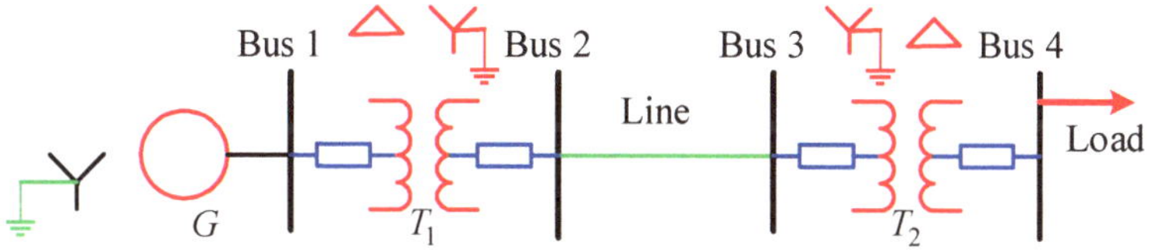

Fig. 2.44 Single-line diagram of a typical power system

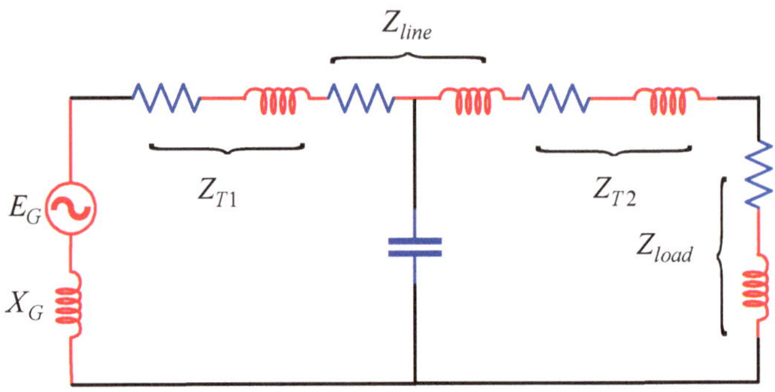

Fig. 2.45 Impedance diagram

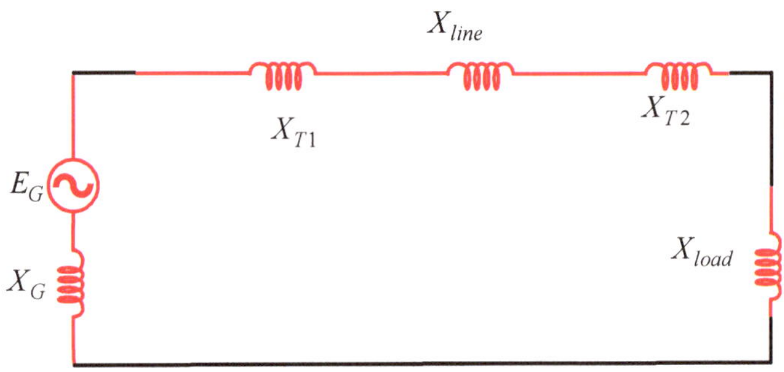

Fig. 2.46 Reactance diagram

2.23 Per Unit System

In a power system, per unit values of voltage, current, power, and impedance are used for analyzing and interpreting data. These per unit values are normalized on a selected base voltage in kV and base apparent power in MVA. This normalization is known as the per unit analysis. There are many advantages to use the per unit values in a power system and these advantages are as follows:

- It gives a clear idea about voltage, current, and impedance.
- The per unit values of voltage, current, and impedance of a transformer is the same, whether referred to as primary or secondary.
- The per unit values are ideal for computer simulation and calculation.
- The per unit impedance of any electrical equipment lies within the narrow numerical range when the equipment ratings are used as base values.
- Manufacturers normally include the necessary information in the nameplate as per unit values.

The per unit value is defined as the ratio of the actual value of the quantity to the base value of the same quantity and it can be expressed as,

$$\text{Per unit value} = \frac{\text{actual value}}{\text{base value}} \tag{2.213}$$

The base value is arbitrarily chosen within the range of voltage and apparent power. The base value is always a real number and per unit value is dimensionless.

2.23.1 Single-Phase System

Consider the base voltage and base power of a single-phase system V_b (1ϕ) and S_b (1ϕ), respectively. The expression of base current (I_b) is expressed as [3],

$$I_b = \frac{S_{b(1\phi)}}{V_{b(LN)}} \tag{2.214}$$

The expression of base impedance (Z_b) is expressed as,

$$Z_b = \frac{V_{b(LN)}}{I_b} \tag{2.215}$$

Substituting Eq. (2.214) into the Eq. (2.215) yields,

$$Z_b = \frac{V_{b(LN)}}{\frac{S_{b(1\phi)}}{V_{b(LN)}}} \tag{2.216}$$

$$Z_b = \frac{V_{b(LN)}^2}{S_{b(1\phi)}} \tag{2.217}$$

2.23.2 Three-Phase System

In a three-phase system, the base current can be calculated as,

$$I_b = \frac{S_{b(3\phi)}}{\sqrt{3}V_{b(LL)}} \tag{2.218}$$

The base impedance can be expressed as,

$$Z_b = \frac{\frac{V_{b(LL)}}{\sqrt{3}}}{I_b} \tag{2.219}$$

Substituting Eq. (2.218) into Eq. (2.219) yields,

$$Z_b = \frac{\frac{V_{b(LL)}}{\sqrt{3}}}{\frac{S_{b(3\phi)}}{\sqrt{3}\,V_{b(LL)}}} \tag{2.220}$$

$$Z_b = \frac{V_{b(LL)}^2}{S_{b(3\phi)}} \tag{2.221}$$

According to Ohm's law, the expression of impedance is,

$$Z = \frac{V}{I} \tag{2.222}$$

Dividing Eq. (2.222) by Eq. (2.215) yields,

$$\frac{Z}{Z_b} = \frac{\frac{V}{I}}{\frac{V_b}{I_b}} = \frac{\frac{V}{V_b}}{\frac{I_b}{I}} \tag{2.223}$$

$$Z_{\text{pu}} = \frac{V_{\text{pu}}}{I_{\text{pu}}} \tag{2.224}$$

where the put unit voltage and current are expressed as,

$$V_{\text{pu}} = \frac{V}{V_b} \tag{2.225}$$

$$I_{\text{pu}} = \frac{I}{I_b} \tag{2.226}$$

In an alternative way, the expression of per unit impedance can be expressed as,

$$Z_{\text{pu}} = \frac{Z}{Z_b} \tag{2.227}$$

Substituting Eq. (2.217) into Eq. (2.227) yields,

$$Z_{\text{pu}} = \frac{Z}{\frac{V_b^2}{S_b}} \tag{2.228}$$

$$Z_{pu} = Z\frac{S_b}{V_b^2} \tag{2.229}$$

According to Eq. (2.229), the per unit impedance for new and old systems can be expressed as,

$$Z_{puo} = Z\frac{S_{bo}}{V_{bo}^2} \tag{2.230}$$

$$Z_{pun} = Z\frac{S_{bn}}{V_{bn}^2} \tag{2.231}$$

Dividing Eq. (2.230) by Eq. (2.231) yields,

$$\frac{Z_{pun}}{Z_{puo}} = \frac{Z\frac{S_{bn}}{V_{bn}^2}}{Z\frac{S_{bo}}{V_{bo}^2}} \tag{2.232}$$

$$\frac{Z_{pun}}{Z_{puo}} = \frac{S_{bn}}{S_{bo}}\left(\frac{V_{bo}}{V_{bn}}\right)^2 \tag{2.233}$$

$$Z_{pun} = Z_{puo}\frac{S_{bn}}{S_{bo}}\left(\frac{V_{bo}}{V_{bn}}\right)^2 \tag{2.234}$$

where

Z_{pun} is the per unit impedance of the new system,
Z_{puo} is the per unit impedance of the old system,
V_{bn} is the base voltage of the new system,
S_{bn} is the base value of the apparent power of the new system,
V_{bo} is the base voltage of the old system,
S_{bo} is the base value of the apparent power of the old system.

Example 2.11
An impedance of $4 + j8\,\Omega$ is connected in series with a voltage source of 220 V, 50 Hz source. Calculate the per unit values of resistance, reactance, and impedance by considering the base values of 100 V and 100 VA.

Solution
The base value of the current is calculated as,

$$I_b = \frac{100}{100} = 1\text{A} \tag{2.235}$$

The base value of the impedance is calculated as,

$$Z_b = \frac{V_b}{I_b} = \frac{100}{1} = 100\,\Omega \quad (2.236)$$

In an alternative way, the base impedance can be calculated as,

$$Z_b = \frac{V_b^2}{S_b} = \frac{100^2}{100} = 100\,\Omega \quad (2.237)$$

Per unit value of the resistance is calculated as,

$$R_{\text{pu}} = \frac{R}{Z_b} = \frac{4}{100} = 0.04 \quad (2.238)$$

Per unit value of the inductance is calculated as,

$$X_{\text{Lpu}} = \frac{X_L}{Z_b} = \frac{8}{100} = 0.08 \quad (2.239)$$

Per unit impedance is calculated as,

$$Z_{\text{pu}} = R_{\text{pu}} + jX_{\text{Lpu}} = 0.04 + j0.08 = 0.089\,\underline{|63.43^\circ}. \quad (2.240)$$

Practice Problem 2.11

The reactance referred to the primary side of a 40 MVA, 11 kV/66 kV transformer is 0.45 Ω. Calculate the per unit reactance referred to as the primary and the secondary sides. Consider the base values as 100 kV and 100 MVA.

Example 2.12

A single-line diagram of a typical power system is shown in Fig. 2.47. The ratings of the components are mentioned below.

Transformer T_1	100 MVA, 132 kV/66 kV, X = 0.02 pu,
Transformer T_2	40 MVA, 66 kV/11 kV, X = 0.01 pu,
Load	50 MW, 10 Mvar
Line	X = 3 Ω.

Three-phase load absorbs power at a lagging power factor of 0.85 from the transmission line. Draw a reactance diagram with per unit values and find the grid to bus voltage.

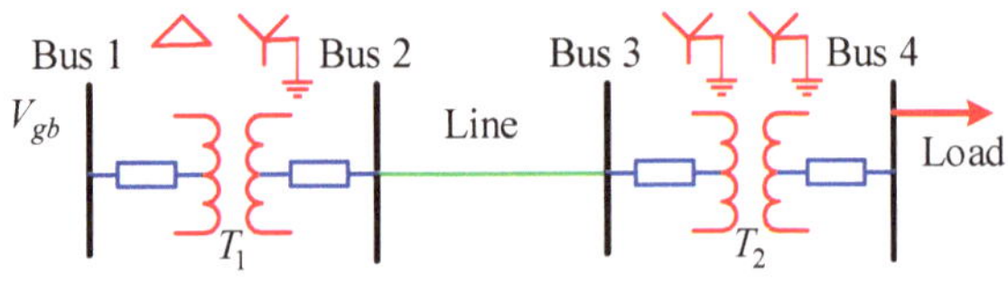

Fig. 2.47 Single-line diagram for Example 2.12

Solution

Let the base power and base voltage are,

$$S_b = 100\,\text{MVA} \tag{2.241}$$

$$V_b = 100\,\text{kV} \tag{2.242}$$

The turns ratio for the first transformer is calculated as,

$$a_1 = \frac{V_1}{V_2} = \frac{132}{66} = 2 \tag{2.243}$$

The base voltage of the line is calculated as,

$$V_{b-\text{line}} = V_{b2} = V_{b3} = \frac{V_b}{a} = \frac{100}{2} = 50\,\text{kV} \tag{2.244}$$

The base impedance of the line is calculated as,

$$Z_{b-\text{line}} = \frac{V_{b-\text{line}}^2}{S_b} = \frac{50^2}{100} = 25\,\Omega \tag{2.245}$$

The per unit reactance of the line is calculated as,

$$X_{\text{pu}-\text{line}} = \frac{3}{25} = 0.12 \tag{2.246}$$

For the first (grid) transformer, the per unit reactance is calculated as,

$$X_{\text{pu}-T1} = 0.02 \times \left(\frac{132}{100}\right)^2 \times \left(\frac{100}{100}\right) = 0.034 \tag{2.247}$$

The base voltage at the load side is calculated as,

$$V_{b-\text{load}} = V_{b4} = \frac{V_{b3}}{a_2} = \frac{11}{66} \times 50 = 8.33\,\text{kV} \tag{2.248}$$

For the second (load) transformer, the per unit reactance is,

$$X_{\text{pu}-T2} = 0.01 \times \left(\frac{11}{8.33}\right)^2 \times \frac{100}{40} = 0.04 \tag{2.249}$$

The base current at the load can be calculated as,

$$I_b = \frac{S_b}{\sqrt{3}\,V_b} = \frac{100 \times 10^6}{\sqrt{3} \times 8.33 \times 10^3} = 693.09\,\text{A} \tag{2.250}$$

The actual current in the load is calculated as,

$$I_L = \frac{50 \times 10^6}{\sqrt{3} \times 11 \times 10^3 \times 0.6} = 3087.43\,\text{A} \tag{2.251}$$

Per unit value of the load current is,

$$I_{\text{Lpu}} = \frac{693.09}{3087.43} = 0.22 \tag{2.252}$$

Per unit value of the voltage at the load terminal (bus 4) is,

$$V_4 = V_L = \frac{8.33}{100} = 0.083\,\text{pu} \tag{2.253}$$

The equivalent circuit is shown in Fig. 2.48 and the grid to bus voltage can be calculated as,

$$V_{gb} = 0.22\,\underline{|\cos -1(0.85)^\circ}. \times j(0.03 + 0.04 + 0.12) + 0.083 = 0.07\,\underline{|30.22^\circ}.\,\text{pu} \tag{2.254}$$

The actual value of the grid to the bus voltage is calculated as

$$V_{gb} = 0.07 \times 100 = 7\,\text{kV} \tag{2.255}$$

Practice Problem 2.12

The primary side reactance of a 100 kVA, 600 V/300 V transformer is 5 Ω. Calculate the per unit value of reactance referred to as the primary side, and to the secondary side. Consider the base quantities are 100 kVA and 100 V.

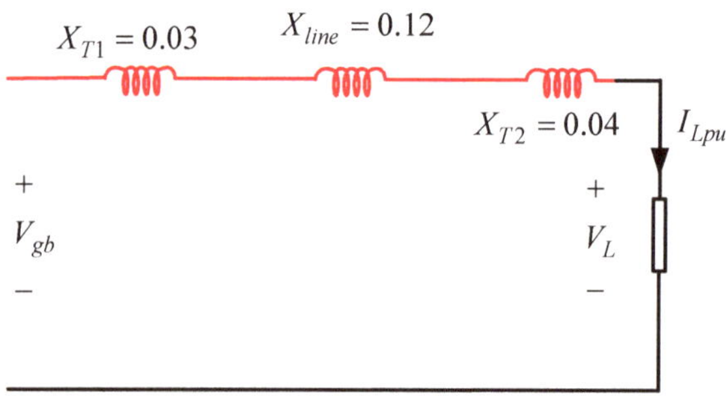

Fig. 2.48 Reactance diagram for Example 2.12

References

1. Wildi T (2006) Electrical machines, drives and power systems, 6th edn. Pearson Education International, Upper Saddle River
2. Salam MA (2012) Fundamentals of electrical machines, 2nd edn. International Ltd., Alpha Science
3. Chapman SJ (2002) Electric machinery and power system fundamentals. McGraw-Hill Higher Education, New York
4. Fitzgerald AE, Kingsley C Jr, Umans SD (2002) Electric machinery, 6th edn. McGraw-Hill Higher Education, New York

Exercise Problems

2.1 A single-phase transformer is having the primary voltage of 240 V. The numbers of turns at the primary and secondary coils are 250 and 50, respectively. Determine the secondary voltage.

2.2 A single-phase transformer is having a secondary voltage of 100 V. The number of turns at the primary and secondary windings is 500 and 50, respectively. Calculate the primary voltage.

2.3 The 100 A current flows in the primary coil of a single-phase transformer. Calculate the secondary current with a turns ratio of 0.05.

2.4 The number of turns at the primary and secondary windings of a single-phase transformer is 500 and 100, respectively. Find the primary current if the secondary current is 10 A.

2.5 The ratio of the primary current to the secondary current of a single-phase transformer is 1:10. Find the voltage ratio. Also, calculate the secondary voltage if the primary voltage is found to 1000 V.

2.6 A single-phase transformer has a turns ratio of 1:4. The secondary coil has 5000 turns, and the voltage is found to be 60 V. Determine the voltage ratio, primary voltage, and a number of turns.

2.7 A single-phase transformer has a voltage ratio of 5:1. The voltage and number of turns at the primary coil are found to be 1100 V and 500, respectively. Calculate the voltage and a secondary number of turns at the secondary coil.

2.8 A single-phase transformer is connected to the 120 V source and draws 4 A current. Calculate the current in the secondary coil when the transformer steps up the voltage to 500 V.

2.9 The number of turns at the primary and secondary coils of a single-phase transformer is found to be 480 and 60, respectively. The transformer draws a current of 0.6 A when connected to 120 V supply. Determine the current and the voltage at the secondary coil.

2.10 The turns ratio of a 5 kVA single-phase transformer is found to be 2. The transformer is connected to a 230 V, 50 Hz supply. If the secondary current of a transformer is 6 A, find the primary current and secondary voltage.

2.11 The secondary number of turns of a 30 kVA, 4400 V/440 V single-phase transformer is found to be 100. Find the number of primary turns, full load primary and secondary currents.

2.12 A single-phase transformer of 5 kVA, 1100 V/230 V, 50 Hz is installed near the domestic area of a country. Find the turns ratio, primary and secondary full load currents.

2.13 The number of turns of the primary winding of a single-phase 50 Hz transformer is found to be 50. Find the value of the core flux, if the induced voltage at this winding is 220 V.

2.14 The number of primary turns of a 60 Hz single-phase transformer is found to be 50. Calculate the induced voltage in this winding, if the value of the core flux is 0.05 Wb.

2.15 A 3300 V/330 V, 50 Hz step-down single-phase transformer has a maximum flux of 0.45 Wb. Calculate the number of primary and secondary turns.

2.16 A 5 kVA, 2200 V/220 V, 50 Hz single-phase step-down transformer has the cross-sectional area is 25 cm^2, and a maximum flux density is 4 Wb/m$^{2.}$ Calculate the primary and secondary turns.

2.17 The number of turns at the primary and secondary of an ideal single-phase transformer is found to be 500 and 250, respectively. The primary of the transformer is connected to a 220 V, 50 Hz source and the secondary coil supplies a current of 5 A to the load. Find the turns ratio, current in the primary, and maximum flux in the core.

2.18 A 4 kVA, 1100 V/230 V, 50 Hz single-phase transformer has the primary number of turns of 500. The cross-sectional area of the transformer core is 85 cm^2. Find the turns ratio, number of turns in the secondary, and maximum flux density in the core.

2.19 A single-phase transformer has the primary and secondary turns of 50 and 500, respectively. The primary winding is connected to a 220 V, 50 Hz supply. Calculate the flux density in the core if the cross-sectional area is 250 cm^2 and induced voltage at the secondary winding.

2.20 The primary coil number of turns of a single-phase 2200 V/220 V, 50 Hz transformer is found to be 1000. Find the area of the core if the maximum flux density of the core is 1.5 Wb/m^2.

2.21 The maximum flux of a single-phase 1100 V/220 V, 50 Hz transformer is found to be 5 mWb. The number of turns and the voltage at the primary winding is found to be 900 and 1100 V, respectively. Determine the frequency of the supply system.

2.22 A 6 kVA, 2200 V/220 V, 50 Hz single-phase transformer has the maximum flux density of 0.6 Wb/m^2 and the induced voltage per turn of the transformer is 10 V. Calculate the primary number of turns, secondary number of turns, and cross-sectional area of the core.

2.23 A single-phase transformer is having the primary voltage per turn is 20 V. Find the magnetic flux density if the cross-sectional area of the core is 0.05 m^2.

2.24 A 1100 V/230 V single-phase transformer has a no-load primary current of 0.5 A and absorbs the power of 350 W from the circuit. Calculate the iron loss and magnetizing components of the no-load current.

2.25 A 3300 V/230 V single-phase transformer draws a no-load current of 0.5 A at a power factor of 0.4 in an open circuit test. Determine the working and magnetizing components of the no-load current.

2.26 A single-phase transformer draws a no-load current of 1.4 A from the 120 V source and absorbs the power of 80 W. The primary winding resistance and leakage reactance are found to be 0.25 Ω and 1.2 Ω, respectively. Find the no-load circuit resistance and reactance.

2.27 Under the no-load condition, a single-phase transformer is connected to a 240 V, 50 Hz supply and draws a current of 0.3 A at a 0.5 power factor. The number of turns on the primary winding is found to be 600. Calculate the maximum value of the flux in the core, magnetizing and working components of the no-load current, and iron loss.

2.28 The no-load current of a 440 V/110 V single-phase transformer is measured 0.8 A and absorbs the power of 255 W. Calculate the working and magnetizing components of the no-load current, copper loss in the primary winding if the primary winding resistance is 1.1 Ω, and core loss.

2.29 A single-phase 1000 V/200 V transformer draws a no-load current of 0.7 A at 0.4 power factor lagging and the secondary winding supplies a current of 85 A at 0.85 power factor lagging. Find the primary current.

2.30 A single-phase 600 V/200 V transformer supplies a no-load current of 45 A at 0.6 power factor lagging to the load, and the transformer draws a no-load current of 3 A at 0.4 power factor lagging. Calculate the primary current and its power factor.

2.31 The primary winding resistance and reactance of a 300 kVA, 2200 V/220 V single-phase transformer are 2 Ω and 9 Ω, respectively. The secondary winding resistance and reactance are found to be 0.02 Ω and 0.1 Ω, respectively. Determine the equivalent impedance referred to as the primary and the secondary.

2.32 The primary winding resistance and reactance of a 200 kVA, 3300 V/220 V single-phase transformer are found to be 5 Ω and 12 Ω, respectively. The same parameters in the secondary winding are found to 0.02 Ω and 0.05 Ω, respectively. Find the equivalent impedance referred to as the primary and the secondary. Also, determine the full load copper loss.

2.33 A 1100 V/400 V single-phase transformer having the resistance of 5 Ω and the reactance of 9 Ω in the primary winding. The secondary winding resistance and reactance are found to be 0.6 Ω and 1.1 Ω, respectively. Find the voltage regulation when the secondary delivers a current of 5 A at a 0.9 lagging power factor.

2.34 A 2200 V/220 V single-phase transformer having the resistance of 6 Ω and the reactance of 16 Ω in the primary coil. The resistance and the reactance in secondary winding are found to be 0.07 Ω and 0.15 Ω, respectively. Find the voltage regulation when the secondary supplies a current of 6 A at a 0.86 power factor leading.

2.35 A 40 kVA single-phase transformer has the iron loss, and full load copper loss is 450 W and 750 W, respectively. Calculate the full load efficiency, output kVA corresponding to maximum efficiency, and maximum efficiency. Consider the power factor of the load is 0.95 lagging.

2.36 A 40 kVA single-phase transformer has an iron loss of 50% of the full load copper loss. The full load copper loss is found to be 850 W. Find the full load efficiency, output kVA corresponding to maximum efficiency and maximum efficiency. Assume the power factor of the load is 0.9 lagging.

2.37 A 25 kVA single-phase transformer has an iron loss and full load copper loss of 300 W and 550 W, respectively. Find the full load efficiency, output kVA corresponding to maximum efficiency, and maximum efficiency. Assume the power factor of the load is 0.6 lagging.

2.38 A 100 kVA, 2200 V/220 V single-phase transformer has the primary and secondary windings parameters of $R_1 = 0.6\,\Omega$, $X_1 = 0.9\,\Omega$, $R_2 = 0.007\,\Omega$, $X_2 = 0.008\,\Omega$. The transformer is operating at a maximum efficiency of 75% of its rated load with 0.9 lagging power factor. Calculate the efficiency of the transformer at full load, and maximum efficiency, if the iron loss is 350 W.

2.39 A 200 kVA, 2400 V/240 V single-phase transformer has the primary and windings parameters $R_1 = 20\,\Omega$, $X_1 = 27\,\Omega$, $R_2 = 0.18\,\Omega$, $X_2 = 0.26\,\Omega$. The transformer is operating at a maximum efficiency of 80% of its rated load with 0.8 pf lagging. Find the efficiency of the transformer at full load, and maximum efficiency if the iron loss is 400 W.

2.40 The low voltage side of a 30 kVA, 2400 V/220 V, 50 Hz single-phase transformer is short-circuited. The test data recorded from the high voltage side are 200 W, 6 A, and 45 V. Find the equivalent resistance, reactance, and impedance referred to the primary, equivalent resistance, reactance, and impedance referred to the secondary, and voltage regulation at 0.6 power factor lagging.

2.41 The low voltage winding of a 25 kVA, 1100 V/220 V, 50 Hz single-phase transformer is short-circuited, and the test data recorded from the high voltage side are 156 W, 4 A and 50 V. Calculate the equivalent resistance, reactance, and impedance referred to the primary, equivalent resistance, reactance, and impedance referred to the secondary, and voltage regulation at 0.85 power factor leading.

2.42 The test data of a 10 kVA, 440 V/220 V, 50 Hz single-phase transformer are as follows:
Open circuit test: 220 V, 1.4 A, 155 W;
Short circuit test: 20.5 V, 15 A, 145 W;
Calculate the approximate equivalent circuit parameters.

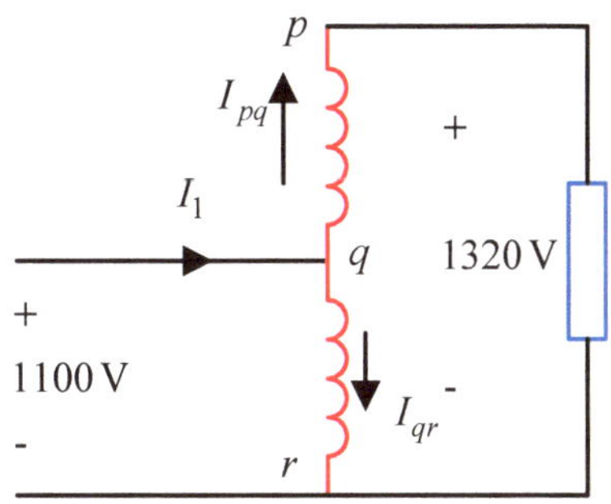

Fig. P2.1 A autotransformer circuit for Problem 2.44

2.43 The open circuit test data of a single-phase transformer are recorded as the 220 V, 1.2 A, and 145 W. Find the no-load circuit parameters.

2.44 A single-phase 100 kVA, 1100 V/220 V transformer is connected as an autotransformer, which is shown in Fig. P2.1. The voltage of the upper portion and the lower portion of the coil are found to be 220 V and 1100 V, respectively. Calculate the kVA rating of the autotransformer.

2.45 A 200 kVA three-phase, 50 Hz core-type transformer is connected as delta–wye and has a line voltage ratio of 1100 V/440 V. The core area and the maximum flux density of the transformer are found to be 0.04 m^2 and 2.3 Wb/m^2, respectively. Calculate the number of turns per phase of the primary and secondary coils.

2.46 The number of turns in the primary and secondary coils is found to be 600 and 150, respectively. The transformer is connected to 11 kV, 50 Hz supply. Find the secondary line voltage when the windings are connected as (a) delta-wye and (b) wye-delta.

2.47 A single-phase load absorbs 120 kVA power from 22 kV busbar. Calculate the base current and base impedance by considering base values at rated MVA and kV.

2.48 A three-phase load absorbs 60 MVA at a 0.8 power factor lagging from 66 kV busbar. Calculate the base current and base impedance. Consider the base values at rated MVA and kV.

2.49 The per phase reactance of a three-phase, 5 kVA, 220 V, Y-connected synchronous generator is found to 12 Ω. Find the per unit reactance by considering the base values at rated MVA and kV.

2.50 The base values of a system are considered 200 kVA and 22 kV. The per unit impedance of that system is 0.05. Determine the actual value of the impedance.

2.51 A 11 kV/66 kV 10 MVA single-phase transformer has the secondary reactance of 6 Ω. Calculate the per unit reactance referred to as primary and secondary.

2.52 The per phase impedance of a three-phase 11 kV transmission line is $0.02 + j0.1$ pu. The transmission line delivers a power of 8 MVA to the load. Find the voltage drop across the line. Consider the base values at rated MVA and kV.

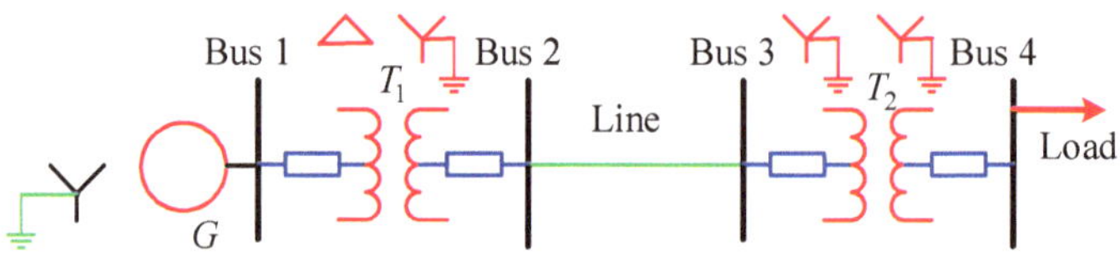

Fig. P2.2 A single-line diagram for Problem 2.53

2.53 Fig. P2.2 shows a single-line diagram of a power system. The related data for generator, transformers, and line are given below:

Generator G	65 MVA, 11 kV, $X = 0.15$,
Transformer T_1	25 MVA, 11 kV/220 kV, $X = 0.12$,
Transformer T_2	35 MVA, 220 kV/11 kV, $X = 0.0.06$,
Line	$X = 12\ \Omega$.

A three-phase load absorbs 25 MVA power at a 0.8 power factor lagging from the busbar 4 at a line voltage of 11 kV. Draw an impedance diagram with all in per unit values. Also, find the generator voltage. Consider base values at rated 80 MVA and 11 kV.

2.54 A single-line diagram of a power system is shown in Fig. P2.3. The related data for generator, transformers, and lines are given below:

Generator G	70 MVA, 11 kV, $X = 0.13$,
Transformer T_1	25 MVA, 11 kV/220 kV, $X = 0.11$,
Transformer T_2	25 MVA, 11 kV/220 kV, $X = 0.11$,
Transformer T_3	25 MVA, 11 kV/110 kV, $X = 0.10$,
Transformer T_4	25 MVA, 11 kV/110 kV, $X = 0.08$.

The series reactance of the lines 1 and 2 are 35 Ω and 40 Ω, respectively. A three-phase load absorbs 20 MVA at a power factor of 0.85 lagging from the bus 4. Draw an impedance diagram with per unit values.

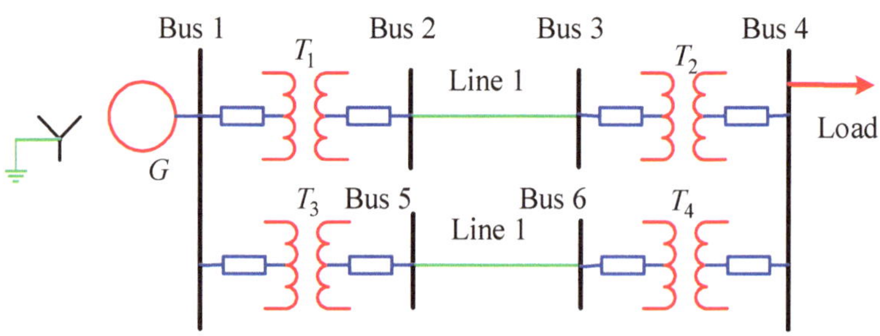

Fig. P2.3 A single-line diagram for Problem 2.54

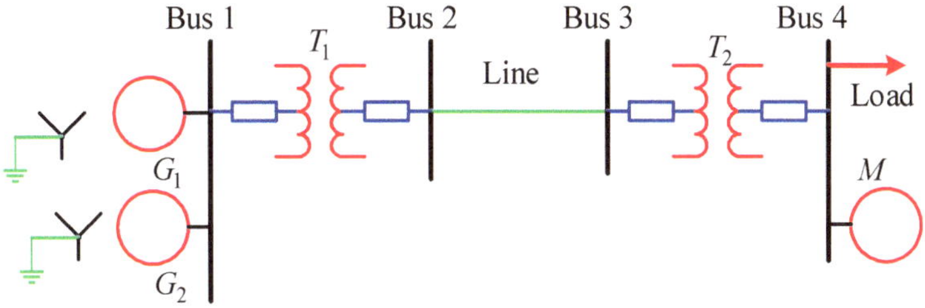

Fig. P2.4 A single-line diagram for Problem 2.55

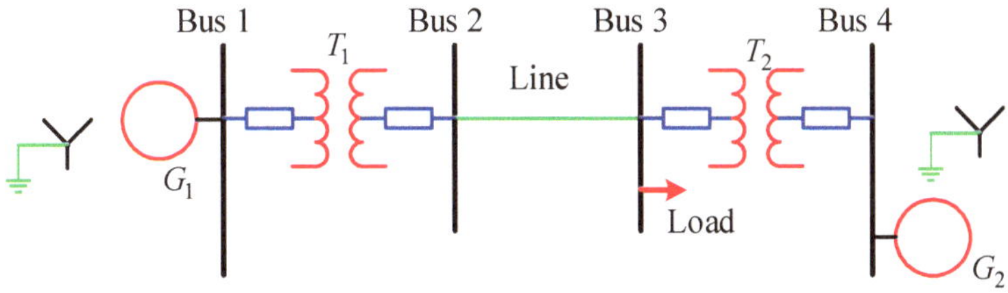

Fig. P2.5 A single-line diagram for Problem 2.56

2.55 Figure P2.4 shows a single-line diagram of a power system. The related data for generator, transformers, and lines are given below:

Generator G_1	15 MVA, 11 kV, $X = 0.12$,
Generator G_2	25 MVA, 11 kV, $X = 0.14$,
Transformer T_1	45 MVA, 11 kV/220 kV, $X = 0.10$,
Transformer T_2	55 MVA, 11 kV/220 kV, $X = 0.08$,
Motor M	10 MVA, 11 kV, $X = 0.04$,
Line	$X = 45\ \Omega$.

A three-phase load absorbs 50 MVA power at a 0.85 lagging power factor from 11 kV supply. Draw an impedance diagram with per unit values.

2.56 A single-line diagram of a power system is shown in Fig. P2.5. The related data for generator, transformers, and lines are given below:

Generator G_1	65 MVA, 11 kV, $X = 0.12$,
Generator G_2	70 MVA, 11 kV, $X = 0.10$,
Transformer T_1	45 MVA, 11 kV/132 kV, $X = 0.10$,
Transformer T_2	55 MVA, 11 kV/132 kV, $X = 0.08$,
Load M	75 MVA, 11 kV, 0.85 pf lag,
Line	$X = 25\ \Omega$.

Draw an impedance diagram by considering base values of 100 MVA and 11 kV.

Chapter 3
Power Generation

3.1 Introduction

Natural gas, water, and fuel are being used for a long time for the generation of electrical power for use in our daily lives. Day by day, the reserve of fossil fuel is decreasing due to heavy usages in power generation. At the same time, the demand for electrical energy is increasing due to the increasing population, increased industrialization, and increasing residential and commercial sectors. Academicians, researchers, and practicing engineers are working to generate electricity through renewable sources such as solar, wind, water, biomass, and nuclear. In solar energy, heat is radiated by the sun and is focused on the solar panel, which in turn generates DC voltage. There are many solar farms across the world generating DC voltage which charges the storage battery. Later, the direct current of the storage battery converts into alternating current through an inverter. Nowadays, wind energy also plays an important role to generate electricity. Wind with a suitable speed is used to turn the windmill, which drives the small generator and produces electrical energy. There is plenty of water available in tropical countries. Water is stored in a suitable place by making a dam. The water with sufficient potential head brings to the turbine house through steel pipes, which drives the turbine. This water turbine drives the three-phase synchronous AC generator, which generates electrical power. In the biomass energy system, gas is produced by rotting vegetation along with organic matter. This gas is then used to burn the water and converted it into steam. Finally, steam is used to generate electrical power.

Md. A. Salam, *Fundamentals of Electrical Power Systems Analysis*,
https://doi.org/10.1007/978-981-15-3212-2_3

3.2 Power Station

A power station or power plant is a plant that generates electricity with a voltage of 11 kV or lower value. This voltage is then stepped up to a higher voltage based on the national transmission voltage. Depending on the available natural energy, power stations can be classified as thermal or steam, hydroelectric, diesel or nuclear power stations.

3.3 Steam Power Station

Steam or thermal power station is an electricity generating station that converts steam into electrical energy. A modern steam power station is rated between 1 and 500 MW, and its efficiency varies from 40 to 50%. This type of power station is usually built in a suitable place where the coal, natural gas, and water are readily available. A simple schematic of a steam power station is shown in Fig. 3.1. In the boiler, natural gas is used to burn the water, which in turn produces steam. The steam is then carried out by a steel pipe to the turbine house. The steam is fallen on the turbine bucket and moves the turbine into the clockwise direction. The turbine shaft is coupled to the rotor shaft of the generator with a mechanical nut-bolt. When the turbine rotates in the clockwise direction, the generator rotor rotates in the same direction as the turbine and produces electricity. Condenser and cooling towers are used to collect exhaust steam from the turbine. The exhaust is then converted to water and is then fed back to the boiler by the feedwater pump.

The block diagram of a steam power station is shown in Fig. 3.2. The heat of the boiler is produced by burning natural gases. The heat input of the boiler is represented by Q_B. The work is accomplished by the feedwater pump when the water is pumped into the boiler, and it is represented as W_P. The work produced by the turbine is W_T.

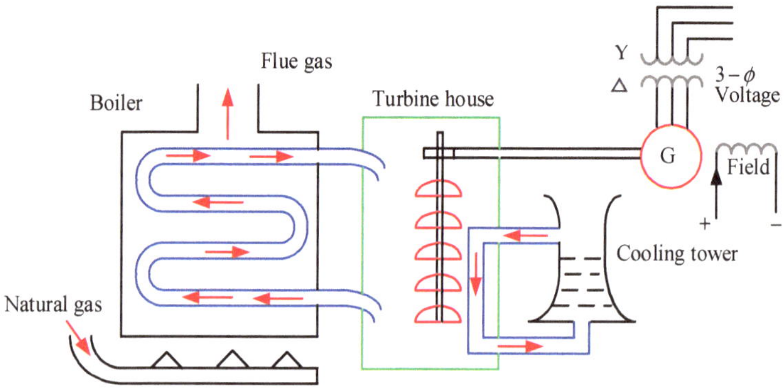

Fig. 3.1 Schematic of a simple steam power station

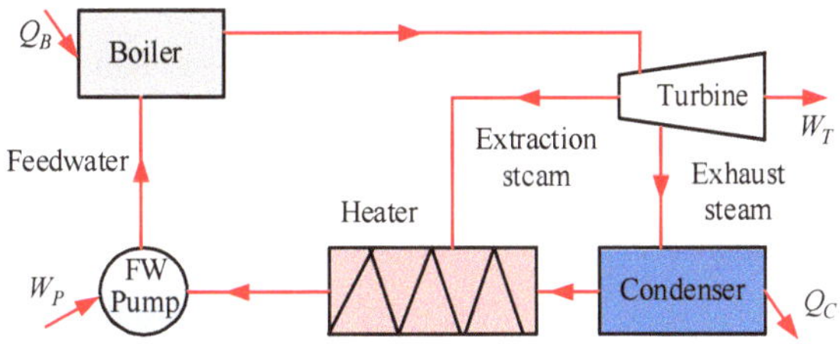

Fig. 3.2 Block diagram of a steam power station

The thermal efficiency is expressed as,

$$\eta = \frac{W_T - W_P}{Q_B} \tag{3.1}$$

3.4 Hydro Power Station

The increasing demand for electricity can be met by using hydropower station. China increases electric power capacity by 8–9% per year to meet growing demand from both industry and other consumers. They constructed a dam for hydropower station on the bank of the Yangtze River which is known as Three Gorges Dam. The total power generation capacity of this hydropower station is 22,500 MW. Currently, the Three Gorges dam is the largest hydropower plant in the world. Hydro Bakun is the largest hydropower station in Malaysia, which is built on the bank of a Balui River. The total power generation capacity of this Hydro Bakun power station is 2400 MW. In the Middle East, near the Red Sea, world's largest hydropower station will be constructed soon, and the total electric generating capacity would be around 50,000 MW. There are 13 generators and 2675 MW generation in a Niagara hydropower station in Canada. A hydropower station is the one which utilizes the energy of falling water to generate electricity. Hydropower station is suitable to build where a large amount of water is available all the time. A dam is constructed across a lake or river to store water as well as to create enough water head. Water is then taken to the turbine house through steel pipes. A water turbine converts the kinetic energy of falling water into mechanical energy. This mechanical energy is used to run the generator, which generates electricity. Two factors mainly determine the amount of electricity generated by a hydroelectric power station. The first factor is the height from the turbine to the water surface, which is known as water head. The second key factor is the volume of water moving through the turbine.

Three forms of energy are involved when water is moving from one place to another place. These are kinetic, pressure, and potential energies. Per kg kinetic energy, pressure energy, and potential energy can be expressed as,

$$E_{kw} = \frac{1}{2}v^2 \text{ Nm/kg} \tag{3.2}$$

$$E_{pw} = \frac{p}{\rho} \text{ Nm/kg} \tag{3.3}$$

$$E_{pw} = gH \text{ Nm/kg} \tag{3.4}$$

where

v is the velocity of water flow in m/s,
p is the pressure of water N/m^2,
ρ is the density of water kg/m^3,
H is the height of water in m.

A simple schematic of a hydropower station is shown in Fig. 3.3. The discharge water from the turbine house is released to the river. The main components of a hydropower station are dam, turbine, generator, transformer, and transmission lines. The brief descriptions of those components are given below.

Dam: It is built on the lake or river to store water in a reservoir. It is also known as the headrace. The dam raises the water level to create the necessary head of the falling water. Therefore, the dam always controls the flow of water through proper equipment.

Penstock: Pipes of large diameter that carry water under pressure from the storage reservoir to the turbines called penstocks. These pipes are usually made of steel or reinforced concrete.

Turbine: In a turbine, different types of buckets or blades are mounted on a wheel called a runner. Water is released from the dam and falls on the turbine blade. The force of falling water pushes the turbine blades, which causes the turbine to rotate.

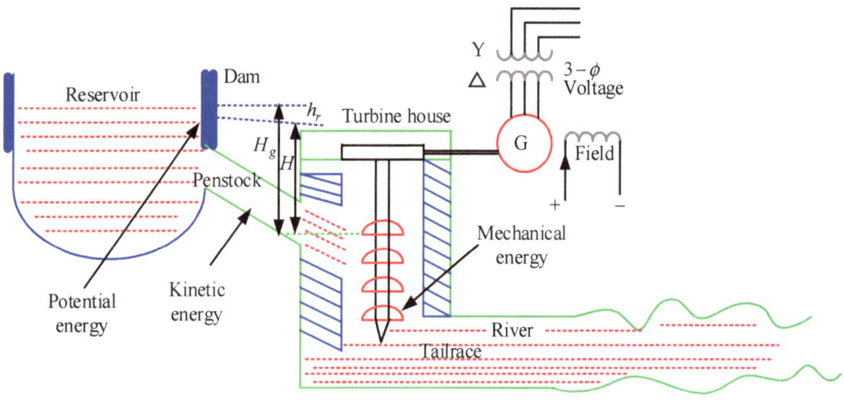

Fig. 3.3 Schematic of a hydropower station

Then, the turbine converts the kinetic energy of falling water into mechanical energy.

Tailrace: A channel that carries water away from the turbine after the water has worked on the turbines is known as a tailrace or river. The water surface in the tailrace is also referred to as tailrace.

Generator: The generator is usually attached to the turbine shaft with the related equipment. The generator field is energized by the DC supply and creates the magnetic field. The turbine rotates the generator, which converts mechanical energy from the turbine into electrical energy.

Transformer: In the power station, a power transformer with a rating from 20 to 50 MVA is used to step up the generating voltage to the national transmission voltage.

Transmission lines: The transmission lines are used to carry out the electrical power from the power station to the nearest grid substation.

3.5 Gross and Net Heads

There are two types of water heads normally used in the hydropower station. They are gross head and net head. The vertical difference between the headrace and tailrace is known as the gross head, and it is represented by H_g. The net or effective head is the actual head available at the inlet of the bucket to work on the turbine. The net head (H) is expressed as,

$$H = H_g - h_L \tag{3.5}$$

where

h_L is the total head loss during the transit of water from the headrace to tailrace which is mainly head loss due to friction.

According to Darcy's equation, the energy loss due to friction can be written as,

$$h_L = f \times \left(\frac{L}{D}\right) \times \left(\frac{v^2}{2g}\right) \tag{3.6}$$

where

h_L is the energy loss due to friction,
L is the length of the flow stream,
D is the diameter of the flow path,
v is the average velocity of flow,
f is the friction factor (dimensionless).

3.6 Efficiency

The efficiency of a hydropower station depends on the input and output. The efficiency is classified as hydraulic, mechanical, and overall.

Hydraulic efficiency: The ratio of the power developed by the runner of a turbine to the power supplied at the inlet of a turbine is known as hydraulic efficiency, and it is represented as η_h. There is a power loss between the striking jet and vane. According to the definition, the hydraulic can be expressed as,

$$\eta_h = \frac{P_R}{P_W} \times 100 \tag{3.7}$$

where

P_R is the runner or turbine power,
P_W is water power.

Mechanical efficiency: The ratio of the power (P_{sh}) available at the shaft to the power (P_R) developed by the runner of a turbine is known as mechanical efficiency, and it is represented as η_m. This efficiency depends on the related mechanical devices that will create a loss of energy between the runner in the annular area, between the nozzle and spear. The mechanical efficiency is expressed as,

$$\eta_m = \frac{P_{\text{sh}}}{P_R} \times 100 \tag{3.8}$$

Overall efficiency: The ratio of the power (P_{sh}) available at the shaft to the power (P_W) supplied at the inlet of a turbine is known as overall efficiency. The overall efficiency is represented as η_O. This depends on both the hydraulic losses and the slips. It also depends on other mechanical problems that will create a loss of energy between the jet power supplied and the power generated at the shaft available for the coupling of the generator.

$$\eta_O = \frac{P_{\text{sh}}}{P_W} \tag{3.9}$$

$$\eta_O = \frac{P_{\text{sh}}}{P_W} \times \frac{P_R}{P_R} \tag{3.10}$$

$$\eta_O = \frac{P_R}{P_W} \times \frac{P_{\text{sh}}}{P_R} \tag{3.11}$$

Substituting Eqs. (3.7) and (3.8) into Eq. (3.10) yields,

$$\eta_O = \eta_h \times \eta_m \tag{3.12}$$

3.7 Pelton Wheel

Pelton wheel or turbine, named after an eminent engineer, is an impulse turbine, wherein the flow is tangential to the runner, and the available energy at the entrance is completely kinetic energy. Further, it is preferred at a very high head and low discharges with low specific speeds. The pressure available at the inlet and the outlet is atmospheric. The main components of a Pelton wheel are shown in Fig. 3.4.

The main components are penstocks, nozzle, spear, runner, casing, and breaking jet. Water is brought to the hydroelectric plant through large penstocks. At the end of penstocks, there is a nozzle, which converts the pressure energy completely into kinetic energy. The kinetic energy will convert the liquid flow into a high-speed jet, which strikes the buckets mounted on the runner, which in turn rotates the runner of

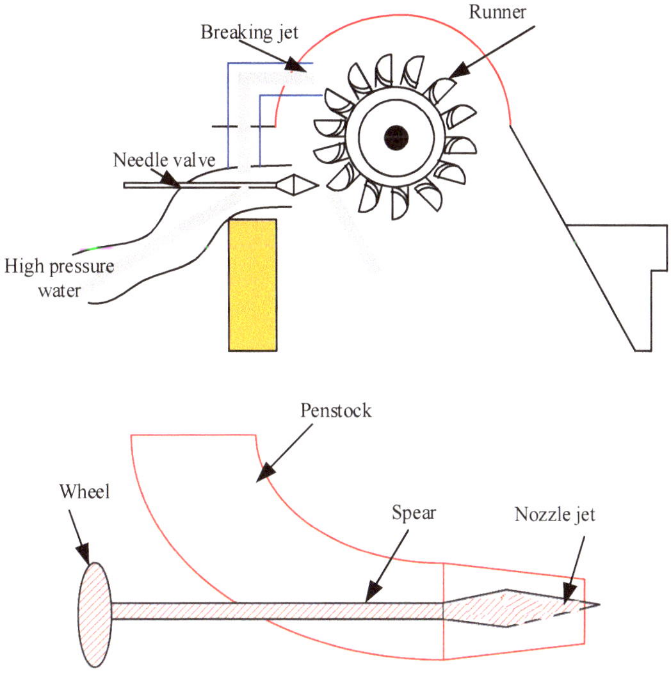

Fig. 3.4 Schematic of a Pelton wheel

the turbine. The amount of water striking is usually controlled by the forward and backward motion of the spear.

The runner is a circular disk mounted on a shaft on the periphery of which some buckets are fixed equally spaced. The buckets are made of cast iron, cast steel, bronze, or stainless steel depending upon the head at the inlet of the turbine. The water jet strikes the bucket on the splitter of the bucket and gets deflected in a proper direction.

The casing is made of cast iron or fabricated steel plates. The main function of the casing is to prevent the splashing of water and to discharge the water into the tailrace.

The amount of water after striking the buckets is completely stopped; the runner goes on rotating for a very long time due to inertia. A small nozzle is provided to stop the runner in a short time, which directs the jet of water on the back of the bucket with which the rotation of the runner is reversed. This jet is called a breaking jet.

3.8 Velocity Triangle

The machine that converts mechanical energy into hydraulic energy is known as a pump, and the machine that converts hydraulic energy into mechanical energy is known as a turbine. Wave is considered as a bent shape as shown in Fig. 3.5. Here, the jet is used to deliver fluid at the inlet with a velocity V_1 with an angle α_1 at the direction plate, and its relative velocity V_{r1} makes a blade angle β_1 with the horizontal.

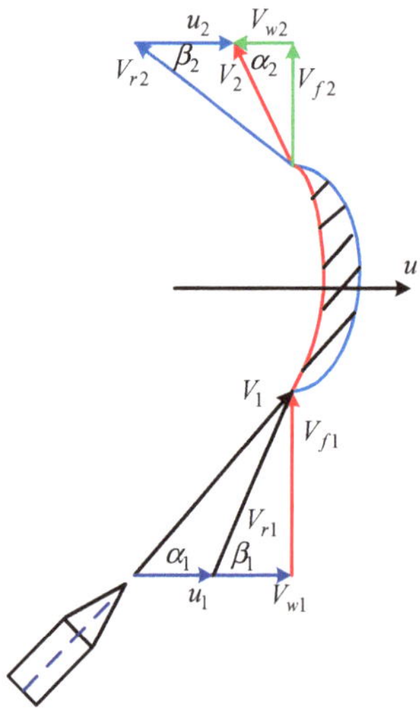

Fig. 3.5 Pelton wheel inlet and outlet triangles

At the inlet, the wall velocity of the jet in the axial direction is $V_{w1,}$ and it can be expressed as,

$$V_{w1} = V_1 \cos \alpha_1 \tag{3.13}$$

Flow component of the jet or water at the inlet is $V_{f1,}$ and it can be expressed as,

$$V_{f1} = V_1 \sin \alpha_1 \tag{3.14}$$

3.9 Hydraulic Efficiency

Jet strikes the splitter of a bucket and gets split into two halves, and they deviate with an angle as shown in Fig. 3.6.

Velocity triangles for the jet striking the bucket at the inlet and outlet are shown in Fig. 3.7. The inlet and outlet parameters are represented with the suffix 1 and 2, respectively.

In Fig. 3.7, consider the linear velocities of the bucket at the inlet, and the outlet is u_1 and u_2, respectively. Again, consider the velocity of water at the inlet of the turbine is V_1. The tangential component of V_1 is V_{w1}. According to Fig. 3.6, the following equation can be written as,

$$V_1 = V_{w1} \tag{3.15}$$

The expression of the relative velocity is,

$$V_{r1} = V_1 - u_1 \tag{3.16}$$

From the outlet velocity triangle as shown in Fig. 3.7, the velocity of water or jet at the outlet of the bucket is $V_{3.}$ The tangential and radial or flow components of V_2 are V_{w2} and V_{f2}, respectively. The deflection angle of the bucket is α, and the exit angle of the bucket is ϕ.

Fig. 3.6 Schematic of a bucket and deviating parts

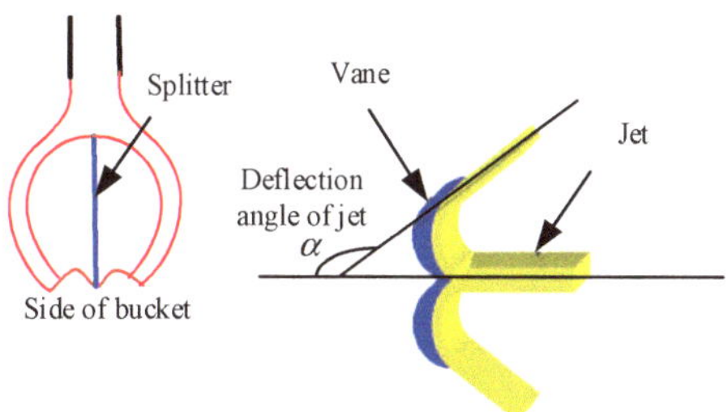

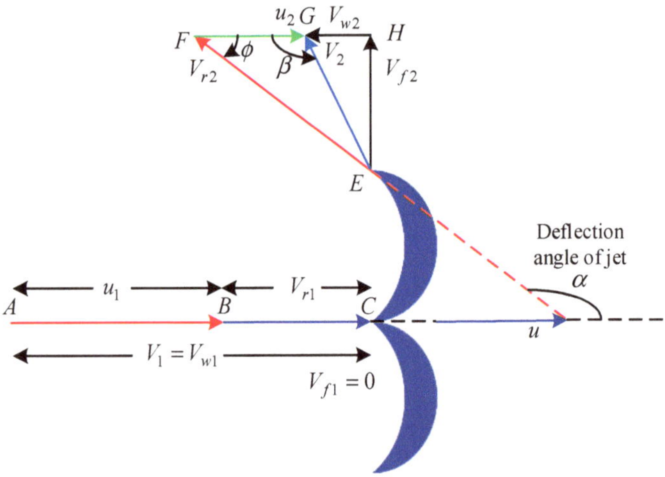

Fig. 3.7 Inlet and outlet triangles of the bucket

According to the impulse-momentum theorem, the force with which the jet strikes the bucket along the direction of vane (blade) is given by,

$$F_x = \text{water flow rate} \times \text{change of velocity in } x - \text{direction} \quad (3.17)$$

$$F_x = m\,(\text{kg/s}) \times [V_{w1} - (-V_{w2})] \quad (3.18)$$

The mass flow rate is defined as,

$$m = \rho(AV_1) \quad (3.19)$$

Substituting Eq. (3.19) into Eq. (3.18) yields,

$$F_x = \rho(AV_1) \times (V_{w1} + V_{w2}) \quad (3.20)$$

The work done per second by the jet of water on the vane is given by the product of force exerted on the vane and the distance moved by the vane in one second. Mathematically, it can be expressed as,

$$\text{WD/s} = F_x \times u \quad (3.21)$$

The work done per second is equal to the runner power, and it can be expressed as,

$$P_R = F_x \times u \quad (3.22)$$

Substituting Eq. (3.20) into Eq. (3.22) yields,

$$P_R = \rho(AV_1) \times (V_{w1} + V_{w2}) \times u \tag{3.23}$$

The input power from the jet or water power is equal to the energy due to the velocity, and it can be expressed as,

$$P_W = \mathrm{KE} = \frac{1}{2} m V_1^2 \tag{3.24}$$

Substituting Eq. (3.19) into Eq. (3.24) yields,

$$P_W = \mathrm{KE} = \frac{1}{2} \rho(AV_1) V_1^2 \tag{3.25}$$

$$P_W = \mathrm{KE} = \frac{1}{2} \rho A V_1^3 \tag{3.26}$$

Then, the hydraulic efficiency is defined as,

$$\eta_h = \frac{P_R}{P_W} \tag{3.27}$$

Substituting Eqs. (3.23) and (3.26) into an Eq. (3.27) yields,

$$\eta_h = \frac{\rho(AV_1) \times (V_{w1} + V_{w2}) \times u}{\frac{1}{2} \rho A V_1^3} \tag{3.28}$$

$$\eta_h = \frac{2u(V_{w1} + V_{w2})}{V_1^2} \tag{3.29}$$

From the inlet velocity triangle, the following equation can be written as,

$$V_{w1} = V_1 \tag{3.30}$$

Assuming no shock and ignoring frictional losses through the blade, then the following equation is expressed as,

$$V_{r1} = V_{r2} = V_1 - u_1 \tag{3.31}$$

In case of Pelton wheel, the inlet and outlet are located at the same radial distance from the center of the runner. Therefore, the linear speed of runner is generalized as,

$$u_1 = u_2 = u \tag{3.32}$$

Substituting Eq. (3.32) into Eq. (3.31) yields,

$$V_{r1} = V_{r2} = V_1 - u \tag{3.33}$$

From the outlet velocity triangle, ΔEFH, the following equation can be written as,

$$\cos\phi = \frac{u_2 + V_{w2}}{V_{r2}} \tag{3.34}$$

$$u_2 + V_{w2} = V_{r2}\cos\phi \tag{3.35}$$

Substituting Eqs. (3.30) and (3.33) into Eq. (3.33) yields,

$$V_{w2} = (V_1 - u)\cos\phi - u \tag{3.36}$$

Substituting Eqs. (3.30) and (3.36) into Eq. (3.29) yields,

$$\eta_h = \frac{2u[V_1 + (V_1 - u)\cos\phi - u]}{V_1^2} \tag{3.37}$$

$$\eta_h = \frac{2u(V_1 - u)(1 + \cos\phi)}{V_1^2} \tag{3.38}$$

Differentiate Eq. (3.38) with respect to u and set equal to zero to obtain the maximum efficiency for a given jet velocity and vane angle.

$$\frac{\mathrm{d}}{\mathrm{d}u}\left[\frac{2u(V_1 - u)(1 + \cos\phi)}{V_1^2}\right] = 0 \tag{3.39}$$

$$\frac{2(1 + \cos\phi)}{V_1^2}\frac{\mathrm{d}}{\mathrm{d}u}u(V_1 - u) = 0 \tag{3.40}$$

$$u(0 - 1) + (V_1 - u) = 0 \tag{3.41}$$

$$u = \frac{V_1}{2} \tag{3.42}$$

From Eq. (3.42), it is seen that the bucket speed is maintained at half the velocity of the jet, and the efficiency will be maximum at this speed. Substituting Eq. (3.42) into Eq. (3.38) yields,

$$\eta_{\text{hmax}} = \frac{2u(2u - u)(1 + \cos\phi)}{(2u)^2} \tag{3.43}$$

$$\eta_{\text{hmax}} = \frac{1}{2}(1 + \cos\phi) \tag{3.44}$$

From Eq. (3.44), it is concluded that the hydraulic efficiency will be maximum when the value of $\cos\phi$ is equal to 1 and the value of the outlet blade angle of the bucket should be 0°.

3.10 Diesel Power Station

A diesel power station uses the diesel engine as a prime mover to generate electricity. A diesel engine is connected with the rotor of the generator. Diesel is burnt inside the engine, which produces mechanical energy. When the diesel engine starts, it runs the generator, which converts mechanical energy into electrical energy. A simple block diagram of a diesel power station is shown in Fig. 3.8. The power rating of a diesel power station is very small, and it is normally installed in the remote area for supplying electricity. The efficiency of the diesel power station is less than 25%. This power station is rarely used due to the high price of the oil and its lower efficiency. It is used as a standby plant, peak load plant, emergency plant, and mobile plant.

3.11 Nuclear Power Station

Nowadays, the nuclear power station is popular for large capacity electricity generation. It provides approximately 17% of the world's electricity. Currently, there are roughly 400 nuclear power stations in the world. More than 100 nuclear power stations are in the USA. France generates approximately 75% of its total electrical energy from nuclear power stations.

The nuclear fuel (uranium) generates a huge amount of nuclear energy through fission of radioactive material and heat by a controlled chain reaction. This heat is used to boil the water, which produces steam. This steam is then used to drive the

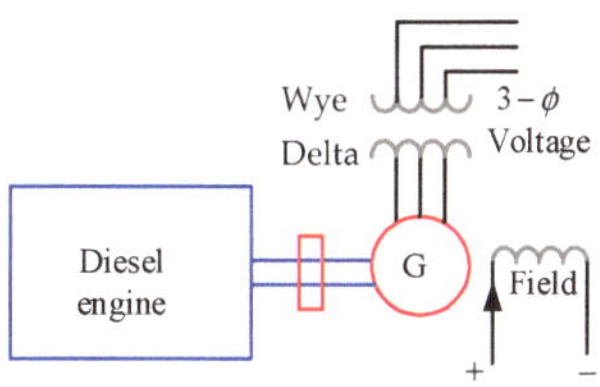

Fig. 3.8 Block diagram of a diesel power station

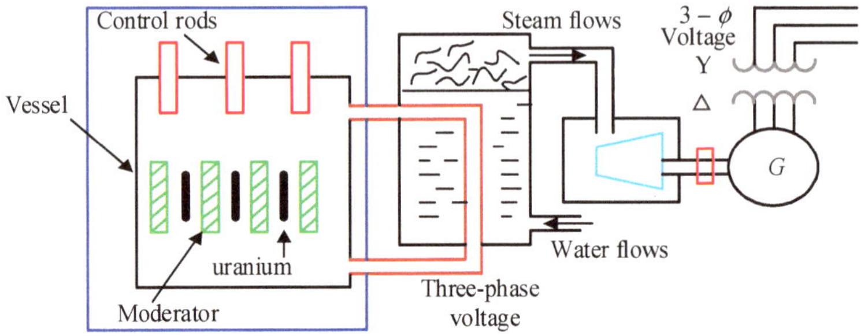

Fig. 3.9 Schematic of a nuclear power station

turbine. Finally, the turbine rotates the generator, which generates electrical energy. The simple schematic of a nuclear power station is shown in Fig. 3.9. The control rod is used to control the chain reaction inside the vessel. The chain reaction starts when the control rods are pushed up from the vessel, and the chain reaction stops when the control rods are pushed down to the vessel. Six types of reactors such as Magnox, AGR, BWR, CANDU, and RBMK in the nuclear power station.

3.12 Variable Load and Different Factors

The power demands of the consumers are varied according to their uses of electrical appliances. Therefore, the load on a power station varies from time to time.

Load Curve: The variation of load on the power station concerning time is known as the load curve. The load curve is obtained by plotting power in Y-axis and time in X-axis. It is the graphical representation between the load in kW and times in hours. When the load curve is plotted for 24 h, then it is known as a daily load curve, and the load curve for one year is known as the annual load curve. Figure 3.10 represents the load curve for 24 h.

There are many advantages of the load curve which are mentioned below.

- It shows the variation of the loads throughout the day.
- The load curve helps to estimate the generation cost.
- The highest point on the load curve represents the maximum demand on that day.
- It helps to select the capacity and number of generating units.
- The area under the load curve represents the exact units generated in the day.

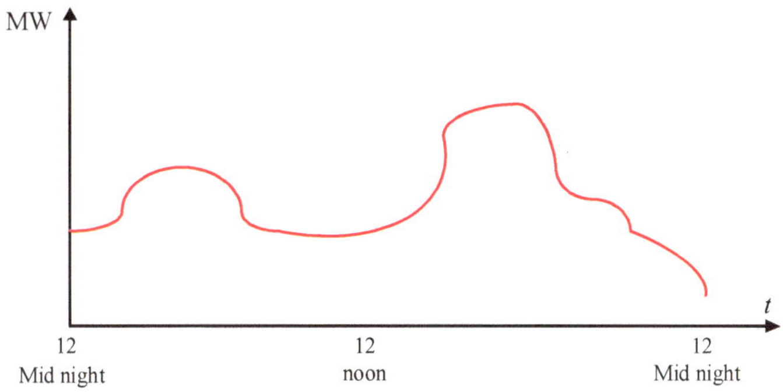

Fig. 3.10 Load curve for 24 h

3.13 Different Terms and Factors of Power Station

There are a few important terms and factors usually used in the power stations. These are connected load, average load, maximum demand, load factor, demand factor, diversity factor, capacity factor, and load duration curve [1].

Connected Load: The sum of ratings of all the connected electrical appliances in the consumer terminals is known as a connected load. If a consumer connected a smart TV of 60 W, a Sony DVD player of 5000 W and two energy savings bulbs of each 7 W across a single-phase system. The sum of the connected load (CL) is calculated as,

$$CL = 60 + 5000 + 2 \times 7 = 5074\,\text{W}$$

Maximum Demand: The maximum demand is the highest amount of load on the power station during a specific time. Maximum demand can be determined from the load curve, and it helps us to install generating units. Normally, all the electrical appliances installed in the consumer terminal are not operated to their fullest capacity at the same time. Therefore, the maximum demand is not equal to the connected load. The maximum demand is always less than the connected load.

Average Load: The average load of the power station occurs at a specific time. It may be daily, monthly, or yearly. Mathematically, these daily, monthly, and yearly average loads can be expressed as,

$$\text{Daily average load} = \frac{\text{No. of kWh generated in a day}}{24\,\text{h}} \tag{3.45}$$

$$\text{Monthly average load} = \frac{\text{No. of kWh generated in a month}}{30 \times 24\,\text{h}} \tag{3.46}$$

$$\text{Monthly average load} = \frac{\text{No. of kWh generated in a year}}{8760\,\text{h}} \tag{3.47}$$

Load Factor: The ratio of the average load to the peak or maximum load is known as the load factor. The load factor is always less than or equal to unity, and it plays an important role in the cost of power generation. For the same maximum demand, the higher load factor reduces the per unit generation cost. Mathematically, the load factor (F_{LF}) can be expressed as,

$$F_{\text{LF}} = \frac{\text{average load}}{\text{peak load}} \tag{3.48}$$

Considering that the power station is operated for T hours, then Eq. (3.48) is modified as,

$$F_{\text{LF}} = \frac{\text{average load} \times T}{\text{peak load} \times T} \tag{3.49}$$

$$F_{\text{LF}} = \frac{\text{units generated}}{\text{peak load} \times T} \tag{3.50}$$

where T is the time. It may be in days, weeks, months, or years. The annual load factor is defined as,

$$F_{\text{aLF}} = \frac{\text{Annual energy generated}}{\text{annual peak load} \times 8760} \tag{3.51}$$

The load factors for different types of load are different. For residential load, it varies from 10% to 15%, and for commercial load, it is around 30%. For small, medium and large industries, it varies as 30% to 50%, 55% to 60%, and 70% to 90%, respectively.

Demand Factor: It is defined as the ratio of the maximum demand to the total connected load of the system. Mathematically, the demand factor (F_{DeF}) can be expressed as,

$$F_{\text{DeF}} = \frac{\text{maximum demand}}{\text{total connected load}} \tag{3.52}$$

The demand factor always changes with hours to hours, and it is less than or equal to one. Lower demand factor indicates less system capacity required to deliver the connected loads. It is considered an indicator of the simultaneous operation of the connected load. The demand factor should be considered for each type of load, especially induction motor load. For residential load, it varies between 50 and 80%. For industrial (induction motor) load, this factor is 75%, and for different types of lighting load, it is equal to one. The relationship between the load factor and the demand factor can be derived as,

$$F_{\mathrm{DeF}} = \frac{\frac{\text{maximum demand}}{\text{average load}} \times \text{average load}}{\text{total connected load}} \tag{3.53}$$

$$F_{\mathrm{DeF}} = \frac{\text{average load}}{\text{total connected load} \times \frac{\text{average load}}{\text{maximum demand}}} \tag{3.54}$$

Substituting Eq. (3.48) into Eq. (3.54) yields,

$$F_{\mathrm{DeF}} = \frac{\text{average load}}{\text{total connected load} \times F_{\mathrm{LF}}} \tag{3.55}$$

$$F_{\mathrm{DeF}} \times F_{\mathrm{LF}} = \frac{\text{average load}}{\text{total connected load}} \tag{3.56}$$

Diversity Factor: The ratio of the sum of the individual maximum demands of the various subsystems of a system to the maximum demand of the whole system is known as a diversity factor. Mathematically, it can be expressed as,

$$F_{\mathrm{DiF}} = \frac{\text{sum of individual maximum demands}}{\text{maximum demand of the whole system}} \tag{3.57}$$

The diversity factor is always greater than or equal to one. This factor is always greater than one because the sum of individual maximum demands is greater than the maximum demand of the system. Diversity occurs in an operating system because all connected loads are not operating simultaneously. The cost of power generation will be lower for higher diversity factors. In the designing or planning stage, the diversity factor is considered one. The diversity factor is used for coordination study and calculation of a load of a particular node of a system. It is also used for determining the size of the transformer based on the total load. For substation, transmission lines, and a whole utility system, diversity factors need to be calculated. According to IEC 60439, the diversity factor for different users is shown in Table 3.1.

Table 3.1 Different appliances with diversity factors

Circuits function	Diversity factor (%)
Lighting	90
Heating and air conditioning	80
Socket-outlets	70
Lifts and catering hoist	
Most powerful motor	100
Second most powerful motor	75
All motors	80

Capacity Factor: The ratio of the actual amount of energy the plant produced during a period to the maximum amount of energy that would have been produced at full capacity. It is often called as the plant factor or the use factor. Capacity factors vary greatly depending on the type of fuel that is used and the design of the plant. The capacity factor is always less than one. There are several reasons for the capacity factor to be less than one. The first reason is that the equipment out of service or operating at the reduced output due to equipment failures and/or routine maintenance. The second reason is that the output generation is reduced because the consumer does not require the electricity or the electricity price is low to make production cost economical. Mathematically, the capacity factor (F_{CF}) is expressed as,

$$F_{CF} = \frac{\text{actual energy generation}}{\text{maximum plant rating}} \tag{3.58}$$

The annual plant factor (F_{aCF}) is defined as,

$$F_{aCF} = \frac{\text{actual annual energy generation}}{\text{maximum plant rating} \times 8760} \tag{3.59}$$

The reserve capacity of the power station is very important for future demand and forecasting, and it can be determined as,

$$\text{Reserve capacity} = \text{plant capacity} - \text{maximum demand} \tag{3.60}$$

Example 3.1
A power station is having the maximum demand and connected load of 30 MW and 60 MW, respectively. The energy generated per annum is found to be 70×10^6 kWh. Calculate the demand factor, average demand, and load factor.

Solution
The value of the demand factor is calculated as,

$$F_{DF} = \frac{\text{maximum demand}}{\text{connected load}} = \frac{30}{60} = 0.5 \tag{3.61}$$

The average load can be determined as,

$$\text{average demand} = \frac{\text{units generated per annum}}{\text{total hours in a year}} = \frac{70 \times 10^6}{8760} = 7990.86\,\text{kW} \tag{3.62}$$

Load factor is calculated as,

$$F_{\mathrm{LF}} = \frac{\text{average load}}{\text{peak load}} = \frac{7990.86}{30 \times 1000} = 0.27 \tag{3.63}$$

Example 3.2
A power station supplies power to three different consumers. The power used by the residential, commercial, and industrial users is 200 kW, 800 kW, and 1200 kW, respectively. The maximum demand and units generated per annum of this power station are 2000 kW and 60×10^5 kWh, respectively. Calculate the diversity factor, average demand, and load factor.

Solution
The value of the diversity factor is,

$$F_{\mathrm{DiF}} = \frac{\text{sum of individual maximum demands}}{\text{maximum demand of the whole system}} = \frac{200 + 800 + 1200}{2000} = 1.1 \tag{3.64}$$

The average load is calculated as,

$$\text{average demand} = \frac{\text{units generated per annum}}{\text{total hours in a year}} = \frac{60 \times 10^5}{8760} = 684.93\,\text{kW} \tag{3.65}$$

Load factor can be determined as,

$$F_{\mathrm{LF}} = \frac{\text{average load}}{\text{peak load}} = \frac{684.93}{2000} = 0.34 \tag{3.66}$$

Practice Problem 3.1
A power station is having the maximum demand and connected load of 20 MW and 45 MW, respectively. The energy generated per annum is found to be 30×10^6 kWh. Find the demand factor, average demand, and load factor.

Practice Problem 3.2
A power station supplies power to three consumers. The 300 kW used by the residential user, 1000 kW used by the commercial user, and 1800 kW used by the industrial user. The maximum demand and units generated of the power station are 2500 kW and 70×10^5 kWh, respectively. Calculate the diversity factor, average demand, and load factor.

Load duration curve: In electricity generation, the load duration curve is used to describe the relationship between power generation and utilization. It is very important for typical days that indicate the power demands at a specific duration of time. It is obtained from the load curve by descending the load magnitudes with respect to time.

The load duration curve indicates how many hours a certain load has been allocated for a day. The load duration curve helps to design a baseload power plant and the peak load power plant. Hydro and diesel power stations are usually used during the peak demand. These power stations are usually used to reduce the generation cost of electricity. Figures 3.11 and 3.12 show a typical load curve and load duration curve for 24 h.

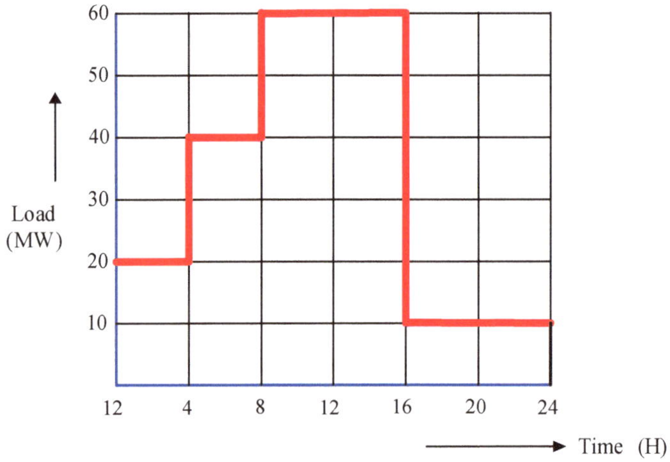

Fig. 3.11 A typical load curve

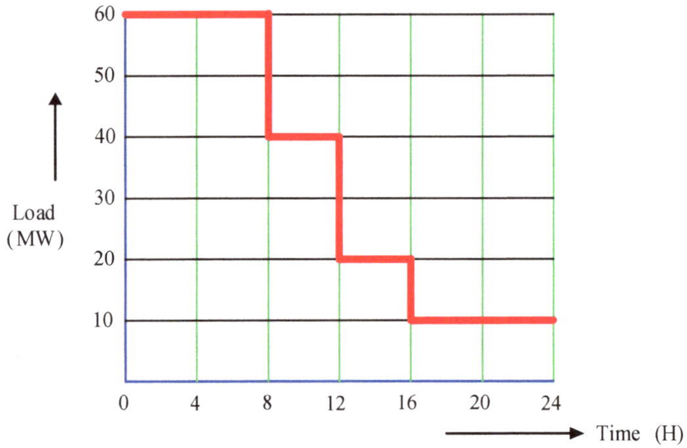

Fig. 3.12 A typical load duration curve

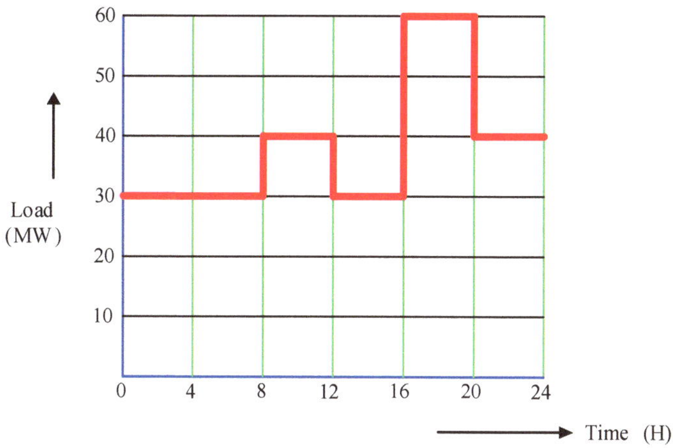

Fig. 3.13 A typical load curve for Example 3.3

Example 3.3
A daily load cycle of a power station is described below.

Time (H):	0–8	8–12	12–16	16–20	20–24
Load (MW):	30	40	30	60	40

Draw the load curve and calculate the maximum demand, units generated per day, average load, and load factor.

Solution
From the given data, the load curve is drawn as shown in Fig. 3.13. From the load duration curve, the following parameters can be determined as,

The maximum demand (MD) of the power station is,

$$MD = 60\,\text{MW} \tag{3.67}$$

Units generated per day (UG_{day}) is,

$$UG_{\text{day}} = 30 \times 8 + 40 \times 4 + 30 \times 4 + 60 \times 4 + 40 \times 4 = 920\,\text{MWh} \tag{3.68}$$

Per day average load (AVL_{day}) is,

$$AVL_{\text{day}} = \frac{920}{24} = 38.33\,\text{MW} \tag{3.69}$$

Load factor can be determined as,

$$F_{LF} = \frac{AVL_{\text{day}}}{MD} = \frac{38.33}{60} = 0.64 \tag{3.70}$$

Practice Problem 3.3
The daily power consumption versus time of a power station is mentioned below.

Time (H):	0–4	4–8	8–16	16–20	20–24
Load (MW):	20	30	60	30	10

Plot the load curve and find the maximum demand, units generated per day, average load, and load factor.

3.14 AC Power Supply System

Electricity is generated at the power station. It is then transferred to the consumers' terminals (load) by the transmission and distribution networks. A very large network that connects power stations to loads is known as an electric power system or electric grid. The electrical lines are used to transfer electrical power from the power station to the loads. The transmission and distribution networks between generating stations and the consumer terminals are classified into transmission and distribution systems. The single-line diagram of a typical AC power system along with step-up and step-down transformers is shown in Fig. 3.14. The vertical black lines represent the busbars. The voltage at the busbar is always constant.

Generating station: A generating station is often known as the power station, power plant, or powerhouse. Most of the power station contains one or more three-phase synchronous generators. Three-phase AC voltage with a magnitude of 11 kV or 13.8 kV is usually generated in the power station by a three-phase synchronous generator, a rotating machine that converts mechanical power into

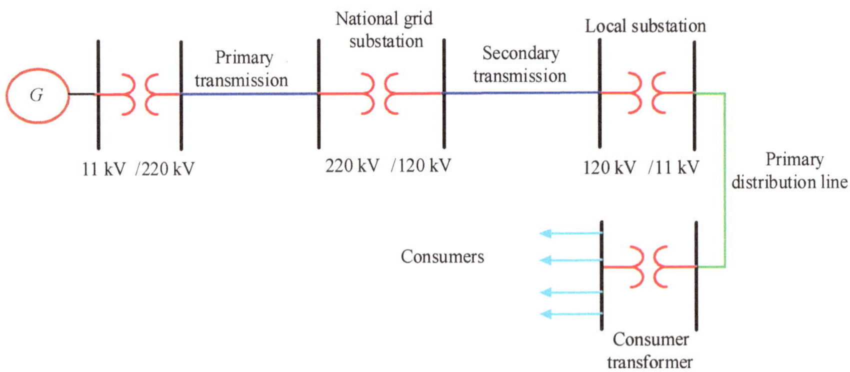

Fig. 3.14 Schematic of a typical AC power system

Table 3.2 Different appliances diversity factors

Voltage classification	Voltage range (kV)	Types
Ultra-high voltage	>765	High voltage transmission >765 kV
Extra-high voltage	345, 500, 765	High voltage transmission
High voltage	115, 138, 161, 230	
Medium voltage	34, 46, 69	Subtransmission
Low voltage	<34	Distribution

electrical power. In Fig. 3.14, the AC generator generates an 11 kV voltage. This generated voltage is then stepped up to a suitable national transmission voltage by a three-phase step-up power transformer [2, 3]. The rating of the power transformer usually varies between 20 and 40 MVA.

Transmission Lines: Transmission lines transmit electricity from one terminal to another terminal. Transmission lines are used for long distances with a higher voltage ranging from 66 to 765 kV, and they connect to the grid substation. Transmission lines are classified as primary and secondary transmission lines. Transmission voltage mainly depends on the total length between the power station and the end terminal of the customers. The national transmission voltage of some Asian countries such as Brunei and Bangladesh are 66 kV and 120 kV, respectively. Based on the voltage levels, the classifications of transmissions and distributions of power system networks are shown in Table 3.2.

Primary Transmission: The voltage of 220 kV or more transmits by the three-phase three wires to the national grid substation through an appropriate steel tower known as primary transmission. The grid substation is usually situated at the outer parts of a city.

Secondary Transmission: At the grid substation, the higher voltage is stepped down to 120 kV or other values by a three-phase transformer. The voltage of 120 kV or less transmits by the three-phase three wires to the various local substations located at different places. At these substations, the voltage is further reduced to either 33 or 11 kV.

Distribution System: The distribution system is classified as a primary distribution and secondary distribution.

Primary Distribution: The overhead lines or underground cables are used as feeders to connect the distribution transformer. The high voltage side of a distribution transformer is known as primary distribution or primaries. Common primary distribution voltages are 3.3, 3.3, 4.16, 6.6, 11, 13.5, 13.8, 25, and 34.5 kV. In Asia, some of the utility companies are used 11 kV as their primary distribution voltage.

Secondary Distribution: The primary distribution lines connect to the distribution transformer, which is usually installed near the consumer's area. This distribution transformer steps down [4] the delivered 11 kV voltage to 440 V line-to-line, and it

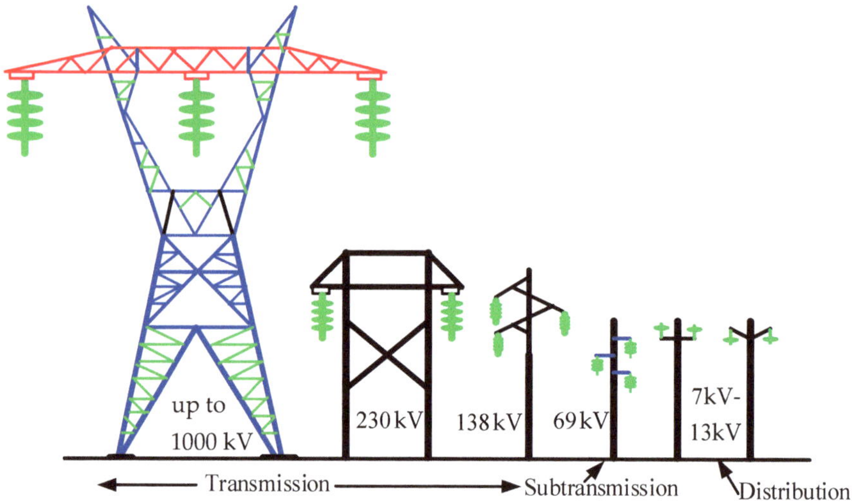

Fig. 3.15 Schematic of transmission towers and distribution poles

distributes by the three-phase four wires system. Therefore, the secondary of a distribution transformer is known as secondary distribution or secondaries. The most common secondary line and phase voltages are 120, 208, 240, 277 and 480 V. Residential and commercial customers usually use the secondary distribution voltage. The secondary distribution consists of feeders, distributors, and service mains. Different types of transmission towers and distribution poles are shown in Fig. 3.15.

3.15 Secondary Distributions and Connections

Secondary distribution voltages are used by the residential, commercial, and light industry. Some of the Asian countries' utility companies use three-phase four wires for power supply to the residential consumers. Their phase voltage and line voltages are 220 V and 440 V, respectively. In North America, both single-phase and three-phase systems are also used based on the connected load. A single-phase three wires are used in the most residential distribution. In this case, the phase and the line voltages are 120 V and 240 V, respectively. 120 V is used for lower-rated electrical appliances, and 240 V can be used for higher-rated electrical usages such as electric vehicle charging. For high-density residential and commercial areas, single-phase four wires are used. In this case, high rating appliances like motors use 208 V. A single-phase three wires and three-phase four wires secondary distribution lines connection are shown in Figs. 3.16 and 3.17, respectively.

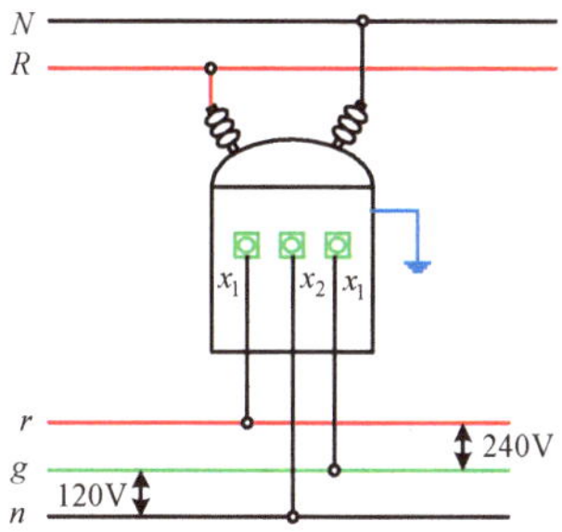

Fig. 3.16 Connection of a single-phase three wires distribution

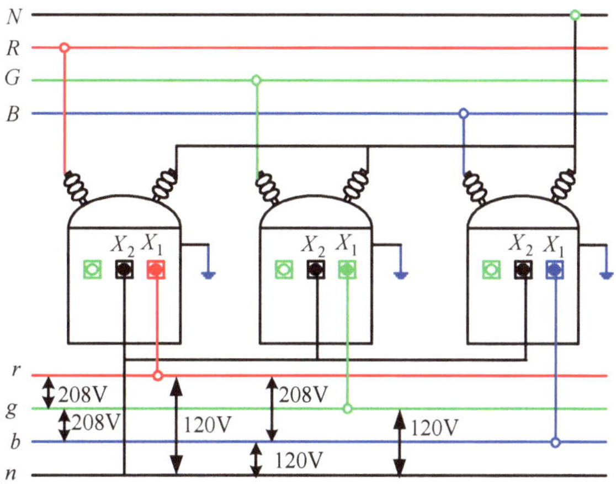

Fig. 3.17 Connection of a three-phase four wires distribution

3.16 Tariff

Electrical energy is delivered to different types of consumers such as residential, commercial, and industrial. The electrical energy consumption at the consumer terminals is recorded by the three-phase or single-phase energy meters. The electrical energy is normally sold to the consumers at a price. This price is known as a tariff. The tariff widely varies from country to country and city to city. Therefore, the tariff is defined as the rate at which electrical energy is supplied to the consumers. The main objectives of the tariff are,

- To recover electricity generation cost.
- To recover cost due to electricity lost during transmission and distribution.
- To recover operation and maintenance cost.
- To earn a reasonable profit on the total investment.

Each system has its own characteristics. The tariff has some important characteristics to identify its performance which is mentioned below.

Return: The tariff should be imposed in such a way that it ensures the proper return from the consumers.

Fairness: The tariff should be fair so that each consumer is able to afford their bills.

Simplicity and Attractive: The tariff should be simple so that an ordinary consumer can easily understand, and it will be attractive if it is simple and user-friendly.

Reasonable Profit: The profit should be marginal and reasonable so that more consumers will use the energy while the investors receive a return.

3.16.1 Classification of Tariff

Tariffs can be classified as the simple tariff, flat rate tariff, block rate tariff, two-part tariff, three-part tariff, and power factor tariff. The brief descriptions of the tariff are mentioned below [5].

Simple Tariff: In this tariff, the fixed rate is imposed based on per unit energy consumption. The price does not vary with a decrease or increase in the number of units used. There is no discrimination between the consumers in a simple tariff.

Flat Rate Tariff: In a flat rate tariff, consumers are grouped into different classes according to diversity and load factors. Each class of consumers is charged at a different uniform rate. The flat rate tariff does not change with the time of use. Here, the cost per unit energy is fixed, and the total charge is dependent on the number of units consumed. The main advantage of this tariff is to prevent the wastage of energy because the consumer has to pay the excess bill. Suppose the charge per kWh is 0.11 cents, and the consumer uses the energy of 200 kWh. The total charge of this tariff is calculated as,

$$TC_{\text{fr}} = 200 \times 0.11 = \$\,22 \tag{3.71}$$

Example 3.4
The maximum demand of a consumer is 5 kW, and the total energy consumption is 6000 kWh. The per annum energy is charged at the rate of 10 cents per kWh for 400 h of the maximum demand plus 5 cents per kWh for additional units. Calculate the annual charges and equivalent flat rate.

Solution
Units consumed for 400 h are,

$$U_{400\text{h}} = 5 \times 400 = 2000\,\text{kWh} \tag{3.72}$$

The charge for 2000 kWh is calculated as,

$$\text{Charge-1} = 2000 \times 0.10 = \text{B\$}\,200 \tag{3.73}$$

The remaining units are determined as,

$$RU = 6000 - 2000 = 4000\,\text{kWh} \tag{3.74}$$

The charge for 4000 kWh is calculated as,

$$\text{Charge-2} = 4000 \times 0.05 = \text{B\$}\,200 \tag{3.75}$$

The total annual charge is determined as,

$$TAC = \text{B\$}\,200 + \text{B\$}\,200 = \text{B\$}\,400 \tag{3.76}$$

The equivalent flat rate is calculated as,

$$Eq_{\text{fr}} = \frac{\text{B\$}\,400}{8760} = \text{B\$}\,0.045 \tag{3.77}$$

Practice Problem 3.4

The maximum demand of the consumer is 4 kW, and the total energy consumption is 7000 kWh. The per annum energy is charged at the rate of 20 cents per kWh for 600 h of the maximum demand plus 10 cents per kWh for additional units. Determine the annual charges and equivalent flat rate.

Block Rate Tariff: The energy consumption is divided into a fixed price per unit block. The price per unit in the first block is usually considered higher rates. Then, the succeeding blocks of energy are charged at progressively reduced rates. In general, the block rate tariff provides lower prices for greater usages to the customer. In this tariff, the main advantage of this tariff is that the cost of energy decreases with the increase in energy consumptions. This tariff does not encourage the customer to use the energy at the off-peak time. Let us consider the x blocks of energy are charged at 0.12 cents and y blocks of energy are charged at 0.25 cents. Therefore, the total charge of this tariff is,

$$TC_{\text{br}} = x \times 0.12 + y \times 0.25 \tag{3.78}$$

Two-part Tariff: This tariff is also known as the Hopkinson demand rate. In this tariff, the energy is charged based upon the consumer's maximum demand and the running charges that depend upon the number of consumed units. Fixed charges are paid irrespective of consumption of electrical energy. Total charges of this tariff can be determined as,

$$TC_{\text{two-part}} = b \times \text{kW} + c \times \text{kWh} \tag{3.79}$$

where

$TC_{\text{two-part}}$ is the total charge for the two-part tariff,
b is the charge per kW of maximum demand,
c is the charge per kWh of total energy consumed.

Example 3.5
The maximum demand of the consumer is 75 kW at 0.55 load factor. The tariff is B $ 80 per kW of maximum demand and 15 cents per kWh. Determine the total units consumed per year, annual charges, and overall cost per kWh.

Solution
The total unit consumed per year is calculated as,

$$TU_{\text{year}} = \text{Maximum demand} \times F_{\text{LF}} \times \text{hours in a year} \tag{3.80}$$

$$TU_{\text{year}} = 75 \times 0.55 \times 8760 = 361350\,\text{kWh} \tag{3.81}$$

The annual charge is calculated as,

$$\text{Annual charge} = 80 \times 75 + 361350 \times 0.15 = \$\,60202.5 \tag{3.82}$$

The overall cost per kWh is determined as,

$$OC_{\text{pkWh}} = \frac{60202.5}{361350} = \$\,0.17 \tag{3.83}$$

Practice Problem 3.5
The maximum demand of a consumer is 55 kW at 40% load factor. The tariff is $ 100 per kW of maximum demand and 10 cents per kWh. Determine the total unit consumed per year, annual charge, and overall cost per kWh.

Maximum Demand Tariff: Maximum demand is the highest amount of electrical energy consumption monitored in a particular period usually for a month. The charge of this tariff is calculated as a charge for the total amount of electricity used plus a demand charge (kW) for the relevant billing period. The demand is a measure of the maximum amount of electricity used at any one time. The chargeable demand in any month is the maximum demand recorded in that month. This type of tariff is used for large consumers.

Three-part Tariff: This tariff is also known as the Doherty rate. In this tariff, the total charge is divided into three parts. These are fixed, semi-fixed, and running charges. This type of tariff is usually applied to large consumers. The total charges of this tariff can be determined as,

$$TC_{\text{three-part}} = a + b \times \text{kW} + c \times \text{kWh} \tag{3.84}$$

where

$TC_{\text{three-part}}$ is the total charge for the two-part tariff,
a is the fixed charge,
b is the charge per kW of maximum demand,
c is the charge per kWh of the total energy consumed.

Power Factor Tariff: This kind of tariff is mainly dependent on the power factor of the consumer. The power factor plays an important role in an AC system. Low power factor increases the line loss which in turn increases the rating of the power system equipment.

References

1. Conejo AJ, Baringo L (2017) Power system operations, 1st edn. Springer
2. Murty PSR (2017) Electrical power systems, 1st edn. Elsevier and Butter-worth Heinemann
3. Salam MA (2012) Fundamentals of electrical machines, 2nd edn. Alpha Science International Ltd., Oxford, UK
4. Fitzgerald AE, Kingsley C Jr, Umans SD (2002) Electric machinery, 6th edn. McGraw-Hill Higher Education, New York, USA
5. Chapman SJ (2002) Electric machinery and power system fundamentals. McGraw-Hill Higher Education, New York, USA

Exercise Problems

3.1 The maximum demand for a power station is 80 MW, and the annual load factor is 0.45. Calculate the total energy generated per annum.

3.2 The total energy generated per annum of a power station is 300×10^6 kWh. Find the load factor if the maximum demand of the power station is 70 MW.

3.3 The maximum demand for a power station is 40 MW and the connected load of 65 MW. The energy generated per annum is 90×10^6 kWh. Calculate the demand factor, average demand, and load factor.

3.4 The maximum demand and the connected load of a power station are 20 MW and 45 MW, respectively. The energy generated per annum is 60×10^6 kWh. Calculate the demand factor, average demand, and load factor.

3.5 A power station supplies power to three different consumers. The power used by the residential user 500 kW, commercial user 1200 kW, and industrial user 1500 kW. The maximum demand and units generated per annum of a power station are 3000 kW and 55×10^5 kWh respectively. Find the diversity factor, average demand, and load factor.

3.6 The three consumers are connected to a power station. The power used by the residential user 300 kW, commercial user and industrial user 1800 kW. The maximum demand and units generated by the power station are 2200 kW and 50×10^5 kWh respectively. Calculate the power used by the commercial user if the diversity factor is 1.2.

3.7 A power station supplies power to the five areas whose maximum demands are 9000 MW, 6000 MW, 5000 MW and 4000 MW respectively. Determine the maximum demand of the power station if the diversity factor is 1.3.

3.8 A power station has the maximum demand of 20 MW. The load factor and a capacity factor of the power station are 0.45 and 0.30, respectively. Calculate the energy generated per annum, plant capacity, and reserve capacity.

3.9 A power station generates per annum energy 65×10^6 kWh, and the maximum demand is 20 MW. Find the capacity factor if the reserve capacity of the power station is 8 MW.

3.10 The daily load cycle of a power station is mentioned below.

Time (H):	0–4	4–8	8–16	16–20	20–24
Load (MW):	25	40	80	40	10

Draw the load curve. Determine the maximum demand, units generated per day, average load, and load factor.

3.11 A daily load cycle of a power station is mentioned below.

Time (H):	0–8	8–16	16–20	20–24
Load (MW):	20	30	60	25

Determine the maximum demand, units generated per day, average load, and load factor.

3.12 A power station supplies power to the consumer A, consumer B and consumer C in the following ways.

Time	Consumer A	Consumer B	Consumer C
12 midnight to 6 a.m.	100 W	No-load	100 W
6 a.m. to 12 noon	500 W	250 W	50 W
12 noon to 4 p.m.	100 W	800 W	300 W
4 p.m. to 10 p.m.	200 W	200 W	100 W
10 p.m. to midnight	50 W	200 W	150 W

Plot the load curve. Determine the maximum demand and diversity factor.

3.13 The maximum demand for a consumer is 10 kW, and the total energy consumption is 8000 kWh. Per annum energy is charged at the rate of 5 cents per kWh for 500 h of the maximum demand plus 3 cents per kWh for additional units. Determine the annual charges and equivalent flat rate.

3.14 The maximum demand of a consumer is 80 kW at a 35% load factor. The tariff is B$ 80 per kW of maximum demand and 5 cents per kWh. Calculate the total units consumed per year, annual charges, and overall cost per kWh.

Chapter 4
Transmission Line Parameters and Analysis

4.1 Introduction

Transmission lines in a power system network are used for transferring power from generating station to the neighboring utility grid substation under all environmental conditions. Transmission lines encompass the electrical properties of resistance, inductance, capacitance and conductance. The parameters inductance and capacitance appear due to the effect of electric and magnetic fields around the conductor. The capacitance is formed between the line and earth of the medium as well as long transmission lines. The effects of resistance, inductance and capacitance, are very important for transmission lines modeling. The transmission circuit consists of conductors, insulators, cross-arms, shield wires, etc. Transmission lines are usually suspended with the help of insulators and towers. The tower is made of wood, steel and reinforced concrete. The transmission and distribution voltages are different in different countries. It can be standardized as 6.6, 11, 33, 66, 69, 120, 220, 500 and 750 kV line to line. Transmission voltage above 220 kV is referred to as extra-high voltage (EHV), and those at 750 kV and above is referred to as ultra-high voltage (UHV). Transmission line conductors are classified as aluminum conductor steel reinforce (ACSR), all-aluminum conductor (AAC), all-aluminum alloy conductor (AAAC), and aluminum conductor alloy reinforce (ACAR). A sample of three-phase transmission lines are shown in Fig. 4.1.

In this chapter, Ampere's circuital law, detail analysis of resistance, inductance, capacitance, bundle conductors will be discussed.

Md. A. Salam, *Fundamentals of Electrical Power Systems Analysis*,
https://doi.org/10.1007/978-981-15-3212-2_4

Fig. 4.1 Three-phase transmission lines

4.2 Ampere's Circuital Law

Ampere's circuital Law represents the integrated magnetic field around a closed path to the current passing through the path. It states that the line integral of the magnetic field intensity around a closed circuit or path is equal to the product of permeability in free space times the total current flowing through the circuit as shown in Fig. 4.2.

Mathematically, the Ampere circuital law is expressed as,

$$\Im = \oint \overrightarrow{B} \cdot \overrightarrow{dl} = \mu_0 I \tag{4.1}$$

$$\Im = \oint \frac{\overrightarrow{B}}{\mu_0} \cdot \overrightarrow{dl} = I \tag{4.2}$$

The magnetic field is directly proportional to magnetic field intensity. It is expressed as $B = \mu_0 H$ and substituting this expression in Eq. (4.2) yields,

$$\Im = \oint \overrightarrow{H} \cdot \overrightarrow{dl} = I \tag{4.3}$$

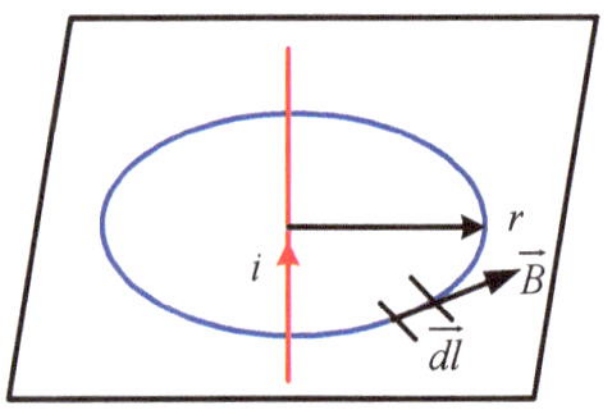

Fig. 4.2 Closed path with the magnetic field

where,

$\Im$ is the magnetomotive force in At,
H is the magnetic field intensity in At per m,
dl is the differential path length in m,
I is the current enclosed in A.

Ampere's law is the most useful method to calculate the magnetic fields of current configurations which have higher symmetry such as long infinity line, circular loop, and cylindrical conductor.

4.3 Line Resistance and Conductance

The transmission line has four parameters which are resistance, inductance, capacitance, and conductance. Resistance is the property of a material that opposes the flow of electricity through the conductor. The resistance generates power loss (I^2R) in the transmission line. It also generates a resistive drop (IR) that affects the voltage regulation of a transmission line. The dc resistance of the conductor is expressed as,

$$R_{dc} = \rho \frac{l}{A} \tag{4.4}$$

where

ρ is the resistivity at 20 °C,
l is the length of the conductor,
A is the area of the conductor.

DC resistance increases linearly with temperature. When direct current flows through a conductor, the current distribution over the cross section of the conductor is uniform, whereas the flow of an alternating current (ac) through a conductor is non-uniform. The outer part of the conductor carries more current than the center of the conductor. Therefore, the current density is higher at the surface of the conductor and it decreases toward the center of the conductor. This results in higher resistance for alternating current than direct current. This behavior is known as skin effect. AC resistance is the ratio of power loss to the square of the current and it is expressed as,

$$R_{ac} = \frac{P_{loss}}{I^2} \tag{4.5}$$

where

P_{loss} is the real power loss of the conductor in watts,
I is the rms value of the current in A.

The effective resistance of the conductor increases due to skin effect at higher frequencies where the skin depth is smaller which reduces the cross-sectional area of the conductor. Conductor loss increases with a higher frequency, which in turn increases the AC resistance. At the frequency of 50 Hz, the ac resistance of the transmission line conductor maybe 5–10% higher than its dc resistance. The resistance of a good conductor such as copper aluminum increases with an increase in temperature, whereas the resistance of electrolyte (*an electrolyte is a substance that produces an electrically conducting solution when dissolved in a polar solvent, such as water*), alloy (*An alloy is a metal, made of the combination of two or more metallic elements*), and insulating material decreases with an increase in temperature. Let us assume that the resistance of a conductor at temperature T and T_0 degrees centigrade is R and R_0, respectively, as shown in Fig. 4.3. In this case, the change in resistance is,

$$\Delta R = R - R_0 \tag{4.6}$$

While the change in temperature is,

$$\Delta T = T - T_0 \tag{4.7}$$

The ratio of change in resistance to the resistance at T_0 degree centigrade is directly proportional to the change in temperature and it can be expressed as,

$$\frac{R - R_0}{R_0} \infty (T - T_0) \tag{4.8}$$

$$\frac{R - R_0}{R_0} = \alpha(T - T_0) \tag{4.9}$$

$$R - R_0 = \alpha R_0(T - T_0) \tag{4.10}$$

$$R = [R_0 + \alpha R_0(T - T_0)] \tag{4.11}$$

$$R = R_0 [1 + \alpha(T - T_0)] \tag{4.12}$$

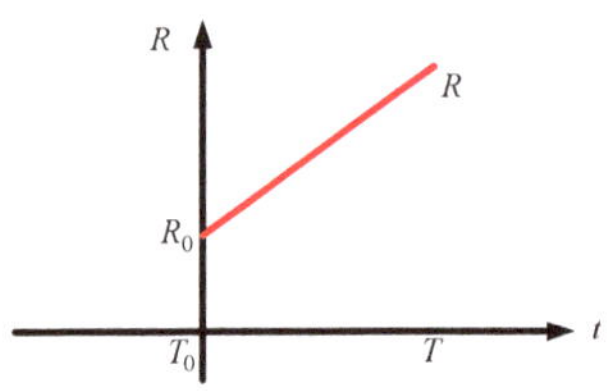

Fig. 4.3 Effect of temperature on resistance

The parameter conductance plays an important role in real power loss between the conductors and between the conductors and earth. In the transmission line, this loss depends on the leakage current in the insulator and corona. The magnitude of the leakage current depends on the amount of salt, dirt, and the presence of the mist or light rain. Leakage current creates a joule heat on the insulator surface and dries the surface, which produces the uneven surface. A higher value of leakage current produces small arcs that create the electric field. When this electric field exceeds its critical value, then the surrounding air of the insulator becomes electrically ionized and provides an electrical path. This phenomenon is known as corona or partial discharge. All the small arcs merge, which form a large arc that creates the flashover on the insulator surface.

Example 4.1
The starting winding resistance of a single-phase induction motor is 10Ω at 10 °C. The temperature coefficient of copper is 0.004/ °C at 0 °C. Calculate the resistance of the starting winding at 20 °C.

Solution
The value of the resistance at 0 °C is calculated as,

$$10 = R_0(1 + 0.004 \times 10) \tag{4.13}$$

$$R_0 = 9.6\,\Omega \tag{4.14}$$

The value of the resistance at 20 °C is calculated as,

$$R_{20} = 9.6(1 + 0.004 \times 20) = 10.4\,\Omega \tag{4.15}$$

Practice Problem 4.1
The value of the resistance of a heater coil is found to be 25Ω at 0 °C. The temperature coefficient of the copper is 0.0043/ °C at 0 °C. Find the resistance of the heater coil at 30 °C.

4.4 Internal Inductance

The inductance is defined as the flux linkage per unit current. Mathematically, it can be expressed as,

$$L = \frac{\psi}{I} \tag{4.16}$$

The inductance of the line can be divided into two parts. One is internal inductance, and the other is external inductance. Internal inductance is due to internal flux linkage when a conductor carries a current. Consider a straight solid cylindrical conductor whose radius is r and carries a current of I ampere as shown in

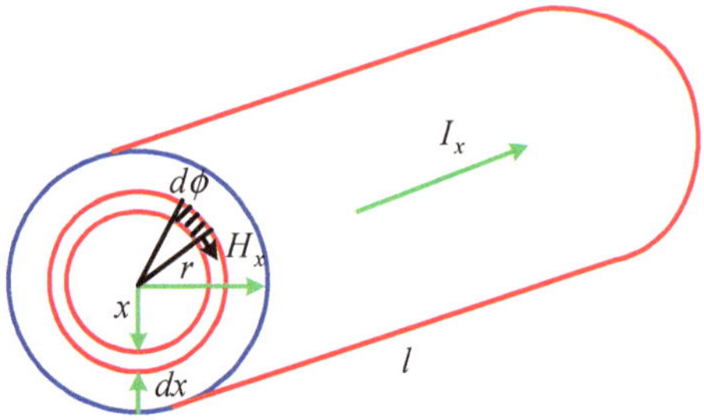

Fig. 4.4 A cylindrical straight conductor with a cross section

Fig. 4.4. An annular ring of thickness dx is drawn at a radius of x. Let the magnetic field intensity at a distance x from the center of the conductor is H_x. This magnetic field intensity is constant throughout the circular ring at a distance x and it is tangent to it. The current enclosed at a radius x is I_x.

Applying Ampere's law to derive the magnetic field intensity around a circle of radius x is,

$$\int_{0}^{2\pi x} H_x \cdot dl = I_x \tag{4.17}$$

$$H_x \times 2\pi x = I_x \tag{4.18}$$

$$H_x = \frac{I_x}{2\pi x} \tag{4.19}$$

The current density at a radius x can be expressed as,

$$J_x = \frac{I_x}{\pi x^2} \tag{4.20}$$

The current density at a radius r is expressed as,

$$J_r = \frac{I}{\pi r^2} \tag{4.21}$$

Neglecting the skin effect and assuming a uniform current density throughout the cross section of the conductor. In this case, the current densities at the two radii can be written as,

$$J_r = J_x \tag{4.22}$$

Substituting Eqs. (4.20) and (4.21) into Eq. (4.22) yields,

$$\frac{I_x}{\pi x^2} = \frac{I}{\pi r^2} \tag{4.23}$$

$$\frac{I_x}{\pi x^2} = \frac{I}{\pi r^2}$$

$$I_x = \frac{x^2 I}{r^2} \tag{4.24}$$

Substituting Eq. (4.24) into Eq. (4.19) yields,

$$H_x = \frac{\frac{x^2 I}{r^2}}{2\pi x} \tag{4.25}$$

$$H_x = \frac{xI}{2\pi r^2} \tag{4.26}$$

The expression of the magnetic flux density at a radius x is,

$$B_x = \mu H_x \tag{4.27}$$

Substituting Eq. (4.26) into Eq. (4.27) yields,

$$B_x = \mu H_x = \frac{\mu_r \mu_0 x I}{2\pi r^2} \tag{4.28}$$

$$B_x = \frac{\mu_r \mu_0 x I}{2\pi r^2} \tag{4.29}$$

Let us consider an elementary area that has a differential thickness dx along the radial direction and a unit length (1 m) along the axial direction is,

$$A = dx \times 1^2 = dx \tag{4.30}$$

The differential flux along the circular strip due to differential thickness be denoted by $d\phi_x$ and this is expressed as,

$$d\phi_x = B_x \times A \tag{4.31}$$

Substituting Eq. (4.30) into Eq. (4.31) yields,

$$d\phi_x = B_x dx \tag{4.32}$$

Substituting Eq. (4.29) into Eq. (4.32) yields

$$d\phi_x = \frac{\mu_r \, \mu_0 \, x \, I}{2\pi r^2} dx \tag{4.33}$$

The entire cross section of the conductor does not link the flux in Eq. (4.33). For a uniform current density, the ratio of the cross-sectional area of the inner circle of radius x to the cross-sectional area of the outer circle of radius r that links the flux $d\phi_x$.

Mathematically, the fraction $\frac{\pi x^2}{\pi r^2}$ of the total current is linked by this flux. Therefore, the differential flux linkage is expressed as,

$$d\psi_x = \frac{\pi x^2}{\pi r^2} d\phi_x \tag{4.34}$$

Substituting Eq. (4.33) into Eq. (4.34) yields,

$$d\psi_x = \frac{\pi x^2}{\pi r^2} \frac{\mu_r \, \mu_0 \, x \, I}{2\pi r^2} dx \tag{4.35}$$

$$d\psi_x = \frac{\mu_r \, \mu_0 \, x^3 \, I}{2\pi r^4} dx \tag{4.36}$$

The internal flux linkage inside of the conductor can be obtained by integrating Eq. (4.36) from 0 to r and it becomes,

$$\psi_x = \int_0^r d\psi_x \tag{4.37}$$

$$\psi_{\text{int}} = \frac{\mu_r \mu_0 I}{2\pi r^4} \int_0^r x^3 dx \tag{4.38}$$

$$\psi_{\text{int}} = \frac{\mu_r \mu_0 I}{2\pi r^4} \times \frac{r^4}{4} \tag{4.39}$$

$$\psi_{\text{int}} = \frac{\mu_r \mu_0 I}{8\pi} \quad \text{Wbt/m} \tag{4.40}$$

Internal inductance of the conductor due to this flux linkage is defined as,

$$L_{\text{int}} = \frac{\psi_{\text{int}}}{I} \tag{4.41}$$

Substituting Eq. (4.40) into Eq. (4.41) yields,

$$L_{\text{int}} = \frac{\mu_r \mu_0}{8\pi} \tag{4.42}$$

Substituting the value of $\mu_0 = 4\pi \times 10^{-7}$ H/m into Eq. (4.21) yields,

$$L_{\text{int}} = \frac{\mu_r}{2} \times 10^{-7} \quad \text{H/m} \tag{4.43}$$

From Eq. (4.43), it is concluded that the internal inductance does not depend on conductor radius.

4.5 External Inductance

An external inductance is due to external flux linkage when a conductor carries a current. Let us consider two conductors with one go which represents the cross and the other one return represents the dot. Again, consider the radius of the conductor is r and the total current is concentrated at the surface of the conductor. It is assumed that the return conductor is placed in a region after the distance D as shown in Fig. 4.5. This distance is longer than the conductor radius. Therefore, no flux lines will be cut by the two conductors. The differential flux for a small region of thickness dx and one meter axial length of the conductor is [1, 2],

$$d\phi_{ext} = B_x \cdot A = \frac{\mu_r \mu_0 I}{2\pi x}(dx \times 1^2) \tag{4.44}$$

Since the flux linkage is close to one conductor, therefore, the differential flux linkage is equal to the differential flux times one and it can be expressed as,

$$d\psi_{ext} = d\phi_x \times 1 \tag{4.45}$$

Substituting Eq. (4.44) into Eq. (4.45) yields,

$$d\psi_{ext} = \frac{\mu_r \mu_0 I}{2\pi x} dx \tag{4.46}$$

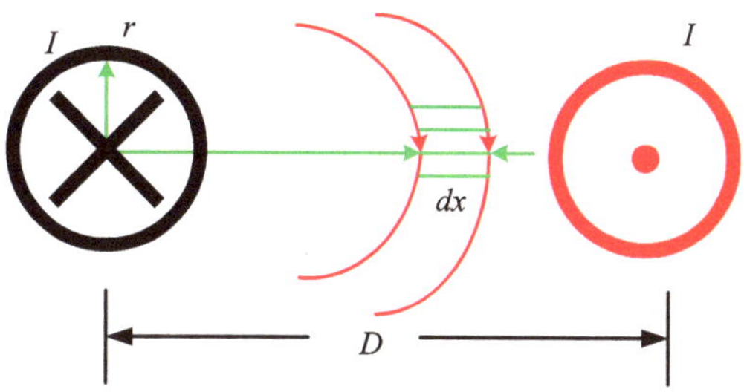

Fig. 4.5 Conductors for external inductance

The total flux linkage due to one conductor can be determined by integrating Eq. (4.46) from r to $D–r$ as,

$$\psi_{ext} = \int_{r}^{D-r} d\psi_x \tag{4.47}$$

Substituting Eq. (4.46) into Eq. (4.47) yields,

$$\psi_{ext} = \int_{r}^{D-r} \frac{\mu_r \mu_0 I}{2\pi} \frac{dx}{x} \tag{4.48}$$

$$\psi_{ext} = \frac{\mu_r \mu_0 I}{2\pi} \ln x \Big|_{r}^{D-r} \tag{4.49}$$

$$\psi_{ext} = \frac{\mu_r \mu_0 I}{2\pi} \ln \frac{D-r}{r} \tag{4.50}$$

Outside distance of the conductor is longer that the conductor radius, i.e., $D \gg r$. In this case, $D - r$ can be reduced to only D and Eq. (4.50) can be modified as,

$$\psi_{ext} = \frac{\mu_r \mu_0 I}{2\pi} \ln \frac{D}{r} \tag{4.51}$$

Substituting Eq. (4.51) into Eq. (4.41) for external inductance, it can be written as,

$$L_{ext} = \frac{\psi_{ext}}{I} = \frac{\mu_r \mu_0}{2\pi} \ln \frac{D}{r} \tag{4.52}$$

The total inductance per meter length is the sum of internal and external inductances and it can be determined as,

$$L = L_{ext} + L_{int} \tag{4.53}$$

Substituting Eqs. (4.42) and (4.52) into Eq. (4.53) yields,

$$L = \frac{\mu_r \mu_0}{8\pi} + \frac{\mu_r \mu_0}{2\pi} \ln \frac{D}{r} \tag{4.54}$$

$$L = \frac{\mu_r \mu_0}{2\pi} \left(\frac{1}{4} + \ln \frac{D}{r} \right) \tag{4.55}$$

$$L = \frac{\mu_r \mu_0}{2\pi} \left(\ln e^{\frac{1}{4}} + \ln \frac{D}{r} \right) \tag{4.56}$$

$$L = \frac{\mu_r \mu_0}{2\pi} \ln \frac{D}{re^{\frac{-1}{4}}} \tag{4.57}$$

$$L = \frac{\mu_r \times 4\pi \times 10^{-7}}{2\pi} \ln \frac{D}{0.7788r} \tag{4.58}$$

$$L = 2 \times 10^{-7} \mu_r \ln \frac{D}{r'} \tag{4.59}$$

where

$r' = 0.7788r$ is the geometric mean radius (GMR).

The GMR can be considered as the radius of a fictitious conductor assumed to have no internal flux linkages but with the same inductance as the actual conductor with the same radius r. The expression of the inductance in the air $(\mu_r = 1)$ is expressed as,

$$L = 2 \times 10^{-7} \ln \frac{D}{r'} \text{ H/m} \tag{4.60}$$

In general, the expression of the inductance is,

$$L = 2 \times 10^{-7} \ln \frac{\text{GMD}}{\text{GMR}} \text{ H/m} \tag{4.61}$$

The geometric mean distance is abbreviated as GMD. Again, consider GMD = $D = D_m$ and GMR = D_s. Equation (4.61) is modified as,

$$L = 2 \times 10^{-7} \ln \frac{D_m}{D_s} \text{ H/m} \tag{4.62}$$

4.6 Concept of GMD and GMR

GMD comes into the picture when there are two or conductors per phase used as in the bundled conductors. The GMD is also used for inductance calculation between the two groups of conductors. In the two groups, the geometric mean distance of a point concerning some points is the geometric mean of the distances between that point and each of the other points. GMD replaces the actual arrangement of the conductors by a hypothetical distance. Therefore, the mutual inductance of the arrangement remains the same. The two groups of conductors which are placed horizontally are shown in Fig. 4.6. The mathematical expression of GMD is,

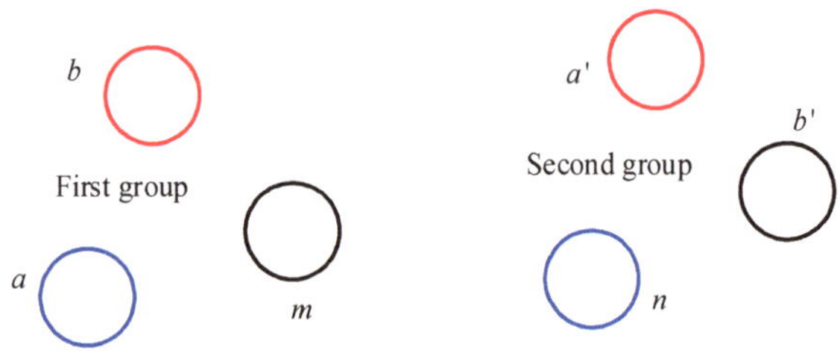

Fig. 4.6 Two groups of conductors

$$\text{GMD} = \sqrt[mn]{(D_{aa'}D_{ab'}D_{an})(D_{ba'}D_{bb'}D_{bn})(D_{ma'}D_{mb'}D_{mn})} \tag{4.63}$$

where

$D_{aa'}$ is the distance between the conductors a and a',
$D_{ab'}$ is the distance between the conductors a and b',
D_{an} is the distance between the conductors a and n,
D_{mn} is the distance between the conductors m and n.

In GMR, the effect of magnetic flux lines is considered outside of the conductor as well as inside. GMR is a hypothetical radius that replaces the actual conductor with a thin-walled hollow conductor of radius equivalent to GMR that the self-inductance of the inductor remains the same. GMR is calculated separately for each phase, and the value of GMR might be different for each phase depending on the conductor size and arrangement. Four conductors are placed in a group as shown in Fig. 4.7 and the expression of GMR can be written as,

$$\text{GMR} = \sqrt[m^2]{(D_{aa}D_{ab}D_{ac}D_{am})(D_{bb}D_{ba}D_{bc}D_{bm})(D_{cc}D_{ca}D_{cb}D_{cm})(D_{mm}D_{ma}D_{mb}D_{mc})} \tag{4.64}$$

where

D_{aa} is the self GMR,
D_{ab} is the distance between the conductors a and b,
D_{ac} is the distance between the conductors a and c,
D_{am} is the distance between the conductors a and m.

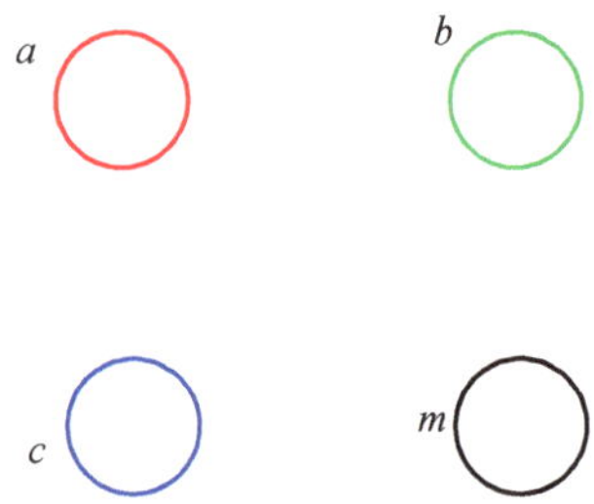

Fig. 4.7 Four conductors in a group

4.7 Inductance of a Single-Phase Line

Consider a single-phase line that consists of two conductors a and b. Let us assume current entering through the conductor a and coming out through the conductor b. These conductors are separated by a distance D as shown in Fig. 4.8. A conductor a with radius, r_a carries a current $I_a = I$ reference out of the page. A conductor b with radius r_b carries a return current $I_b = -I$. Therefore, the sum of these two currents is zero. The internal and external inductance of the conductor a is,

$$L_{a(\text{int})} = \frac{1}{2} \times 10^{-7}\ \text{H/m} \tag{4.65}$$

$$L_{a(ext)} = 2 \times 10^{-7} \ln\frac{D}{r_a}\ \text{H/m} \tag{4.66}$$

The total inductance of the conductor a is,

$$L_a = \frac{1}{2} \times 10^{-7} + 2 \times 10^{-7} \ln\frac{D}{r_a} \tag{4.67}$$

$$L_a = 2 \times 10^{-7}\left(\frac{1}{4} + \ln\frac{D}{r_a}\right) \tag{4.68}$$

$$L_a = 2 \times 10^{-7}\left(\ln e^{\frac{1}{4}} + \ln\frac{D}{r_a}\right) \tag{4.69}$$

$$L_a = 2 \times 10^{-7} \ln\frac{D}{r_a e^{-\frac{1}{4}}} \tag{4.70}$$

$$L_a = 2 \times 10^{-7} \ln\frac{D}{0.7788 r_a} \tag{4.71}$$

$$L_a = 2 \times 10^{-7} \ln\frac{D}{r'_a} \tag{4.72}$$

In general, Eq. (4.72) can be expressed as,

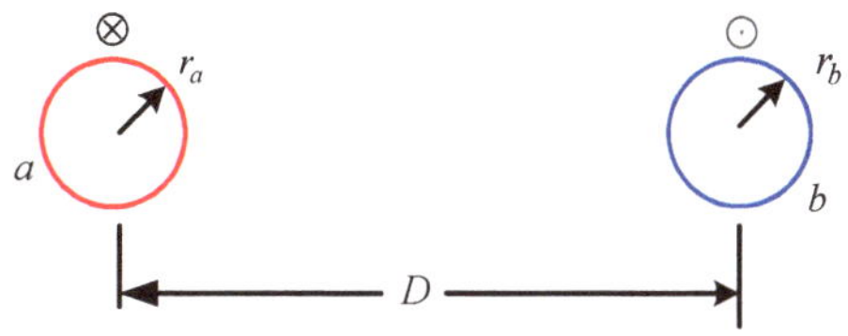

Fig. 4.8 Two conductors with separation distance

$$L_a = 2 \times 10^{-7} \ln \frac{\text{GMD}}{\text{GMR}_a} \tag{4.73}$$

Substituting r'_a by GMR_a and D by GMD in Eq. (4.72) yields,

$$L_a = 2 \times 10^{-7} \left(\ln \frac{1}{r'_a} + \ln \frac{D}{1} \right) \tag{4.74}$$

According to Eq. (4.72), the similar expression for the inductance of the conductor b can be written as,

$$L_b = 2 \times 10^{-7} \ln \frac{D}{r'_b} \tag{4.75}$$

In general, Eq. (4.75) can be written as,

$$L_b = 2 \times 10^{-7} \ln \frac{\text{GMD}}{\text{GMR}_b} \tag{4.76}$$

Equation (4.75) can also be modified as,

$$L_b = 2 \times 10^{-7} \left(\ln \frac{1}{r'_b} + \ln \frac{D}{1} \right) \tag{4.77}$$

where

$r'_a = 0.7788 r_a$ is the GMR of the conductor a,
$r'_b = 0.7788 r_b$ is the GMR of the conductor b,
D is the geometric mean distance (GMD) between the conductors.

Therefore, total inductance of the single-phase line can be determined as,

$$L_T = L_a + L_b \tag{4.78}$$

Substituting Eqs. (4.72) and (4.75) into Eq. (4.78) yields,

$$L_T = 2 \times 10^{-7} \ln \frac{D}{r'_a} + 2 \times 10^{-7} \ln \frac{D}{r'_b} \tag{4.79}$$

$$L_T = 2 \times 10^{-7} \ln \frac{D^2}{r'_a r'_b} \tag{4.80}$$

$$L_T = 2 \times 10^{-7} \times 2 \times \frac{1}{2} \ln \frac{D^2}{r'_a r'_b} \tag{4.81}$$

$$L_T = 4 \times 10^{-7} \ln\left(\frac{D^2}{r'_a r'_b}\right)^{\frac{1}{2}} \tag{4.82}$$

$$L_T = 4 \times 10^{-7} \ln\frac{D}{\sqrt{r'_a r'_b}} \tag{4.83}$$

If two conductors are identical, then the following condition can be written as,

$$r'_a = r'_b = r' \tag{4.84}$$

Equation (4.83) can be modified as,

$$L_T = 4 \times 10^{-7} \ln\frac{D}{r'}\ \text{H/m} \tag{4.85}$$

In general, Eq. (4.85) can be expressed as,

$$L_T = 4 \times 10^{-7} \ln\frac{D_m}{D_s}\ \text{H/m} \tag{4.86}$$

where

D_m is the geometric mean distance,
D_s is the geometric mean radius.

Example 4.2
Two circuits of a single-phase 50 Hz line are shown in Fig. 4.9. The radius of each conductor of the circuit A and circuit B is 0.2 cm and 0.4 cm, respectively. Calculate the GMD, GMR for each circuit, the inductance for each circuit, and the total inductance per meter.

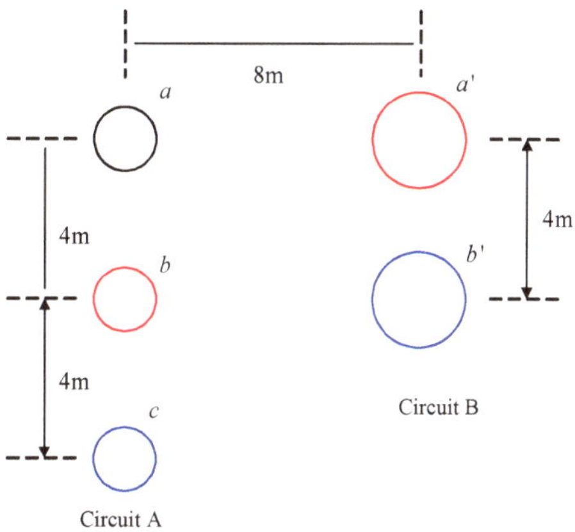

Fig. 4.9 Circuit for Example 4.2

Solution

In the circuit as shown in Fig. 4.9, the corresponding distances are calculated as,

$$D_{aa'} = 8\,\text{m} \tag{4.87}$$

$$D_{bb'} = 8\,\text{m} \tag{4.88}$$

$$D_{ab'} = D_{ba'} = D_{cb'} = \sqrt{8^2 + 4^2} = 8.94\,\text{m} \tag{4.89}$$

$$D_{ca'} = \sqrt{8^2 + 8^2} = 11.31\,\text{m} \tag{4.90}$$

The number of conductors for each circuit is,

$$m = 3,\ n' = 2$$

The value of the GMD can be calculated as,

$$\text{GMD} = \sqrt[mn']{(D_{aa'}D_{ab'})(D_{ba'}D_{bb'})(D_{ca'}D_{cb'})} \tag{4.91}$$

$$\text{GMD} = \sqrt[6]{(8 \times 8.94)(8.94 \times 8)(11.31 \times 8.94)} = 8.96\,\text{m} \tag{4.92}$$

Geometric mean radius for circuit A is calculated as,

$$\text{GMR}_\text{A} = \sqrt[3^2]{(D_{aa}D_{ab}D_{ac})(D_{bb}D_{ba}D_{bc})(D_{cc}D_{ca}D_{cb})} \tag{4.93}$$

$$\text{GMR}_\text{A} = \sqrt[3^2]{(0.2 \times 10^{-2} \times 0.7788)^3 \times 4^4 \times 8^2} = 0.734\,\text{m} \tag{4.94}$$

The geometric mean radius for circuit B is calculated as,

$$\text{GMR}_\text{B} = \sqrt[2^2]{(D_{a'a'}D_{a'b'})(D_{b'b'}D_{b'a'})} \tag{4.95}$$

$$\text{GMR}_\text{B} = \sqrt{4}(0.4 \times 10^{-2} \times 0.7788)^2 \times 4^2 = 0.353\,\text{m} \tag{4.96}$$

Inductance for circuit A is determined as,

$$L_\text{A} = 2 \times 10^{-7} \ln \frac{\text{GMD}}{\text{GMR}_\text{A}} = 2 \times 10^{-7} \ln \frac{8.96}{0.734} = 5 \times 10^{-7}\ \text{H/m} \tag{4.97}$$

Inductance for circuit B is calculated as,

$$L_\text{B} = 2 \times 10^{-7} \ln \frac{\text{GMD}}{\text{GMR}_\text{B}} = 2 \times 10^{-7} \ln \frac{8.96}{0.353} = 6.47 \times 10^{-7}\ \text{H/m} \tag{4.98}$$

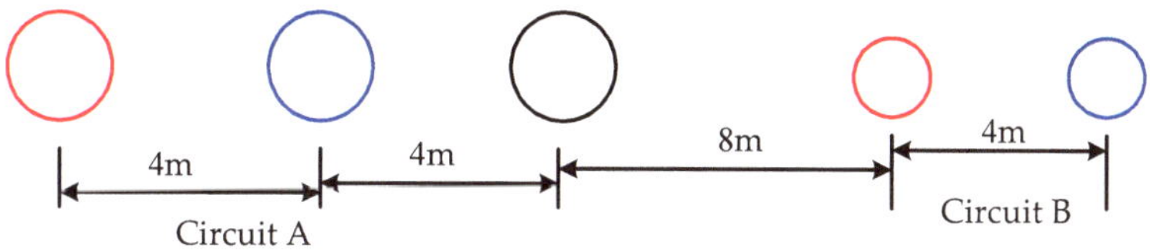

Fig. 4.10 Circuit for Practice Problem 4.2

The total inductance per meter is determined as,

$$L_T = L_A + L_B = 11.47 \times 10^{-7}\ \text{H/m} \tag{4.99}$$

Practice Problem 4.2
A go circuit of a single-phase transmission line is composed of the three solid wires of 0.4 cm radius. The return circuit is composed of the two solid wires of 0.2 cm radius. The arrangement of the conductors is shown in Fig. 4.10. Calculate the GMD, GMR for each circuit, the inductance for each circuit, and the total inductance per meter.

4.8 Self- and Mutual Inductance

The rate of change of current in the coil induces the voltage is known as self-inductance. The rate of change of current in the nearby coil is known as mutual inductance. Inductance inductors of the two conductors are shown in Fig. 4.11. Here, L_{aa} and L_{bb} are the self-inductances of two conductors. Also, L_{ab} and L_{ba} are their mutual inductances. The flux linkage due to self-inductance of the conductor a is expressed as,

$$\psi_{as} = L_{aa} I_a \tag{4.100}$$

The flux linkage due to mutual inductance of the same conductor is expressed as,

$$\psi_{am} = L_{ab} I_b \tag{4.101}$$

Fig. 4.11 Separate inductors

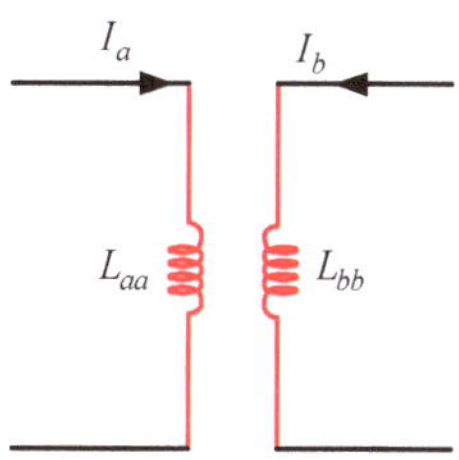

The total flux linkage of the conductor a is the sum of the flux linkage due to self and mutual it can be expressed as,

$$\psi_a = \psi_{as} + \psi_{am} \tag{4.102}$$

Substituting Eqs. (4.100) and (4.101) into Eq. (4.102) yields,

$$\psi_a = L_{aa}I_a + L_{ab}I_b \tag{4.103}$$

Similarly, the total flux linkage of the conductor b can be written as,

$$\psi_b = \psi_{bs} + \psi_{bm} = L_{bb}I_b + L_{ba}I_a \tag{4.104}$$

Since the two currents are in opposite direction, it can be written as,

$$I_a = -I_b \tag{4.105}$$

Substituting Eq. (4.105) into Eq. (4.103) yields,

$$\psi_a = (L_{aa} - L_{ab})I_a \tag{4.106}$$

Again, substituting Eq. (4.105) into Eq. (4.104) yields,

$$\psi_b = (L_{bb} - L_{ba})I_b \tag{4.107}$$

Comparing Eqs. (4.74) and (4.77), following expressions for self-inductance and mutual inductance can be written as,

$$L_{aa} = 2 \times 10^{-7} \ln \frac{1}{r'_a} \tag{4.108}$$

$$L_{bb} = 2 \times 10^{-7} \ln \frac{1}{r'_b} \tag{4.109}$$

$$L_{ab} = L_{ba} = 2 \times 10^{-7} \ln \frac{1}{D} \tag{4.110}$$

where

L_{aa} is the self-inductance of the conductor a,
L_{bb} is the self-inductance of the conductor b,
L_{ab} and L_{ba} are the mutual inductances.

In general, the flux linkage of the ith conductor is expressed as,

$$\psi_i = L_{ii}I_i + \sum_{j=1}^{n} L_{ij}I_j \qquad i \neq j \tag{4.111}$$

Substituting the expressions of self-inductance and mutual inductance in Eq. (4.111) yields,

$$\psi_i = 2 \times 10^{-7}\left(I_i \ln\frac{1}{r_{i'}} + \sum_{j=1}^{n} I_j \ln\frac{1}{D_{ij}}\right) \qquad i \neq j \tag{4.112}$$

4.9 Inductance of Three-Phase Lines

Conductors in the three-phase transmission lines are usually placed as either a symmetrical spacing or an unsymmetrical spacing. The analysis of symmetrical spacing and the unsymmetrical spacing of conductors of the three-phase transmission lines are discussed.

4.9.1 Symmetrical Spacing Conductors

Three conductors a, b, and c carry current I_a, I_b, and I_c, respectively. These conductors are placed at the corners of an equilateral triangle as shown in Fig. 4.12 and it can be written as,

$$D_{12} = D_{23} = D_{31} = D \tag{4.113}$$

For balanced condition, the sum of the currents in the three conductors is equal to zero and it can be expressed as,

$$I_a + I_b + I_c = 0 \tag{4.114}$$

$$I_b + I_c = -I_a \tag{4.115}$$

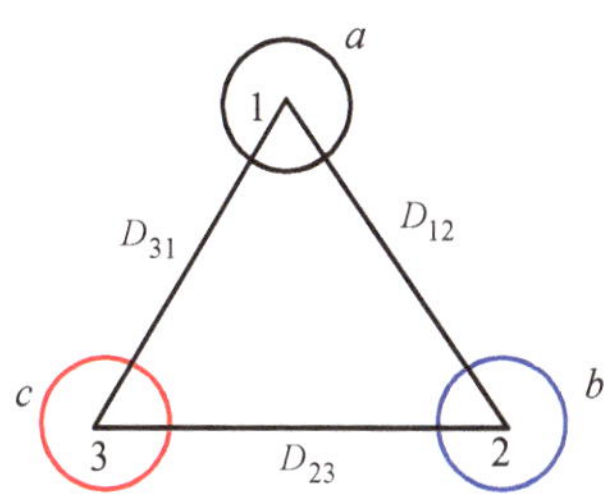

Fig. 4.12 Conductors placed at an equilateral triangle

According to Eq. (4.112), the total flux linkage of the conductor a is,

$$\psi_a = 2 \times 10^{-7}\left(I_a \ln\frac{1}{r'_a} + I_b \ln\frac{1}{D_{12}} + I_c \ln\frac{1}{D_{31}}\right) \tag{4.116}$$

Substituting Eqs. (4.113) and (4.115) into Eq. (4.116) yields,

$$\psi_a = 2 \times 10^{-7} I_a \ln\frac{1}{r'_a} + 2 \times 10^{-7}(I_b + I_c)\ln\frac{1}{D} \tag{4.117}$$

Substituting Eq. (4.115) into Eq. (4.117) yields,

$$\psi_a = 2 \times 10^{-7} I_a \ln\frac{1}{r'_a} - 2 \times 10^{-7} I_a \ln\frac{1}{D} \tag{4.118}$$

$$\psi_a = 2 \times 10^{-7} I_a \ln\frac{\frac{1}{r'_a}}{\frac{1}{D}} \tag{4.119}$$

$$\psi_a = 2 \times 10^{-7} I_a \ln\frac{D}{r'_a} \tag{4.120}$$

Line inductance of the conductor a is expressed as,

$$L_a = \frac{\psi_a}{I_a} \tag{4.121}$$

Substituting Eq. (4.120) into Eq. (4.121) yields,

$$L_a = 2 \times 10^{-7} \ln\frac{D}{r'_a} \tag{4.122}$$

Similarly, the line inductances of the conductors b and c are,

$$L_b = 2 \times 10^{-7} \ln\frac{D}{r'_b} \tag{4.123}$$

$$L_c = 2 \times 10^{-7} \ln\frac{D}{r'_c} \tag{4.124}$$

4.9.2 Unsymmetrical Spacing Conductors

Three conductors a, b, and c carry current I_a, I_b, and I_c, respectively. These conductors having the same radius r are placed unsymmetrically as shown in Fig. 4.13, and it can be written as,

$$D_{12} \neq D_{23} \neq D_{31} \tag{4.125}$$

According to Eq. (4.112), the expression of the flux linkage of conductor a due to I_a, I_b and I_c can be written as,

$$\psi_a = 2 \times 10^{-7}\left(I_a \ln\frac{1}{r'} + I_b \ln\frac{1}{D_{12}} + I_c \ln\frac{1}{D_{31}}\right) \tag{4.126}$$

Similarly, the expressions for the flux linkages of the conductors b and c are,

$$\psi_b = 2 \times 10^{-7}\left(I_a \ln\frac{1}{D_{12}} + I_b \ln\frac{1}{r'} + I_c \ln\frac{1}{D_{23}}\right) \tag{4.127}$$

$$\psi_c = 2 \times 10^{-7}\left(I_a \ln\frac{1}{D_{31}} + I_b \ln\frac{1}{D_{23}} + I_c \ln\frac{1}{r'}\right) \tag{4.128}$$

Again, consider the current I_a is the reference phasor acts in OA direction. The reference phasor rotates counterclockwise as shown in Fig. 4.14. Multiplying the phasor I_a by a and the new phasor aI_a acts in the OB line. Again, multiplying the phasor aI_a by a and the new phasor a^2I_a rotates with an angle of 240° counterclockwise which represents by the line OC. It can be expressed as,

$$a^2I = I \angle 240^\circ \tag{4.129}$$

$$a^2 = 1 \angle 240^\circ = -0.5 - j0.866 \tag{4.130}$$

$$\frac{1}{a^2} = \frac{1}{1 \angle 240^\circ} = 1 \angle 120^\circ = a \tag{4.131}$$

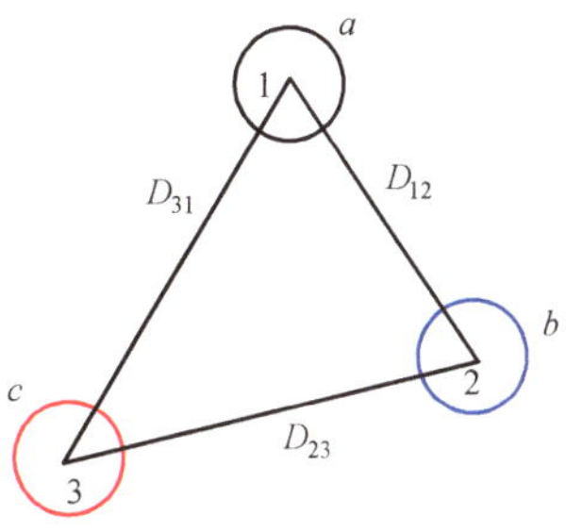

Fig. 4.13 Unsymmetrical spacing conductors

Fig. 4.14 Representation of current phasors

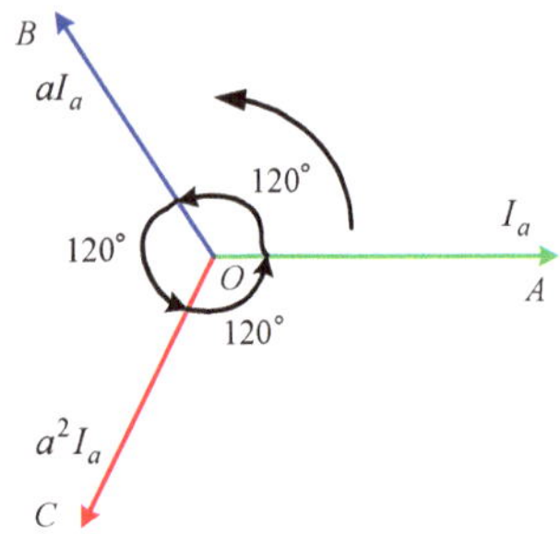

According to the positive phase sequence, the following equations can be written as,

$$I_a = I_a \angle 0^\circ \tag{4.132}$$

$$I_b = I_a \angle 240^\circ = a^2 I_a \tag{4.133}$$

$$I_c = I_a \angle 120^\circ = aI_a \tag{4.134}$$

Substituting Eqs. (4.133) and (4.134) into Eq. (4.126) yields,

$$\psi_a = 2 \times 10^{-7}\left(I_a \ln\frac{1}{r'} + a^2 I_a \ln\frac{1}{D_{12}} + aI_a \ln\frac{1}{D_{31}}\right) \tag{4.135}$$

From Eq. (4.135), the expression of inductance of the conductor a can be derived as,

$$L_a = \frac{\psi_a}{I_a} = 2 \times 10^{-7}\left(\ln\frac{1}{r'} + a^2 \ln\frac{1}{D_{12}} + a \ln\frac{1}{D_{31}}\right) \tag{4.136}$$

Substituting Eq. (4.133) into Eq. (4.134) yields,

$$I_c = I_a \angle 120^\circ = a\frac{I_b}{a^2} = \frac{I_b}{a} \tag{4.137}$$

Substituting Eqs. (4.133) and (4.137) into Eq. (4.127) yields the expression for the inductance of the conductor b as,

$$L_b = \frac{\psi_b}{I_b} = 2 \times 10^{-7}\left(\ln\frac{1}{r'} + \frac{1}{a^2}\ln\frac{1}{D_{12}} + a\frac{1}{a^2}\ln\frac{1}{D_{23}}\right) \tag{4.138}$$

$$L_b = 2 \times 10^{-7}\left(\ln\frac{1}{r'} + a\ln\frac{1}{D_{12}} + \frac{1}{a}\ln\frac{1}{D_{23}}\right) \tag{4.139}$$

$$L_b = 2 \times 10^{-7}\left(\ln\frac{1}{r'} + a\ln\frac{1}{D_{12}} + a^2\ln\frac{1}{D_{23}}\right) \tag{4.140}$$

Substituting Eq. (4.134) into Eq. (4.133) yields,

$$I_b = a^2\frac{I_c}{a} = aI_c \tag{4.141}$$

The expression for the inductance of the conductor c can be expressed as,

$$L_c = \frac{\psi_c}{I_c} \tag{4.142}$$

Substituting Eq. (4.128) into Eq. (4.142) yields,

$$L_c = \frac{2 \times 10^{-7}}{I_c}\left(I_a\ln\frac{1}{D_{31}} + I_b\ln\frac{1}{D_{23}} + I_c\ln\frac{1}{r'}\right) \tag{4.143}$$

Substituting Eqs. (4.141) and (4.134) into Eq. (4.143) yields,

$$L_c = \frac{2 \times 10^{-7}}{I_c}\left(I_c\ln\frac{1}{r'} + \frac{I_c}{a}\ln\frac{1}{D_{31}} + aI_c\ln\frac{1}{D_{23}}\right) \tag{4.144}$$

$$L_c = 2 \times 10^{-7}\left(\ln\frac{1}{r'} + \frac{1}{a}\ln\frac{1}{D_{31}} + a\ln\frac{1}{D_{23}}\right) \tag{4.145}$$

The average value of the inductance can be determined as,

$$L = \frac{L_a + L_b + L_c}{3} \tag{4.146}$$

Substituting Eqs. (4.136), (4.140), and (4.145) into Eq. (4.146) yields,

$$L = \frac{2 \times 10^{-7}}{3}\left[3\ln\frac{1}{r'} + \ln\frac{1}{D_{12}}(a^2 + a) + \ln\frac{1}{D_{23}}(a^2 + a) + \ln\frac{1}{D_{31}}(a^2 + a)\right] \tag{4.147}$$

$$L = \frac{2 \times 10^{-7}}{3}\left[3\ln\frac{1}{r'} + (a^2 + a)\left(\ln\frac{1}{D_{12}} + \ln\frac{1}{D_{23}} + \ln\frac{1}{D_{31}}\right)\right] \tag{4.148}$$

$$L = \frac{2 \times 10^{-7}}{3}\left[3\ln\frac{1}{r'} + (-1)\left(\ln\frac{1}{D_{12}D_{23}D_{31}}\right)\right] \tag{4.149}$$

$$L = 2 \times 10^{-7}\left[\ln\frac{1}{r'} + \left(-\frac{1}{3}\right)\left(\ln\frac{1}{D_{12}D_{23}D_{31}}\right)\right] \tag{4.150}$$

$$L = 2 \times 10^{-7}\left[\ln\frac{1}{r'} - \ln\frac{1}{\sqrt[3]{D_{12}D_{23}D_{31}}}\right] \tag{4.151}$$

$$L = 2 \times 10^{-7}\ln\frac{\sqrt[3]{D_{12}D_{23}D_{31}}}{r'} \quad \text{H/m} \tag{4.152}$$

$$L = 2 \times 10^{-7}\ln\frac{D_m}{D_s} \quad \text{H/m} \tag{4.153}$$

where the geometric mean distance D_m for this arrangement is,

$$D_m = \sqrt[3]{D_{12}D_{23}D_{31}} \tag{4.154}$$

Example 4.3
The conductors of a three-phase, 50 Hz, 120 km long transmission line are shown in Fig. 4.15. The diameter of each conductor is 0.6 cm. Calculate the inductance in H/m and inductive reactance per phase.

Solution
The value of the GMR of the conductor is,

$$D_s = 0.7788r = 0.7788 \times 0.3 = 0.233 \text{ cm} \tag{4.155}$$

The value of the GMD is,

$$D_m = \sqrt[3]{D_{ab}D_{bc}D_{ca}} = \sqrt[3]{2 \times 2 \times 3} = 2.29 \text{ m} \tag{4.156}$$

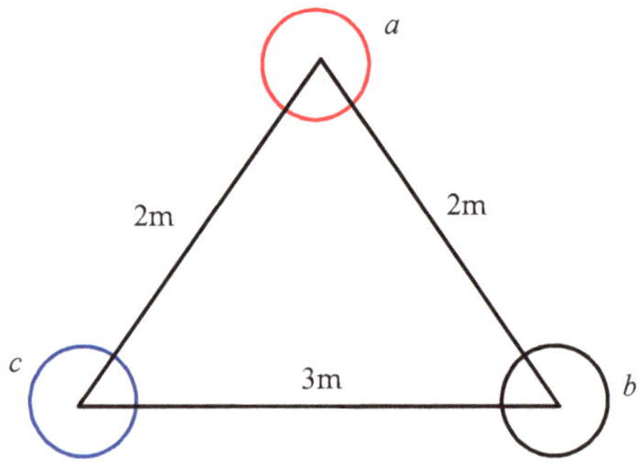

Fig. 4.15 Conductors arrangement for Example 4.3

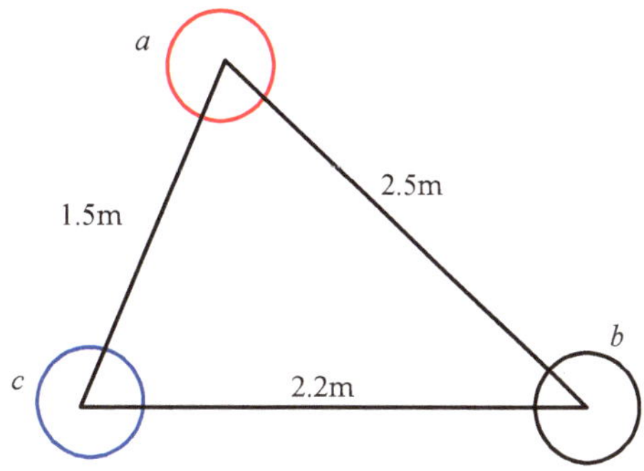

Fig. 4.16 Conductors arrangement for Practice Problem 4.3

The value of the inductance per phase is,

$$L = 2 \times 10^{-7} \ln \frac{D_m}{D_s} = 2 \times 10^{-7} \ln \frac{229}{0.233} \tag{4.157}$$

$$L = 1.38 \times 10^{-6}\ \mathrm{H/m} \tag{4.158}$$

The value of the inductance per km is,

$$L = 1.38 \times 10^{-3}\ \mathrm{H/km} \tag{4.159}$$

The value of the inductance per phase is calculated as,

$$L = 1.38 \times 10^{-3} \times 120 = 0.17\ \mathrm{H} \tag{4.160}$$

The value of the inductive reactance per phase is determined as,

$$X_L = 2\pi \times 50 \times 0.17 = 53.41\ \Omega \tag{4.161}$$

Practice Problem 4.3
The conductors' arrangement of a three-phase, 50 Hz, 80 km long transmission line is shown in Fig. 4.16. The diameter of each conductor is 1 cm. Determine the inductance in H/m and inductive reactance per phase.

4.10 Transposition of Conductors

The phase conductors of transmission lines cannot maintain symmetrical spacing along the entire length because of the construction of transmission towers on uneven lands. Therefore, the inductance will be different in an unsymmetrical spacing conductor with a corresponding unbalanced voltage for each conductor. It is very difficult to find the inductance when the conductors are placed in an unsymmetrical spacing. Due to unsymmetrical spacing, flux linkages and

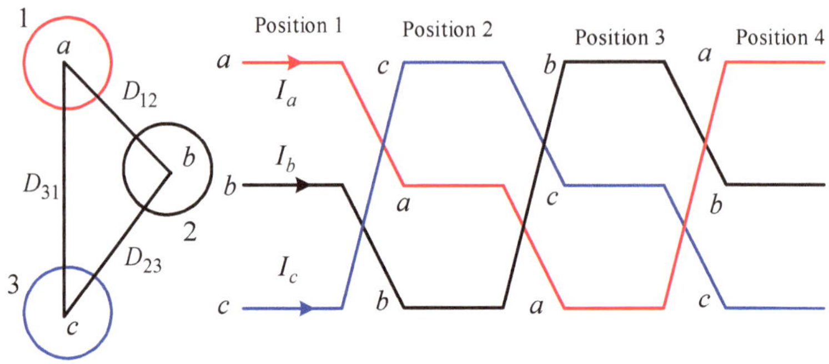

Fig. 4.17 Position of conductors in different steps

inductances of each phase are not the same, and the magnetic field to the conductors is not zero. Due to this magnetic field, the voltage induced in adjacent conductors, especially in telephone lines, may result in interference. The conductors of transmission lines need to be transposed to reduce this interference effect to a minimum. In transposition, the balanced condition of the three-phase lines can be obtained by changing the position of the conductors at regular intervals along the line so that each conductor will come to the original position along with another conductor over an equal distance. This kind of exchange of the position of the conductors is known as transposition. The conductor's position at different steps is shown in Fig. 4.17. At position 1, the conductor a is in position 1, the conductor b is in position 2 and the conductor c is in position 4. The total flux linkage of the conductor a when it is at position 1 is,

$$\psi_{a1} = 2 \times 10^{-7}\left(I_a \ln\frac{1}{r'} + I_b \ln\frac{1}{D_{12}} + I_c \ln\frac{1}{D_{31}}\right) \quad (4.162)$$

At position 2, the conductor a is in position 2, the conductor b is in position 3, and the conductor c is in position 1. The total flux linkage of the conductor a when it is at position 2 is,

$$\psi_{a2} = 2 \times 10^{-7}\left(I_a \ln\frac{1}{r'} + I_b \ln\frac{1}{D_{23}} + I_c \ln\frac{1}{D_{12}}\right) \quad (4.163)$$

At position 3, the conductor a is in position 3, the conductor b is in position 1, and the conductor c is in position 2. The total flux linkage of the conductor a when it is at position 3 is,

$$\psi_{a3} = 2 \times 10^{-7}\left(I_a \ln\frac{1}{r'} + I_b \ln\frac{1}{D_{31}} + I_c \ln\frac{1}{D_{23}}\right) \quad (4.164)$$

The average value of the flux linkage of the conductor a is calculated as,

$$\psi_a = \frac{\psi_{a1} + \psi_{a2} + \psi_{a3}}{3} \tag{4.165}$$

Substituting Eqs. (4.162), (4.163), and (4.164) into Eq. (4.165) yields,

$$\psi_a = \frac{2 \times 10^{-7}}{3}\left(3I_a \ln\frac{1}{r'} + I_b \ln\frac{1}{D_{12}D_{23}D_{31}} + I_c \ln\frac{1}{D_{12}D_{23}D_{31}}\right) \tag{4.166}$$

For a balanced load, the sum of the currents is equal to zero and it can be expressed as,

$$I_a + I_b + I_c = 0 \tag{4.167}$$

$$I_b + I_c = -I_a \tag{4.168}$$

Substituting Eq. (4.168) into Eq. (4.166) yields,

$$\psi_a = 2 \times 10^{-7}\left(I_a \ln\frac{1}{r'} - \frac{I_a}{3}\ln\frac{1}{D_{12}D_{23}D_{31}}\right) \tag{4.169}$$

$$\psi_a = 2 \times 10^{-7} I_a\left(\ln\frac{1}{r'} - \frac{1}{3}\ln\frac{1}{D_{12}D_{23}D_{31}}\right) \tag{4.170}$$

From Eq. (4.170), the expression of inductance for the conductor a can be expressed as,

$$L_a = \frac{\psi_a}{I_a} = 2 \times 10^{-7}\left(\ln\frac{1}{r'} - \frac{1}{3}\ln\frac{1}{D_{12}D_{23}D_{31}}\right) \tag{4.171}$$

$$L_a = 2 \times 10^{-7}\left(\ln\frac{1}{r'} - \ln\frac{1}{\sqrt[3]{D_{12}D_{23}D_{31}}}\right) \tag{4.172}$$

$$L_a = 2 \times 10^{-7}\left(\ln\frac{1}{r'} + \ln\sqrt[3]{D_{12}D_{23}D_{31}}\right) \tag{4.173}$$

$$L_a = 2 \times 10^{-7} \ln\frac{\sqrt[3]{D_{12}D_{23}D_{31}}}{r'} \tag{4.174}$$

$$L_a = 2 \times 10^{-7} \ln\frac{D_m}{D_s} \tag{4.175}$$

The inductance per phase can be expressed as,

$$X_{Lphase} = 2\pi f \times 2 \times 10^{-7} \ln \frac{D_m}{D_s} \tag{4.176}$$

where the geometric mean distance is,

$$D_m = \sqrt[3]{D_{12} D_{23} D_{31}}$$

4.11 Bundled Conductors

In long transmission lines, corona discharge is common. Corona occurs when the surface potential gradient or electric field of a conductor exceeds the dielectric strength of the surrounding air. This causes ionization of the area near the conductor and creates hissing sounds and arcs. Corona produces power loss and causes interference with communication lines or channels. In order to reduce the corona effect, hollow round conductors are used. For economical constraints, it is feasible to use more than one conductor per phase when the voltages greater than 220 kV. The electric field or high voltage surface gradient is usually reduced considerably by considering two or more conductors per phase in close proximity. This arrangement is known as the bundling of conductors. The conductors are bundled in groups of two, three, or four. The advantages of bundle conductors are as follows:

- Self-GMD of conductors increased as radius increased
- The inductance of conductors decreases which lead to reducing the line voltage drop and improves line performance
- Reduce the voltage gradient in the vicinity of the line
- Reduce the corona loss in the line
- Reduce radio interference in the communication lines or channels
- Provide higher capacitance and lower inductance
- Provide higher surge impedance loading $(SIL = \sqrt{(L/C)})$ which leads to maximum power transferability.

Depending on the voltage magnitude of the transmission lines, the bundled conductor consists of two, three, and four as shown in Fig. 4.18.

The expression of GMR for the two conductors per bundle is,

$$D_s^{b2} = \sqrt[4]{D_s \times d \times D_s \times d} \tag{4.177}$$

$$D_s^{b2} = \sqrt[4]{D_s^2 \times d^2} \tag{4.178}$$

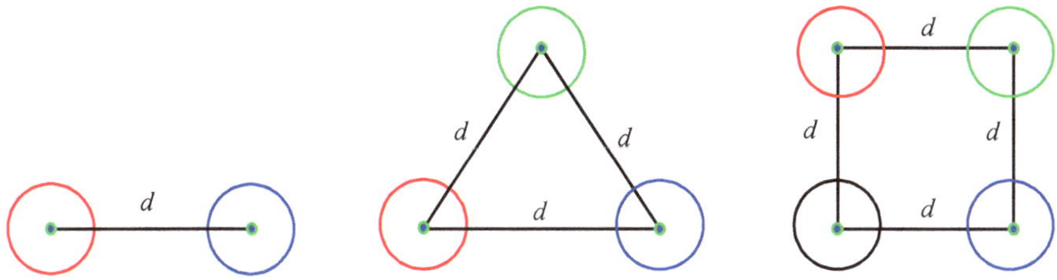

Fig. 4.18 Arrangements of bundled conductors

$$D_s^{b2} = \sqrt{D_s \times d} \tag{4.179}$$

The expression of GMR for the three conductors per bundle is,

$$D_s^{b3} = \sqrt[9]{(D_s \times d \times d)(D_s \times d \times d)(D_s \times d \times d)} \tag{4.180}$$

$$D_s^{b3} = \sqrt[9]{(D_s \times d \times d)^3} \tag{4.181}$$

$$D_s^{b3} = \sqrt[3]{D_s \times d^2} \tag{4.182}$$

The expression of GMR for four conductors per bundle is,

$$D_s^{b4} = \sqrt[16]{(D_s \times d \times d \times \sqrt{2}d)(D_s \times d \times d \times \sqrt{2}d)(D_s \times d \times d \times \sqrt{2}d)(D_s \times d \times d \times \sqrt{2}d)} \tag{4.183}$$

$$D_s^{b4} = \sqrt[16]{(D_s \times d \times d \times \sqrt{2}d)^4} \tag{4.184}$$

$$D_s^{b4} = 1.09\sqrt[4]{D_s \times d^3} \tag{4.185}$$

Example 4.4
The length of a three-phase single circuit 260 kV transmission line is 100 km. This line is composed of the three bundle conductors as shown in Fig. 4.19. The radius of each conductor is 0.8 cm and the distance $d = 0.3$ m. Determine the inductance in H/km and inductance per phase.

Solution
The value of the GMR can be determined as,

$$D_s^{b3} = \sqrt[3]{D_s \times d^2} = \sqrt[3]{0.7788 \times 0.8 \times 10^{-2} \times (0.3)^2} = 0.082\,\text{m} \tag{4.186}$$

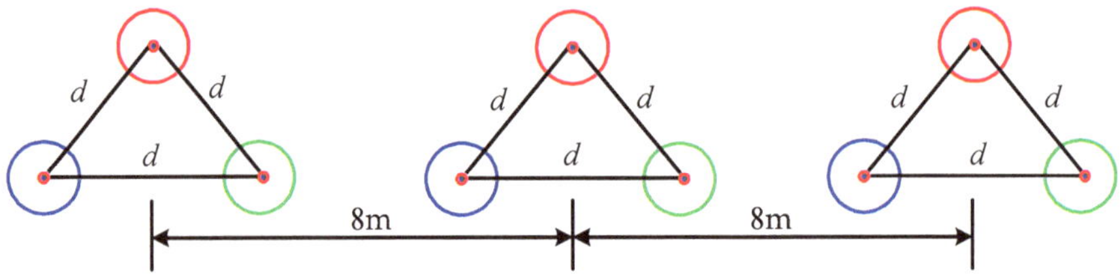

Fig. 4.19 Three conductors bundle

The value of the GMD is calculated as,

$$D_m = \sqrt[3]{8 \times 8 \times 16} = 10.08 \text{ m} \tag{4.187}$$

The value of the inductance is calculated as,

$$L = 2 \times 10^{-7} \ln \frac{D_m}{D_s^{b3}} = 2 \times 10^{-7} \ln \frac{10.08}{0.082} \quad \text{H/m} \tag{4.188}$$

$$L = 9.62 \times 10^{-7} \text{ H/m} \tag{4.189}$$

$$L = 9.62 \times 10^{-4} \text{ H/km} \tag{4.190}$$

The value of the inductance per phase for 100 km is calculated as,

$$L = 9.62 \times 10^{-4} \times 100 = 0.096 \text{ H} \tag{4.191}$$

Practice Problem 4.4

The two bundle conductors per phase are used for a 120 km long three-phase single circuit 230 kV, 50 Hz transmission line as shown in Fig. 4.20. The radius of each conductor is 0.4 cm and the distance $d = 0.4$ m. Calculate the inductance in H/km and inductance per phase.

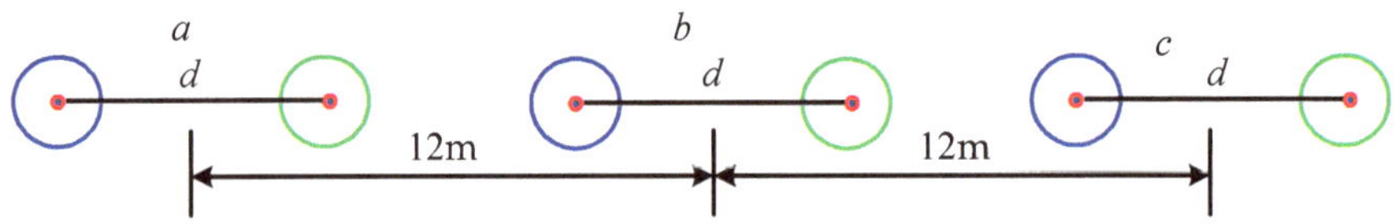

Fig. 4.20 Two conductors bundle

4.12 Line Capacitance

The capacitance between the conductors is defined as the charge per unit potential. Mathematically, it can be expressed as,

$$C = \frac{q}{V} \tag{4.192}$$

The capacitance between parallel conductors is normally constant. The capacitance, which is formed in between two lines, is considered as a shunt. The effect of capacitance is very small for a transmission line that is less than 80 km and therefore is often neglected but is considered for over 80 km.

Generally, Gauss's law is considered for analyzing capacitance. This law states that within a closed surface, the total electric charge equals the total electric flux emerging from the surface. Mathematically, Gauss's law can be expressed as,

$$\oint_A D \,.\, dA = Q \tag{4.193}$$

where,

D is the electric flux density in $\mathrm{C/m^2}$,
Q is the total electric charge on the conductor in coulombs (C).
dA is the unit vector normal to the surface in $\mathrm{m^2}$.

Consider a round conductor whose radius is r and carries charge q C/m as shown in Fig. 4.20. Applying Gauss's law to Fig. 4.21 to find the electric flux density at a radius x and l m length of the cylindrical conductor is,

$$DA = Q \tag{4.194}$$

$$D \times 2\pi x l = ql \tag{4.195}$$

$$D = \frac{q}{2\pi x} \tag{4.196}$$

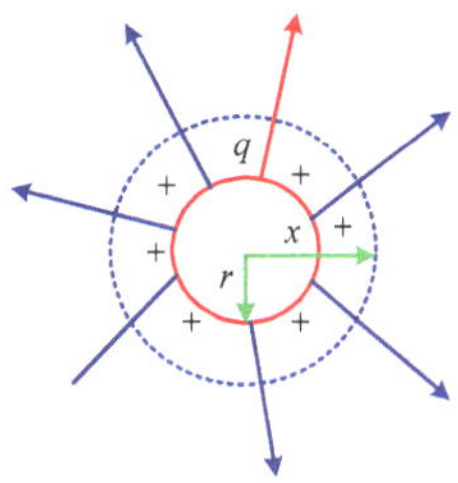

Fig. 4.21 A conductor with a charge

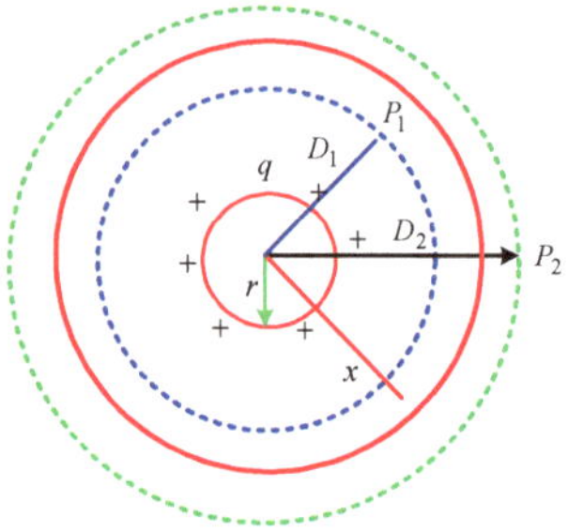

Fig. 4.22 Two points with different distances

where

l is the length of the cylindrical conductor,
q is the charge on the conductor C/m.

The electric field is related to the electric flux density and it can be expressed as,

$$E = \frac{D}{\varepsilon = \varepsilon_0 \varepsilon_r} \tag{4.197}$$

where

ε_0 is the permittivity of free space and its value is equal to 8.854×10^{-12} F/m,
ε_r is the relative permittivity of the space and its value is equal to 1 for the air.

Substituting Eq. (4.196) into Eq. (4.197) yields,

$$E = \frac{q}{2\pi\varepsilon x} \tag{4.198}$$

Again, consider a long straight conductor that carries a positive charge q C/m as shown in Fig. 4.22. Assuming that the uniformly distributed charge concentrated at the center of the conductor due to its equipotential surface, the points P_1 and P_2 are located at distances D_1 and D_2, respectively, from the center of the conductor.

The potential difference (V_{12}) between the two points is the work done by moving a unit charge between the two points. Therefore, the voltage drop between the two points can be determined by integrating the field intensity from D_1 to D_2 over a radial path between the equipotential surfaces as,

$$V_{12} = \int_{D_1}^{D_2} E dx \tag{4.199}$$

Substituting Eq. (4.198) into Eq. (4.199) yields,

$$V_{12} = \int_{D_1}^{D_2} \frac{q}{2\pi\varepsilon x} dx \tag{4.200}$$

$$V_{12} = \frac{q}{2\pi\varepsilon} \ln \frac{D_2}{D_1} \tag{4.201}$$

The line capacitance can be expressed as,

$$C = \frac{q}{V_{12}} = \frac{2\pi\varepsilon}{\ln \frac{D_2}{D_1}} \tag{4.202}$$

4.13 Capacitance of Single-Phase Line

Consider a single-phase line consists of two parallel round conductors of the radius r_1 and r_2 as shown in Fig. 4.23. The distance D separates the two conductors. Let us assume conductor 1 carries a charge q_1 C/m, and conductor 2 carries a charge q_2 C/m. The conductor's separation distance is much larger than the radius of the conductors, and the height of the conductors is much larger than the separation distance to avoid the ground effect.

The potential difference between conductor 1 and conductor 2 due to charge q_1 is,

$$V_{12}(q_1) = \frac{q_1}{2\pi\varepsilon} \ln \frac{D_{12}}{r_1} \tag{4.203}$$

The potential difference between conductor 2 and conductor 1 due to charge q_2 is,

$$V_{21}(q_2) = \frac{q_2}{2\pi\varepsilon} \ln \frac{D_{21}}{r_2} \tag{4.204}$$

$$V_{12}(q_2) = \frac{q_2}{2\pi\varepsilon} \ln \frac{r_2}{D_{21}} \tag{4.205}$$

Applying principle of superposition to write the following equation,

$$V_{12} = V_{12}(q_1) + V_{12}(q_2) \tag{4.206}$$

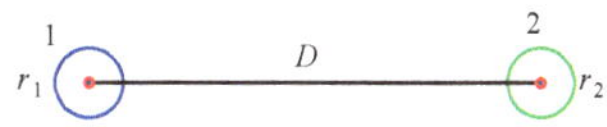

Fig. 4.23 Two parallel conductors

Substituting Eqs. (4.203) and (4.205) into Eq. (4.206) yields,

$$V_{12} = \frac{q_1}{2\pi\varepsilon} \ln \frac{D_{12}}{r_1} + \frac{q_2}{2\pi\varepsilon} \ln \frac{r_2}{D_{21}} \tag{4.207}$$

$$V_{12} = \frac{q_1}{2\pi\varepsilon} \ln \frac{D_{12}}{D_{11}} + \frac{q_2}{2\pi\varepsilon} \ln \frac{D_{22}}{D_{21}} \tag{4.208}$$

In general, the voltage difference between conductors k and i based on Eq. (4.207) can be written as,

$$V_{ki} = \frac{1}{2\pi\varepsilon} \sum_{m=1}^{N} q_m \ln \frac{D_{im}}{D_{km}} \tag{4.209}$$

Let us consider the following assumptions for simplification,

$$r_1(D_{11}) = r_2(D_{22}) = r \tag{4.210}$$

$$D_{12} = D_{21} = D \tag{4.211}$$

$$q_1 = -q_2 = q \tag{4.212}$$

Substituting Eqs. (4.210), (4.211), and (4.212) into Eq. (4.207) yields,

$$V_{12} = \frac{q}{2\pi\varepsilon} \ln \frac{D}{r} - \frac{q}{2\pi\varepsilon} \ln \frac{r}{D} \tag{4.213}$$

$$V_{12} = \frac{q}{2\pi\varepsilon} \ln \frac{D}{r} + \frac{q}{2\pi\varepsilon} \ln \frac{D}{r} \tag{4.214}$$

$$V_{12} = \frac{q}{2\pi\varepsilon} \ln \left(\frac{D}{r}\right)^2 \tag{4.215}$$

$$V_{12} = \frac{q}{\pi\varepsilon} \ln \frac{D}{r} \tag{4.216}$$

The expression between the conductors is expressed as,

$$C_{12} = \frac{q}{V_{12}} \tag{4.217}$$

Substituting Eq. (4.216) into Eq. (4.217) yields,

$$C_{12} = \frac{q}{\frac{q}{\pi\varepsilon} \ln \frac{D}{r}} \tag{4.218}$$

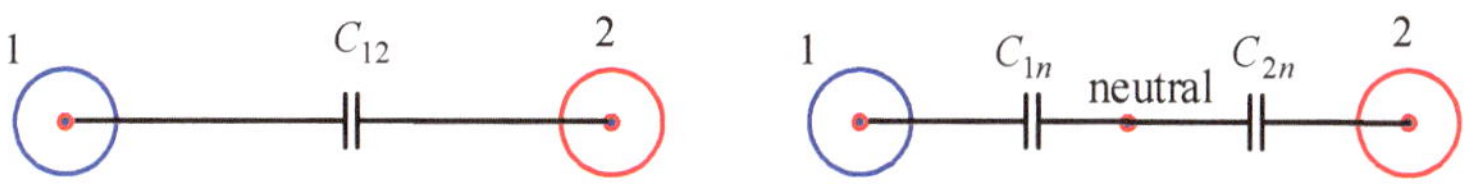

Fig. 4.24 Capacitance between two conductors and capacitance between neutral to the conductor

$$C_{12} = \frac{\pi\varepsilon}{\ln\frac{D}{r}} = \frac{\pi\varepsilon_0\varepsilon_r}{\ln\frac{D}{r}} \tag{4.219}$$

For an air medium, Eq. (4.219) can be modified as,

$$C_{12} = \frac{\pi\varepsilon_0}{\ln\frac{D}{r}} \tag{4.220}$$

For transmission line modeling, the capacitance is defined between the conductor and the neutral point as shown in Fig. 4.24. The potential difference between one conductor to neutral is equal to one-half of the potential difference between two conductors. Therefore, the capacitance to neutral is expressed as,

$$C_n = C_{1n} = C_{2n} = 2C_{12} \tag{4.221}$$

Substituting Eq. (4.220) into Eq. (4.221) yields,

$$C_n = \frac{2\pi\varepsilon_0}{\ln\frac{D}{r}} = \frac{2\pi \times 8.854 \times 10^{-12}}{\ln\frac{D}{r}} \tag{4.222}$$

$$C_n = \frac{5.56 \times 10^{-11}}{\ln\frac{D}{r}}\ \text{F/m} \tag{4.223}$$

$$C_n = \frac{0.0556}{\ln\frac{D}{r}}\ \mu\text{F/km} \tag{4.224}$$

Example 4.5

The radius of the two parallel conductors of a 25 km long single-phase transmission line is 0.2 cm. The separation distance between the conductors is found to be 2.2 m. Calculate the capacitance in F/m and μF.

Solution

Here, the values of the radius and separation distance are,

$$r = 0.2\,\text{cm} \tag{4.225}$$

$$D = 2.2\,\text{m} = 220\,\text{cm} \tag{4.226}$$

The value of the line capacitance can be determined as,

$$C = \frac{\pi \varepsilon_0}{\ln \frac{D}{r}} = \frac{\pi \times 8.854 \times 10^{-12}}{\ln \frac{220}{0.2}} = 3.97 \times 10^{-12}\ \text{F/m} \tag{4.227}$$

$$C = 3.97 \times 10^{-12} \times 25 \times 10^3 = 0.099\ \mu\text{F} \tag{4.228}$$

Practice Problem 4.5

A 135 km long single-phase transmission line consists of two parallel conductors. The radius of each conductor is 1.5 mm. The separation between the conductors is found to be 2 m. Consider the line voltage is 33 kV, 50 Hz. Find the capacitance in μF and charging current.

4.14 Capacitance of Three-Phase Lines

In three-phase transmission lines, the three conductors are arranged either in equal or unequal spacing. Detailed analysis of equal spacing and unequal spacing of conductors has been discussed separately.

4.14.1 Capacitance of Three-Phase Lines with Equal Spacing Conductors

Consider the three conductors with an equal radius r and separated with an equal distance D as shown in Fig. 4.25. For a balanced three-phase system, the sum of charges is zero, and it can be expressed as,

$$q_a + q_b + q_c = 0 \tag{4.229}$$

$$q_b + q_c = -q_a \tag{4.230}$$

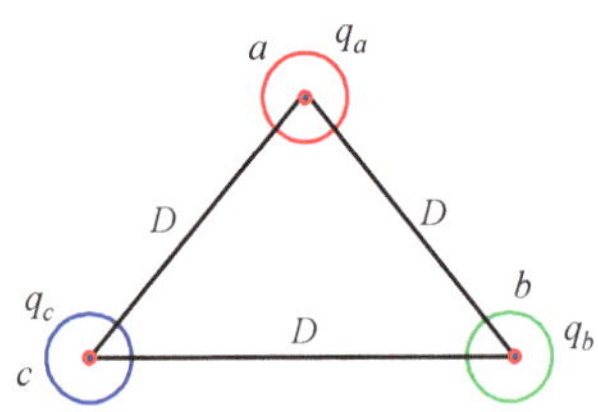

Fig. 4.25 Conductors with equal spacing

The voltage at the conductor a due to self- and other conductors b and c can be written as,

$$V_a = \frac{1}{2\pi\varepsilon}\left(q_a \ln\frac{1}{r} + q_b \ln\frac{1}{D} + q_c \ln\frac{1}{D}\right) \tag{4.231}$$

$$V_a = \frac{1}{2\pi\varepsilon}\left[q_a \ln\frac{1}{r} + (q_b + q_c) \ln\frac{1}{D}\right] \tag{4.232}$$

Substituting Eq. (4.230) into Eq. (4.232) yields,

$$V_a = \frac{1}{2\pi\varepsilon}\left[q_a \ln\frac{1}{r} - q_a \ln\frac{1}{D}\right] \tag{4.233}$$

$$V_a = V_{an} = \frac{q_a}{2\pi\varepsilon} \ln\frac{D}{r} \tag{4.234}$$

The capacitance from line to neutral can be expressed as,

$$C_{an} = \frac{q_a}{V_{an}} = \frac{2\pi\varepsilon}{\ln\frac{D}{r}} \tag{4.235}$$

$$C_{an} = \frac{2\pi \times 1 \times 8.854 \times 10^{-12}}{\ln\frac{D}{r}}\ \text{F/m} \tag{4.236}$$

$$C_{an} = \frac{0.0556}{\ln\frac{D}{r}}\ \mu\text{F/km} \tag{4.237}$$

From Eq. (4.237), it is concluded that the value of the capacitance can be calculated by substituting the separation distance and radius.

Example 4.6

A three-phase 33 kV, 50 Hz transmission line is having a length of 115 km. The transmission line consists of the three conductors with an equal radius of 1.5 cm, and the conductors are arranged as an equilateral triangle with a separation distance of 2 m. Calculate the capacitance per phase and charging current.

Solution

The separation distance between the two conductors is,

$$D = 2\,\text{m} = 200\,\text{cm} \tag{4.238}$$

The radius of the conductor is,

$$r = 1.5\,\text{cm} \tag{4.239}$$

The value of the capacitance per phase is calculated as,

$$C = \frac{0.0556}{\ln\frac{D}{r}} \times 115 = \frac{0.0556}{\ln\frac{200}{1.5}} \times 115 = 1.31\ \mu\text{F} \tag{4.240}$$

The value of the charging current can be determined as,

$$I_c = \frac{\frac{V_L}{\sqrt{3}}}{X_c} = \frac{33000}{\sqrt{3}} \times 2\pi \times 50 \times 1.31 \times 10^{-6} = 7.84\ \text{A} \tag{4.241}$$

Practice Problem 4.6

A 120 km long three-phase transmission line consists of the three conductors with an equal radius of 12 mm. The conductors are arranged as an equilateral triangle with a separation distance of 150 cm. Calculate the capacitance per phase and charging current if the line voltage and frequency are 33 kV and 50 Hz, respectively.

4.14.2 Capacitance of Three-Phase Lines with Unequal Spacing Conductors

It is very difficult to find the capacitance inductance when the conductors are placed in an unequal spacing. The conductors' positions at different steps are shown in Fig. 4.26. According to Eq. (4.209), at step 1, the voltage difference between the conductors a and b is written as,

$$V_{ab1} = \frac{1}{2\pi\varepsilon}\left(q_a \ln\frac{D_{12}}{D_{11} = r} + q_b \ln\frac{r = D_{22}}{D_{12}} + q_c \ln\frac{D_{23}}{D_{13}}\right) \tag{4.242}$$

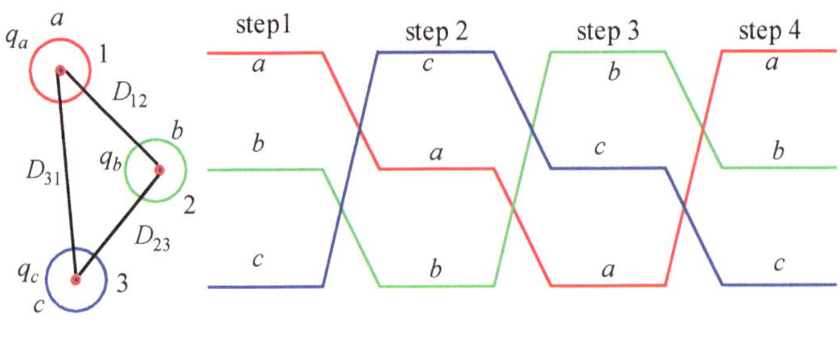

Fig. 4.26 Position of conductors at different steps

Similarly, at step 2, the voltage difference between the conductors a and b is written as,

$$V_{ab2} = \frac{1}{2\pi\varepsilon}\left(q_a \ln\frac{D_{23}}{r} + q_b \ln\frac{r}{D_{23}} + q_c \ln\frac{D_{13}}{D_{12}}\right) \tag{4.243}$$

Also, at step 3, the voltage difference between the conductors a and b is expressed as,

$$V_{ab3} = \frac{1}{2\pi\varepsilon}\left(q_a \ln\frac{D_{13}}{r} + q_b \ln\frac{r}{D_{13}} + q_c \ln\frac{D_{12}}{D_{23}}\right) \tag{4.244}$$

The average voltage between the conductors a and b is written as,

$$V_{ab} = \frac{V_{ab1} + V_{ab2} + V_{ab3}}{3} \tag{4.245}$$

Substituting Eqs. (4.242), (4.243), and (4.244) into Eq. (4.245) yields,

$$V_{ab} = \frac{1}{3 \times 2\pi\varepsilon}\left(q_a \ln\frac{D_{12}D_{23}D_{31}}{r} + q_b \ln\frac{r}{D_{12}D_{23}D_{31}} + q_c \ln\frac{D_{12}D_{23}D_{31}}{D_{12}D_{23}D_{31}}\right) \tag{4.246}$$

$$V_{ab} = \frac{1}{3 \times 2\pi\varepsilon}\left(q_a \ln\frac{D_{12}D_{23}D_{31}}{r} + q_b \ln\frac{r}{D_{12}D_{23}D_{31}}\right) \tag{4.247}$$

$$V_{ab} = \frac{1}{2\pi\varepsilon}\left(q_a \ln\frac{\sqrt[3]{D_{12}D_{23}D_{31}}}{r} + q_b \ln\frac{r}{\sqrt[3]{D_{12}D_{23}D_{31}}}\right) \tag{4.248}$$

Similarly, the average voltage between the conductors a and c is written as,

$$V_{ac} = \frac{1}{2\pi\varepsilon}\left(q_a \ln\frac{\sqrt[3]{D_{12}D_{23}D_{31}}}{r} + q_c \ln\frac{r}{\sqrt[3]{D_{12}D_{23}D_{31}}}\right) \tag{4.249}$$

Adding Eqs. (4.248) and (4.249) yields,

$$V_{ab} + V_{ac} = \frac{1}{2\pi\varepsilon}\left(2q_a \ln\frac{\sqrt[3]{D_{12}D_{23}D_{31}}}{r} + (q_b + q_c) \ln\frac{r}{\sqrt[3]{D_{12}D_{23}D_{31}}}\right) \tag{4.250}$$

Substituting Eq. (4.229) into Eq. (4.250) yields,

$$V_{ab} + V_{ac} = \frac{1}{2\pi\varepsilon}\left(2q_a \ln\frac{\sqrt[3]{D_{12}D_{23}D_{31}}}{r} - q_a \ln\frac{r}{\sqrt[3]{D_{12}D_{23}D_{31}}}\right) \tag{4.251}$$

$$V_{ab} + V_{ac} = \frac{1}{2\pi\varepsilon}\left(2q_a \ln\frac{\sqrt[3]{D_{12}D_{23}D_{31}}}{r} + q_a \ln\frac{\sqrt[3]{D_{12}D_{23}D_{31}}}{r}\right) \tag{4.252}$$

$$V_{ab} + V_{ac} = \frac{1}{2\pi\varepsilon} 3q_a \ln\frac{\sqrt[3]{D_{12}D_{23}D_{31}}}{r} \tag{4.253}$$

In a balanced three-phase system, the following relations can be written as,

$$V_{ab} = V_{an} \angle 0^\circ - V_{an} \angle -120^\circ \tag{4.254}$$

$$V_{ac} = V_{an} \angle 0^\circ - V_{an} \angle 120^\circ \tag{4.255}$$

Adding Eqs. (4.254) and (4.255) yields,

$$V_{ab} + V_{ac} = V_{an} \angle 0^\circ - V_{an} \angle -120^\circ + V_{an} \angle 0^\circ - V_{an} \angle 120^\circ \tag{4.256}$$

$$V_{ab} + V_{ac} = 3V_{an} \tag{4.257}$$

Substituting Eq. (4.253) into Eq. (4.257) yields,

$$3V_{an} = \frac{1}{2\pi\varepsilon} 3q_a \ln\frac{\sqrt[3]{D_{12}D_{23}D_{31}}}{r} \tag{4.258}$$

$$V_{an} = \frac{1}{2\pi\varepsilon} q_a \ln\frac{\sqrt[3]{D_{12}D_{23}D_{31}}}{r} \tag{4.259}$$

The capacitance from line to neutral is expressed as,

$$C_{an} = \frac{q_a}{V_{an}} = \frac{2\pi\varepsilon}{\ln\frac{\sqrt[3]{D_{12}D_{23}D_{31}}}{r}} \tag{4.260}$$

$$C_{an} = \frac{2\pi \times 1 \times 8.854 \times 10^{-12}}{\ln\frac{\sqrt[3]{D_{12}D_{23}D_{31}}}{r}}\ \text{F/m} \tag{4.261}$$

$$C_{an} = \frac{0.0556}{\ln\frac{GMD}{r}}\ \mu\text{F/km} \tag{4.262}$$

where,

$$GMD = \sqrt[3]{D_{12}D_{23}D_{31}} \tag{4.263}$$

Example 4.7
Figure 4.27 shows an arrangement of the three conductors of a 40 km long three-phase transmission line. The diameter of each conductor is 2.5 cm. Find the capacitance per phase.

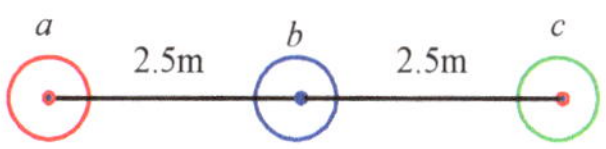

Fig. 4.27 Conductors arrangement for Example 4.7

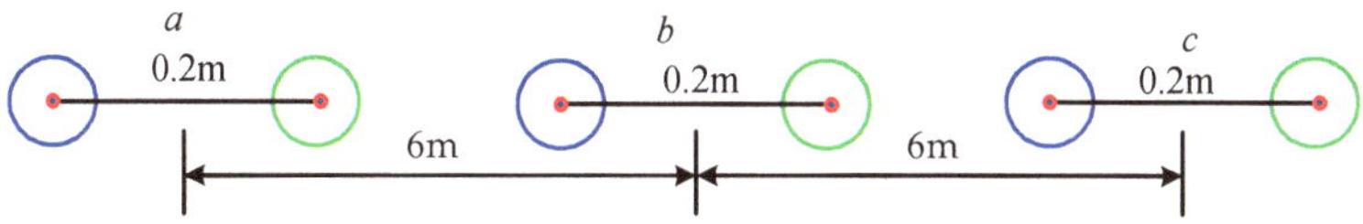

Fig. 4.28 Conductors arrangement for Practice Problem 4.7

Solution
The value of the *GMD* is calculated as,

$$GMD = \sqrt[3]{D_{12}D_{23}D_{31}} = \sqrt[3]{2.5 \times 2.5 \times 5} = 3.15\,\text{m} = 315\,\text{cm} \tag{4.264}$$

The radius of the conductor is calculated as,

$$r = \frac{2.5}{2} = 1.25\,\text{cm} \tag{4.265}$$

The per phase capacitance can be determined as,

$$C_{an} = \frac{0.0556}{\ln\frac{GMD}{r}}\ \mu\text{F/km} = \frac{0.0556}{\ln\frac{315}{1.25}} \times 40 = 0.402\ \mu\text{F} \tag{4.266}$$

Practice Problem 4.7
A 20 km long three-phase transposed line consists of two bundle conductors as shown in Fig. 4.28. The radius of each conductor in the bundle is 0.9 cm. Calculate the capacitance per phase and capacitive reactance.

4.15 Effect of Earth on Capacitance

The method in the image is used to calculate the effect of the earth on the capacitance of a conductor. The conductor with a positive charge placed at a certain height from earth induces the negative charge on the earth's surface.

The electric flux lines originate from the positive charge conductor and finish at the negative charge conductor in the earth. The earth is considered to be conducting and an equipotential plane to an infinite extent. The electric flux lines and an equipotential surface are perpendicular to each other. Since the earth is an equipotential plane, which is possible only if it is assumed the presence of an imaginary conductor below the surface of the earth at a depth equal to the height of

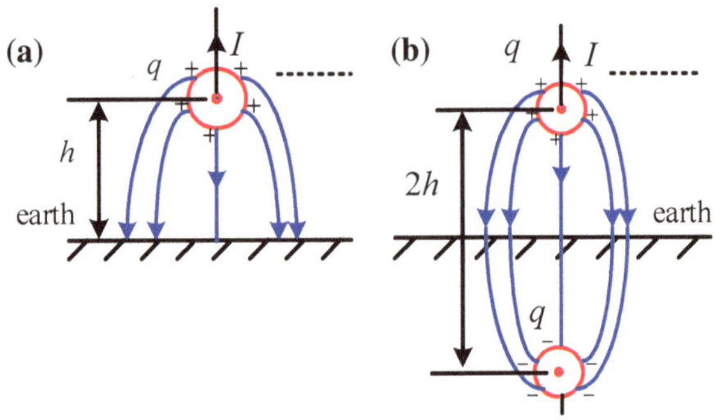

Fig. 4.29 Real and image conductors

the actual conductor above the surface on the earth. This imaginary conductor is called the image of the actual conductor. Figure 4.29a shows a conductor with a positive charge q C/m above a height h from the earth's surface. The positive charges pass to the earth's surface as indicated by the arrows. Figure 4.29b shows that the earth is replaced by a negative charge conductor, place below the original conductor.

4.16 Capacitance of Single Conductor to Earth

A conductor with a positive charge is placed at a height h m from the earth's surface. The image conductor is placed at the same height below the original conductor as shown in Fig. 4.30. According to Eq. (4.130), the capacitance for a single-phase conductor can be written as,

$$C_{1earth} = \frac{\pi \varepsilon_0}{\ln \frac{2h}{r}} \tag{4.267}$$

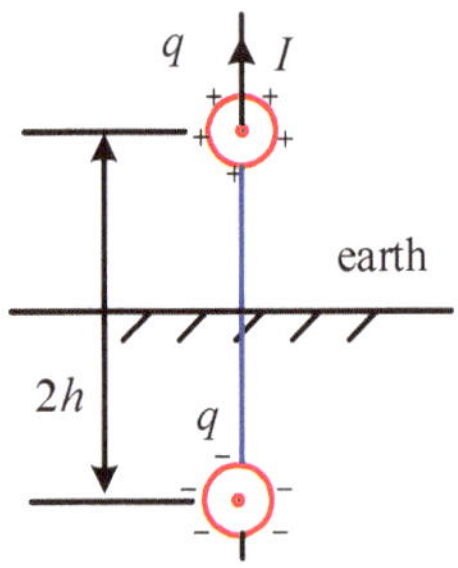

Fig. 4.30 Single conductor to earth

According to Eq. (4.260), the capacitance with reference to earth is expressed as,

$$C_{earth} = \frac{2\pi\varepsilon_0}{\ln\frac{2h}{r}} \tag{4.268}$$

4.17 Single-Phase Line Capacitance with Effect of Earth

The two conductors 1 and 2 of a single-phase run parallel by maintaining the same height from the earth. The conductors 3 and 4 are the image conductors of the conductors 1 and 2, respectively, as shown in Fig. 4.31. The radius of the conductors 1 and 2 is r. The conductor 1 contains the positive charge, and the conductor 2 contains the negative charge.

According to Eq. (4.209), the voltage difference between the two conductors ($k = 1$, $i = 2$) can be written as,

$$V_{12} = \frac{1}{2\pi\varepsilon}\sum_{m=1}^{4} q_m \ln\frac{D_{2m}}{D_{1m}} \tag{4.269}$$

Equation (4.269) can be expanded for four conductors as,

$$V_{12} = \frac{1}{2\pi\varepsilon}\left[q_1 \ln\frac{D_{21}}{D_{11}} + q_2 \ln\frac{D_{22}}{D_{12}} + q_3 \ln\frac{D_{23}}{D_{13}} + q_4 \ln\frac{D_{24}}{D_{14}}\right] \tag{4.270}$$

From Fig. 4.31, the following equations can be written as,

$$D_{11} = D_{22} = r \tag{4.271}$$

$$D_{12} = D_{21} = D \tag{4.272}$$

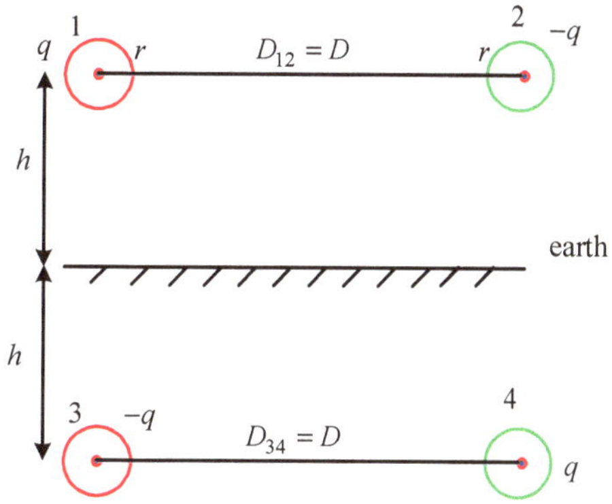

Fig. 4.31 Original and image conductors of a single-phase line

$$D_{13} = D_{24} = 2h \tag{4.273}$$

$$D_{14} = D_{23} = \sqrt{4h^2 + D^2} \tag{4.274}$$

$$\begin{aligned} q_1 = q, \quad q_2 = -q \\ q_3 = -q, \quad q_4 = q \end{aligned} \tag{4.275}$$

Substituting equations Eqs. (4.271) and (4.275) into Eq. (4.270) yields,

$$V_{12} = \frac{1}{2\pi\varepsilon}\left[q\ln\frac{D}{r} - q\ln\frac{r}{D} - q\ln\frac{\sqrt{4h^2+D^2}}{2h} + q\ln\frac{2h}{\sqrt{4h^2+D^2}}\right] \tag{4.276}$$

$$V_{12} = \frac{1}{2\pi\varepsilon}\left[q\ln\frac{D}{r} + q\ln\frac{D}{r} + q\ln\frac{2h}{\sqrt{4h^2+D^2}} + q\ln\frac{2h}{\sqrt{4h^2+D^2}}\right] \tag{4.277}$$

$$V_{12} = \frac{q}{\pi\varepsilon}\left[\ln\frac{2hD}{r\sqrt{4h^2+D^2}}\right] \tag{4.278}$$

$$C_{12} = \frac{q}{V_{12}} = \frac{\pi\varepsilon}{\ln\frac{2hD}{r\sqrt{4h^2+D^2}}} \tag{4.279}$$

$$C_{12} = \frac{\pi \times 8.854 \times 10^{-12}}{\ln\frac{D}{r\sqrt{1+\left(\frac{D}{2h}\right)^2}}}\ \text{F/m} \tag{4.280}$$

$$C_{12} = \frac{0.0278}{\ln\frac{D}{r\sqrt{1+\left(\frac{D}{2h}\right)^2}}}\quad \mu\text{F/km} \tag{4.281}$$

Consider Fig. 4.32 to derive generalized relationship between the capacitance and the earth. Comparing Figs. 4.30 and 4.31, the following relations can be written as,

$$D_{13} = H_{xy} \tag{4.282}$$

$$D_{14} = H_{xy} \tag{4.283}$$

$$D_{23} = H_{yx} \tag{4.284}$$

Substituting Eqs. (4.271), (4.272), (4.275), (4.276), (4.277), and (4.278) into Eq. (4.279) yields,

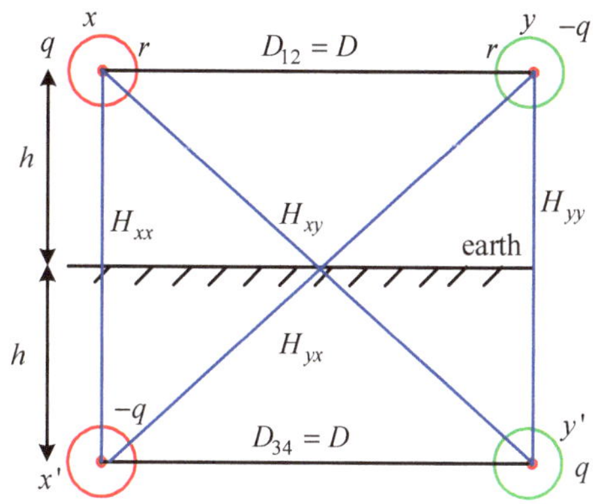

Fig. 4.32 Original and image conductors of a single-phase line for generalized cases

$$V_{12} = \frac{1}{2\pi\varepsilon}\left[q\ln\frac{D_{21} = D}{D_{11} = r} - q\ln\frac{D_{22} = r}{D_{12} = D} - q\ln\frac{H_{yx}}{H_{xx}} + q\ln\frac{H_{yy}}{H_{xy}}\right] \tag{4.285}$$

$$V_{12} = \frac{1}{2\pi\varepsilon}\left[2q\ln\frac{D}{r} - q\ln\frac{H_{yx}H_{xy}}{H_{xx}H_{yy}}\right] \tag{4.286}$$

The distances $H_{yx} = H_{xy}$ and $H_{yy} = H_{xx}$, and then Eq. (4.280) can be modified as,

$$V_{12} = \frac{1}{2\pi\varepsilon}\left[2q\ln\frac{D}{r} - q\ln\frac{H_{yx}H_{xy}}{H_{xx}H_{yy}}\right] \tag{4.287}$$

$$V_{12} = \frac{q}{\pi\varepsilon}\left[\ln\frac{D}{r} - \ln\frac{H_{xy}}{H_{xx}}\right] \tag{4.288}$$

The general expression of line-to-line capacitance with the effect of the earth is expressed as,

$$C_{12} = \frac{\pi\varepsilon}{\left[\ln\frac{D}{r} - \ln\frac{H_{xy}}{H_{xx}}\right]} \quad \text{F/m} \tag{4.289}$$

Then, the expression of neutral capacitance is,

$$C_n = 2C_{12} = \frac{2\pi\varepsilon}{\left[\ln\frac{D}{r} - \ln\frac{H_{xy}}{H_{xx}}\right]} \quad \text{F/m} \tag{4.290}$$

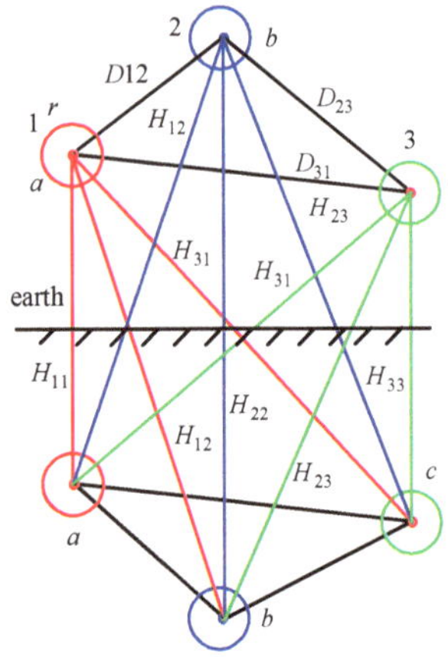

Fig. 4.33 Original and image conductors of three-phase lines

4.18 Three-Phase Line Capacitance with Effect of Earth

Consider the three conductors *a*, *b*, and *c* are placed above the ground at points 1, 2, and 3, respectively, as shown in Fig. 4.33. These conductors carry charges q_a, q_b, and q_c C/m, respectively. The image conductors are placed below the main conductors and carry negative charges. Conductors are transposed as shown in Fig. 4.34.

According to Eq. (4.279), the voltage difference between the conductors *a* and *b* when conductors *a*, *b*, and *c* are at positions 1, 2, and 3, respectively, as shown in Fig. 4.34 can be written as [3],

$$V_{ab1}=\frac{1}{2\pi\varepsilon}\left[q_a\left(\ln\frac{D_{21}}{D_{11}}-\ln\frac{H_{21}}{H_{11}}\right)+q_b\left(\ln\frac{D_{22}}{D_{12}}-\ln\frac{H_{22}}{H_{12}}\right)+q_c\left(\ln\frac{D_{23}}{D_{13}}-\ln\frac{H_{23}}{H_{13}}\right)\right] \tag{4.291}$$

$$V_{ab1}=\frac{1}{2\pi\varepsilon}\left[q_a\left(\ln\frac{D_{12}}{r}-\ln\frac{H_{12}}{H_{11}}\right)+q_b\left(\ln\frac{r}{D_{12}}-\ln\frac{H_{22}}{H_{12}}\right)+q_c\left(\ln\frac{D_{23}}{D_{31}}-\ln\frac{H_{23}}{H_{31}}\right)\right] \tag{4.292}$$

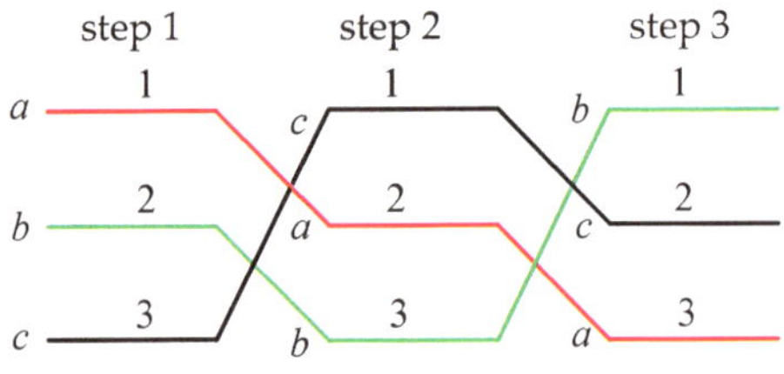

Fig. 4.34 Transposed conductors

Similarly, the voltage difference between the conductors a and b when conductors a, b, and c are at positions 2, 3, and 1, respectively, as shown in Fig. 4.34 can be written as,

$$V_{ab2} = \frac{1}{2\pi\varepsilon}\left[q_a\left(\ln\frac{D_{23}}{r} - \ln\frac{H_{23}}{H_{22}}\right) + q_b\left(\ln\frac{r}{D_{23}} - \ln\frac{H_{33}}{H_{23}}\right) + q_c\left(\ln\frac{D_{31}}{D_{12}} - \ln\frac{H_{31}}{H_{12}}\right)\right] \tag{4.293}$$

Again, the voltage difference between the conductors a and b when conductors a, b, and c are at positions 3, 1, and 2, respectively, as shown in Fig. 4.34 can be written as,

$$V_{ab3} = \frac{1}{2\pi\varepsilon}\left[q_a\left(\ln\frac{D_{31}}{r} - \ln\frac{H_{31}}{H_{33}}\right) + q_b\left(\ln\frac{r}{D_{31}} - \ln\frac{H_{11}}{H_{31}}\right) + q_c\left(\ln\frac{D_{12}}{D_{23}} - \ln\frac{H_{12}}{H_{23}}\right)\right] \tag{4.294}$$

The average value of the voltage can be written as,

$$V_{ab} = \frac{V_{ab1} + V_{ab2} + V_{ab3}}{3} \tag{4.295}$$

Substituting Eqs. (4.285) and (4.287) into Eq. (4.288) yields,

$$V_{ab} = \frac{1}{6\pi\varepsilon}\left[\begin{array}{l} q_a\left(\ln\frac{D_{12}}{r} + \ln\frac{D_{23}}{r} + \ln\frac{D_{31}}{r}\right) + q_b\left(\ln\frac{r}{D_{12}} + \ln\frac{r}{D_{23}} + \ln\frac{r}{D_{31}}\right) \\ -q_a\left(\ln\frac{H_{12}}{H_{11}} + \ln\frac{H_{23}}{H_{22}} + \ln\frac{H_{31}}{H_{33}}\right) - q_b\left(\ln\frac{H_{22}}{H_{12}} + \ln\frac{H_{33}}{H_{23}} + \ln\frac{H_{11}}{H_{31}}\right) \\ + q_c\ln\frac{D_{12}D_{23}D_{31}}{D_{12}D_{23}D_{31}} \end{array}\right] \tag{4.296}$$

$$V_{ab} = \frac{1}{6\pi\varepsilon}\left[q_a\ln\frac{D_{12}D_{23}D_{31}}{r^3} + q_b\ln\frac{r^3}{D_{12}D_{23}D_{31}} - q_a\ln\frac{H_{12}H_{23}H_{31}}{H_{11}H_{22}H_{33}} - q_b\ln\frac{H_{11}H_{22}H_{33}}{H_{12}H_{23}H_{31}}\right] \tag{4.297}$$

$$V_{ab} = \frac{1}{6\pi\varepsilon}\left[q_a\ln\frac{D_{12}D_{23}D_{31}}{r^3} + q_b\ln\frac{r^3}{D_{12}D_{23}D_{31}} - q_a\ln\frac{H_{12}H_{23}H_{31}}{H_{11}H_{22}H_{33}} - q_b\ln\frac{H_{11}H_{22}H_{33}}{H_{12}H_{23}H_{31}}\right] \tag{4.298}$$

$$V_{ab} = \frac{1}{2\pi\varepsilon}\left[\begin{array}{l} q_a\ln\left(\frac{D_{12}D_{23}D_{31}}{r^3}\right)^{\frac{1}{3}} + q_b\ln\left(\frac{r^3}{D_{12}D_{23}D_{31}}\right)^{\frac{1}{3}} - q_a\ln\left(\frac{H_{12}H_{23}H_{31}}{H_{11}H_{22}H_{33}}\right)^{\frac{1}{3}} \\ -q_b\ln\left(\frac{H_{11}H_{22}H_{33}}{H_{12}H_{23}H_{31}}\right)^{\frac{1}{3}} \end{array}\right] \tag{4.299}$$

Substituting $\sqrt[3]{D_{12}D_{23}D_{31}} = D_{eq}$ in Eq. (4.290) yields,

$$V_{ab} = \frac{1}{2\pi\varepsilon}\left[q_a \ln\frac{D_{eq}}{r} + q_b \ln\frac{r}{D_{eq}} - q_a \ln\frac{\sqrt[3]{H_{12}H_{23}H_{31}}}{\sqrt[3]{H_{11}H_{22}H_{33}}} - q_b \ln\frac{\sqrt[3]{H_{11}H_{22}H_{33}}}{\sqrt[3]{H_{12}H_{23}H_{31}}}\right] \tag{4.300}$$

Similarly, the voltage between a and c can be written as,

$$V_{ac} = \frac{1}{2\pi\varepsilon}\left[q_a \ln\frac{D_{eq}}{r} + q_c \ln\frac{r}{D_{eq}} - q_a \ln\frac{\sqrt[3]{H_{12}H_{23}H_{31}}}{\sqrt[3]{H_{11}H_{22}H_{33}}} - q_c \ln\frac{\sqrt[3]{H_{11}H_{22}H_{33}}}{\sqrt[3]{H_{12}H_{23}H_{31}}}\right] \tag{4.301}$$

Substituting Eqs. (4.290) and (4.291) into Eq. (4.257) yields,

$$3V_{an} = \frac{1}{2\pi\varepsilon}\cdot\left[\begin{matrix} q_a \ln\frac{D_{eq}}{r} + q_b \ln\frac{r}{D_{eq}} - q_a \ln\frac{\sqrt[3]{H_{12}H_{23}H_{31}}}{\sqrt[3]{H_{11}H_{22}H_{33}}} - q_b \ln\frac{\sqrt[3]{H_{11}H_{22}H_{33}}}{\sqrt[3]{H_{12}H_{23}H_{31}}} \\ q_a \ln\frac{D_{eq}}{r} + q_c \ln\frac{r}{D_{eq}} - q_a \ln\frac{\sqrt[3]{H_{12}H_{23}H_{31}}}{\sqrt[3]{H_{11}H_{22}H_{33}}} - q_c \ln\frac{\sqrt[3]{H_{11}H_{22}H_{33}}}{\sqrt[3]{H_{12}H_{23}H_{31}}} \end{matrix}\right] \tag{4.302}$$

$$3V_{an} = \frac{1}{2\pi\varepsilon}\left[2q_a \ln\frac{D_{eq}}{r} + (q_b + q_c) \ln\frac{r}{D_{eq}} - 2q_a \ln\frac{\sqrt[3]{H_{12}H_{23}H_{31}}}{\sqrt[3]{H_{11}H_{22}H_{33}}} - (q_b + q_c) \ln\frac{\sqrt[3]{H_{11}H_{22}H_{33}}}{\sqrt[3]{H_{12}H_{23}H_{31}}}\right] \tag{4.303}$$

Substituting Eq. (4.230) into Eq. (4.293) yields,

$$3V_{an} = \frac{1}{2\pi\varepsilon}\left[2q_a \ln\frac{D_{eq}}{r} - q_a \ln\frac{r}{D_{eq}} - 2q_a \ln\frac{\sqrt[3]{H_{12}H_{23}H_{31}}}{\sqrt[3]{H_{11}H_{22}H_{33}}} + q_a \ln\frac{\sqrt[3]{H_{11}H_{22}H_{33}}}{\sqrt[3]{H_{12}H_{23}H_{31}}}\right] \tag{4.304}$$

$$3V_{an} = \frac{1}{2\pi\varepsilon}\left[3q_a \ln\frac{D_{eq}}{r} - 3q_a \ln\frac{\sqrt[3]{H_{12}H_{23}H_{31}}}{\sqrt[3]{H_{11}H_{22}H_{33}}}\right] \tag{4.305}$$

$$V_{an} = \frac{1}{2\pi\varepsilon}\left[q_a \ln\frac{D_{eq}}{r} - q_a \ln\frac{\sqrt[3]{H_{12}H_{23}H_{31}}}{\sqrt[3]{H_{11}H_{22}H_{33}}}\right] \tag{4.306}$$

The expression of the capacitance can be written as,

$$C_n = \frac{2\pi\varepsilon}{q_a \ln\frac{D_{eq}}{r} - q_a \ln\frac{\sqrt[3]{H_{12}H_{23}H_{31}}}{\sqrt[3]{H_{11}H_{22}H_{33}}}} \quad \text{F/m} \tag{4.307}$$

Example 4.8

The two conductors of single-phase transmission lines are spaced 3 m apart and 7 m above the ground. The radius of the conductor is 0.02 m. Find the capacitance to neutral per km with and without the effect of the earth.

Solution

The value of the capacitance with the effect of the earth is calculated as,

$$C_{12} = \frac{0.0278}{\ln \frac{3}{(0.02)\sqrt{1+\left(\frac{3}{14}\right)^2}}} = 5.58 \times 10^{-3}\ \mu\text{F/km} \tag{4.308}$$

The value of the capacitance to neutral with the effect of the earth is determined as,

$$C_{1n} = C_{2n} = 2C_{12} = 2 \times 5.58 \times 10^{-3} = 0.011\ \mu\text{F/km} \tag{4.309}$$

According to Eq. (4.220), the value of the capacitance to neutral without the effect of earth can be calculated as,

$$C_{1n} = C_{2n} = 2C_{12} = \frac{2\pi\varepsilon_0}{\ln \frac{D}{r}} = \frac{0.0556}{\ln \frac{3}{0.02}} = 0.011\ \mu\text{F/km} \tag{4.310}$$

Practice Problem 4.8

The two conductors of single-phase transmission lines are spaced 4 m apart and 8 m above the ground. Determine the radius of the conductor if the value of the capacitance with the earth is found to be 0.02 μF/km.

4.19 Effect of Bundling in Capacitance

The effect of bundling is also remarkable in the calculation of capacitance in the transmission lines. Let d be the separation distance between two bundle conductors and the expressions of geometric mean radius (R^b) for different types of bundle conductors are also summarized here based on bundling in inductance.

For two-conductor bundle, the geometric mean radius is,

$$R^{b2} = \sqrt{r \times d} \tag{4.311}$$

For three-conductor bundle, the geometric mean radius is expressed as,

$$R^{b3} = \sqrt[3]{r \times d^2} \tag{4.312}$$

For four-conductor bundle, the geometric mean radius is written as,

$$R^{b4} = 1.09\sqrt[4]{r \times d^3} \tag{4.313}$$

References

1. Wildi T (2005) Electrical machines, drives and power systems, 6th edn. Pearson Education, USA, pp 1–934
2. Burgen AR, Vittal V (1999) Power systems analysis, 2nd edn. Pearson Education, USA, pp 1–632
3. Stevenson WD (1982) Elements of power system analysis, 4th edn. McGraw Hill Higher Education, USA, pp 1–436

Exercise Problems

4.1 The temperature coefficient of the copper of the transmission line is $0.0043/^{\circ}\text{C}$ at 0 °C and The resistance of a transmission line at 10 °C is 15 Ω. Calculate the line impedance at 25 °C.

4.2 The temperature coefficient of a copper of a distribution transmission line is $0.004/^{\circ}\text{C}$ at 0 °C. Calculate the distribution line impedance at 20 °C if the resistance of a distribution line at 0 °C is 15 Ω.

4.3 A system draws 80 A current through a 20 km long transmission line and the loss of the line is found to be 40 kW. Calculate the radius of the conductor if the resistivity of the conductor is given by $1.55 \times 10^{-8}\ \Omega^{-m}$.

4.4 A single-phase transmission line is 20 km long and consists of the two-round conductors. The diameter of each conductor is found to be 0.6 cm and is separated from each other by 50 cm. Calculate the value of the inductance.

4.5 Figure P4.1 shows a stranded conductor consists of the three identical conductors whose radius is given as *r*. Calculate the GMR.

4.6 A stranded conductor which consists of two identical conductors as shown in Fig. P4.2. Calculate the GMR if the radius of each conductor is *r*.

4.7 The four identical conductors are used to build a stranded conductor as shown in Fig. P4.3 and the radius of each conductor is *r*. Determine the GMR.

Fig. P4.1 Conductors for Exercise Problem 4.5

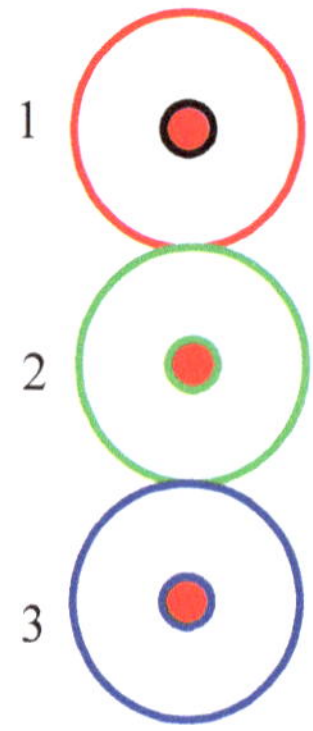

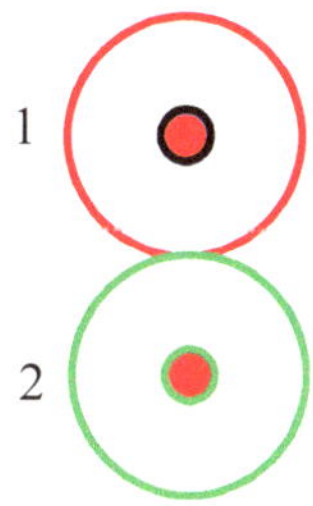

Fig. P4.2 Conductors for Exercise Problem 4.6

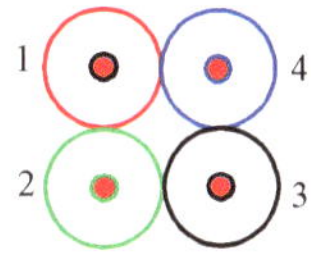

Fig. P4.3 Conductors for Exercise Problem 4.7

4.8 Figure P4.4 shows a single-phase line which consists of go and return conductors. The radius of each go conductor is 0.2 cm, and each return conductor is 0.3 cm. Calculate the GMD, GMR for each circuit and the total inductance per km.

4.9 Figure P4.5 shows the conductors' arrangement of a single-phase double circuit transmission line. The radius of each conductor is 1.5 cm. Calculate GMD and the total inductance per km.

4.10 Figure P4.6 shows the conductors' arrangement of a single-phase transmission line. The parallel conductors 1 and 2 forms go the path, and the conductors 1′ and 2′ forms return path. The radius of each conductor is 1.8 cm. Calculate the GMD, GMR and the total inductance per km.

4.11 Figure P4.7 shows the arrangement of a three-phase transmission line where the conductors are placed horizontally. The radius of each conductor is 0.8 cm. Determine the inductance per km.

4.12 Figure P4.8 represents three flat conductors of a three-phase 50 Hz transposed transmission. The line inductance is 12×10^{-7} H/m and the radius of each conductor is 2 cm. Calculate the separation distance *d*.

4.13 A double circuit three-phase transmission line is shown in Fig. P4.9. The diameter of each conductor is 2 cm. Find the inductance per km.

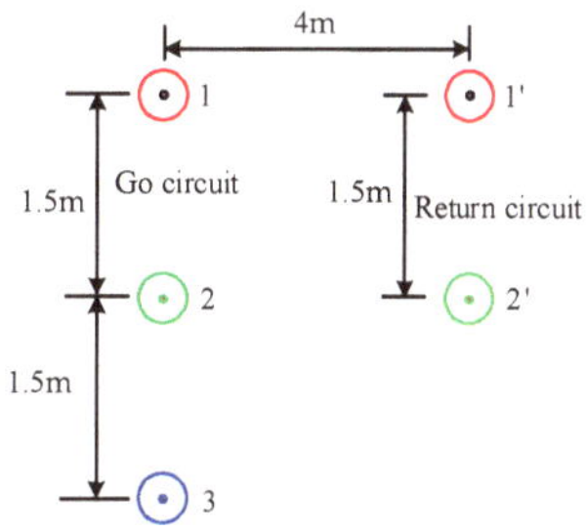

Fig. P4.4 Conductors for Exercise Problem 4.8

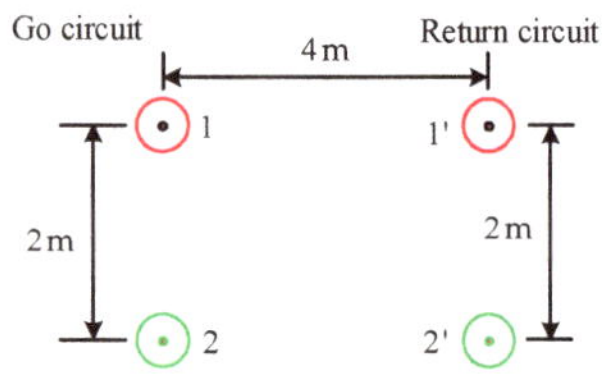

Fig. P4.5 Conductors for Exercise Problem 4.9

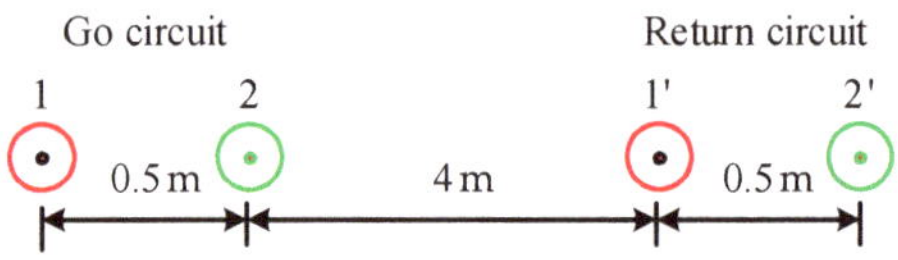

Fig. P4.6 Conductors for Exercise Problem 4.10

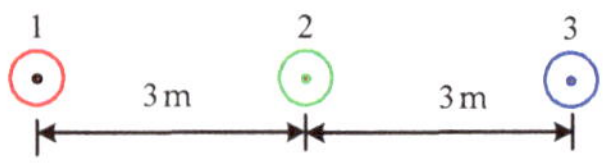

Fig. P4.7 Conductors for Exercise Problem 4.11

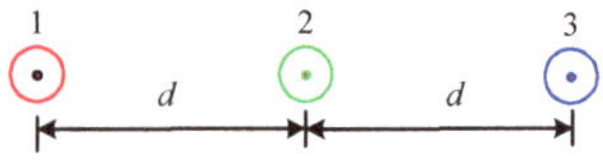

Fig. P4.8 Conductors for Exercise Problem 4.12

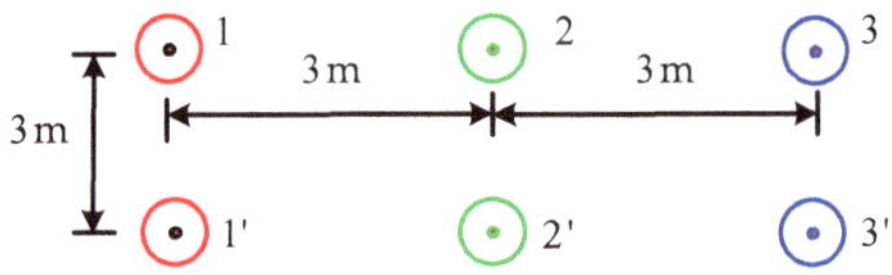

Fig. P4.9 Conductors for Exercise Problem 4.13

4.14 The conductors of a double circuit three-phase transmission line are shown in Fig. P4.10. Calculate the inductance per km if the radius of each conductor is 2 cm.

4.15 Figure P4.11 shows a three-phase line with two conductors per bundle. The radius of each conductor is 1.5 cm. Calculate the inductance in H/km and the inductance in H if the line is 240 km long.

4.16 Figure P4.12 a three-phase transmission line that consists of three conductors per bundle per phase. The radius of the conductor is 1.5 cm, and the phase spacing between the conductors of a bundle is 0.5 m. Find the line inductance in H/km.

4.17 A three-phase transmission line with a double circuit is shown in Fig. P4.13. The radius of each conductor is 2.5 cm. Calculate the inductance per km.

4.18 Figure P4.14 shows a 120 km long three-phase transmission line with four conductors per bundle. The radius of each conductor is 1.26 cm. Calculate the inductance in H and the inductive reactance per phase (Fig. P4.13).

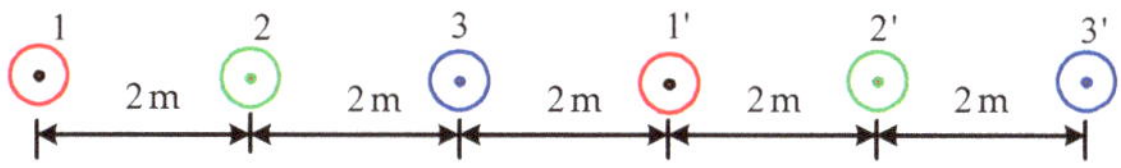

Fig. P4.10 Conductors for Exercise Problem 4.14

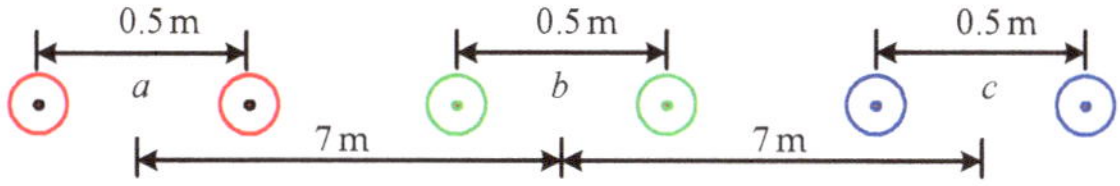

Fig. P4.11 Conductors for Exercise Problem 4.15

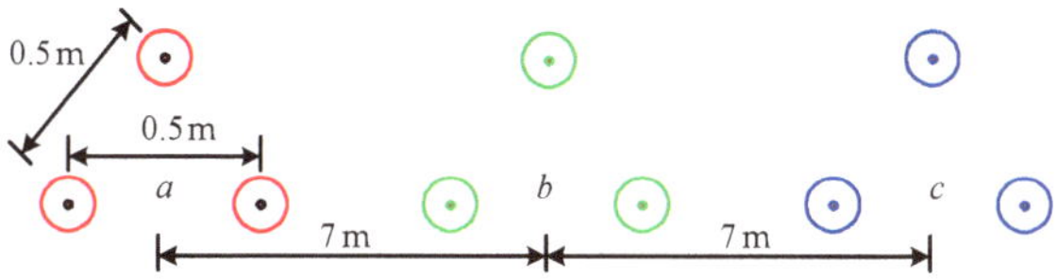

Fig. P4.12 Conductors for Exercise Problem 4.16

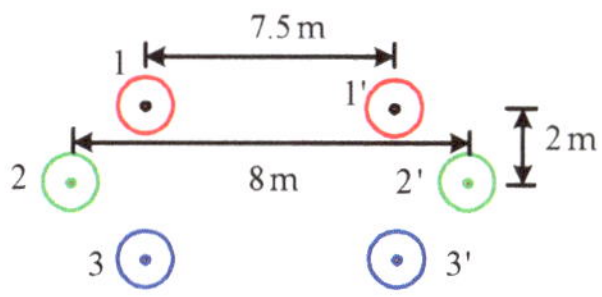

Fig. P4.13 Conductors for Exercise Problem 4.17

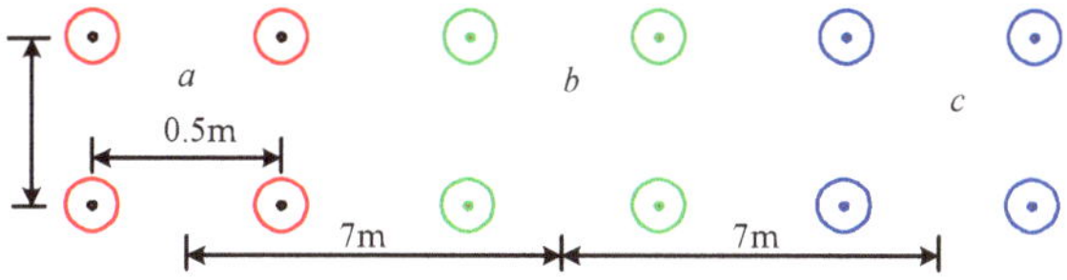

Fig. P4.14 Conductors for Exercise Problem 4.18

4.19 A 150 km long single-phase transmission line is having a capacitance of 0.86 μF and the separation distance between the conductor is given as 1.5 m. Calculate the radius of the conductor.

4.20 The radius of a 120 km long single-phase transmission line conductor is given as 2 cm, and the capacitance of the line is found to be 0.84 μF. Find the separation distance between the conductors.

4.21 Figure P4.15 shows the conductors' arrangement of a 120 kV and 130 km three-phase transmission lines. The conductors are transposed, and the radius of the conductor is 1.3 cm. Calculate the capacitance in μF and per phase charging current.

4.22 A 220 km three-phase line with two conductors per bundle is shown in Fig. P4.16. The diameter of each conductor is 4 cm, and the phase spacing between the bundles is 0.5 m. Find the capacitance in μF per km and per phase current if the line voltage is 132 kV.

4.23 The three conductors per bundle of a 120 km long balanced three-phase line is shown in Fig. P4.17. The horizontal phase spacing between the bundles is

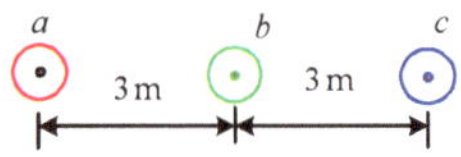

Fig. P4.15 Conductors for Exercise Problem 4.21

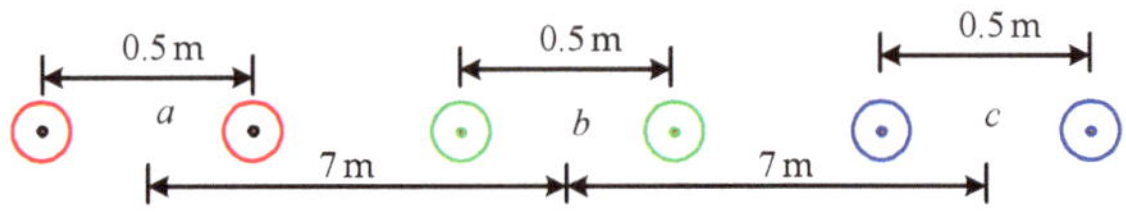

Fig. P4.16 Conductors for Exercise problem 4.22

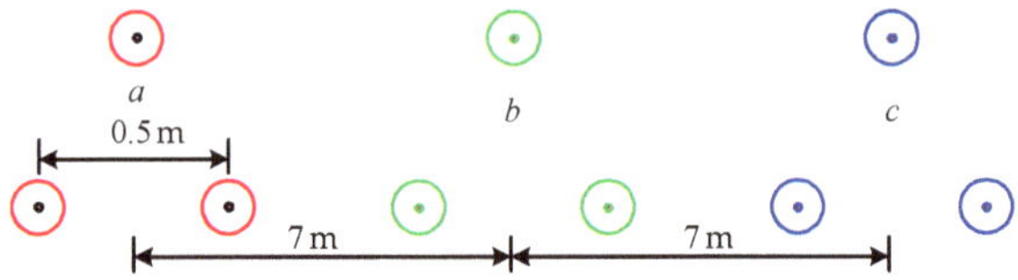

Fig. P4.17 Conductors for exercise Problem 4.23

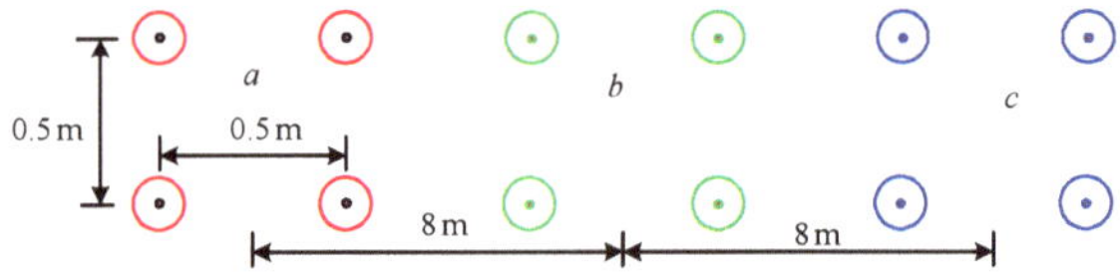

Fig. P4.270 Conductors for exercise Problem 4.24

7 m, and the spacing between the conductors of a bundle is 0.5 m. Assume the line is uniformly transposed, and the radius of the conductor is 1.4 cm. Calculate the per phase capacitance, capacitive reactance and per phase line current if the line voltage is 132 kV with a frequency of 50 Hz.

4.24 A 150 km long three-phase transmission line with four conductors per bundle is shown in Fig. P4.18. The horizontal phase spacing between the bundles is 8 m, and the spacing between the conductors of a bundle is 0.5 m. Assume the line is uniformly transposed, and the radius of the conductor is 0.6 cm. Find the per phase capacitance and the capacitive reactance if the frequency is 60 Hz.

Chapter 5
Modeling and Performance of Transmission Lines

5.1 Introduction

Electric power is generated at the power station and stepped up to a higher voltage known as national transmission voltage at the substation. This higher voltage is transferred from the substation to the grid substation by transmission lines. This voltage is again stepped down to a lower voltage (11 kV) and brought to the consumer terminals by a distribution line. The end of the distribution line where the load is connected is called the receiving end, whereas the beginning of the distribution line, where the source voltage is connected, is called the sending end. The line voltage drops, and losses, efficiency, and voltage regulation identify the performance of the transmission and distribution lines. These parameters are greatly influenced by the resistance, inductance, and capacitance. In this chapter, different types of transmission lines, voltage regulation, transmission efficiency, surge impedance loading, *ABCD* parameters, Ferranti effect, ground wires, corona discharge, and traveling waves will be discussed.

5.2 Classification of Transmission Lines

In the transmission lines, resistance, inductance, and capacitance are uniformly distributed over its whole length. The resistance and the inductance form the series impedance of the line, whereas the capacitance forms the shunt path through the length of the line. Therefore, the effect of capacitance in the transmission line is more complex for calculation and simplification. Depending on the effect of capacitance, the transmission lines are classified as short, medium, and long.

Short transmission line: The length of the line is up to 80 km is known as a short transmission line. In this line, the voltage is less than 20 kV. Therefore, the effect of capacitance is smaller due to its short length, and it is neglected.

Md. A. Salam, *Fundamentals of Electrical Power Systems Analysis*,
https://doi.org/10.1007/978-981-15-3212-2_5

Medium transmission line: The length of the line is more than 80 km but is less than 160 km is known as a medium transmission line. The voltage of this line is more than 20 kV but less than 100 kV. In the medium transmission line, the effect of capacitance is considered.

Long transmission line: The length of the transmission line is more than 160 km, and the voltage rating is more than 100 kV is known as a long transmission line. All the line parameters are considered uniformly distributed throughout the length of the line.

5.3 Efficiency and Voltage Regulation

The efficiency and voltage regulation is used to study the performance of a transmission line. The power received at the receiving end is always less than the sending end power. It is due to the losses of the transmission line. The ratio of receiving end power to the sending end power is known as the efficiency of the transmission line. Mathematically, it can be expressed as,

$$\eta = \frac{\text{Power delivered at the receiving end}}{\text{Power sent from the sending end}} \times 100 \tag{5.1}$$

$$\eta = \frac{P_r}{P_s} \times 100 = \frac{V_r I_r \cos\phi_r}{V_s I_s \cos\phi_s} \times 100 \tag{5.2}$$

where

$P_r = V_r I_r \cos\phi_r$ is the receiving end power,
$P_s = V_s I_s \cos\phi_s$ is the sending end power,
$V_r, I_r, \cos\phi_r$ are the receiving end parameters,
$V_s, I_s, \cos\phi_s$ are the sending end parameters.

The receiving end voltage is normally less than the sending end voltage. It is due to the resistive and inductive voltage drops of the line. The change in receiving end voltage from no-load to full load is known as the voltage regulation of the line. It is expressed as the percent of the full load voltage. Mathematically, the voltage regulation in percent can be expressed as,

$$VR = \frac{V_{rn} - V_{rf}}{V_{rf}} \times 100 \tag{5.3}$$

where

V_{rn} is the receiving end voltage under no-load condition,
V_{rf} is the receiving end voltage under full load condition.

5.4 Analysis of Short Transmission Line

In short transmission lines, the effect of capacitance is neglected, whereas the effects of resistance and inductance are considered for analysis. The equivalent circuit of a single-phase short transmission line is shown in Fig. 5.1. The impedance of the line can be determined as,

$$Z = R + jX_L \tag{5.4}$$

The expression of voltage per phase at the sending end is,

$$V_s = V_r + I_r Z \tag{5.5}$$

The sending end current is the same as the receiving end current, and it is written as,

$$I_s = I_r \tag{5.6}$$

The equivalent circuit of a short transmission line is compared with the four terminals' two-port network as shown in Fig. 5.2. Equations (5.5) and (5.6) can be represented by the generalized circuit constants that are commonly known as *ABCD* constants. The followings equations relate the voltage and current in the sending end and receiving end as,

$$V_s = AV_r + BI_r \tag{5.7}$$

$$I_s = CV_r + DI_r \tag{5.8}$$

Equations (5.7) and (5.8) can be expressed in the matrix form as,

$$\begin{bmatrix} V_s \\ I_s \end{bmatrix} = \begin{bmatrix} A & B \\ C & D \end{bmatrix} \begin{bmatrix} V_r \\ I_r \end{bmatrix} \tag{5.9}$$

The *ABCD* parameters can be determined by considering open circuit and short circuit at the receiving end. For an open circuit ($I_r = 0$), Eqs. (5.7) and (5.8) can be written as,

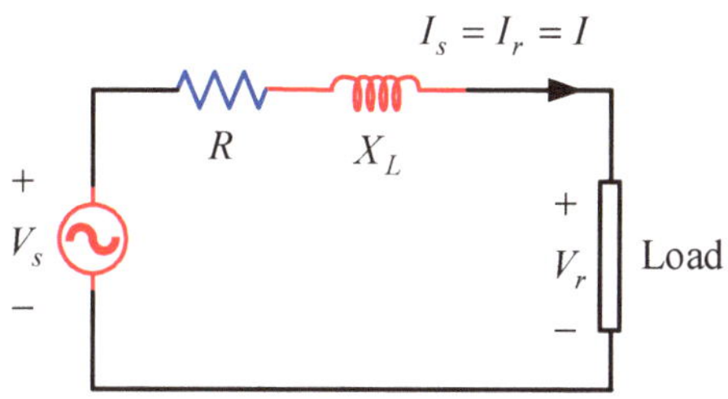

Fig. 5.1 Equivalent circuit of single-phase short transmission line

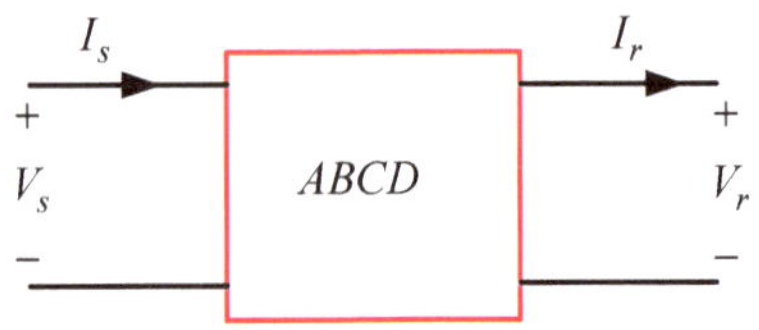

Fig. 5.2 Two-port network

$$V_s = AV_r \tag{5.10}$$

$$A = \left.\frac{V_s}{V_r}\right|_{I_r=0} \tag{5.11}$$

$$I_s = CV_r \tag{5.12}$$

$$C = \left.\frac{I_s}{V_r}\right|_{I_r=0} \tag{5.13}$$

From Eqs. (5.12) and (5.13), it is concluded that the A is a dimension less and C has a dimension mho (℧).

Again, for a short circuit ($V_r = 0$), Eqs. (5.7) and (5.8) can be modified as,

$$V_s = BI_r \tag{5.14}$$

$$B = \left.\frac{V_s}{I_r}\right|_{V_r=0} \tag{5.15}$$

$$I_s = DI_r \tag{5.16}$$

$$D = \left.\frac{I_s}{I_r}\right|_{V_r=0} \tag{5.17}$$

From Eqs. (5.15) and (5.17), it is concluded that the B has a dimension Ohm (Ω) and D is a dimension less.

Now, comparing Eq. (5.7) with Eq. (5.5) yields,

$$A = 1 \tag{5.18}$$

$$B = Z \tag{5.19}$$

Again, comparing Eq. (5.6) with (5.8) yields,

$$C = 0 \tag{5.20}$$

$$D = 1 \tag{5.21}$$

Now, applying the following,

$$AD - BC = 1 \times 1 - Z \times 0 = 1 \tag{5.22}$$

The phasor diagram of the short transmission line with a lagging power factor is shown in Fig. 5.3, where the line current is considered as a reference vector.

Here, δ is considered as the load angle, i.e., the angle between the sending end and the receiving end voltages. From Fig. 5.3, the following relations can be written as,

$$dc = bf = V_r \sin\phi_r \tag{5.23}$$

$$ab = V_r \cos\phi_r \tag{5.24}$$

$$bc = fd = IR \tag{5.25}$$

$$de = IX_L \tag{5.26}$$

$$ae = V_s \tag{5.27}$$

From the right-angle triangle Δace in Fig. 5.3, the following relation can be written as,

$$ae^2 = ac^2 + ec^2 \tag{5.28}$$

$$ae^2 = (ab + bc)^2 + (cd + de)^2 \tag{5.29}$$

Substituting equations from (5.23) to (5.27) into Eq. (5.29) yields,

$$V_s^2 = (V_r \cos\phi_r + IR)^2 + (V_r \sin\phi_r + IX_L)^2 \tag{5.30}$$

$$V_s = \sqrt{(V_r \cos\phi_r + IR)^2 + (V_r \sin\phi_r + IX_L)^2} \tag{5.31}$$

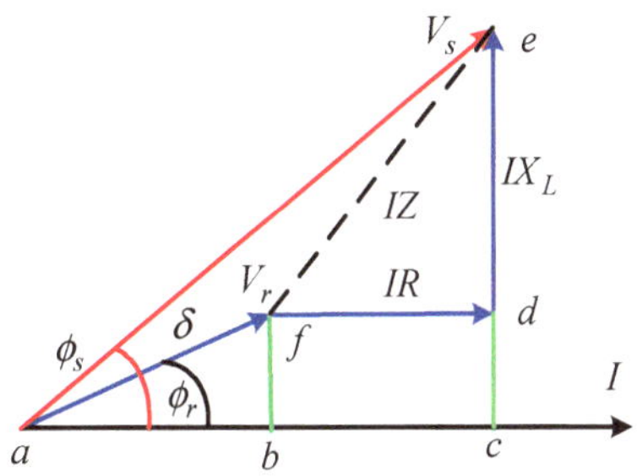

Fig. 5.3 Phasor diagram for lagging power factor

$$V_s = \sqrt{V_r^2(\cos^2\phi_r + \sin^2\phi_r) + 2V_rIR\cos\phi_r + 2V_rIX_L\sin\phi_r + I^2(R^2 + X_L^2)} \tag{5.32}$$

$$V_s = V_r\sqrt{1 + \frac{2IR}{V_r}\cos\phi_r + \frac{2IX_L}{V_r}\sin\phi_r + \frac{I^2}{V_r^2}(R^2 + X_L^2)} \tag{5.33}$$

The ratio of current to voltage is low in the last term, and its square is also very low. Therefore, the last term in Eq. (5.33) can be neglected. Equation (5.33) is modified as,

$$V_s = V_r\sqrt{1 + \frac{2IR}{V_r}\cos\phi_r + \frac{2IX_L}{V_r}\sin\phi_r} \tag{5.34}$$

Expanding Eq. (5.34) using binomial expansion and neglecting the higher terms and it can be expressed as,

$$V_s = V_r\left[1 + \frac{1}{2}\left(\frac{2IR}{V_r}\cos\phi_r + \frac{2IX_L}{V_r}\sin\phi_r\right)\right] \tag{5.35}$$

$$V_s = V_r + IR\cos\phi_r + IX_L\sin\phi_r \tag{5.36}$$

According to Eq. (5.3), the percentage of voltage regulation in terms of receiving end voltage is,

$$\%VR = \frac{V_s - V_r}{V_r} \times 100 \tag{5.37}$$

Substituting Eq. (5.36) into Eq. (5.37) yields,

$$\%VR = \frac{IR\cos\phi_r + IX_L\sin\phi_r}{V_r} \times 100 \tag{5.38}$$

Again, considering leading power factor for analyzing voltage regulation. Figure 5.4 shows a phasor diagram where the line current leads the receiving end voltage. From Fig. 5.4, the following relations can be written as,

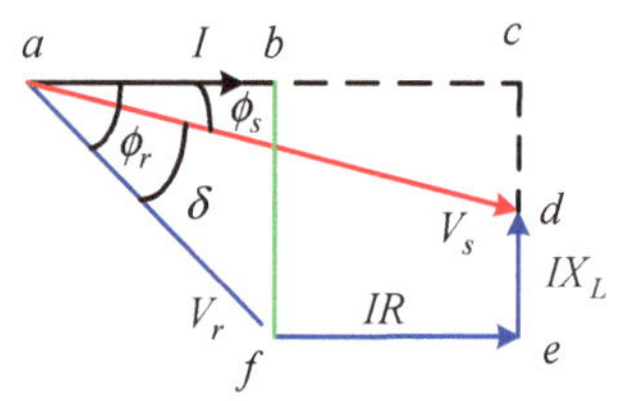

Fig. 5.4 Phasor diagram for leading power factor

$$ec = bf = V_r \sin \phi_r \tag{5.39}$$

$$ab = V_r \cos \phi_r \tag{5.40}$$

$$bc = fe = IR \tag{5.41}$$

$$de = IX_L \tag{5.42}$$

$$ad = V_s \tag{5.43}$$

From Fig. 5.4, the following relation can be written as,

$$ad^2 = ac^2 + dc^2 \tag{5.44}$$

$$ad^2 = (ab + bc)^2 + (ec - de)^2 \tag{5.45}$$

Substituting equations from (5.39) to (5.43) into Eq. (5.45) yields,

$$V_s^2 = (V_r \cos \phi_r + IR)^2 + (V_r \sin \phi_r - IX_L)^2 \tag{5.46}$$

$$V_s = \sqrt{(V_r \cos \phi_r + IR)^2 + (V_r \sin \phi_r - IX_L)^2} \tag{5.47}$$

Again, from Fig. 5.3, the load angle can be written as,

$$\tan(\phi_r + \delta) = \frac{ec}{ac} = \left(\frac{V_r \sin \phi_r + IX_L}{V_r \cos \phi_r + IR} \right) \tag{5.48}$$

$$\phi_r + \delta = \tan^{-1} \left(\frac{V_r \sin \phi_r + IX_L}{V_r \cos \phi_r + IR} \right) \tag{5.49}$$

However, the sending end angle from Fig. 5.3 can be written as,

$$\phi_s = \phi_r + \delta \tag{5.50}$$

$$\delta = \phi_s - \phi_r \tag{5.51}$$

Again, considering receiving end voltage as a reference vector where the current lags the voltage as shown in Fig. 5.5. The sending end voltage from Fig. 5.5 can be written as,

$$V_s = V_r + I \underline{\left|-\phi_r\right.} \times Z \tag{5.52}$$

$$V_s = V_r + I(\cos \phi_r - j \sin \phi_r)(R + jX_L) \tag{5.53}$$

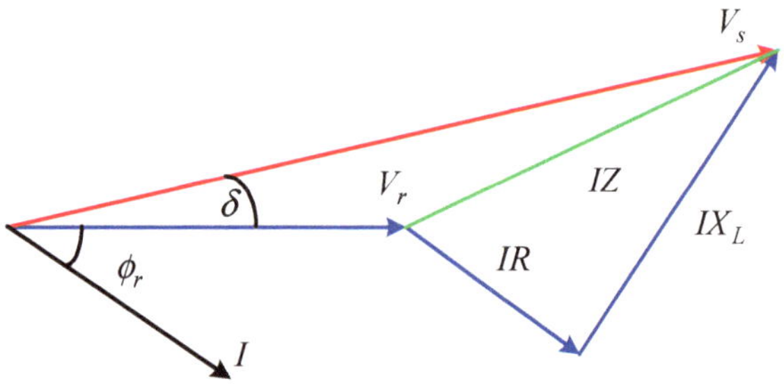

Fig. 5.5 Phasor diagram with V_r reference

$$V_s = V_r + IR\cos\phi_r + IX_L\sin\phi_r + j(IX_L\cos\phi_r - IR\sin\phi_r) \tag{5.54}$$

$$V_s + j0 = V_r + IR\cos\phi_r + IX_L\sin\phi_r + j(IX_L\cos\phi_r - IR\sin\phi_r) \tag{5.55}$$

Equating the real part of Eq. (5.55) yields,

$$V_s = V_r + IR\cos\phi_r + IX_L\sin\phi_r \tag{5.56}$$

From Fig. 5.5, the sending end angle can be written as,

$$\cos\phi_s = \cos(\phi_r + \delta) = \frac{V_r\cos\phi_r + IR}{V_s} \tag{5.57}$$

$$\phi_s = (\phi_r + \delta) = \cos^{-1}\left(\frac{V_r\cos\phi_r + IR}{V_s}\right) \tag{5.58}$$

Again, substituting Eq. (5.56) into Eq. (5.53) to express the voltage regulation for a lagging power factor as,

$$\%VR = \frac{IR\cos\phi_r + IX_L\sin\phi_r}{V_r} \times 100 \tag{5.59}$$

Similarly, the voltage regulation for a leading power factor can be expressed as,

$$VR = \frac{IR\cos\phi_r - IX_L\sin\phi_r}{V_r} \times 100 \tag{5.60}$$

Example 5.1
A single-phase short transmission line delivers power to a load of 12 kW and 0.8 lagging power factor. The terminal voltage across the load is 440 V, and the total transmission line resistance and reactance are 5Ω and 10Ω, respectively. Calculate the sending end voltage, sending end power factor, and transmission efficiency.

Solution
The value of the impedance is,

$$Z = 5 + j10 = 11.2\ \underline{|63.4^\circ}\ \Omega \tag{5.61}$$

The value of the line current is calculated as,

$$I = \frac{12000}{440 \times 0.8}\ \underline{|-36.86^\circ} = 34.1\ \underline{|-36.86^\circ}\ \mathrm{A} \tag{5.62}$$

The value of the sending end voltage is determined as,

$$V_s = V_r + IZ = 440 + 34.1\ \underline{|-36.86^\circ} \times 11.2\ \underline{|63.4^\circ} = 800\ \underline{|12.31^\circ}\ \mathrm{V} \tag{5.63}$$

The power factor angle at the sending end is calculated as,

$$\phi_s = 12.31 + 36.86 = 49.17^\circ \tag{5.64}$$

Sending end power factor is calculated as,

$$\cos\phi_s = \cos 49.17^\circ = 0.65 \tag{5.65}$$

The transmission line loss is determined as,

$$P_l = I^2R = 34.1^2 \times 5 = 5814.05\ \mathrm{W} \tag{5.66}$$

The power at the sending end is calculated as,

$$P_s = 12000 + 5814.05 = 17814.05\ \mathrm{W} \tag{5.67}$$

The transmission efficiency is calculated as,

$$\eta = \frac{12000}{17814.05} \times 100 = 67.36\% \tag{5.68}$$

Practice Problem 5.1
A load of 4000 kW with a 0.86 lagging power factor receives power by a three-phase, wye-connected transmission lines. Consider the line voltage at the load terminal is 30 kV. The total transmission line resistance and reactance are 3 Ω and 12 Ω per phase, respectively. Calculate the sending end voltage, voltage regulation, and transmission efficiency.

5.5 Medium Transmission Line

In the medium transmission line, the effects of shunt capacitance are considered throughout the whole length, and it plays an important role to determine other parameters [1–3]. The shunt capacitance is assumed to be lumped in the middle of the line, and the impedance is divided equally on either side of the shunt capacitance. This configuration is known as the nominal *T*-model as shown in Fig. 5.6. Applying KVL to the circuit in Fig. 5.6 to find out the voltage across the capacitance and it becomes,

$$V_c = V_r + I_r \frac{Z}{2} \tag{5.69}$$

The current through the capacitance is written as,

$$I_c = j\omega C V_c = Y V_c \tag{5.70}$$

Where the shunt admittance is,

$$Y = j\omega C \tag{5.71}$$

Applying KCL to Fig. 5.6 to determine the sending end current as,

$$I_s = I_r + I_c \tag{5.72}$$

Substituting Eq. (5.69) into Eq. (5.70) yields,

$$I_c = Y\left(V_r + I_r \frac{Z}{2}\right) \tag{5.73}$$

Substituting Eq. (5.73) into Eq. (5.72) yields,

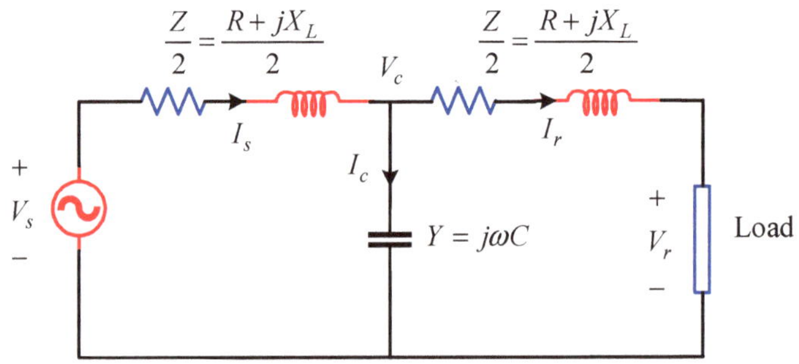

Fig. 5.6 Circuit for nominal *T*-model

$$I_s = I_r + Y\left(V_r + I_r\frac{Z}{2}\right) \tag{5.74}$$

$$I_s = YV_r + \left(1 + \frac{ZY}{2}\right)I_r \tag{5.75}$$

Applying KVL to Fig. 5.6 to determine the expression of sending end voltage as,

$$V_s = V_c + I_s\frac{Z}{2} \tag{5.76}$$

Substituting Eqs. (5.69) and (5.75) into Eq. (5.76) yields,

$$V_s = V_r + \frac{Z}{2}I_r + \left\{YV_r + \left(1 + \frac{ZY}{2}\right)I_r\right\}\frac{Z}{2} \tag{5.77}$$

$$V_s = V_r + \frac{Z}{2}I_r + \frac{YZ}{2}V_r + \frac{Z}{2}I_r + \frac{Z^2Y}{4}I_r \tag{5.78}$$

$$V_s = \left(1 + \frac{YZ}{2}\right)V_r + \left(1 + \frac{ZY}{4}\right)ZI_r \tag{5.79}$$

Comparing Eq. (5.79) with Eq. (5.6) and Eq. (5.75) with Eq. (5.7) and the expression of *ABCD* parameters can be written as,

$$A = 1 + \frac{YZ}{2} \tag{5.80}$$

$$B = Z\left(1 + \frac{YZ}{4}\right) \tag{5.81}$$

$$C = Y \tag{5.82}$$

$$D = 1 + \frac{YZ}{2} \tag{5.83}$$

$$AD - BC = \left(1 + \frac{YZ}{2}\right)^2 - YZ\left(1 + \frac{YZ}{4}\right) = 1 + YZ + \left(\frac{YZ}{2}\right)^2 - YZ - \left(\frac{YZ}{2}\right)^2 \tag{5.84}$$

$$AD - BC = 1 \tag{5.85}$$

When the total shunt admittance is divided into two halves, and each half is placed at the sending end and the receiving end, while the total circuit impedance is

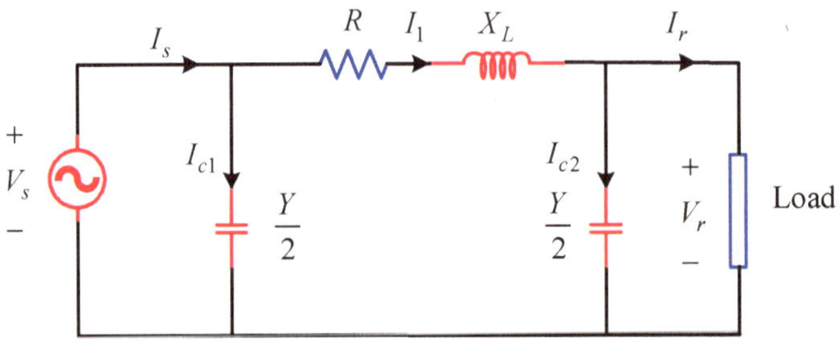

Fig. 5.7 Circuit for nominal π-model

placed in between the two, is known as the nominal π-model. The π-model is shown in Fig. 5.7.

Applying KCL to Fig. 5.7 yields,

$$I_1 = I_{c2} + I_r = \frac{Y}{2}V_r + I_r \tag{5.86}$$

Also, applying KVL to Fig. 5.7 yields,

$$V_s = V_r + ZI_1 \tag{5.87}$$

Substituting Eq. (5.86) into Eq. (5.87) yields,

$$V_s = V_r + Z\left(\frac{Y}{2}V_r + I_r\right) \tag{5.88}$$

$$V_s = \left(1 + \frac{YZ}{2}\right)V_r + ZI_r \tag{5.89}$$

Applying KCL to Fig. 5.7 yields,

$$I_s = I_1 + \frac{Y}{2}V_s \tag{5.90}$$

Substituting Eqs. (5.86) and (5.89) into Eq. (5.90) yields,

$$I_s = I_r + \frac{Y}{2}V_r + \frac{Y}{2}\left\{\left(1 + \frac{YZ}{2}\right)V_r + ZI_r\right\} \tag{5.91}$$

$$I_s = \left(\frac{Y}{2} + \frac{Y}{2} + \frac{Y^2Z}{4}\right)V_r + \left(1 + \frac{YZ}{2}\right)I_r \tag{5.92}$$

$$I_s = Y\left(1 + \frac{YZ}{4}\right)V_r + \left(1 + \frac{YZ}{2}\right)I_r \tag{5.93}$$

Comparing Eq. (5.89) with Eq. (5.6) and Eq. (5.93) with Eq. (5.7) yields,

$$A = 1 + \frac{YZ}{2} \tag{5.94}$$

$$B = Z \tag{5.95}$$

$$C = Y\left(1 + \frac{YZ}{4}\right) \tag{5.96}$$

$$D = 1 + \frac{YZ}{2} \tag{5.97}$$

The following relations can be written as,

$$AD - BC = \left(1 + \frac{YZ}{2}\right)^2 - YZ\left(1 + \frac{YZ}{4}\right) = 1 \tag{5.98}$$

Example 5.2
A 30 MW and 0.95 lagging power factor balanced load receives power through 100 km long three-phase 50 Hz transmission lines. The load terminal line voltage is 66 kV. The transmission line parameters per phase are $R = 0.2\Omega/\text{km}$, $X_L = 0.5\Omega/\text{km}$, $Y = 0.06 \times 10^{-4}\mho/\text{km}$. Calculate the sending end voltage, voltage regulation, and transmission efficiency. Consider the nominal T-model.

Solution
Per phase receiving end voltage is calculated as,

$$V_r = \frac{66000}{\sqrt{3}} = 38105.12\,\text{V} \tag{5.99}$$

The magnitude of the load current is calculated as,

$$I_r = \frac{30 \times 1000000}{\sqrt{3} \times 66 \times 1000 \times 0.95} = 276.2\,\text{A} \tag{5.100}$$

The impedance per phase is calculated as,

$$Z = (0.2 + j0.5) \times 100 = 20 + j50 = 53.85\,\underline{|68.2^\circ}\,\Omega \tag{5.101}$$

The admittance per phase is determined as,

$$Y = j0.06 \times 10^{-4} \times 100 = j6 \times 10^{-4}\,\text{S} \tag{5.102}$$

The voltage across the capacitance is calculated as,

$$V_c = 38105.12 + 276.2\,\underline{|-18.2^\circ} \times \frac{53.85\,\underline{|68.2^\circ}}{2} = 43262.05\,\underline{|7.6^\circ}\,\text{V} \tag{5.103}$$

The current through the capacitor is calculated as,

$$I_c = jYV_c = j6 \times 10^{-4} \times 43262.05\,\underline{|7.6} = 25.96\,\underline{|97.6^\circ}\,\text{A} \tag{5.104}$$

The sending end current is calculated as,

$$I_s = I_c + I_r = 25.96\,\underline{|97.6^\circ} + 276.2\,\underline{|-18.2^\circ} = 265.93\,\underline{|-13.16^\circ}\,\text{A} \tag{5.105}$$

The sending end voltage is calculated as,

$$V_s = I_s\frac{Z}{2} + V_c = \frac{53.85\,\underline{|68.2^\circ}}{2} \times 265.93\,\underline{|-13.16^\circ} + 43262.05\,\underline{|7.6^\circ} \tag{5.106}$$

$$V_s = 48393.15\,\underline{|13.86^\circ}\,\text{V} \tag{5.107}$$

Alternative approach:
The values of the constants are calculated as,

$$A = D = 1 + \frac{YZ}{2} = 1 + \frac{(20 + j50)(6 \times 10^{-4}\,\underline{|90^\circ})}{2} = 0.99\,\underline{|0.35^\circ} \tag{5.108}$$

$$B = Z\left(1 + \frac{YZ}{4}\right) = (20 + j50)\left(1 + \frac{(20 + j50)(6 \times 10^{-4}\,\underline{|90^\circ})}{4}\right) \tag{5.109}$$

$$B = 53.45\,\underline{|68.37^\circ}\,\Omega \tag{5.110}$$

$$C = Y = 6 \times 10^{-4}\,\underline{|90^\circ}\,\text{S} \tag{5.111}$$

The value of the sending end voltage is determined as,

$$\begin{aligned} V_s &= AV_r + BI_r = 0.99\,\underline{|0.35^\circ} \times 38105.12 + 53.45\,\underline{|68.37^\circ} \times 276.2\,\underline{|-18.2^\circ} \\ &= 48576.57\,\underline{|13.78^\circ}\,\text{V} \end{aligned} \tag{5.112}$$

The value of the sending end current can be calculated as,

$$I_s = CV_r + DI_r = 6 \times 10^{-4} \angle 90^\circ \times 38105.12 + 0.99 \angle 0.35^\circ \times 276.2 \angle -18.2^\circ$$
$$= 267.32 \angle -13.18^\circ \text{ A} \tag{5.113}$$

The receiving end voltage under no-load condition is calculated as,

$$V_{rn} = \frac{48576.57}{0.99} = 49067.24 \text{ V} \tag{5.114}$$

The voltage regulation is calculated as,

$$VR = \frac{V_{rn} - V_{rf}}{V_{rf}} \times 100 = \frac{49067.24 - 38105.12}{38105.12} \times 100 = 28.77\% \tag{5.115}$$

The transmission line loss is calculated as,

$$P_{Line} = 3\left(I_s^2 \frac{R}{2} + I_r^2 \frac{R}{2}\right) = 3\left(267.32^2 \times 10 + 276.2^2 \times 10\right) = 4.43 \text{ MW} \tag{5.116}$$

The transmission efficiency is determined as,

$$\eta = \frac{30}{30 + 4.43} \times 100 = 87.13\% \tag{5.117}$$

Example 5.3
A 120 km long three-phase 50 Hz transmission line is having per phase per km resistance, reactance, and susceptance 0.2 Ω, 0.7 Ω, and 0.08×10^{-4} ℧, respectively. The line delivers a power of 25 MW at 0.85 power factor lagging to the receiving end side, and its terminal voltage is 66 kV. Find the sending end voltage, sending end power factor, and voltage regulation by considering a nominal π-model.

Solution

The total values of the line parameters per phase are calculated as,

$$R = 0.2 \times 120 = 24\,\Omega \tag{5.118}$$

$$X_L = 0.7 \times 120 = 84\,\Omega \tag{5.119}$$

$$Y = 0.08 \times 10^{-4} \times 120 = 0.00096 \text{ S} \tag{5.120}$$

The receiving end voltage per phase is calculated as,

$$V_r = \frac{66 \times 1000}{\sqrt{3}} = 38105.12 \text{ V} \tag{5.121}$$

The value of the load current is calculated as,

$$I_r = \frac{25 \times 10^6}{\sqrt{3} \times 66 \times 10^3 \times 0.85} = 201.63\,\text{A} \tag{5.122}$$

The current through the line is calculated as,

$$I_1 = I_{c2} + I_r = j\frac{0.00096}{2} \times 38105.12 + 201.63\,\angle{-31.78^\circ} = 192.63\,\angle{-27.15^\circ}\,\text{A} \tag{5.123}$$

The sending end voltage is calculated as,

$$V_s = V_r + ZI_1 = 38105.12 + (24 + j84) \times 192.63\,\angle{-27.15^\circ} = 51102.03\,\angle{13.91^\circ}\,\text{V} \tag{5.124}$$

The sending end current can be calculated as,

$$\begin{aligned} I_s &= I_1 + \frac{Y}{2}V_s = 192.63\,\angle{-27.15^\circ} + j\frac{0.00096}{2} \times 51102.03\,\angle{13.91^\circ} \\ &= 177.48\,\angle{-21.17^\circ}\,\text{A} \end{aligned} \tag{5.125}$$

Alternative approach:
The values of the constants are,

$$A = D = 1 + \frac{YZ}{2} = 1 + \frac{(24 + j84)(0.00096\,\angle{90^\circ})}{2} = 0.96\,\angle{0.69^\circ} \tag{5.126}$$

$$B = Z = 24 + j84 = 87.36\,\angle{74.05^\circ}\,\Omega \tag{5.127}$$

$$\begin{aligned} C &= Y\left(1 + \frac{YZ}{4}\right) = 0.00096\,\angle{90^\circ}\left(1 + \frac{(0.00096\,\angle{90^\circ})(24 + j84)}{4}\right) \\ &= 0.00094\,\angle{90.34^\circ}\,\text{S} \end{aligned} \tag{5.128}$$

The value of the sending end voltage is calculated as,

$$\begin{aligned} V_s &= AV_r + BI_r = 0.96\,\angle{0.69^\circ} \times 38105.12 + 87.36\,\angle{74.05^\circ} \times 201.63\,\angle{-31.78^\circ} \\ &= 51111.81\,\angle{13.91^\circ}\,\text{V} \end{aligned} \tag{5.129}$$

The value of the sending end current can be calculated as,

$$\begin{aligned} I_s &= CV_r + DI_r = 0.00094\,\angle{90.34^\circ} \times 38105.12 + 0.96\,\angle{0.69^\circ} \times 201.63\,\angle{-31.78^\circ} \\ &= 177.54\,\angle{-21.18^\circ}\,\text{A} \end{aligned} \tag{5.130}$$

The sending end power factor is calculated as,

$$\cos\phi_s = \cos(13.91^\circ + 21.18^\circ) = 0.82 \tag{5.131}$$

The receiving end voltage under no-load condition is calculated as,

$$V_{rn} = \frac{51102.03}{0.96} = 53231.28\,\text{V} \tag{5.132}$$

The voltage regulation can be calculated as,

$$VR = \frac{V_{rn} - V_{rf}}{V_{rf}} \times 100 = \frac{53231.28 - 38105.12}{38105.12} \times 100 = 39.7\,\% \tag{5.133}$$

Practice Problem 5.2

A 220 km long three-phase 50 Hz transmission line delivers 25 MW power at a 0.85 power factor lagging to a balanced load, and its line voltage is 33 kV. The transmission line parameters are given as, $R = 0.12\Omega/\text{km}/\text{phase}, X_L = 0.4\Omega/\text{km}/\text{phase}$, $Y = 0.09 \times 10^{-5}\mho/\text{km}/\text{phase}$.

Calculate the sending end voltage and voltage regulation. Consider a nominal T-model.

Practice Problem 5.3

A 100 km long three-phase 50 Hz transmission line per phase per km resistance, reactance, and susceptance is $0.1\,\Omega$, $0.5\,\Omega$, and $10 \times 10^{-6}\,\mho$, respectively. The line delivers 20 MW power at a 0.95 power factor lagging to the receiving end whose terminal voltage is 33 kV. Find the sending end voltage and power factor by considering a nominal π-model.

5.6 Long Transmission Line

In a long transmission line, the line parameters are distributed uniformly throughout the length of the line. As a result, the voltages and the current vary from point to point on the line. The one phase of a long transmission line with a π-model is shown in Fig. 5.8. The following parameters are defined as,

z	is the series impedance per unit length,
y	is the shunt admittance per unit length,
l	is the total length of the line,
$Z = zl$	is the total line impedance,
$Y = yl$	is the total admittance of the line,
$V(x)$	is the voltage at a location of x on the line,
$I(x)$	is the current in the location of x on the line,
$V(x + \Delta x)$	is the voltage at a location of $x + \Delta x$ on the line,
$I(x + \Delta x)$	is the current in the location of $x + \Delta x$ on the line

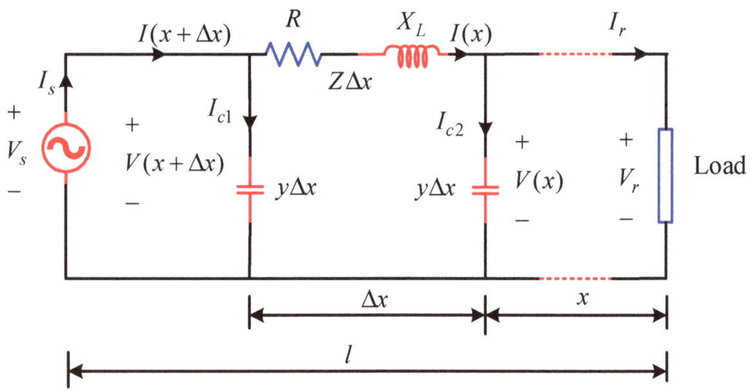

Fig. 5.8 Long transmission line with parameters

Applying KVL to the circuit in Fig. 5.8 yields,

$$V(x + \Delta x) = V(x) + z\Delta x I(x) \tag{5.134}$$

$$\frac{V(x + \Delta x) - V(x)}{\Delta x} = zI(x) \tag{5.135}$$

Taking limit as $\Delta x \rightarrow 0$ to Eq. (5.135) yields,

$$\lim_{\Delta x \rightarrow 0} \frac{V(x + \Delta x) - V(x)}{\Delta x} = zI(x) \tag{5.136}$$

$$\frac{dV(x)}{dx} = zI(x) \tag{5.137}$$

Applying KCL to the circuit in Fig. 5.8 yields,

$$I(x + \Delta x) = I(x) + y\Delta x V(x + \Delta x) \tag{5.138}$$

$$\frac{I(x + \Delta x) - I(x)}{\Delta x} = yV(x + \Delta x) \tag{5.139}$$

Taking limit as $\Delta x \rightarrow 0$ to Eq. (5.139) yields,

$$\lim_{\Delta x \rightarrow 0} \frac{I(x + \Delta x) - I(x)}{\Delta x} = y \lim_{\Delta x \rightarrow 0} V(x + \Delta x) \tag{5.140}$$

$$\frac{dI(x)}{dx} = yV(x) \tag{5.141}$$

Differentiating Eq. (5.137) with respect to x yields,

$$\frac{d^2V(x)}{dx^2} = z\frac{dI(x)}{dx} \tag{5.142}$$

Substituting Eq. (5.141) into Eq. (5.142) yields,

$$\frac{d^2V(x)}{dx^2} = yzV(x) \tag{5.143}$$

$$\frac{d^2V(x)}{dx^2} - \gamma^2 V(x) = 0 \tag{5.144}$$

Considering the following relation,

$$\gamma^2 = yz \tag{5.145}$$

$$\gamma = \alpha + j\beta = \sqrt{yz} \tag{5.146}$$

Where γ is known as propagation constant. The real part α is known as attenuation constant, and the imaginary part β is known as phase constant.

Let $D = \frac{d}{dx}$, and then, Eq. (5.144) becomes,

$$D^2 - \gamma^2 = 0 \tag{5.147}$$

$$D = \pm\gamma \tag{5.148}$$

The solution of Eq. (5.144) can be written as,

$$V(x) = Ae^{\gamma x} + Be^{-\gamma x} \tag{5.149}$$

Substituting Eq. (5.149) into Eq. (5.137) yields,

$$I(x) = \frac{1}{z}\frac{d}{dx}(Ae^{\gamma x} + Be^{-\gamma x}) \tag{5.150}$$

$$I(x) = \frac{\gamma}{z}(Ae^{\gamma x} - Be^{-\gamma x}) = \frac{\sqrt{yz}}{z}(Ae^{\gamma x} - Be^{-\gamma x}) \tag{5.151}$$

$$I(x) = \sqrt{\frac{y}{z}}(Ae^{\gamma x} - Be^{-\gamma x}) \tag{5.152}$$

$$I(x) = \frac{1}{Z_c}(Ae^{\gamma x} - Be^{-\gamma x}) \tag{5.153}$$

The characteristics impedance of the line is expressed as,

$$Z_c = \sqrt{\frac{z}{y}} \tag{5.154}$$

The coefficients A and B can be determined by the initial conditions at the receiving end of the line. These conditions are,

At

$$x = 0, V(0) = V_r \tag{5.155}$$

At

$$x = 0, I(0) = I_r \tag{5.156}$$

Substituting Eq. (5.155) into Eq. (5.149) yields,

$$A + B = V_r \tag{5.157}$$

Substituting Eq. (5.156) into Eq. (5.153) yields,

$$A - B = I_r Z_c \tag{5.158}$$

The expressions of the constants are determined as,

$$A = \frac{V_r + I_r Z_c}{2} \tag{5.159}$$

$$B = \frac{V_r - I_r Z_c}{2} \tag{5.160}$$

Substituting Eqs. (5.159) and (5.160) into Eqs. (5.149) yields,

$$V(x) = \frac{V_r + I_r Z_C}{2} e^{\gamma x} + \frac{V_r - I_r Z_C}{2} e^{-\gamma x} \tag{5.161}$$

$$V(x) = \left(\frac{e^{\gamma x} + e^{-\gamma x}}{2}\right) V_r + Z_c \left(\frac{e^{\gamma x} - e^{-\gamma x}}{2}\right) I_r \tag{5.162}$$

Substituting Eqs. (5.159) and (5.160) into Eqs. (5.153) yields,

$$I(x) = \left(\frac{e^{\gamma x} - e^{-\gamma x}}{2Z_c}\right) V_r + \left(\frac{e^{\gamma x} + e^{-\gamma x}}{2}\right) I_r \tag{5.163}$$

The hyperbolic functions are defined as,

$$\sinh\phi = \frac{e^{\phi} - e^{-\phi}}{2} \tag{5.164}$$

$$\cosh\phi = \frac{e^{\phi} + e^{-\phi}}{2} \tag{5.165}$$

Substituting Eqs. (5.164) and (5.165) into Eqs. (5.162) and (5.163) yields,

$$V(x) = V_r \cosh\gamma x + Z_c I_r \sinh\gamma x \tag{5.166}$$

$$I(x) = \frac{V_r}{Z_c}\sinh\gamma x + I_r \cosh\gamma x \tag{5.167}$$

The voltage and current at the sending end can be determined by setting the following conditions,

$$x = l,\ V(l) = V_s \tag{5.168}$$

$$x = l,\ \ I(l) = I_s \tag{5.169}$$

Substituting Eq. (5.168) into Eq. (5.166) yields,

$$V_s = V_r \cosh\gamma l + Z_c I_r \sinh\gamma l \tag{5.170}$$

Substituting Eq. (5.169) into Eq. (5.167) yields,

$$I_s = \frac{V_r}{Z_c}\sinh\gamma l + I_r \cosh\gamma l \tag{5.171}$$

Comparing Eq. (5.170) with Eq. (5.6) and Eq. (5.171) with Eq. (5.7),

$$A = D = \cosh\gamma l \tag{5.172}$$

$$B = Z_c \sinh\gamma l \tag{5.173}$$

$$C = \frac{\sin\gamma l}{Z_c} \tag{5.174}$$

The following expression can be written as,

$$e^{\gamma l} = e^{(\alpha + j\beta)l} = e^{\alpha l} e^{j\beta l} = e^{\alpha l}\ \underline{|\beta l} \tag{5.175}$$

The following expressions can be modified as,

$$\cosh \gamma l = \frac{e^{\gamma l} + e^{-\gamma l}}{2} = \frac{1}{2}\left(e^{\alpha l}\,\underline{|\beta l} + e^{-\alpha l}\,\underline{|-\beta l}\right) \tag{5.176}$$

$$\sinh \gamma l = \frac{e^{\gamma l} - e^{-\gamma l}}{2} = \frac{1}{2}\left(e^{\alpha l}\,\underline{|\beta l} - e^{-\alpha l}\,\underline{|-\beta l}\right) \tag{5.177}$$

$$AD - BC = \cosh^2 \gamma l - Z_c \sinh \gamma l \times \frac{\sinh \gamma l}{Z_c} \tag{5.178}$$

$$AD - BC = \cosh^2 \gamma l - \sinh^2 \gamma l = 1 \tag{5.179}$$

For a lossless transmission line, the real part of the propagation constant is zero, i.e., $\alpha = 0$. The propagation constant becomes,

$$\gamma = 0 + j\beta = j\beta \tag{5.180}$$

Again, the hyperbolic functions can be modified as,

$$\cosh \gamma x = \cosh j\beta x = \frac{e^{j\beta x} + e^{-j\beta x}}{2} = \cos \beta x \tag{5.181}$$

$$\sinh \gamma x = \sinh j\beta x = \frac{e^{j\beta x} - e^{-j\beta x}}{2} = j \sin \beta x \tag{5.182}$$

Substituting Eqs. (5.182) and (5.183) into Eqs. (5.166) and (5.167) yields,

$$V(x) = V_r \cos \beta x + jZ_c I_r \sin \beta x \tag{5.183}$$

$$I(x) = j\frac{V_r}{Z_c} \sin \beta x + I_r \cos \beta x \tag{5.184}$$

Figure 5.9 shows an equivalent π-model for a long transmission line. According to Eqs. (5.89) and (5.93), the following relations can be written as,

$$V_s = \left(1 + \frac{Y'Z'}{2}\right) V_r + Z' I_r \tag{5.185}$$

$$I_s = Y'\left(1 + \frac{Y'Z'}{4}\right) V_r + \left(1 + \frac{Y'Z'}{2}\right) I_r \tag{5.186}$$

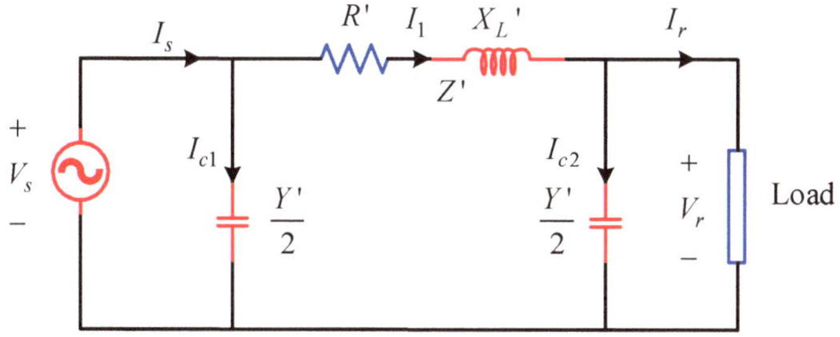

Fig. 5.9 Nominal π-model circuit for a long transmission line

Comparing Eq. (5.185) with Eq. (5.170) and Eq. (5.186) with the Eq. (5.171) yields,

$$1 + \frac{Y'Z'}{2} = \cosh \gamma l \quad (5.187)$$

$$Z' = Z_c \sinh \gamma l \quad (5.188)$$

$$Y'\left(1 + \frac{Y'Z'}{4}\right) = \frac{\sinh \gamma l}{Z_c} \quad (5.189)$$

$$1 + \frac{Y'Z'}{2} = \cosh \gamma l \quad (5.190)$$

Equation (5.188) can be rearranged as,

$$Z' = Z_c \sinh \gamma l = \sqrt{\frac{z}{y}} \frac{\sinh \gamma l}{\sqrt{yz}} \sqrt{yz} \quad (5.191)$$

$$Z' = \sqrt{\frac{z}{y}} l \frac{\sinh \gamma l}{l \sqrt{yz}} \sqrt{yz} \quad (5.192)$$

$$Z' = \sqrt{\frac{z}{y}} l \frac{\sinh \gamma l}{\gamma l} \sqrt{yz} \quad (5.193)$$

$$Z' = z l \frac{\sinh \gamma l}{\gamma l} \quad (5.194)$$

$$Z' = Z \frac{\sinh \gamma l}{\gamma l} \quad (5.195)$$

Equation (5.190) can be rearranged as,

$$1+\frac{Y'Z'}{2}=\cosh\gamma l \tag{5.196}$$

$$\frac{Y'Z'}{2}=\cosh\gamma l-1 \tag{5.197}$$

$$\frac{Y'}{2}Z\frac{\sinh\gamma l}{\gamma l}=\cosh\gamma l-1 \tag{5.198}$$

$$\frac{Y'}{2}=\frac{\gamma l}{Z}\frac{\cosh\gamma l-1}{\sinh\gamma l} \tag{5.199}$$

$$\frac{Y'}{2}=\frac{\gamma l}{Z}\frac{\cosh\gamma l-1}{\sinh\gamma l} \tag{5.200}$$

$$\frac{Y'}{2}=\frac{\sqrt{yz}l}{zl}\tanh\frac{\gamma l}{2} \tag{5.201}$$

$$\frac{Y'}{2}=\sqrt{\frac{y}{z}}\tanh\frac{\gamma l}{2} \tag{5.202}$$

$$\frac{Y'}{2}=\frac{1}{Z_c}\tanh\frac{\gamma l}{2} \tag{5.203}$$

Equation (5.203) can be expressed as,

$$\frac{Y'}{2}=\sqrt{\frac{y}{z}}\frac{\tanh\frac{\gamma l}{2}}{\frac{\gamma l}{2}}\frac{\gamma l}{2} \tag{5.204}$$

$$\frac{Y'}{2}=\sqrt{\frac{y}{z}}\frac{\sqrt{yz}}{2}l\frac{\tanh\frac{\gamma l}{2}}{\frac{\gamma l}{2}} \tag{5.205}$$

$$\frac{Y'}{2}=\frac{y}{2}l\frac{\tanh\frac{\gamma l}{2}}{\frac{\gamma l}{2}} \tag{5.206}$$

$$\frac{Y'}{2}=\frac{Y}{2}\frac{\tanh\frac{\gamma l}{2}}{\frac{\gamma l}{2}} \tag{5.207}$$

$$Y'=Y\frac{\tanh\frac{\gamma l}{2}}{\frac{\gamma l}{2}} \tag{5.208}$$

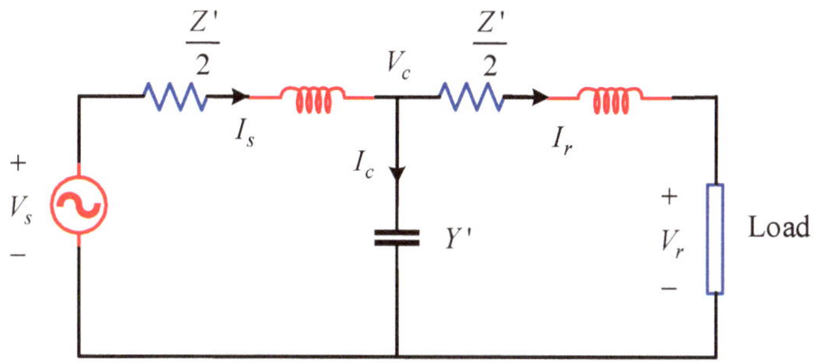

Fig. 5.10 Nominal *T*-model circuit for a long transmission line

Figure 5.10 shows an equivalent circuit for T-model of a long transmission line. According to Eqs. (5.79) and (5.75), the following relations can be written as,

$$V_s = \left(1 + \frac{Y'Z'}{2}\right)V_r + Z'\left(1 + \frac{Z'Y'}{4}\right)I_r \tag{5.209}$$

$$I_s = Y'V_r + Y'\left(1 + \frac{Z'Y'}{2}\right)I_r \tag{5.210}$$

Comparing Eqs. (5.170) and (5.209) yields,

$$1 + \frac{Y'Z'}{2} = \cosh\gamma l \tag{5.211}$$

$$Z'\left(1 + \frac{Z'Y'}{4}\right) = Z_c \sinh\gamma l \tag{5.212}$$

Again, comparing Eq. (5.171) with Eq. (5.210) yields,

$$Y' = \frac{1}{Z_c}\sinh\gamma l \tag{5.213}$$

$$1 + \frac{Y'Z'}{2} = \cosh\gamma l \tag{5.214}$$

Substituting Eq. (5.213) into Eq. (5.214) yields,

$$1 + \frac{Z'}{2}\frac{1}{Z_c}\sinh\gamma l = \cosh\gamma l \tag{5.215}$$

$$\frac{Z'}{2}\frac{1}{Z_c} = \frac{\cosh \gamma l - 1}{\sinh \gamma l} \tag{5.216}$$

$$\frac{Z'}{2} = \frac{2\sinh^2 \frac{\gamma l}{2}}{2\sinh \frac{\gamma l}{2} \cosh \frac{\gamma l}{2}} = Z_c \tanh \frac{\gamma l}{2} \tag{5.217}$$

$$\frac{Z'}{2} = \sqrt{\frac{z}{y}}\frac{1}{2\sqrt{yz}}\sqrt{yz}\frac{\tanh \frac{\gamma l}{2}}{\frac{l}{2}}\, l \tag{5.218}$$

$$\frac{Z'}{2} = \frac{zl}{2}\frac{\tanh \frac{\gamma l}{2}}{\frac{\gamma l}{2}} \tag{5.219}$$

$$\frac{Z'}{2} = \frac{Z}{2}\frac{\tanh \frac{\gamma l}{2}}{\frac{\gamma l}{2}} \tag{5.220}$$

Example 5.4
A load 25 MW at a 0.95 power factor lagging receives power by a 220 km long three-phase 50 Hz transmission line. The transmission line parameters are given by r/phase/km = 0.11Ω, x/phase/km = 0.23Ω, y/phase/km = 1.4×10^{-6}℧. Calculate the sending end voltage and current if the receiving end voltage is 120 kV.

Solution
The total resistance per phase is calculated as,

$$R = 0.11 \times 220 = 24.2\Omega \tag{5.221}$$

The total reactance per phase is calculated as,

$$X_L = 0.23 \times 220 = 50.6\Omega \tag{5.222}$$

Per phase impedance of the line is calculated as,

$$Z = 24.2 + j50.6 = 56.08\,\underline{|64.44^\circ}\ \Omega \tag{5.223}$$

The total shunt admittance is calculated as,

$$Y = j1.4 \times 10^{-6} \times 220 = 0.00028\,\underline{|90^\circ}\ ℧ \tag{5.224}$$

The receiving end voltage per phase is calculated as,

$$V_r = \frac{120 \times 1000}{\sqrt{3}} = 69282.03\ \text{V} \tag{5.225}$$

The value of the load current is determined as,

$$I_r = \frac{25 \times 10^6}{\sqrt{3} \times 120 \times 10^3 \times 0.95} = 126.6\,\mathrm{A} \tag{5.226}$$

The following parameters can be calculated as,

$$\begin{aligned} \gamma l &= \sqrt{yz}\, l = \sqrt{(0.11 + j0.23) \times 1.4 \times 10^{-6} \underline{|90^\circ}} \times 220 \\ &= \left(3.57 \times 10^{-7} \underline{|154.44^\circ}\right)^{0.5} \times 220 = 0.13 \underline{|77.22^\circ} = 0.03 + j0.13 \end{aligned} \tag{5.227}$$

$$e^{\gamma l} = e^{0.03} \underline{|0.13^\circ} = 1.03 \underline{|0.13^\circ} \tag{5.228}$$

$$e^{-\gamma l} = e^{-0.03} \underline{|-0.13^\circ} = 0.97 \underline{|-0.13^\circ} \tag{5.229}$$

The values of the hyperbolic functions are determined as,

$$\cosh \gamma l = \frac{1}{2}\left(1.03 \underline{|0.13^\circ} + 0.97 \underline{|-0.13^\circ}\right) = 1 \underline{|3.9^\circ} \tag{5.230}$$

$$\sinh \gamma l = \frac{1}{2}\left(1.03 \underline{|0.13^\circ} - 0.97 \underline{|-0.13^\circ}\right) = 0.03 \underline{|4.33^\circ} \tag{5.231}$$

The value of the characteristics impedance is calculated as,

$$Z_c = \sqrt{\frac{z}{y}} = \sqrt{\frac{0.25 \underline{|64.44^\circ}}{1.4 \times 10^{-6} \underline{|90^\circ}}} = 422.58 \underline{|-12.78^\circ}\,\Omega \tag{5.232}$$

The sending end voltage per phase can be determined as,

$$\begin{aligned} V_s &= V_r \cosh \gamma l + I_r Z_c \sinh \gamma l \\ &= 69282.03 \times 1 \underline{|3.9^\circ} + 126.6 \times 422.58 \underline{|-12.78^\circ} \times 0.03 \underline{|4.33^\circ} \end{aligned} \tag{5.233}$$

$$V_s = 70850.68 \underline{|3.62^\circ}\,\mathrm{V} \tag{5.234}$$

The sending end current can be calculated as,

$$\begin{aligned} I_s &= \frac{V_r}{Z_c} \sinh \gamma l + I_r \cosh \gamma l = \frac{69282.03}{422.58 \underline{|-12.78^\circ}} \times 0.03 \underline{|4.33^\circ} + 126.6 \times 1 \underline{|3.9^\circ} \\ &= 131.39 \underline{|4.39^\circ}\,\mathrm{A} \end{aligned} \tag{5.235}$$

Practice Problem 5.4

The line parameters of a 240 km long transmission r/phase/km = 0.03Ω, x/phase/km = 0.15Ω, y/phase/km = 0.99×10^{-6} ℧. Transmission line delivers a power to a 25 MW at a 0.85 lagging power factor. Calculate the sending end line voltage and current if the receiving end voltage is 80 kV.

5.7 Surge Impedance Loading

Surge impedance loading is an important parameter to study the power system when an issue arises related to prediction of maximum loading capability of transmission line. The maximum three-phase active power transfer capability of a transmission line is called the surge impedance loading (*SIL*). A long transmission lines have the distributed lumped inductance and capacitance. When a transmission lines are energized, capacitance feeds the reactive power to the line, and the inductance absorbs the reactive power. The amount of reactive power in MVAR range depends on the capacitive reactance and the energized line voltage. Mathematically, the expression of MVAR produced is written as,

$$\mathrm{MVAR}_C = \frac{\mathrm{kV}^2}{X_C} \tag{5.236}$$

Transmission line also uses reactive power to support their magnetic field. The strength of the magnetic field depends on the magnitude of the current and its natural reactance. The expression of the reactive power uses or absorbs by the transmission line is,

$$\mathrm{MVAR}_L = I^2 X_L \tag{5.237}$$

In surge impedance loading, reactive power production is equal to reactive power uses by the transmission line. This reactive power balance relation is written as,

$$I^2 X_L = \frac{V^2}{X_C} \tag{5.238}$$

$$X_L X_C = \frac{V^2}{I^2} \tag{5.239}$$

$$\frac{\omega L}{\omega C} = \frac{V^2}{I^2} \tag{5.240}$$

$$\sqrt{\frac{\omega L}{\omega C}} = \frac{V}{I} \tag{5.241}$$

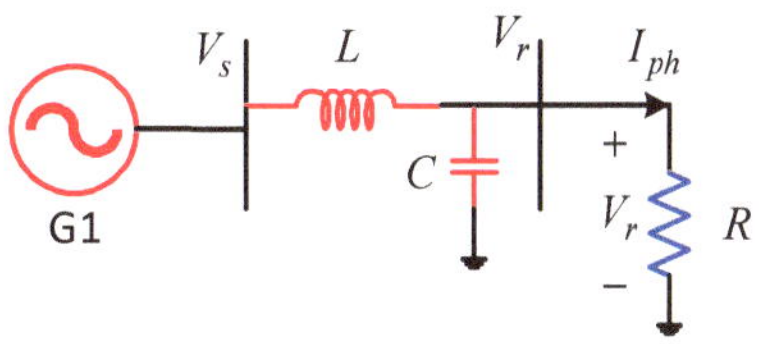

Fig. 5.11 Circuit for SIL

$$\frac{V}{I} = \sqrt{\frac{L}{C}} \tag{5.242}$$

$$Z_c = \sqrt{\frac{L}{C}} \tag{5.243}$$

Where Z_c is the surge impedance or characteristic impedance.

The voltage is supplied to the load at a unity power factor using a transmission line as shown in Fig. 5.11. The expression of *SIL* is written as,

$$SIL = 3V_{ph}I_{ph}\cos\phi \tag{5.244}$$

For a resistive load, the power factor is unity. The phase current is expressed as,

$$I_{ph} = \frac{V_{ph}}{R} = \frac{V_L}{\sqrt{3}Z_L} \tag{5.245}$$

Substituting Eq. (5.245) into Eq. (5.244) yields,

$$SIL = 3\frac{V_L}{\sqrt{3}} \times \frac{V_L}{\sqrt{3}Z_L} \times 1 \tag{5.246}$$

$$SIL = \frac{V_L^2}{Z_L} = \frac{V_L^2}{Z_c} \tag{5.247}$$

Substituting Eq. (5.243) into Eq. (5.247) yields,

$$SIL = V_L^2\sqrt{\frac{C}{L}} \tag{5.248}$$

From Eq. (5.248), the following relations are summarized as,

$$SIL \propto V_L^2 \tag{5.249}$$

$$SIL \propto \sqrt{C} \tag{5.250}$$

$$SIL \infty \frac{1}{\sqrt{L}} \tag{5.251}$$

Alternative approach: The expression of receiving end current is,

$$I_r = \frac{V_r}{Z_c} \tag{5.252}$$

Substituting Eq. (5.252) into Eq. (5.183) yields,

$$V(x) = V_r \cos \beta x + jZ_c \frac{V_r}{Z_c} \sin \beta x \tag{5.253}$$

$$V(x) = V_r(\cos \beta x + j \sin \beta x) \tag{5.254}$$

$$V(x) = V_r \underline{|\beta x} \tag{5.255}$$

Again, substituting Eq. (5.252) into Eq. (5.184) yields,

$$I(x) = j\frac{V_r}{Z_c} \sin \beta x + \frac{V_r}{Z_c} \cos \beta x \tag{5.256}$$

$$I(x) = \frac{V_r}{Z_c}(\cos \beta x + j \sin \beta x) \tag{5.257}$$

$$I(x) = \frac{V_r}{Z_c} \underline{|\beta x} \tag{5.258}$$

The complex power flow at any point x of the transmission line is written as,

$$S = V(x)I(x)^* \tag{5.259}$$

Substituting Eqs. (5.255) and (5.258) into Eq. (5.259) yields,

$$S = V_r \underline{|\beta x} \times \frac{V_r}{Z_c} \underline{|-\beta x} = \frac{|V_r|^2}{Z_c} \tag{5.260}$$

At rated line voltage, the *SIL* is defined as the square of the voltage divided by the surge impedance. Mathematically, it can be expressed as,

$$SIL = \frac{V_L^2}{Z_c} \tag{5.261}$$

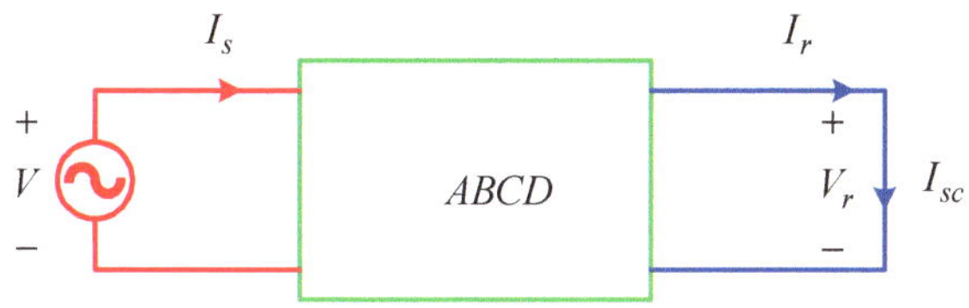

Fig. 5.12 Two-port circuit with receiving end short circuit

5.8 *ABCD* Parameters and Measurements

The sending end voltage and current of the transmission lines in terms of *ABCD* parameters are represented as,

$$V_s = AV_r + BI_r \tag{5.262}$$

$$I_s = CV_r + DI_r \tag{5.263}$$

Initially, a voltage source with a magnitude of V is connected at the sending end terminals and short circuit the receiving end terminals as shown in Fig. 5.12. Here, the receiving end voltage is zero, and Eq. (5.262) is modified as,

$$V = A \times 0 + BI_{sc} \tag{5.264}$$

$$I_{sc} = \frac{V}{B} \tag{5.265}$$

Again, short circuit the sending end terminals and insert a voltage source at the receiving end terminals as shown in Fig. 5.13. Substituting $V_s = 0$ in Eq. (5.262) yields,

$$0 = AV + BI_r \tag{5.266}$$

$$I_r = -\frac{AV}{B} \tag{5.267}$$

According to Fig. 5.12, the sending end current can be written as,

$$I_s = -I_{sc} = CV + DI_r \tag{5.268}$$

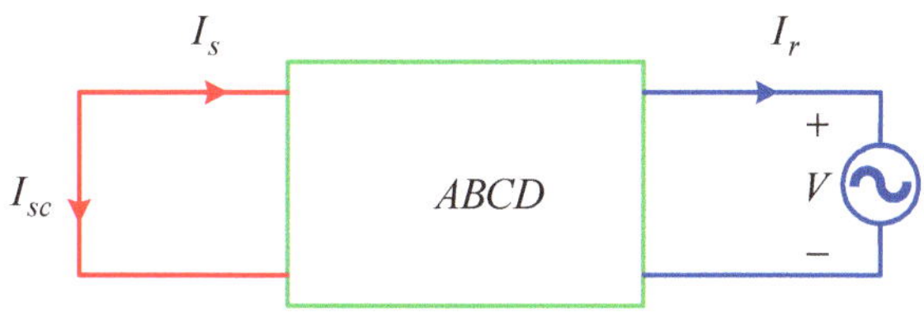

Fig. 5.13 Two-port circuit with sending end short circuit

Substituting Eqs. (5.265) and (5.267) into Eq. (5.268) yields,

$$-\frac{V}{B} = CV - D\frac{AV}{B} \tag{5.269}$$

$$-\frac{1}{B} = C - D\frac{A}{B} \tag{5.270}$$

$$AD - BC = 1 \tag{5.271}$$

Multiplying Eq. (5.262) by D and Eq. (5.263) by B yields,

$$DV_s = ADV_r + BDI_r \tag{5.272}$$

$$BI_s = BCV_r + BDI_r \tag{5.273}$$

Subtracting Eq. (5.273) from Eq. (5.272) yields,

$$(AD - BC)V_r = DV_s - BI_s \tag{5.274}$$

Substituting Eq. (5.271) into Eq. (5.274) yields,

$$V_r = DV_s - BI_s \tag{5.275}$$

Again, multiplying Eq. (5.262) by C and Eq. (5.263) by A yields,

$$CV_s = ACV_r + BCI_r \tag{5.276}$$

$$AI_s = ACV_r + ADI_r \tag{5.277}$$

Subtracting Eq. (5.277) from Eq. (5.276) yields,

$$(BC - AD)I_r = CV_s - AI_s \tag{5.278}$$

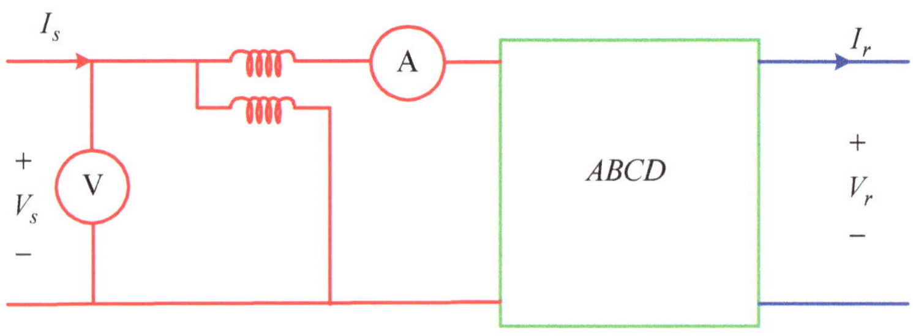

Fig. 5.14 Circuit for open circuit test

Substituting Eq. (5.271) into Eq. (5.278) yields,

$$-I_r = CV_s - AI_s \tag{5.279}$$

$$I_r = -CV_s + AI_s \tag{5.280}$$

The *ABCD* parameters can be determined using the concept of short circuit impedance and open circuit impedance. Initially, open the receiving end terminals and put the measuring instruments in the sending end side as shown in Fig. 5.14. For an open circuit test, setting $I_r = 0$ to Eqs. (5.262) and (5.263) yields,

$$V_s = AV_r \tag{5.290}$$

$$A = \frac{V_s}{V_r} \tag{5.291}$$

$$I_s = CV_r \tag{5.292}$$

$$C = \frac{I_s}{V_r} \tag{5.293}$$

From Eqs. (5.291) and (5.292), the expression of open circuit impedance can be written as,

$$Z_{oc} = \left(\frac{V_s}{I_s}\right)_{oc} = \frac{A}{C} \tag{5.294}$$

Again, short circuit the receiving end terminals and put all the measuring instruments in the sending end side as shown in Fig. 5.15. For a short circuit test, put $V_r = 0$ in Eqs. (5.262) and (5.263) yields,

$$V_s = BI_r \tag{5.295}$$

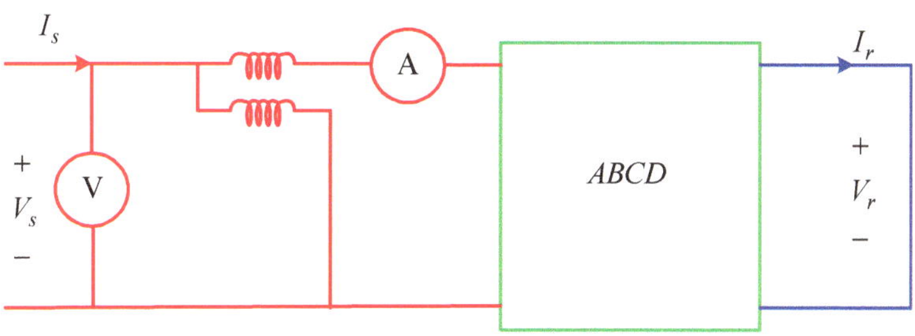

Fig. 5.15 Circuit for short circuit test

$$B = \frac{V_s}{I_r} \tag{5.296}$$

$$I_s = DI_r \tag{5.297}$$

$$D = \frac{I_s}{I_r} \tag{5.298}$$

From Eqs. (5.296) and (5.298), the expression of short circuit impedance can be written as,

$$Z_{sc} = \left(\frac{V_s}{I_s}\right)_{sc} = \frac{B}{D} = \frac{B}{A} \tag{5.299}$$

Substituting $A = D$ into Eq. (5.271) yields,

$$A^2 - BC = 1 \tag{5.300}$$

Subtracting Eq. (5.299) from Eq. (5.294) yields,

$$Z_{oc} - Z_{sc} = \frac{A}{C} - \frac{B}{A} = \frac{A^2 - BC}{AC} \tag{5.301}$$

$$Z_{oc} - Z_{sc} = \frac{1}{AC} \tag{5.302}$$

Dividing Eq. (5.294) by Eq. (5.302) yields,

$$\frac{Z_{oc}}{Z_{oc} - Z_{sc}} = \frac{\frac{A}{C}}{\frac{1}{AC}} = A^2 \tag{5.303}$$

$$A = \sqrt{\frac{Z_{oc}}{Z_{oc} - Z_{sc}}} \tag{5.304}$$

Substituting Eq. (5.304) into Eq. (5.299) yields,

$$B = Z_{sc}\sqrt{\frac{Z_{oc}}{Z_{oc} - Z_{sc}}} \tag{5.305}$$

Substituting Eq. (5.304) into Eq. (5.294) yields,

$$C = \frac{1}{Z_{oc}}\sqrt{\frac{Z_{oc}}{Z_{oc} - Z_{sc}}} \tag{5.306}$$

Example 5.5

Figure 5.16 shows a *Y*-model circuit. Calculate A, B, C, and D parameters.

Solution

Put V_s in the sending end side and the receiving end is open circuit as shown in Fig. 5.17. Applying KVL to the circuit in Fig. 5.17 yields,

$$V_s = V_r \tag{5.307}$$

The parameter *A* is determined as,

$$A = \frac{V_s}{V_r} = 1 \tag{5.308}$$

Again, applying KVL to the circuit in Fig. 5.17 yields,

$$-V_r + 10I_s = 0 \tag{5.309}$$

$$C = \frac{I_s}{V_r} = \frac{1}{10} = 0.1\ \mho \tag{5.310}$$

Again, short circuit the receiving end to find the parameters *B* and *D* as shown in Fig. 5.18.

Applying KCL to the circuit in Fig. 5.18 yields,

$$I_s = I_r \tag{5.311}$$

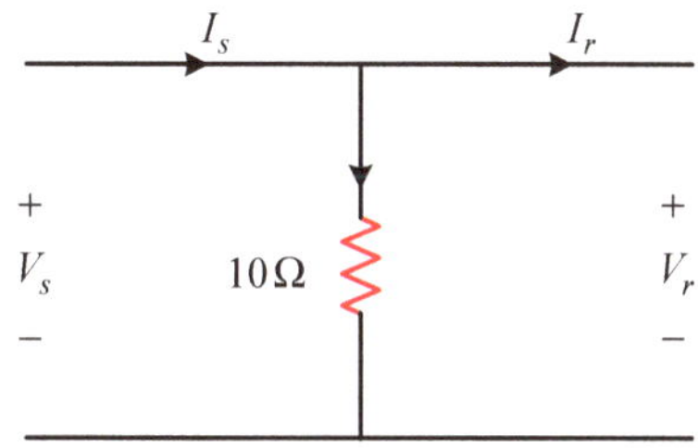

Fig. 5.16 Circuit for Example 5.5

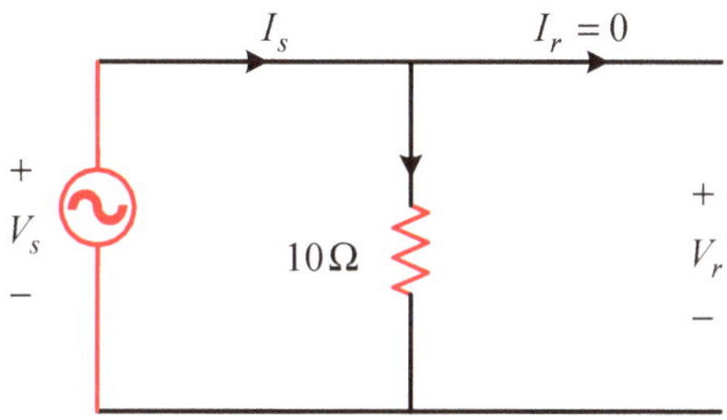

Fig. 5.17 Circuit with receiving end open circuit

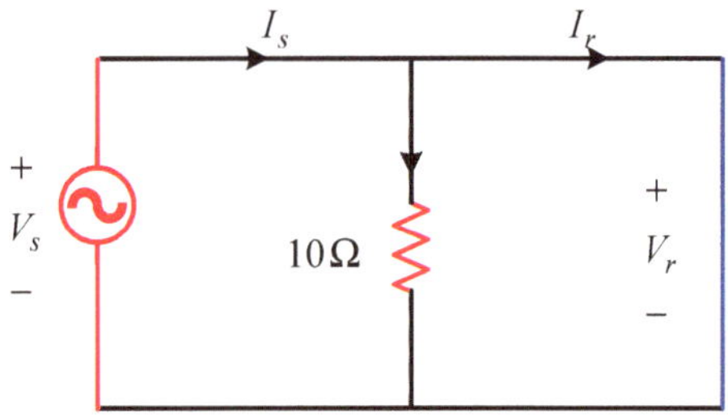

Fig. 5.18 Circuit with receiving end short circuit

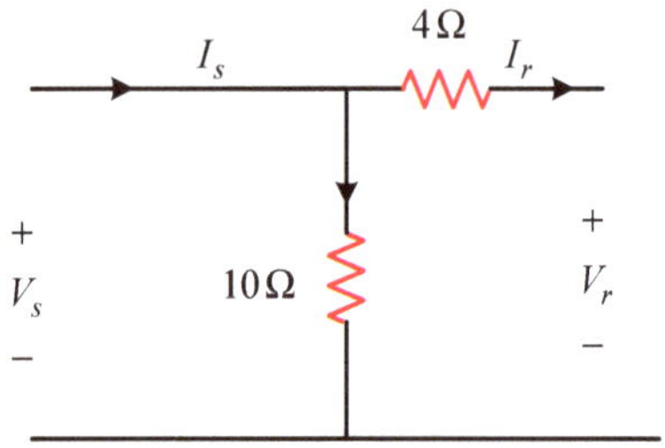

Fig. 5.19 Circuit for Practice Problem 5.5

Then, the parameters are calculated as,

$$D = \frac{I_s}{I_r} = 1 \tag{5.312}$$

$$B = \frac{V_s}{I_r} = \frac{0}{I_s} = 0 \tag{5.313}$$

Practice Problem 5.5
Calculate the parameters A, B, C, and D of the circuit as shown in Fig. 5.19.

5.9 Series Transmission Networks

Figure 5.20 shows transmission networks where the output of the first network is connected as an input to the second network. This arrangement is known as a tandem network. The tandem network is normally used in the communication circuit.

The sending end voltage and current in the first network can be written as,

$$V_s = A_1 V + B_1 I \tag{5.314}$$

$$I_s = C_1 V + D_1 I \tag{5.315}$$

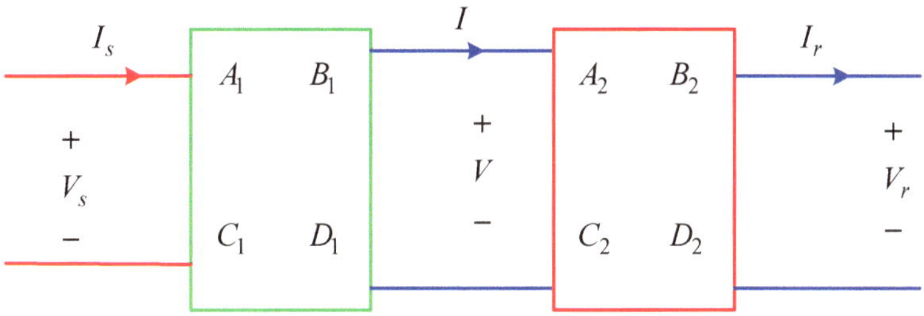

Fig. 5.20 Transmission networks in series

Equations (5.314) and (5.315) can be expressed in the matrix form as,

$$\begin{bmatrix} V_s \\ I_s \end{bmatrix} = \begin{bmatrix} A_1 & B_1 \\ C_1 & D_1 \end{bmatrix} \begin{bmatrix} V \\ I \end{bmatrix} \tag{5.316}$$

The sending end voltage and current in the second network can be written as,

$$V = A_2 V_r + B_2 I_r \tag{5.317}$$

$$I = C_2 V_r + D_2 I_r \tag{5.318}$$

Equations (5.317) and (5.318) can be expressed in the matrix form as,

$$\begin{bmatrix} V \\ I \end{bmatrix} = \begin{bmatrix} A_2 & B_2 \\ C_2 & D_2 \end{bmatrix} \begin{bmatrix} V_r \\ I_r \end{bmatrix} \tag{5.319}$$

Substituting Eq. (5.319) into Eq. (5.316) yields,

$$\begin{bmatrix} V_s \\ I_s \end{bmatrix} = \begin{bmatrix} A_1 & B_1 \\ C_1 & D_1 \end{bmatrix} \begin{bmatrix} A_2 & B_2 \\ C_2 & D_2 \end{bmatrix} \begin{bmatrix} V_r \\ I_r \end{bmatrix} \tag{5.320}$$

$$\begin{bmatrix} V_s \\ I_s \end{bmatrix} = \begin{bmatrix} A_1A_2 + B_1C_2 & A_1B_2 + B_1D_2 \\ C_1A_2 + D_1C_2 & C_1B_2 + D_1D_2 \end{bmatrix} \begin{bmatrix} V_r \\ I_r \end{bmatrix} \tag{5.321}$$

$$\begin{bmatrix} V_s \\ I_s \end{bmatrix} = \begin{bmatrix} A_s & B_s \\ C_s & D_s \end{bmatrix} \begin{bmatrix} V_r \\ I_r \end{bmatrix} \tag{5.322}$$

The equivalent parameters are represented as,

$$A_s = A_1A_2 + B_1C_2 \tag{5.323}$$

$$B_s = A_1B_2 + B_1D_2 \tag{5.324}$$

$$C_s = C_1A_2 + D_1C_2 \tag{5.325}$$

$$D_s = C_1B_2 + D_1D_2 \tag{5.326}$$

Example 5.6
The transmission lines are connected in series and deliver 120A current from the receiving end terminal with a 0.85 power factor lagging and 100 kV line voltage. The *ABCD* parameters are given as $A_1 = D_1 = 0.75\,\angle 4^\circ$, $B_1 = 24\,\angle 35^\circ\ \Omega$, $C_1 = 0.0004\,\angle 86^\circ\ \mho$, $A_2 = D_2 = 0.87\,\angle 5^\circ$, $B_2 = 35\,\angle 42^\circ\ \Omega$, $C_2 = 0.0008\,\angle 89^\circ\ \mho$. Calculate the sending end voltage and current.

Solution
The equivalent constants can be determined as,

$$A = A_1A_2 + B_1C_2 = 0.75\,\angle 4^\circ \times 0.87\,\angle 5^\circ + 24\,\angle 35^\circ \times 0.0008\,\angle 89^\circ = 0.64\,\angle 10.55^\circ \tag{5.327}$$

$$B = A_1B_2 + B_1D_2 = 0.75\,\angle 4^\circ \times 35\,\angle 42^\circ + 24\,\angle 35^\circ \times 0.87\,\angle 5^\circ = 50.73\,\angle 40.17^\circ\ \Omega \tag{5.328}$$

$$\begin{aligned} C &= C_1A_2 + D_1C_2 = 0.0004\,\angle 86^\circ \times 0.87\,\angle 5^\circ + 0.75\,\angle 4^\circ \times 0.0008\,\angle 89^\circ \\ &= 0.0009\,\angle 92.27^\circ\ \mho \end{aligned} \tag{5.329}$$

$$\begin{aligned} D &= C_1B_2 + D_1D_2 = 0.0004\,\angle 86^\circ \times 35\,\angle 42^\circ + 0.75\,\angle 4^\circ \times 0.87\,\angle 5^\circ \\ &= 0.65\,\angle 10.09^\circ \end{aligned} \tag{5.330}$$

The per phase receiving end voltage is calculated as,

$$V_r = \frac{100 \times 1000}{\sqrt{3}} = 57735.03\,\text{V} \tag{5.331}$$

The power factor is calculated as,

$$\cos\phi_r = 0.85, \phi_r = 31.79^\circ \tag{5.332}$$

The receiving end current is expressed as,

$$I_r = 120\,\angle -31.79^\circ\ \text{A} \tag{5.333}$$

The sending end voltage per phase can be determined as,

$$\begin{aligned} V_s &= AV_r + BI_r = 0.64\,\angle 10.55^\circ \times 57735.03 + 50.73\,\angle 40.17^\circ \times 120\,\angle -31.79^\circ \\ &= 43034.27\,\angle 10.24^\circ\ \text{V} \end{aligned} \tag{5.334}$$

The line voltage at the sending end is calculated as,

$$V_{sL} = \sqrt{3} \times 43034.27 = 74.54\,\text{kV} \tag{5.335}$$

The sending end current can be determined as,

$$\begin{aligned} I_s &= CV_r + DI_r = 0.0009 \angle 92.27^\circ \times 57735.03 + 0.65 \angle 10.09^\circ \times 120 \angle -31.79^\circ \\ &= 74.1 \angle 18.15^\circ\ \text{A} \end{aligned} \tag{5.336}$$

Practice Problem 5.6

A 120 kV line voltage with a 0.85 lagging power factor receiving end terminals receives 200 A current through series transmission networks. The ABCD parameters are given as $A_1 = D_1 = 0.55 \angle 5^\circ$, $B_1 = 20 \angle 25^\circ\ \Omega$, $C_1 = 0.0003 \angle 87^\circ\ \mho$, $A_2 = D_2 = 0.72 \angle 7^\circ$, $B_2 = 32 \angle 36^\circ\ \Omega$, $C_2 = 0.0005 \angle 89^\circ\ \mho$. Find the sending end voltage.

5.10 Parallel Transmission Networks

Transmission lines are sometimes connected in parallel for convenience. Figure 5.21 shows a circuit where the two transmission networks are connected in parallel. From Fig. 5.21, the following voltage and current equations can be written as,

$$V_s = V_{s1} = V_{s2} \tag{5.337}$$

$$V_r = V_{r1} = V_{r2} \tag{5.338}$$

$$I_s = I_{s1} + I_{s2} \tag{5.339}$$

$$I_r = I_{r1} + I_{r2} \tag{5.340}$$

The expressions of the sending end voltages for the two parallel networks are written as,

$$V_{s1} = A_1 V_{r1} + B_1 I_{r1} \tag{5.341}$$

$$V_{s2} = A_2 V_{r2} + B_2 I_{r2} \tag{5.342}$$

Substituting Eqs. (5.337) and (5.338) into Eqs. (5.341) and (5.342) yields,

$$V_s = A_1 V_r + B_1 I_{r1} \tag{5.343}$$

$$V_s = A_2 V_r + B_2 I_{r2} \tag{5.344}$$

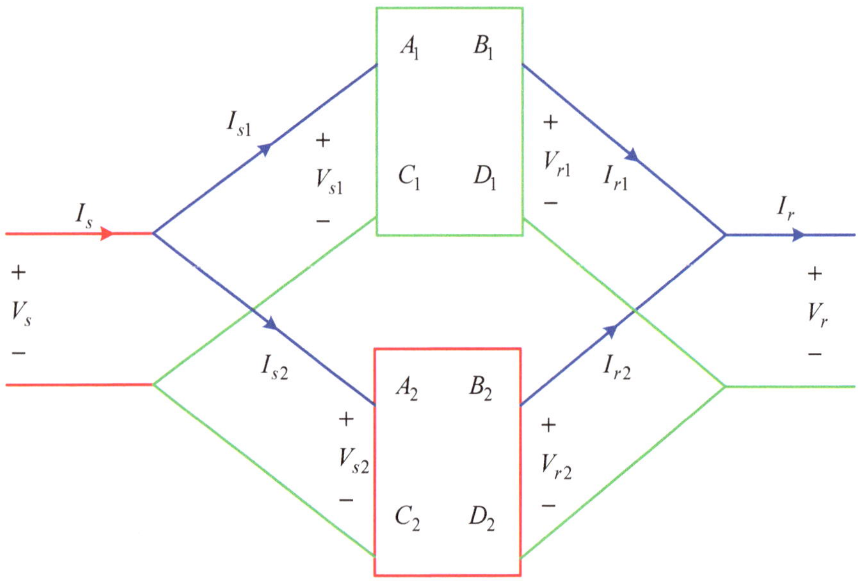

Fig. 5.21 Two transmission parallel networks

Multiplying Eq. (5.343) by B_2 and Eq. (5.344) by B_1, and adding the results yields,

$$(B_1 + B_2)V_s = (A_1B_2 + A_2B_1)V_r + B_1B_2(I_{r1} + I_{r1}) \tag{5.345}$$

$$V_s = \frac{A_1B_2 + A_2B_1}{B_1 + B_2}V_r + \frac{B_1B_2}{B_1 + B_2}(I_{r1} + I_{r1}) \tag{5.346}$$

Again, substituting Eq. (5.340) into Eq. (5.346) yields,

$$V_s = \frac{A_1B_2 + A_2B_1}{B_1 + B_2}V_r + \frac{B_1B_2}{B_1 + B_2}I_r \tag{5.347}$$

$$V_s = A_pV_r + B_pI_r \tag{5.348}$$

Where the expressions of general A and B parameters are written as,

$$A_p = \frac{A_1B_2 + A_2B_1}{B_1 + B_2} \tag{5.349}$$

$$B_p = \frac{B_1B_2}{B_1 + B_2} \tag{5.350}$$

The expressions of the sending end currents in the parallel networks are written as,

$$I_{s1} = C_1 V_r + D_1 I_{r1} \tag{5.351}$$

$$I_{s2} = C_2 V_r + D_2 I_{r2} \tag{5.352}$$

Substituting Eqs. (5.351) and (5.352) into Eq. (5.339) yields,

$$I_s = (C_1 + C_2)V_r + D_1 I_{r1} + D_2 I_{r2} \tag{5.353}$$

Subtracting Eq. (5.344) from Eq. (5.343) yields,

$$(A_1 - A_2)V_r + B_1 I_{r1} - B_2 I_{r2} = 0 \tag{5.354}$$

Again, substituting Eq. (5.340) into Eq. (5.353) yields,

$$I_s = (C_1 + C_2)V_r + D_1 I_{r1} + D_2 I_r - D_2 I_{r1} \tag{5.355}$$

$$I_s = (C_1 + C_2)V_r + (D_1 - D_2)I_{r1} + D_2 I_r \tag{5.356}$$

Substituting Eq. (5.340) into Eq. (5.354) yields,

$$(A_1 - A_2)V_r + B_1 I_{r1} - B_2 I_r + B_2 I_{r1} = 0 \tag{3.357}$$

$$(B_1 + B_2)I_{r1} = (A_2 - A_1)V_r + B_2 I_r \tag{5.358}$$

$$I_{r1} = \frac{A_2 - A_1}{B_1 + B_2}V_r + \frac{B_2}{B_1 + B_2}I_r \tag{5.359}$$

Again, substituting Eq. (5.359) into Eq. (5.356) yields,

$$I_s = (C_1 + C_2)V_r + (D_1 - D_2)\left[\frac{A_2 - A_1}{B_1 + B_2}V_r + \frac{B_2}{B_1 + B_2}I_r\right] + D_2 I_r \tag{5.360}$$

$$I_s = \left[C_1 + C_2 + \frac{(D_1 - D_2)(A_2 - A_1)}{B_1 + B_2}\right]V_r + \left[D_2 + (D_1 - D_2)\frac{B_2}{B_1 + B_2}\right]I_r \tag{5.361}$$

$$I_s = \left[C_1 + C_2 + \frac{(D_2 - D_1)(A_1 - A_2)}{B_1 + B_2}\right]V_r + \left[\frac{B_1 D_2 + B_2 D_2 + B_2 D_1 - B_2 D_2}{B_1 + B_2}\right]I_r \tag{5.362}$$

$$I_s = \left[C_1 + C_2 + \frac{(D_2 - D_1)(A_1 - A_2)}{B_1 + B_2}\right] V_r + \left[\frac{B_1 D_2 + B_2 D_1}{B_1 + B_2}\right] I_r \tag{5.363}$$

$$I_s = C_p V_r + D_p I_r \tag{5.364}$$

Where the expressions of the general C and D parameters are expressed as,

$$C_p = C_1 + C_2 + \frac{(D_2 - D_1)(A_1 - A_2)}{B_1 + B_2} \tag{5.365}$$

$$D_p = \frac{B_1 D_2 + B_2 D_1}{B_1 + B_2} \tag{5.366}$$

Example 5.7
The parameters of parallel transmission networks are given by $A_1 = D_1 = 0.65\angle 3^\circ$, $B_1 = 34\angle 30^\circ\ \Omega$, $C_1 = 0.0002\angle 89^\circ\ \mho$, $A_2 = D_2 = 0.76\angle 4^\circ$, $B_2 = 45\angle 48^\circ\ \Omega$, $C_2 = 0.0004\angle 90^\circ\ \mho$. Calculate the equivalent parameters.

Solution
The equivalent constants can be determined as,

$$A_p = \frac{A_1 B_2 + A_2 B_1}{B_1 + B_2} = \frac{0.65\angle 3^\circ \times 45\angle 48^\circ + 0.76\angle 4^\circ \times 34\angle 30^\circ}{34\angle 30^\circ + 45\angle 48^\circ} = 0.7\angle 2.77^\circ \tag{5.367}$$

$$B_p = \frac{B_1 B_2}{B_1 + B_2} = \frac{34\angle 30^\circ \times 45\angle 48^\circ}{34\angle 30^\circ + 45\angle 48^\circ} = 19.60\angle 37.74^\circ\ \Omega \tag{5.368}$$

$$\begin{aligned} C_p &= C_1 + C_2 + \frac{(D_2 - D_1)(A_1 - A_2)}{B_1 + B_2} = 0.0002\angle 89^\circ + 0.0004\angle 90^\circ \\ &+ \frac{(0.76\angle 4^\circ - 0.65\angle 3^\circ)(0.65\angle 3^\circ - 0.76\angle 4^\circ)}{34\angle 30^\circ + 45\angle 48^\circ} = 0.00067\angle 102.36^\circ \end{aligned} \tag{5.369}$$

$$D_p = \frac{B_1 D_2 + B_2 D_1}{B_1 + B_2} = \frac{0.65\angle 3^\circ \times 45\angle 48^\circ + 0.76\angle 4^\circ \times 34\angle 30^\circ}{34\angle 30^\circ + 45\angle 48^\circ} = 0.7\angle 2.77^\circ \tag{5.370}$$

Practice Problem 5.7
The parameters of parallel transmission networks are given by $A_1 = D_1 = 0.61\angle 4^\circ$, $B_1 = 24\angle 32^\circ\ \Omega$, $C_1 = 0.0004\angle 86^\circ\ \mho$, $A_2 = D_2 = 0.71\angle 5^\circ$, $B_2 = 35\angle 40^\circ\ \Omega$, $C_2 = 0.0006\angle 89^\circ\ \mho$. Calculate the equivalent parameters.

5.11 Ferranti Effect

In 1890, Sebastian Ziani de Ferranti, a British Electrical Engineer introduced a theory that the receiving end voltage is higher than the sending end voltage of the medium or the long transmission lines in case of light load or no-load condition. This phenomenon is known as the Ferranti effect. A high amount of capacitance and inductance distributed across the entire length of a long transmission line. The current drawn by the distributed capacitance is greater than the current drawn by the load at the receiving end of the line during the light or no-load. This charging current of the capacitor leads to a voltage drop across the inductor, which is in phase with the sending end voltage. This voltage drop keeps increasing as it moves toward the receiving end terminals. As a result, the receiving end voltage is greater than the sending end voltage leading to phenomena called the Ferranti effect. Therefore, both the inductance and the capacitance are responsible for producing the Ferranti effect. The distributed capacitance and inductance of a long transmission line are shown in Fig. 5.22.

The distributed parameters of a long transmission line are represented as lumped parameters as shown in Fig. 5.23. The shunt capacitance is concentrated at the receiving end. Figure 5.24 shows the phasor diagram where the receiving end voltage is considered as a reference phasor. The charging current (I_c) leads the receiving end voltage by 90 degrees. The sending end voltage can be written as,

$$V_s = V_r + I_c(R + jX_L) \tag{5.371}$$

Substituting the charging current $I_c = V_r(j\omega C)$ and $X_L = \omega L$ in Eq. (5.371) yields,

$$V_s = V_r + j\omega C V_r(R + j\omega L) \tag{5.372}$$

$$V_s = V_r - \omega^2 LCV_r + j\omega CV_r R \tag{5.373}$$

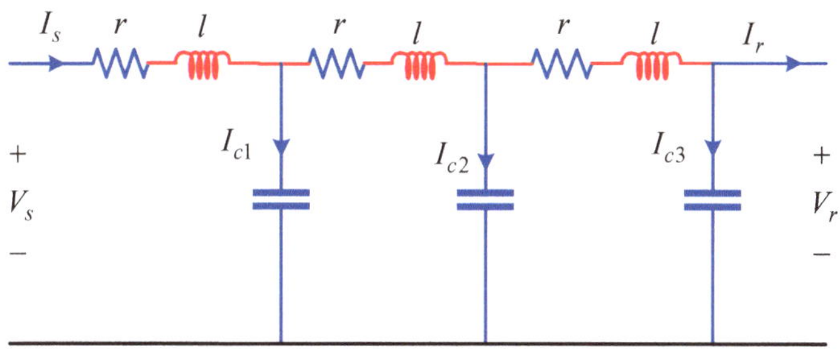

Fig. 5.22 Long transmission line with distributed parameters

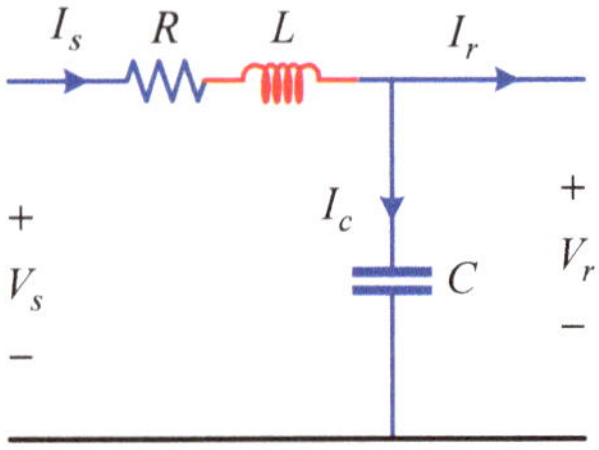

Fig. 5.23 Long transmission line with lumped parameters

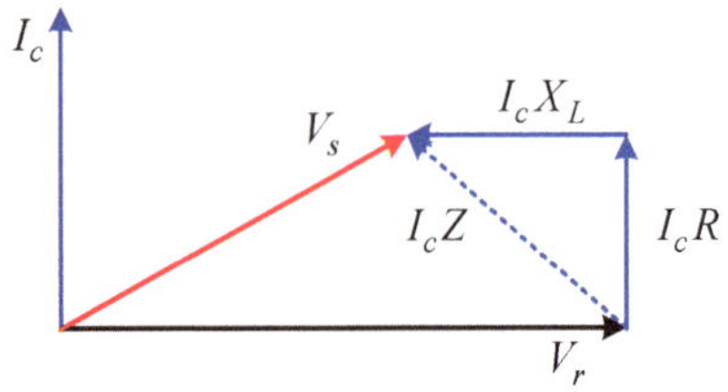

Fig. 5.24 Phasor diagram of a long transmission line with lumped parameters

From Fig. 5.24, the following relation can be written as,

$$V_s^2 = (V_r - I_c X_L)^2 + (I_c R)^2 \tag{5.374}$$

The effect of resistance in a long transmission line is negligible, and Eqs. (5.373) and (5.374) are modified as,

$$V_s = V_r - \omega^2 LCV_r \tag{5.375}$$

$$V_s = V_r - I_c X_L \tag{5.376}$$

$$V_r = V_s + I_c X_L \tag{5.377}$$

Again, considering a nominal π-model of a medium transmission line to explain the Ferranti effect. For the open circuit line, putting $I_r = 0$ in Eq. (5.89) yields,

$$V_s = \left(1 + \frac{YZ}{2}\right) V_r \tag{5.378}$$

$$V_s - V_r = \frac{YZ}{2} V_r \tag{5.379}$$

Substituting the expression of impedance and admittance in Eq. (5.379) yields,

$$V_s - V_r = \frac{(j\omega Cl)(r + j\omega L)l}{2} V_r \tag{5.380}$$

Substitute $r = 0$ in Eq. (5.380) yields,

$$V_s - V_r = \frac{-\omega^2 Cl^2}{2} LV_r \quad (5.381)$$

$$V_s = \left(1 - \frac{\omega^2 Cl^2}{2} L\right) V_r \quad (5.382)$$

5.12 Ground Wires and Corona Discharge

Ground wires: The ground wires are bare conductors installed at the top of transmission towers. The overhead transmission lines are normally protected from the lightning strikes by the ground wires. The ground wires normally do not carry current. Therefore, ground wires are made of steel and are directly connected to the ground at each transmission tower and distribution pole. The ground wires help to pass the high current that receives during lightning strikes.

Corona discharge: When a low alternating voltage is applied across the two conductors whose diameters are very small compared to their separation distance, there is no apparent change of the surrounding atmospheric air of the conductors. When the applied voltage exceeds a certain value, called critical disruptive voltage, the conductors are surrounded by a violet glow. Therefore, the corona is a partial discharge that occurs at the surface of the power conductor when the electric field exceeds the breakdown strength of the surrounding air. The electric field around the surface of the conductor increases if the voltage between the conductors is increased. Due to high electric fields, the surrounding air molecules become ionized. This causes hissing sound and causes a glow on the line. This phenomenon is known as corona discharge. Coronas may be positive or negative. The positive corona or negative corona is normally determined by the polarity of the voltage on the power conductor.

5.13 Traveling Waves

The voltage and current on a long transmission line always satisfy the relation of the wave equation. The effects of resistance and conductance are neglected for a long transmission line. In this case, the model is dependent on the inductance and capacitance. The transmission line without resistance and conductance is known as a lossless transmission line [4, 5]. Considering a long transmission equation as shown in Fig. 5.25 for deriving traveling wave equations, applying KVL to the circuit in Fig. 5.25 yields,

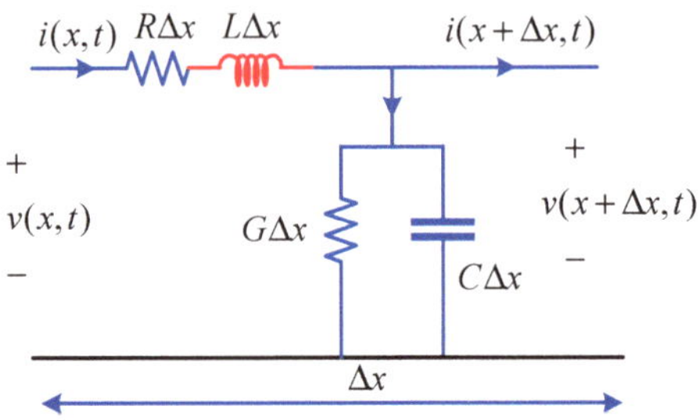

Fig. 5.25 Part of a long transmission line with lumped parameters

$$-v(x,t) + R\Delta x i(x,t) + L\Delta x \frac{\partial i(x,t)}{\partial t} + v(x+\Delta x,t) = 0 \tag{5.383}$$

$$\frac{v(x+\Delta x,t) - v(x,t)}{\Delta x} = -Ri(x,t) - L\frac{\partial i(x,t)}{\partial t} \tag{5.384}$$

Taking limit as $\Delta x \to 0$, Eq. (5.384) is modified as,

$$Lt_{\Delta x \to 0} \frac{v(x+\Delta x,t) - v(x,t)}{\Delta x} = -Ri(x,t) - L\frac{\partial i(x,t)}{\partial t} \tag{5.385}$$

$$\frac{\partial v(x,t)}{\partial x} = -Ri(x,t) - L\frac{\partial i(x,t)}{\partial t} \tag{5.386}$$

Applying KCL to the circuit in Fig. 5.25 yields,

$$i(x,t) = G\Delta x v(x,t) + C\Delta x \frac{\partial v(x,t)}{\partial t} + i(x+\Delta x,t) \tag{5.387}$$

$$\frac{i(x+\Delta x,t) - i(x,t)}{\Delta x} = -Gv(x,t) - C\frac{\partial v(x,t)}{\partial t} \tag{5.388}$$

Taking limit as $\Delta x \to 0$, Eq. (5.388) is modified as,

$$Lt_{\Delta x \to 0} \frac{i(x+\Delta x,t) - i(x,t)}{\Delta x} = -Gv(x,t) - C\frac{\partial v(x,t)}{\partial t} \tag{5.389}$$

$$\frac{\partial i(x,t)}{\partial x} = -Gv(x,t) - C\frac{\partial v(x,t)}{\partial t} \tag{5.390}$$

Considering the voltage and current in complex forms as

$$v(x,t) = \mathrm{Re}\left[V(x)e^{j\omega t}\right] \tag{5.391}$$

$$i(x,t) = \mathrm{Re}\left[I(x)e^{j\omega t}\right] \tag{5.392}$$

Substituting Eq. (5.391) into Eq. (5.386) yields,

$$\frac{\partial V(x)}{\partial x} = -(R + j\omega L)I(x) \tag{5.393}$$

Again, substituting Eq. (5.392) into Eq. (5.390) yields,

$$\frac{\partial I(x)}{\partial x} = -(G + j\omega C)V(x) \tag{5.394}$$

For a long transmission line, $R = 0$ and $G = 0$, Eqs. (5.393) and (5.394) are modified as,

$$\frac{\partial V(x)}{\partial x} = -j\omega L I(x) \tag{5.395}$$

$$\frac{\partial I(x)}{\partial x} = -j\omega C V(x) \tag{5.396}$$

Differentiating Eqs. (5.395) and (5.396) with respect to x yields,

$$\frac{\partial^2 V(x)}{\partial x^2} + \omega^2 LC\, V(x) = 0 \tag{5.397}$$

$$\frac{\partial^2 I(x)}{\partial x^2} + \omega^2 LC\, I(x) = 0 \tag{5.398}$$

Considering the expression of phase constant as

$$\beta = \omega\sqrt{LC} \tag{5.399}$$

The general solutions of Eqs. (5.397) and (5.398) are,

$$V(x) = V_0^+ e^{-\beta x} + V_0^- e^{-\beta x} \tag{5.400}$$

$$I(x) = I_0^+ e^{-\beta x} + I_0^- e^{-\beta x} \tag{5.401}$$

Here, the superscripts positive (+) and negative (−) represent the wave traveling in the $+x$ and $-x$ directions along the transmission line.

The wavelength (λ) of a signal is defined as the distance between the two peaks over 2π radians, and it can be expressed as,

$$\beta\lambda = 2\pi \tag{5.402}$$

$$\lambda = \frac{2\pi}{\beta} \tag{5.403}$$

Substituting Eq. (5.399) into Eq. (5.403) yields,

$$\lambda = \frac{1}{f\sqrt{LC}} \tag{5.404}$$

Since the signal is oscillating in time at rate ω rad sec, the propagation velocity of the wave is expressed as,

$$v = \frac{\omega}{\beta} \tag{5.405}$$

Substituting Eq. (5.399) into Eq. (5.405) yields,

$$v = \frac{1}{\sqrt{LC}} \tag{5.406}$$

Example 5.8 The per phase per km line inductance and capacitance of 200 km long three-phase 500 kV, 50 Hz transmission line are 0.87 mH and 0.012 μF. Assume a lossless line, and the line supplies power from 500 kV line to the load of 600 MW at a 0.85 power factor lagging. Determine the phase constant velocity of the wave propagation, wavelength, surge impedance, per phase sending end voltage, sending end current, three-phase sending end power, and voltage regulation.

Solution
The value of the phase constant is calculated as,

$$\beta = \omega\sqrt{LC} = 2\pi \times 50\sqrt{0.87 \times 10^{-3} \times 0.012 \times 10^{-6}} = 0.001\,\text{rad/km} \tag{5.407}$$

The velocity of the wave propagation is calculated as,

$$v = \frac{1}{\sqrt{LC}} = \frac{1}{\sqrt{0.87 \times 10^{-3} \times 0.012 \times 10^{-6}}} = 3.09 \times 10^5\,\text{km/s} \tag{5.408}$$

The value of the wavelength is as,

$$\lambda = \frac{v}{f} = \frac{3.09 \times 10^5}{50} = 6189.84\,\text{km} \tag{5.409}$$

The value of the surge impedance is determined as,

$$Z_c = \sqrt{\frac{L}{C}} = \sqrt{\frac{0.87 \times 10^{-3}}{0.012 \times 10^{-6}}} = 269.26\,\Omega \tag{5.410}$$

The receiving end current per phase is calculated as,

$$I_r = \frac{600 \times 1000}{\sqrt{3} \times 500 \times 0.85} \underline{|-31.79^\circ} = 815.08 \underline{|-31.79^\circ}\,\text{A} \tag{5.412}$$

Per phase receiving end voltage is calculated as,

$$V_r = \frac{500000}{\sqrt{3}} = 288675.13\,\text{V} \tag{5.413}$$

The value of the following parameter is calculated as,

$$\beta l = 0.001 \times 200 = 0.2\,\text{rad} = 11.46^\circ \tag{5.414}$$

The sending end voltage can be determined as,

$$\begin{aligned} V_s &= V_r \cos\beta l + jZ_c I_r \sin\beta l \\ &= (\cos 11.46) \times 288675.13 + (j269.26 \times \sin 11.46)815.08 \underline{|-31.79^\circ} \end{aligned} \tag{5.415}$$

$$\begin{aligned} V_s &= (0.980) \times 288675.13 + (j269.26 \times \sin 11.46)815.08 \underline{|-31.79^\circ} \\ &= 308110.32 \underline{|6.91^\circ}\,\text{V} \end{aligned} \tag{5.416}$$

The sending end current is calculated as,

$$\begin{aligned} I_s &= \left(j\frac{1}{Z_c}\sin\beta l\right)V_r + (\cos\beta l)I_r = \left(j\frac{1}{269.26}\sin 11.46\right) \times 288675.13 \\ &+ (\cos 11.46) \times 815.08 \underline{|-31.79^\circ} = 710.09 \underline{|-17.02^\circ}\,\text{A} \end{aligned} \tag{5.417}$$

Three-phase sending end power is calculated as,

$$\begin{aligned} S_s &= 3V_s I_s^* = 3 \times 308110.32 \underline{|6.91^\circ} \times 710.09 \underline{|17.02^\circ} \\ &= 599.94\,\text{MW} + 266.23\,\text{MVAR} \end{aligned} \tag{5.418}$$

The percentage voltage regulation can be determined as,

$$VR = \frac{\frac{V_s}{A} - V_r}{V_r} \times 100 \frac{\frac{308110.32}{0.98} - 288675.13}{288675.13} \times 100 = 8.91\% \tag{5.419}$$

Practice Problem 5.8

A three-phase 400 kV 50 Hz long transmission line is having a length of 170 km. The per phase per km line inductance and capacitance are 0.96 mH and 0.025 μF. Assume a lossless line, and the line supplies power to the load of 300 MW at a 0.95 power factor lagging. Calculate the phase constant, the velocity of the wave propagation, wavelength, surge impedance, and sending end phase voltage.

References

1. Glover JD, Sarma MS, Overbye TJ (2017) Power system analysis and design, 6th ed, Cengage Learning, pp 1–942, USA
2. Nagsarkar TK, Sukhija MS (2014) Power system analysis, 6th ed, Oxford University Press, pp 1–726
3. Grainger JJ, Stevenson WD, Stevenson WD (2015) Power systems analysis, 2nd Education, McGraw-Hill Education, USA
4. Wildi T (2005) Electrical machines, drives and power systems, 6th Edn, Pearson Education, pp 1–934, USA
5. Bergen AR, Vijay V (1999) Power systems analysis, 2nd Edn., Pearson Education, pp 1–632, USA

Exercise Problems

5.1 The per phase series impedance of a single-phase short transmission line is $5+j16\,\Omega$, and the line delivers a power of 10 kW at a power factor of 0.95 lagging from the 240 V receiving end terminals. Calculate the sending end voltage, sending end power factor, and transmission efficiency.

5.2 A long single-phase 24 km long short transmission line delivers a power of 200 kW at a 0.85 leading power factor to the load from the 11 kV terminals. The per phase per km resistance and reactance of the line are $0.2\,\Omega$ and $0.5\,\Omega$, respectively. Find the sending end voltage, voltage regulation, and efficiency.

5.3 A 10 km long single-phase short transmission line delivers a power of 180 kW at a 0.90 lagging power factor to the load from the 6.6 kV receiving end terminals. The per phase per km resistance and reactance of the line are $0.1\,\Omega$ and $0.4\,\Omega$, respectively. Calculate the sending end voltage, voltage regulation, and efficiency.

5.4 A load of 140 kW receives power from the 11 kV single-phase short transmission line with a unity power factor. The per phase impedance of the line is $3+j8\,\Omega$. Find the sending end voltage and efficiency.

5.5 The sending end voltage of a three-phase short transmission line is 11 kV and delivers a power of 1200 kW at a 0.9 power factor lagging to a three-phase load. The impedance of the line is found to be $3+j5\,\Omega$. Calculate the receiving end voltage, line current, and efficiency.

5.6 The per phase impedance of an 11 kV three-phase short transmission line is found to be $10+j18\,\Omega$. A three-phase load of 4 MW at a 0.95 power factor lagging is connected at the receiving end of the line. Find the sending end voltage per phase and load angle.

5.7 The per phase per km resistance and reactance of the three-phase 11 kV and 50 km short transmission line are $0.16\,\Omega$ and $0.3\,\Omega$, respectively. The line delivers a three-phase load of 280 MVA at a 0.85 power factor lagging. Calculate the sending end voltage, real power, and sending end complex power.

5.8 The impedance of a three-phase short transmission line is $4+j10\,\Omega$. This line delivers power to the load at a 0.85 power factor lagging. The per phase sending ending voltage and receiving end voltage are 11 kV and 6.6 kV, respectively. Calculate the line current, per phase load power, and sending end power factor.

5.9 A load of 20 MW at a 0.85 lagging power factor is connected with a three-phase 66 kV medium transmission line. The line has a per phase impedance of $1.2+j5\,\Omega$ and a per phase shunt admittance $j0.0003$ S. Use the nominal *T*-model to determine the *A*, *B*, *C*, *D* parameters, sending end line voltage, sending end current, and sending end power factor.

5.10 The line has an impedance and a shunt admittance of a three-phase 33 kV, and 90 km medium transmission line is $0.1+j0.6\,\Omega$ per km and $j2.4\times10^{-6}$ S per km, respectively. The line delivers a power of 24 MW at a 0.90 power factor leading to the load. Use the nominal *T*-model to determine the *A*, *B*, *C*, *D* parameters, sending end line voltage, sending end current, and sending end power factor.

5.11 The line has an impedance, and a shunt admittance of a three-phase 66 kV and 100 km long medium transmission line are $0.2+j0.8\,\Omega$ per km and $j3.5\times10^{-6}$ S per km, respectively. The line delivers a power of 30 MW at a 0.95 power factor lagging to the load. Use the nominal π-model to determine the *A*, *B*, *C*, *D* parameters, sending end line voltage, sending end current, and sending end power factor.

5.12 The per km line impedance and admittance of a three-phase 100 kV and 120 km long medium transmission line are $0.3+j0.5\,\Omega$ and $j1.9\times10^{-6}$ S, respectively. The line delivers a power of 25 MW at a 0.85 power factor lagging to the load. Use a nominal T-model to determine the sending end line voltage, sending end current, and transmission efficiency.

5.13 A 130 km long three-phase 50 Hz medium transmission line delivers a power of 30 MW at a power factor of 0.90 lagging to a balanced load whose line voltage is 66 kV. The per phase per km transmission line parameters are $R=0.3\,\Omega, X_L=0.6\,\Omega$, $Y=0.06\times10^{-4}$ ℧. Calculate the sending end phase voltage and voltage regulation by considering the nominal *T*-model.

5.14 The per phase per km parameters of a 250 km long transmission line are $r=0.04\,\Omega, x=0.19\,\Omega$, $y=0.09\times10^{-4}$ ℧. The transmission line delivers a power of 40 MW at a 0.90 lagging power factor. Calculate the per phase sending end line voltage and sending end current if the receiving end voltage is 110 kV.

5.15 The parameters of a series of transmission networks are given by $A_1=D_1=0.85\,\angle 6^\circ$, $B_1=30\,\angle 38^\circ\,\Omega$, $C_1=0.0003\,\angle 82^\circ$ ℧, $A_2=D_2=0.89\,\angle 8^\circ$, $B_2=37\,\angle 40^\circ\,\Omega$, $C_2=0.0006\,\angle 87^\circ$ ℧.

The line delivers a current of 90 A from the 110 kV receiving end terminal at a 0.95 power factor lagging. Calculate the sending end voltage and current.

5.16 The parameters of a parallel transmission network are given by $A_1 = D_1 = 0.56\angle 6^\circ$, $B_1 = 44\angle 35^\circ\ \Omega$, $C_1 = 0.0004\angle 82^\circ\ \mho$, $A_2 = D_2 = 0.77\angle 5^\circ$, $B_2 = 35\angle 44^\circ\ \Omega$, $C_2 = 0.0007\angle 89^\circ\ \mho$. Calculate the equivalent parameters.

5.17 The per phase per km line inductance and capacitance of 180 km 50 Hz long transmission line are 0.79 mH and 0.018 μF. Assume a lossless line and the line supplies power to the load of 600 MW at a 0.85 power factor lagging from the 500 kV line. Calculate the phase constant, the velocity of the wave propagation, wavelength, and surge impedance.

Chapter 6
Symmetrical and Unsymmetrical Faults

6.1 Introduction

The goal of any power utility company is to run its power system network under balanced condition. The power system network is said to be balanced when it is operating in normal-load condition. This normal operating condition of the power system can be disrupted due to adverse weather conditions such as heavy wind, lightning strikes, or due to other factors such as birds shorting out the lines, vehicles collide with transmission line poles or towers by accident, or trees fall on the transmission lines. The lightning strike on the transmission line may generate a very high transient voltage, which exceeds the basic insulation voltage level of the transmission lines. This event triggers the flashover from the resultant high magnitude of the current that passes through the transmission tower to the ground. This condition of the transmission lines is known as a short circuit condition, and the fault associated with this phenomenon is known as a short circuit fault. In a short circuit situation, a very low impedance path is created either in between two transmission lines or in between a transmission line and ground. In this case, the resulting high magnitude current imposes a heavy duty on the circuit breaker and other controlling equipment. The short circuit faults are classified as symmetrical and unsymmetrical faults. In this chapter, symmetrical and unsymmetrical faults, symmetrical components, zero sequence components of the machines, and classification unsymmetrical faults will be discussed.

Md. A. Salam, *Fundamentals of Electrical Power Systems Analysis*,
https://doi.org/10.1007/978-981-15-3212-2_6

6.2 Symmetrical Faults

The symmetrical faults are often known as balanced faults. In the case of balanced faults, three lines are affected equally, and the system remains in a balanced condition. These types of faults are rare in the power system, and it contributes 2–5% of the total fault.

These faults are easy to analyze. The symmetrical faults are classified as three line-to-ground fault (LLLG) and three-line fault (LLL). The connection diagrams of symmetrical faults are shown in Fig. 6.1. If the fault impedance $Z_f = 0$, then the fault is known as a solid or bolted fault.

The balanced three-phase fault occurs suddenly at synchronous generator terminals. Initially, the magnitude of the short circuit current is high. Later on, the magnitude of this current is reduced. This short circuit current is divided into three periods, namely subtransient, transient, and steady-state periods. The trace of a short circuit stator current with subtransient, transient, and steady-state periods is shown in Fig. 6.2. In the subtransient period, the short circuit current reduces rapidly, and it is lasting for a few cycles.

In the transient period, the reduction of the short circuit current is more moderate, and it continues for more cycles. In the final state, the oscillation of the current is stable, and the currents in these regions are used to define the various reactances of the synchronous generator if the armature resistance is neglected. The rms value of the ac current flows in the generator during the subtransient period is known as subtransient current. This current is represented by I'', and it is generated by the damper windings of the generator. The ratio of generated voltage to the subtransient current is known as subtransient reactance, and it is denoted by X_d''. The subtransient reactance can be expressed as,

$$I'' = \frac{oc}{\sqrt{2}} = \frac{E_g}{X_d''} \tag{6.1}$$

The rms value of the AC current flows in the generator during the transient period is known as transient current. This current is represented by I', and it is generated by the transient DC component of the field current of the generator during a fault. The ratio of generated voltage to the transient current is known as transient reactance, and it is denoted by X_d'. The subtransient reactance can be expressed as,

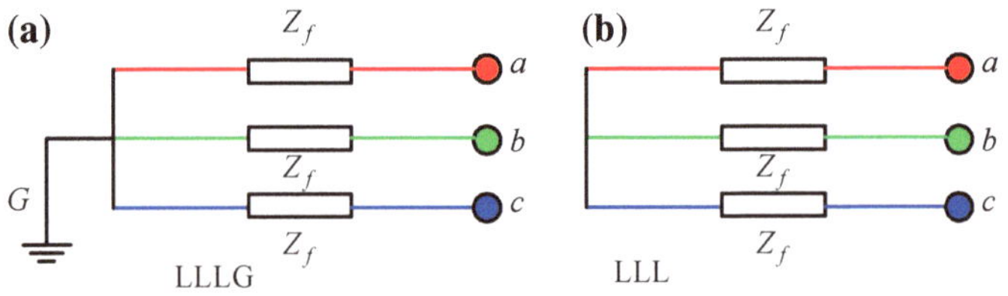

Fig. 6.1 Symmetrical faults' connection diagram

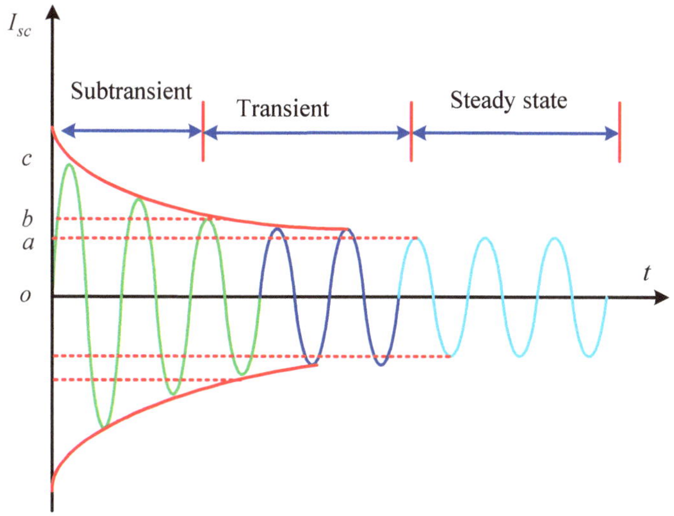

Fig. 6.2 Short circuit current waveform

$$I' = \frac{ob}{\sqrt{2}} = \frac{E_g}{X'_d} \tag{6.2}$$

After a transient period, the fault current reaches a stable condition which is known as steady-state current, and it is denoted by I. The synchronous reactance or direct axis reactance is the ratio of induced voltage to the steady-state current, and it can be expressed as,

$$I = \frac{oa}{\sqrt{2}} = \frac{E_g}{X_d} \tag{6.3}$$

The rms value of the AC fault current in a synchronous generator varies over time, and it can be expressed as,

$$I(t) = (I'' - I')e^{-\frac{t}{T''}} + (I' - I)e^{-\frac{t}{T'}} + I \tag{6.4}$$

where

T'' is the time constant for the subtransient period,

T' is the time constant for the transient period

Example 6.1
A three-phase 100 MVA, 11 kV, 50 Hz synchronous generator is used in the power station, and a three-phase fault occurs at the generator terminals. The per unit reactances of the generator are given as $X'' = 0.16$, $X' = 0.25$, and $X = 1.05$. The

subtransient and transient time constants are 0.05 s and 1.04 s, respectively. Assume the initial DC component of the current is 50% of the initial AC component of the current. Calculate the subtransient current, transient current, steady-state current, total current at the beginning of the fault, and AC component of the current as a function of time.

Solution

Consider the base values are 100 MVA and 11 kV. The value of the base current is calculated as,

$$I_b = \frac{100 \times 10^6}{\sqrt{3} \times 11 \times 10^3} = 5248.63 \text{ A} \tag{6.5}$$

The value of the subtransient current or initial AC component of the current is calculated as,

$$I'' = \frac{1}{0.16} = 6.25 \text{ pu} \tag{6.6}$$

$$I'' = 6.25 \times 5248.63 = 32803.93 \text{ A} \tag{6.7}$$

The value of the transient current is calculated as,

$$I' = \frac{1}{0.25} = 4 \text{ pu} \tag{6.8}$$

$$I' = 4 \times 5248.63 = 20994.52 \text{ A} \tag{6.9}$$

The value of the steady-state current is calculated as,

$$I = \frac{1}{1.05} = 0.95 \text{ pu} \tag{6.10}$$

$$I = 0.95 \times 5248.63 = 4986.19 \text{ A} \tag{6.11}$$

The total current at the beginning of the fault is calculated as,

$$I_{\text{total}} = 1.5I'' = 1.5 \times 32803.93 = 49205.89 \text{ A} \tag{6.12}$$

The value of the AC component of the current can be determined as,

$$\begin{aligned} I(t) &= (I'' - I')e^{-\frac{t}{T''}} + (I' - I)e^{-\frac{t}{T'}} + I \\ &= 11809.41e^{-\frac{t}{0.05}} + 16008.33e^{-\frac{t}{1.04}} + 4986.19 \text{ A} \end{aligned} \tag{6.13}$$

Practice Problem 6.1

A three-phase 90 MVA, 11 kV, 50 Hz generator is installed at the power station, and the three-phase fault occurs at the generator terminals. The per unit reactances of the generator are given as $X'' = 0.11$, $X' = 0.20$, and $X = 1.12$. Assume the initial DC component of the current is 50% of the initial AC component of the current. Find the subtransient current, transient current, steady-state current, and total current at the beginning of the fault.

6.3 Calculation of Short Circuit Current and kVA

A short circuit in a power system occurs due to failure of the insulation of the equipment during lightning or switching activities. This short circuit current is several times higher than the normal rating current, which in turn damages the system severely. Sometimes, a high impedance fault current is not enough to run the relay and circuit breaker or blow the related fuses. Therefore, there is a prime importance to determine the short circuit current in the power system. The transformer and transmission line is normally represented by the leakage reactance and series reactance, respectively. The voltage source and the reactance represent the synchronous generator. Initially, the percentage of reactance needs to be defined to derive the expression of the short circuit current. The voltage drop due to reactance at rated current is expressed in percent of the rated voltage as,

$$\%X_p = \frac{IX}{V} \times 100 \tag{6.14}$$

Equation (6.14) can be rearranged as,

$$\frac{X}{V} = \frac{\%X_p}{100 \times I} \tag{6.15}$$

$$\frac{V}{X} = \frac{100 \times I}{\%X_p} \tag{6.16}$$

Again, consider a short circuit reactance X. For a rated voltage, the short circuit current can be determined as,

$$I_{\text{sc}} = \frac{V}{X} \tag{6.17}$$

Substituting Eq. (6.16) into Eq. (6.17) yields,

$$I_{\text{sc}} = I \times \frac{100}{\%X_p} \tag{6.18}$$

For given base VA and voltage, the expression of base impedance is written as,

$$Z_b = \frac{V_b^2}{S_b} = \frac{V_b^2}{V_b I_b} \tag{6.19}$$

The right side of Eq. (6.19) can be rearranged as,

$$Z_b = \frac{\left(\frac{V_b}{1000}\right)^2 \times 1000^2}{\frac{V_b I_b}{1000} \times 1000} \tag{6.20}$$

$$Z_b = \frac{(\text{base kV})^2 \times 1000^2}{(\text{base kVA}) \times 1000} \tag{6.21}$$

$$Z_b = \frac{(\text{base kV})^2 \times 1000}{(\text{base kVA})} \tag{6.22}$$

The expression of per unit impedance can be written as,

$$Z_{\text{pu}} = \frac{Z_{\text{actual}}}{Z_b} \tag{6.23}$$

Substituting Eq. (6.22) into Eq. (6.23) yields,

$$Z_{\text{pu}} = \frac{Z_{\text{actual}}}{\frac{(\text{base kV})^2 \times 1000}{(\text{base kVA})}} \tag{6.24}$$

$$Z_{\text{pu}} = \frac{Z_{\text{actual}}(\text{base kVA})}{(\text{base kV})^2 \times 1000} \tag{6.25}$$

From Eq. (6.25), the per unit impedance in percentage can be represented as,

$$\%Z_{\text{pu}} = \frac{Z_{\text{actual}}(\text{base kVA})}{(\text{base kV})^2 \times 1000} \times 100 \tag{6.26}$$

$$\%\,Z_{\text{pu}} = \frac{Z_{\text{actual}}(\text{base kVA})}{(\text{base kV})^2 \times 10} \tag{6.27}$$

The short circuit kVA at any bus m can be expressed as,

$$\text{kVA}_{\text{sc}} = \sqrt{3} \times V_{Lm} I_F \tag{6.28}$$

where

V_{Lm} is the line voltage in kV,
I_F is the fault current in A.

The per unit three-phase fault current at any bus m can be determined as,

$$I_F(pu) = \frac{V_m}{X_m} \tag{6.29}$$

where

V_m is the prefault bus voltage in pu,
X_m is the total reactance up to the point of fault in pu.

Consider S_b is the base kVA and V_b is the base voltage in kV, and the base current can be determined as,

$$I_b = \frac{S_b}{\sqrt{3}V_b} \tag{6.30}$$

The expression of fault current in ampere is expressed as,

$$I_F = I_F(pu) \times I_b \tag{6.31}$$

Substituting Eqs. (6.29) and (6.30) into Eq. (6.31) yields,

$$I_F = \frac{V_m}{X_m} \times \frac{S_b}{\sqrt{3}V_b} \tag{6.32}$$

Substituting Eq. (6.32) into Eq. (6.28) yields,

$$\text{kVA}_{\text{sc}} = \sqrt{3} \times V_{Lm} \times \frac{V_m}{X_m} \times \frac{S_b}{\sqrt{3}V_b} \tag{6.33}$$

$$\text{kVA}_{\text{sc}} = \frac{V_m V_{Lm} S_b}{X_m V_b} \tag{6.34}$$

If the base voltage (V_b) is equal to the line voltage (V_{Lm}), Eq. (6.34) becomes,

$$\text{kVA}_{\text{sc}} = \frac{V_m S_b}{X_m} \tag{6.35}$$

The value of the prefault bus voltage (V_m) is usually considered as 1 pu and Eq. (6.35) becomes,

$$\text{kVA}_{\text{sc}} = \frac{S_b}{X_m} \tag{6.36}$$

In terms of percentage reactance, Eq. (6.36) can be expressed as,

$$\text{kVA}_{\text{sc}} = \text{Base kVA} \times \frac{100}{\%X_m} \tag{6.37}$$

Example 6.2
Figure 6.3 shows a single-line diagram of a power system. The ratings of the equipment are as follows.

Generator	20 MVA, 11 kV, $X = 0.20$
Transformer	10 MVA, 11/66 kV, $X = 0.08$
Line	$R_L = 0.4\ \Omega$, $X_L = 1.6\ \Omega$
Load	20 MW, 40 MVAR

A three-phase balanced fault occurs at the load bus. Calculate the short circuit kVA and short circuit current.

Solution
Consider the base values are 20 MVA and 11 kV. The new percentage reactance of the generator, transformer, and line can be determined as,

$$X_G = \frac{20}{20} \times 0.20 = 0.20 \tag{6.38}$$

$$X_T = \frac{20}{10} \times \ 0.08 = 0.16 \tag{6.39}$$

New base in the secondary side of the transformer is,

$$V_{b2} = \frac{V_{b1}}{a} = 11 \times \frac{66}{11} = 66\text{ kV} \tag{6.40}$$

$$\%R_L = \frac{R_{\text{Lactual}}(\text{base kVA})}{(\text{base kV})^2 \times 10} = \frac{0.4 \times 20000}{66^2 \times 10} = 0.18 \tag{6.41}$$

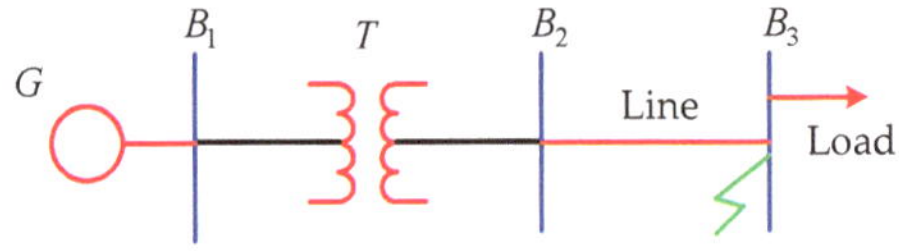

Fig. 6.3 Single-line diagram for Example 6.2

$$\%X_L = \frac{X_{\text{Lactual}}(\text{base kVA})}{(\text{base kV})^2 \times 10} = \frac{1.6 \times 20000}{66^2 \times 10} = 0.73 \tag{6.42}$$

The total reactance up to the fault is calculated as,

$$X_{\text{total}} = 20 + 16 + 0.73 = 36.73\% \tag{6.43}$$

The value of the short circuit kVA can be calculated as,

$$\text{kVA}_{\text{sc}} = \text{Base kVA} \times \frac{100}{\%X} = 20 \times 1000 \times \frac{100}{36.73} = 54.45 \times 10^3 \tag{6.44}$$

The value of the short circuit MVA is,

$$\text{MVA}_{\text{sc}} = 54.45 \tag{6.45}$$

The full load current with 20 MVA base and 66 kV base voltage is calculated as,

$$I_F = \frac{20000}{\sqrt{3} \times 66} = 174.95\ \text{A} \tag{6.46}$$

The short circuit current can be determined as,

$$I_{\text{sc}} = I_F \times \frac{100}{\%X} = 174.95 \times \frac{100}{36.73} = 476.32\ \text{A} \tag{6.47}$$

The single-line diagram is drawn in the CYME power system software. Then data are given input by selecting each equipment. Then press the run button, and select the parameters as the prefault base voltage and the machine impedance as the steady state. Finally, apply fault at bus 3 and run the simulation. The simulation results are shown in Fig. 6.4.

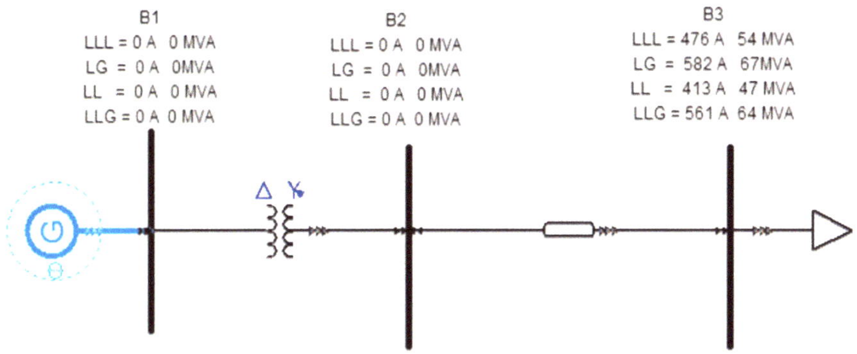

Fig. 6.4 CYME simulation results for Example 6.2

From Fig. 6.4, it is seen that the three-phase fault (LLL) current is,

$$I_{\mathrm{scCYME}} = 476\,\mathrm{A} \tag{6.48}$$

The value of the short circuit MVA is,

$$\mathrm{MVA_{sc}} = 54 \tag{6.49}$$

Example 6.3

Figure 6.5 shows a single-line diagram of a power system. A three-phase balanced fault occurs at the generator bus. The ratings of the equipment are as follows.

Generator 1	20 MVA, 11 kV, $X = 30\%$
Generator 2	40 MVA, 11 kV, $X = 20\%$
Transformer	15 MVA, 11/33 kV, $X = 10\%$
Line	$X_L = 20\ \Omega$
Load	25 MW, 55 MVAR

Calculate the short circuit kVA and the short circuit current.

Solution

Consider the base values are 40 MVA and 11 kV. The new percentage reactance of the generators, transformer, and line can be determined as,

$$X_{G1} = \frac{40}{20} \times 0.30 = 60\% \tag{6.50}$$

$$X_{G2} = \frac{40}{40} \times 0.20 = 20\% \tag{6.51}$$

$$X_T = \frac{40}{15} \times \left(\frac{33}{33}\right)^2 \times 0.10 = 26.67\% \tag{6.52}$$

$$\%X_L = \frac{X_{\mathrm{Lactual}}(\text{base kVA})}{(\text{base kV})^2 \times 10} = \frac{20 \times 40000}{33^2 \times 10} = 73.46 \tag{6.53}$$

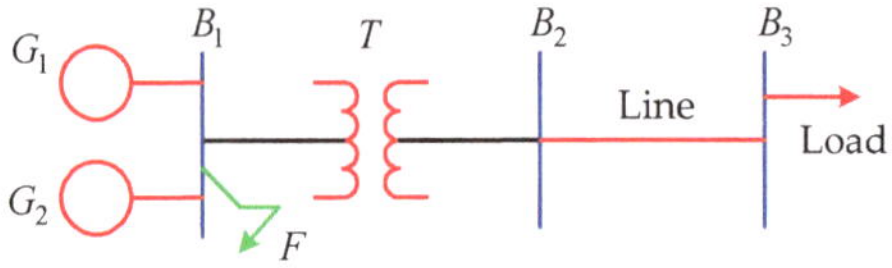

Fig. 6.5 Single-line diagram for Example 6.3

Alternative approach:
The value of the new base voltage is calculated as,

$$V_{b1} = 11 \times \frac{33}{11} = 33\ \text{kV} \tag{6.54}$$

The value of the base impedance is calculated as,

$$Z_b = \frac{V_{b1}^2}{S_b} = \frac{33^2}{40} = 27.23\ \Omega \tag{6.55}$$

The value of the per unit line reactance is,

$$X_{\text{line}} = \frac{20}{27.23} = 0.7344 \tag{6.56}$$

To calculate the reactance up to the fault point, the reactance diagram is considered, which is shown in Fig. 6.6.

The total reactance up to the fault point can be determined as,

$$\%X = ((73.46 + 26.67)\|20)\|60 = 13.04 \tag{6.57}$$

The short circuit kVA can be determined as,

$$\text{kVA}_{\text{sc}} = \text{Base kVA} \times \frac{100}{\%X} = 40 \times 1000 \times \frac{100}{13.04} = 306748 \tag{6.58}$$

The value of the short circuit MVA is,

$$\text{MVA}_{\text{sc}} = 306.75 \tag{6.59}$$

The value of the base current is calculated as,

$$I_b = \frac{40 \times 10^6}{\sqrt{3} \times 11 \times 1000} = 2099.45\ \text{A} \tag{6.60}$$

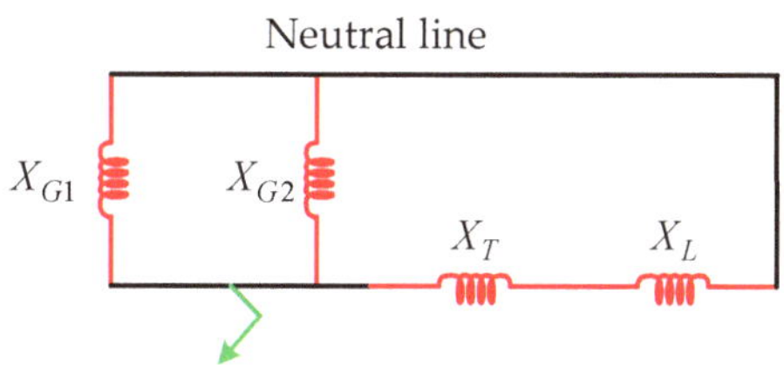

Fig. 6.6 Reactance diagram for Example 6.3

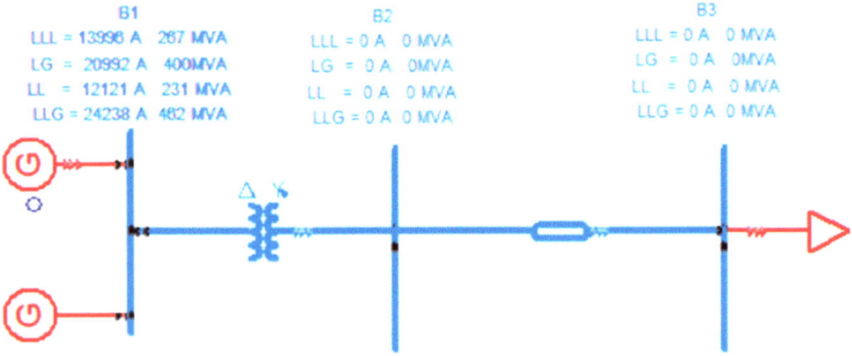

Fig. 6.7 CYME simulation results for Example 6.3

The short circuit current can be determined as,

$$I_{sc} = I_b \times \frac{100}{\%X} = 2099.45 \times \frac{100}{13.04} = 16100\,\text{A}$$

The simulation results are shown in Fig. 6.7. From Fig. 6.7, it is seen that the three-phase fault (LLL) current is,

$$I_{scCYME} = 13996\,\text{A} \tag{6.61}$$

The value of the short circuit MVA is,

$$\text{MVA}_{sc} = 267 \tag{6.62}$$

Practice Problem 6.2

A single-line diagram of a power system is shown in Fig. 6.8. A three-phase balanced fault occurs at the load bus. The ratings of the equipment are as follows.

Generator	10 MVA, 11 kV, $X = 20\%$
Transformer	5 MVA, 11/33 kV, $X = 8\%$
Line	$X = 2\ \Omega$
Load	30 MW, 60 MVAR

Calculate the short circuit kVA and short circuit current.

Fig. 6.8 Single-line diagram for Practice Problem 6.2

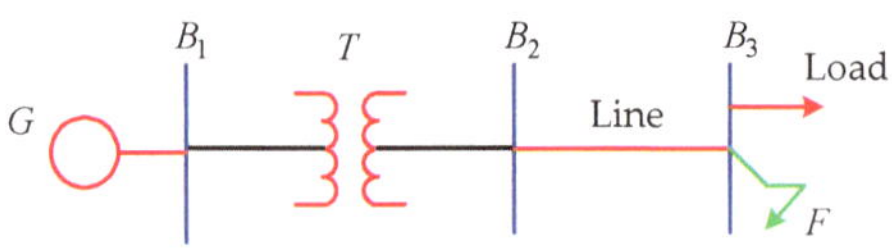

Practice Problem 6.3

Figure 6.9 shows a single-line diagram where three generators are connected to a common busbar. A three-phase balanced fault occurs at the generator bus. The ratings of the equipment are as follows.

Generator 1	50 MVA, 13 kV, $X = 30\%$
Generator 2	40 MVA, 13 kV, $X = 20\%$
Generator 2	30 MVA, 13 kV, $X = 10\%$
Load	30 MW, 50 MVAR

Find the short circuit kVA and short circuit current.

Example 6.4

Figure 6.10 shows a single-line diagram of a power system. The ratings of the equipment are as follows.

Generator G_1	100 MVA, 20 kV, $X = 0.20$ pu
Generator G_2	80 MVA, 20 kV, $X = 0.20$ pu
Generator G_3	60 MVA, 20 kV, $X = 0.12$ pu
Transformer T_1	100 MVA, 20/138 kV, primary delta and secondary earthed wye, $X = 0.08$ pu
Transformer T_2	50 MVA, 20/138 kV, primary delta and secondary wye, $X = 0.04$ pu
Line	100 MVA, $X_L = 20\ \Omega$

A three-phase balanced fault occurs at bus 1. Calculate the short circuit kVA and short circuit current. Compare the result with IPSA software simulation.

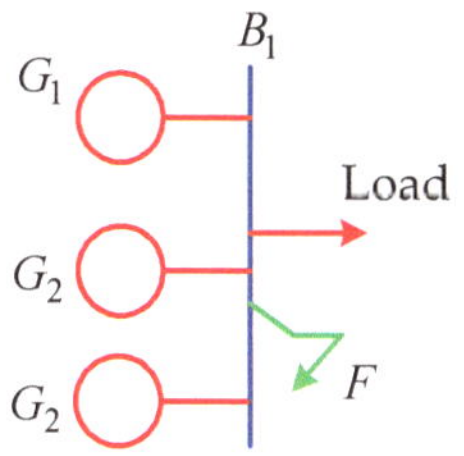

Fig. 6.9 Single-line diagram for Practice Problem 6.3

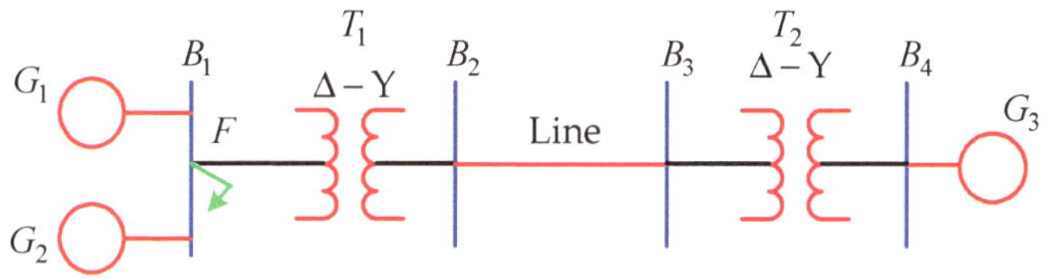

Fig. 6.10 Single-line diagram for Practice Problem 6.4

Solution

Consider the base values 100 MVA and 20 kV. The new percentage reactance of the generators, transformers, and line can be determined as,

$$X_{G1} = \frac{100}{100} \times 0.20 = 0.2\,\text{pu} \tag{6.63}$$

$$X_{G1} = \frac{100}{100} \times 0.20 = 0.2\,\text{pu} \tag{6.64}$$

$$X_{G2} = \frac{100}{80} \times 0.20 = 0.25\,\text{pu} \tag{6.65}$$

$$X_{G3} = \frac{100}{60} \times 0.12 = 0.20\,\text{pu} \tag{6.66}$$

$$X_{T1} = \frac{100}{100} \times 0.08 = 0.08\,\text{pu} \tag{6.67}$$

$$X_{T2} = \frac{100}{50} \times 0.04 = 0.08\,\text{pu} \tag{6.68}$$

The base voltage for the line is calculated as,

$$V_{b1} = \frac{V_b}{a} = 20 \times \frac{138}{20} = 138\,\text{kV} \tag{6.69}$$

The value of the base impedance is calculated as,

$$Z_b = \frac{V_{b1}^2}{S_b} = \frac{138^2}{100} = 190.44\,\Omega \tag{6.70}$$

The per unit line reactance is calculated as,

$$X_{\text{line}} = \frac{X_{\text{line}}}{Z_b} = \frac{20}{190.44} = 0.105 \tag{6.71}$$

The reactance diagram is considered which is shown in Fig. 6.11, to calculate the reactance up to the fault point,

The equivalent reactance for the generator 1 and generator 2 is calculated as,

$$X_{ge} = \frac{0.2 \times 0.25}{0.45} = 0.111\,\text{pu} \tag{6.72}$$

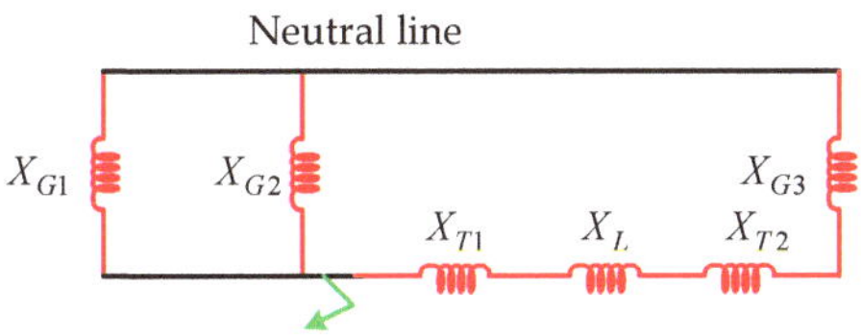

Fig. 6.11 Reactance diagram for Example 6.4

The total reactance up to the fault point can be determined as,

$$X = (0.08 + 0.08 + 0.105 + 0.2) \parallel 0.111 \tag{6.73}$$

$$X = \frac{0.465 \times 0.111}{0.576} = 0.0896\,\text{pu} \tag{6.74}$$

The value of the short circuit kVA can be determined as,

$$\text{kVA}_{\text{sc}} = \text{Base kVA} \times \frac{100}{\%X} = 100 \times 1000 \times \frac{1}{0.0896} = 1116071.429 \tag{6.75}$$

The short circuit MVA is calculated as,

$$\text{MVA}_{\text{sc}} = 1116.071 \tag{6.76}$$

The value of the base current is calculated as,

$$I_b = \frac{100 \times 10^6}{\sqrt{3} \times 20 \times 1000} = 2886.751\,\text{A} \tag{6.77}$$

The value of the short circuit current can be determined as,

$$I_{\text{sc}} = I_b \times \frac{100}{\%X} = 2886.751 \times \frac{1}{0.0896} = 32218.20\,\text{A} \tag{6.78}$$

The simulation is done by the software IPSA, and the result is shown in Fig. 6.12. The balanced three-phase fault current is,

$$I_{fIPSA} = 32188.831\,\text{A} \tag{6.79}$$

The simulation is done by the software CYME, and the result is shown in Fig. 6.13. The balanced three-phase fault current is,

$$I_{fCYME} = 26062\,\text{A} \tag{6.80}$$

$$\text{MVA}_{\text{scCYME}} = 903 \tag{6.81}$$

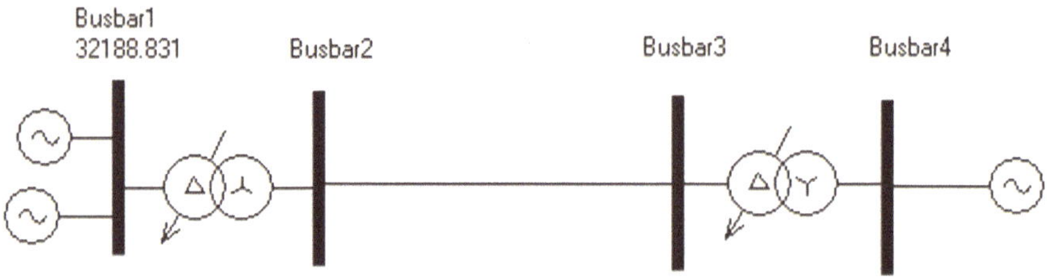

Fig. 6.12 IPSA simulation results for Example 6.4

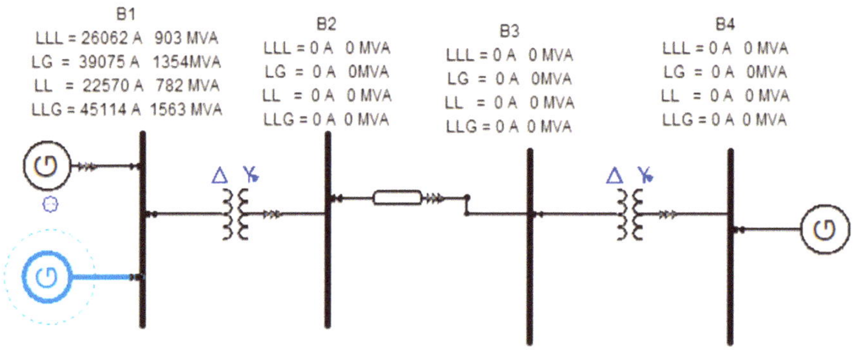

Fig. 6.13 CYME simulation results for Example 6.4

Practice Problem 6.4

Figure 6.14 shows a single-line diagram of a power system. The ratings of the equipment are as follows.

Generator G_1	100 MVA, 11 kV, $X = 0.20\,\text{pu}$
Generator G_2	90 MVA, 11 kV, $X = 0.15\,\text{pu}$
Generator G_3	80 MVA, 11 kV, $X = 0.12\,\text{pu}$
Generator G_4	70 MVA, 11 kV, $X = 0.10\,\text{pu}$
Transformer T_1	100 MVA, 11/33 kV, primary delta and secondary earthed wye, $X = 0.12$
Transformer T_2	90 MVA, 11/33 kV, primary delta and secondary earthed wye, $X = 0.11$
Line	$X_L = 28\,\Omega$

A three-phase balanced fault occurs at bus 3. Calculate the short circuit kVA and short circuit current.

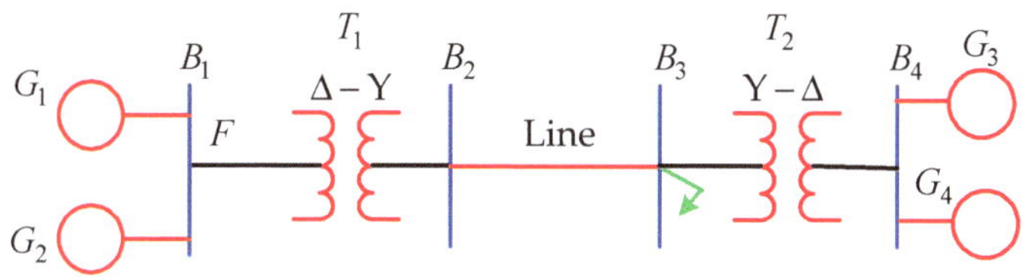

Fig. 6.14 Single-line diagram for Practice Problem 6.4

6.4 Unsymmetrical Faults

The faults in the power system network which disturb the balanced condition of the network are known as unsymmetrical faults. The unsymmetrical faults are classified as a single-line-to-ground faults (SLG), double-line-to-ground faults (DLG), and line-to-line faults. More than 90% of faults which occur in a power system are single-line-to-ground faults. The connection diagrams of different types of unsymmetrical faults are shown in Fig. 6.15.

6.5 Symmetrical Components

The knowledge of phase sequence is very important to analyze the symmetrical components. The phase sequence is defined as the order in which they reach the maximum value of the voltage. In 1918, C. L. Fortescue, an American Scientist, stated that three separate balanced phasors can replace the three-phase unbalanced phasors of a three-phase system.

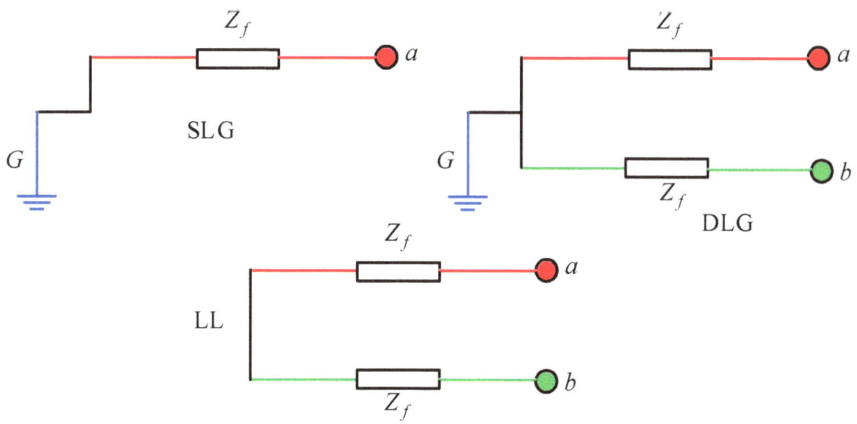

Fig. 6.15 Unsymmetrical faults' connection diagram

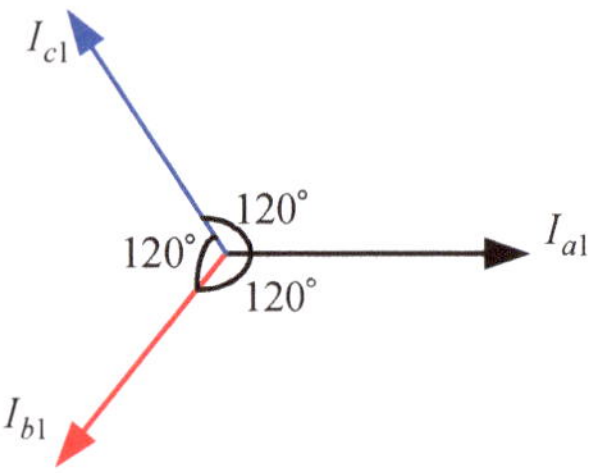

Fig. 6.16 Representation of positive sequence components

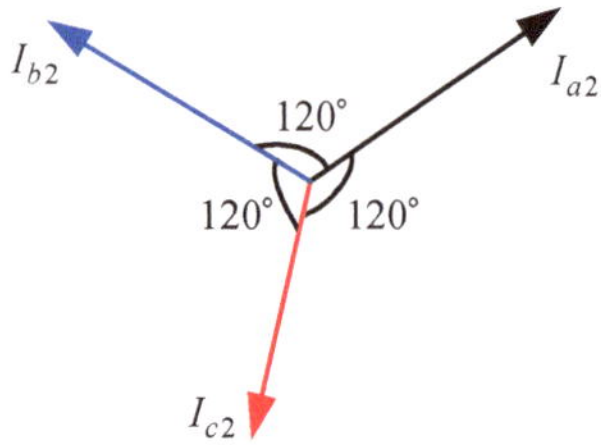

Fig. 6.17 Representation of negative sequence components

The symmetrical components are classified as positive sequence components, negative sequence components, and zero sequence components.

The components of the system have three phasors with equal magnitude but are displaced from each other by 120°, and they follow the *abc* phase sequence. Phase *b* lags the phase *a* by 120°, and phase *c* lags the phase *b* by 120°. The neutral current of the positive sequence components is zero. The positive sequence components I_{a1}, I_{b1}, and I_{c1} are shown in Fig. 6.16. In this system, the components have three phasors with an equal magnitude but are displaced from each other by 120°, and they maintain the *acb* phase sequence. Phase *c* lags the phase *a* by 120°, and phase *b* lags the phase *c* by 120°. The negative sequence components are similar to the positive sequence components, except the phase order is reversed. The neutral current of the negative sequence components is zero. The negative sequence components I_{a2}, I_{b2}, and I_{c2} are shown in Fig. 6.17.

The components have three phasors with equal magnitude, but zero displacements are known as zero sequence components. The zero sequence components are in phase with each other and have the neutral current. The zero sequence components of the current I_{a0}, I_{b0}, and I_{c0} are shown in Fig. 6.18. The unsymmetrical components of the currents I_a, I_b, and I_c can be derived from the phasor diagram as shown in Fig. 6.18.

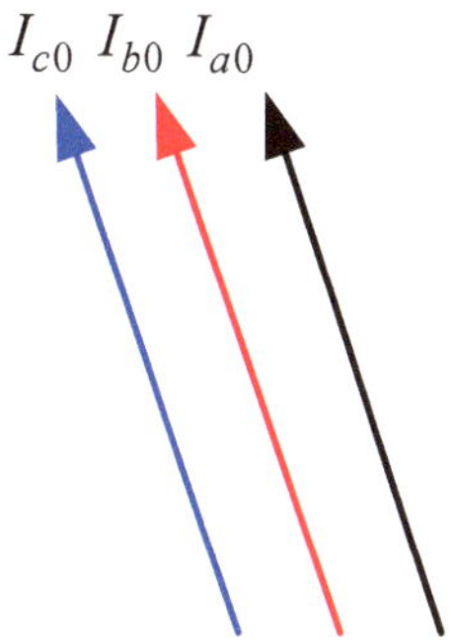

Fig. 6.18 Representation of zero sequence components

6.6 Representation of Symmetrical Components

Figure 6.19 represents a phasor diagram where the line *OA* represents the current phasor *I*. This phasor, after being multiplied by the operator *a*, gives the new phasor *aI* which is represented by the line *OB*. The phasor *aI* in the diagram is leading (in the counterclockwise direction) the phasor by 120°, which can be expressed as,

$$aI = I \angle 120^\circ \tag{6.82}$$

$$a = 1 \angle 120^\circ = -0.5 + j0.866 \tag{6.83}$$

Similarly, multiplying the phasor *aI* by *a* which gives a new phasor a^2I, which is represented by the line *OC*. This new phasor a^2I is leading phasor *I* by 240° in the phasor diagram, which can be mathematically expressed as,

$$a^2I = I \angle 240^\circ \tag{6.84}$$

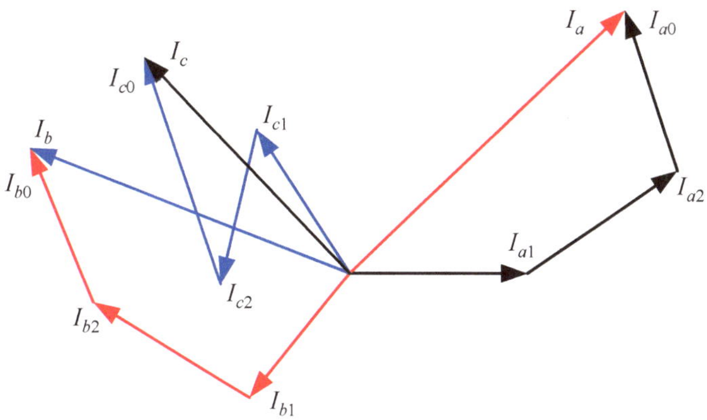

Fig. 6.19 Representation of unsymmetrical and symmetrical components of current

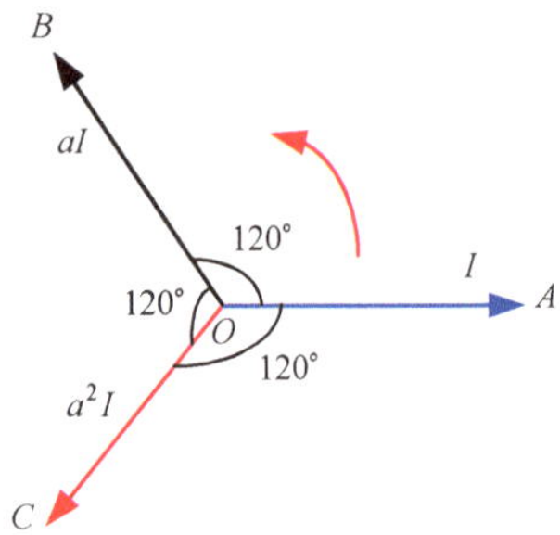

Fig. 6.20 Phasors with counterclockwise direction

$$a^2 = 1 \underline{|240^\circ} = -0.5 - j0.866 \tag{6.85}$$

A similar approach can lead us to the following expression:

$$a^3 = 1 \underline{|360^\circ} = 1 \tag{6.86}$$

Adding Eqs. (6.83), (6.85), and (6.86) yields,

$$1 + a^2 + a^3 = 0 \tag{6.87}$$

Comparing Fig. 6.16 with Fig. 6.20, the positive phase sequence components of the current can be represented as,

$$I_{a1} = I_{a1} \underline{|0^\circ} \tag{6.88}$$

$$I_{b1} = I_{a1} \underline{|240^\circ} = a^2 I_{a1} \tag{6.89}$$

$$I_{c1} = I_{a1} \underline{|120^\circ} = aI_{a1} \tag{6.90}$$

Again, comparing Fig. 6.17 with Fig. 6.20, the negative phase sequence components of the current can be represented as,

$$I_{a2} = I_{a2} \underline{|0^\circ} \tag{6.91}$$

$$I_{b2} = I_{a2} \underline{|120^\circ} = aI_{a2} \tag{6.92}$$

$$I_{c2} = I_{a2} \underline{|240^\circ} = a^2 I_{a2} \tag{6.93}$$

The magnitudes of the zero phase sequence components are the same, and it can be written as,

$$I_{a0} = I_{b0} = I_{c0} \tag{6.94}$$

From Fig. 6.19, the unsymmetrical currents can be represented by the symmetrical components of current as,

$$I_a = I_{a0} + I_{a1} + I_{a2} \tag{6.95}$$

$$I_b = I_{b0} + I_{b1} + I_{b2} \tag{6.96}$$

$$I_c = I_{c0} + I_{c1} + I_{c2} \tag{6.97}$$

where the suffixes 0, 1, and 2 indicate the zero sequence, positive sequence, and negative sequence components, respectively. Equations (6.96) and (6.97) can be replaced by the symmetrical components of the current of phase a. Substituting Eqs. (6.89), (6.92), and (6.94) into Eq. (6.96) yields,

$$I_b = I_{a0} + a^2 I_{a1} + a I_{a2} \tag{6.98}$$

Again, substituting Eqs. (6.90), (6.93) and (6.94) into the Eq. (6.97) yields,

$$I_c = I_{a0} + a I_{a1} + a^2 I_{a2} \tag{6.99}$$

Equations (6.95), (6.98), and (6.99) can be rearranged in the matrix form as,

$$\begin{bmatrix} I_a \\ I_b \\ I_c \end{bmatrix} = \begin{bmatrix} 1 & 1 & 1 \\ 1 & a^2 & a \\ 1 & a & a^2 \end{bmatrix} \begin{bmatrix} I_{a0} \\ I_{a1} \\ I_{a2} \end{bmatrix} \tag{6.100}$$

In Eq. (6.100), let's represent the following symmetrical component transformation matrix as A, where

$$A = \begin{bmatrix} 1 & 1 & 1 \\ 1 & a^2 & a \\ 1 & a & a^2 \end{bmatrix} \tag{6.101}$$

Equation (6.100) can be modified as,

$$\begin{bmatrix} I_a \\ I_b \\ I_c \end{bmatrix} = A \begin{bmatrix} I_{a0} \\ I_{a1} \\ I_{a2} \end{bmatrix} \tag{6.102}$$

Taking the inverse of A, Eq. (6.102) can be written as,

$$\begin{bmatrix} I_{a0} \\ I_{a1} \\ I_{a2} \end{bmatrix} = A^{-1} \begin{bmatrix} I_a \\ I_b \\ I_c \end{bmatrix} \tag{6.103}$$

From Eq. (6.103), the inverse of the A can be derived as,

$$A^{-1} = \frac{1}{3}\begin{bmatrix} 1 & 1 & 1 \\ 1 & a & a^2 \\ 1 & a^2 & a \end{bmatrix} \tag{6.104}$$

Substituting Eq. (6.104) into Eq. (6.103) yields,

$$\begin{bmatrix} I_{a0} \\ I_{a1} \\ I_{a2} \end{bmatrix} = \frac{1}{3}\begin{bmatrix} 1 & 1 & 1 \\ 1 & a & a^2 \\ 1 & a^2 & a \end{bmatrix}\begin{bmatrix} I_a \\ I_b \\ I_c \end{bmatrix} \tag{6.105}$$

From Eq. (6.105), the following expressions for the current components can be found,

$$I_{a0} = \frac{1}{3}(I_a + I_b + I_c) \tag{6.106}$$

$$I_{a1} = \frac{1}{3}\left(I_a + aI_b + a^2 I_c\right) \tag{6.107}$$

$$I_{a2} = \frac{1}{3}\left(I_a + a^2 I_b + aI_c\right) \tag{6.108}$$

Similarly, the mathematical expressions for voltage components can be written as,

$$V_{a0} = \frac{1}{3}(V_a + V_b + V_c) \tag{6.109}$$

$$V_{a1} = \frac{1}{3}\left(V_a + aV_b + a^2 V_c\right) \tag{6.110}$$

$$V_{a2} = \frac{1}{3}\left(V_a + a^2 V_b + aV_c\right) \tag{6.111}$$

In a three-phase Y-connection system, the magnitude of the neutral current can be calculated as,

$$I_n = I_a + I_b + I_c \tag{6.112}$$

Substituting Eq. (6.112) into Eq. (6.106) yields,

$$I_{a0} = \frac{1}{3} I_n \tag{6.113}$$

$$I_n = 3I_{a0} = 3I_0 \tag{6.114}$$

Example 6.5
A three-phase system is having phase voltages $V_a = 90\,\underline{|0^\circ}$ kV, $V_b = 66\,\underline{|-100^\circ}$ kV, and $V_c = 22\,\underline{|85^\circ}$ kV. Calculate the symmetrical voltage components of for phases *a*, *b*, and *c*.

Solution
The magnitude of the zero sequence voltage component is calculated as,

$$V_{a0} = \frac{1}{3}\left(90 + 66\,\underline{|-100^\circ} + 22\,\underline{|85^\circ}\right) = 30.42\,\underline{|-28.17^\circ}\text{ kV} \tag{6.115}$$

The magnitude of the positive sequence voltage component is calculated as,

$$V_{a1} = \frac{1}{3}\left(90 + 66\,\underline{|-100^\circ + 120^\circ} + 22\,\underline{|85^\circ + 240^\circ}\right) = 56.21\,\underline{|3.35^\circ}\text{ kV} \tag{6.116}$$

The magnitude of the negative sequence voltage component is,

$$\begin{aligned} V_{a2} &= \frac{1}{3}\left(V_a + a^2V_b + aV_c\right) = \frac{1}{3}\left(90 + 66\,\underline{|-100^\circ + 240^\circ} + 22\,\underline{|85^\circ + 120^\circ}\right) \\ &= 12.69\,\underline{|59.51^\circ}\text{ kV} \end{aligned} \tag{6.117}$$

For phase *b*:
The zero sequence voltage component is calculated as,

$$V_{b0} = 30.42\,\underline{|-28.17^\circ}\text{ kV} \tag{6.118}$$

The positive sequence voltage component is calculated as,

$$V_{b1} = a^2V_{a1} = 56.21\,\underline{|240^\circ + 3.35^\circ} = 56.21\,\underline{|243.35^\circ}\text{ kV} \tag{6.119}$$

The negative sequence voltage component is calculated as,

$$V_{b2} = V_{a2}\,\underline{|120^\circ} = 12.69\,\underline{|120^\circ + 59.51^\circ} = 12.69\,\underline{|179.51^\circ}\text{ kV} \tag{6.120}$$

For phase *c*:
The zero sequence voltage component is,

$$V_{c0} = 30.42\,\underline{|-28.17^\circ}\text{ kV} \tag{6.121}$$

The positive sequence voltage component is determined as,

$$V_{c1} = V_{a1}\,\underline{|120^\circ} = 56.21\,\underline{|120^\circ + 3.35^\circ} = 56.21\,\underline{|123.35^\circ}\text{ kV} \tag{6.122}$$

The negative sequence voltage component is calculated as,

$$V_{c2} = V_{a2}\,\angle 240^\circ = 12.69\,\angle 240^\circ + 59.51^\circ = 12.69\,\angle 299.51^\circ \text{ kV} \quad (6.123)$$

Example 6.6
The symmetrical voltage components of a three-phase system are given by $V_{a0} = 94\,\angle 150^\circ$ kV, $V_{a1} = 56\,\angle 80^\circ$ kV, and $V_{a2} = 122\,\angle 55^\circ$ kV for phase a. Find the three-phase unbalanced voltages.

Solution
The positive phase sequence voltage components can be determined as,

$$V_{b1} = V_{a1}\,\angle 240^\circ = 56\,\angle 80 + 240^\circ = 56\,\angle 320^\circ \text{ kV} \quad (6.124)$$

$$V_{c1} = V_{a1}\,\angle 120^\circ = 56\,\angle 80^\circ + 120^\circ = 56\,\angle 200^\circ \text{ kV} \quad (6.125)$$

The negative phase sequence voltage components are calculated as,

$$V_{b2} = V_{a2}\,\angle 120^\circ = 122\,\angle 55^\circ + 120^\circ = 122\,\angle 175^\circ \text{ kV} \quad (6.126)$$

$$V_{c2} = V_{a2}\,\angle 240^\circ = 122\,\angle 55^\circ + 240^\circ = 122\,\angle 295^\circ \text{ kV} \quad (6.127)$$

The zero phase sequence voltage components can be calculated as,

$$V_{a0} = V_{b0} = V_{c0} = 94\,\angle 150^\circ \text{ kV} \quad (6.128)$$

Unbalance three-phase voltage components can be determined as,

$$V_a = 94\,\angle 150^\circ + 56\,\angle 80^\circ + 122\,\angle 55^\circ = 202.09\,\angle 90.48^\circ \text{ kV} \quad (6.129)$$

$$V_a = 94\,\angle 150^\circ + 56\,\angle 320^\circ + 122\,\angle 175^\circ = 161.50\,\angle 172.5^\circ \text{ kV} \quad (6.130)$$

$$V_c = 94\,\angle 150^\circ + 56\,\angle 200^\circ + 122\,\angle 295^\circ = 116.81\,\angle -134.91^\circ \text{ kV} \quad (6.131)$$

Practice Problem 6.5
A Y-connected three-phase load draws current from the source as $I_a = 15\,\angle 0^\circ$ A, $I_b = 25\,\angle 20^\circ$ A, and $I_c = 30\,\angle 40^\circ$ A. Calculate the symmetrical current components for the phases a, b, and c.

Practice Problem 6.6
A three-phase system has the following symmetrical current components: Positive sequence current components are $I_{a1} = 6\,\angle 10^\circ$ A and $I_{b1} = 6\,\angle 250^\circ$ A. The negative sequence current components are $I_{a2} = 5\,\angle 53.13^\circ$ A, $I_{b2} = 5\,\angle 173.13^\circ$ A, and $I_{c2} = 5\,\angle 293.13^\circ$ A. The zero sequence current components are $I_{a0} = I_{b0} = I_{c0} = 4\,\angle 35^\circ$ A. Calculate the unsymmetrical current components.

6.7 Complex Power in Symmetrical Components

The product of the voltage and the conjugate of the current is known as complex power, and it is denoted by the capital letter *S*. The expression of the complex power is,

$$\mathbf{S} = P + jQ = VI^* = VI\cos\phi + VI\sin\phi \tag{6.132}$$

Based on Eq. (6.132), the expression of the complex power for the three-phase lines can be written as,

$$P + jQ = V_a I_a^* + V_b I_b^* + V_c I_c^* \tag{6.133}$$

Equation (6.133) can be rearranged in the matrix form as,

$$P + jQ = [V_a + V_b + V_c]\begin{bmatrix} I_a \\ I_b \\ I_c \end{bmatrix}^* = \begin{bmatrix} V_a \\ V_b \\ V_c \end{bmatrix}^T \begin{bmatrix} I_a \\ I_b \\ I_c \end{bmatrix}^* \tag{6.134}$$

According to Eq. (6.100), the unbalanced voltages can be written as,

$$\begin{bmatrix} V_a \\ V_b \\ V_c \end{bmatrix} = \begin{bmatrix} 1 & 1 & 1 \\ 1 & a^2 & a \\ 1 & a & a^2 \end{bmatrix}\begin{bmatrix} V_{a0} \\ V_{a1} \\ V_{a2} \end{bmatrix} = AV_{012} \tag{6.135}$$

where

$$\begin{bmatrix} V_{a0} \\ V_{a1} \\ V_{a2} \end{bmatrix} = V_{012} \tag{6.136}$$

Taking transpose of Eq. (6.135) yields,

$$\begin{bmatrix} V_a \\ V_b \\ V_c \end{bmatrix}^T = (AV_{012})^T = A^T V_{012}^T \tag{6.137}$$

$$\begin{bmatrix} V_a \\ V_b \\ V_c \end{bmatrix}^T = \begin{bmatrix} 1 & 1 & 1 \\ 1 & a^2 & a \\ 1 & a & a^2 \end{bmatrix}\begin{bmatrix} V_{a0} & V_{a1} & V_{a2} \end{bmatrix} \tag{6.138}$$

Taking conjugate of Eq. (6.100) yields,

$$\begin{bmatrix} I_a \\ I_b \\ I_c \end{bmatrix}^* = \begin{bmatrix} 1 & 1 & 1 \\ 1 & a^2 & a \\ 1 & a & a^2 \end{bmatrix}^* \begin{bmatrix} I_{a0} \\ I_{a1} \\ I_{a2} \end{bmatrix}^* = \begin{bmatrix} 1 & 1 & 1 \\ 1 & a & a^2 \\ 1 & a^2 & a \end{bmatrix} \begin{bmatrix} I_{a0} \\ I_{a1} \\ I_{a2} \end{bmatrix}^* \tag{6.139}$$

Substituting Eqs. (6.137) and (6.138) into Eq. (6.134) yields,

$$P + jQ = \begin{bmatrix} 1 & 1 & 1 \\ 1 & a^2 & a \\ 1 & a & a^2 \end{bmatrix} \begin{bmatrix} V_{a0} & V_{b0} & V_{c0} \end{bmatrix} \begin{bmatrix} 1 & 1 & 1 \\ 1 & a & a^2 \\ 1 & a^2 & a \end{bmatrix} \begin{bmatrix} I_{a0} \\ I_{a1} \\ I_{a2} \end{bmatrix}^* \tag{6.140}$$

$$P + jQ = \begin{bmatrix} V_{a0} & V_{a1} & V_{a2} \end{bmatrix} \begin{bmatrix} 1 & 1 & 1 \\ 1 & a^2 & a \\ 1 & a & a^2 \end{bmatrix} \begin{bmatrix} 1 & 1 & 1 \\ 1 & a & a^2 \\ 1 & a^2 & a \end{bmatrix} \begin{bmatrix} I_{a0} \\ I_{a1} \\ I_{a2} \end{bmatrix}^* \tag{6.141}$$

Equation (6.141) can be modified as,

$$P + jQ = \begin{bmatrix} V_{a0} & V_{a1} & V_{a2} \end{bmatrix} \begin{bmatrix} 1+1+1 & 1+a+a^2 & 1+a^2+a \\ 1+a^2+a & 1+a^3+a^3 & 1+a^4+a^2 \\ 1+a+a^2 & 1+a^2+a^4 & 1+a^3+a^3 \end{bmatrix} \begin{bmatrix} I_{a0} \\ I_{a1} \\ I_{a2} \end{bmatrix}^* \tag{6.142}$$

$$P + jQ = \begin{bmatrix} V_{a0} & V_{a1} & V_{a2} \end{bmatrix} \begin{bmatrix} 3 & 0 & 0 \\ 0 & 3 & 0 \\ 0 & 0 & 3 \end{bmatrix} \begin{bmatrix} I_{a0} \\ I_{a1} \\ I_{a2} \end{bmatrix}^* \tag{6.143}$$

$$P + jQ = 3\begin{bmatrix} V_{a0} & V_{a1} & V_{a2} \end{bmatrix} \begin{bmatrix} I_{a0} \\ I_{a1} \\ I_{a2} \end{bmatrix}^* \tag{6.144}$$

$$P + jQ = 3\left[V_{a0}I_{a0}^* + V_{a1}I_{a1}^* + V_{a2}I_{a2}^*\right] \tag{6.145}$$

Equation (6.145) can be used to find the real power and reactive power from the symmetrical components of voltage and current. According to Eq. (6.132), the expressions of real and reactive power from Eq. (6.145) can be written as,

$$P = 3V_{a0}I_{a0}\cos\phi_0 + 3V_{a1}I_{a1}\cos\phi_1 + 3V_{a2}I_{a2}\cos\phi_2 \tag{6.146}$$

$$Q = 3V_{a0}I_{a0}\sin\phi_0 + 3V_{a1}I_{a1}\sin\phi_1 + 3V_{a2}I_{a2}\sin\phi_2 \tag{6.147}$$

6.8 Sequence Impedance of Power System Equipment

The sequence impedances of a power system equipment are defined as the impedance offered by the equipment to the flow of sequence (positive or negative or zero) current through it. These sequence impedances are zero, positive, and negative sequence impedances. The zero sequence impedance of equipment is defined as the impedance offered by the equipment to the flow of the zero sequence current, and it is represented by Z_0. The impedance offered by the power system equipment to the flow of the positive sequence current is known as the positive sequence impedance, and it is denoted by Z_1. The impedance offered by the power system equipment to the flow of the negative sequence current is known as the negative sequence impedance, and it is denoted by Z_2. In the case of a synchronous machine, positive sequence impedance is equal to the synchronous impedance of the machine, whereas the negative sequence impedance is much less than the positive sequence impedance. If the zero sequence impedance is not given, then its value is assumed to be equal to the positive sequence impedance. For the transformer, positive sequence impedance, negative sequence impedance, and zero sequence impedances are equal. In the case of the transmission line, positive sequence impedance and negative sequence impedances are equal. The zero sequence impedance is much higher than the positive sequence impedance or the negative sequence impedance. The balanced Y-connected load and the neutral impedance are shown in Fig. 6.21. The current in the neutral point is,

$$I_n = I_a + I_b + I_c \tag{6.148}$$

The voltage between the phase a and the ground point is,

$$V_{ag} = I_a Z_y + Z_n I_n \tag{6.149}$$

Fig. 6.21 Balanced Y-connected load

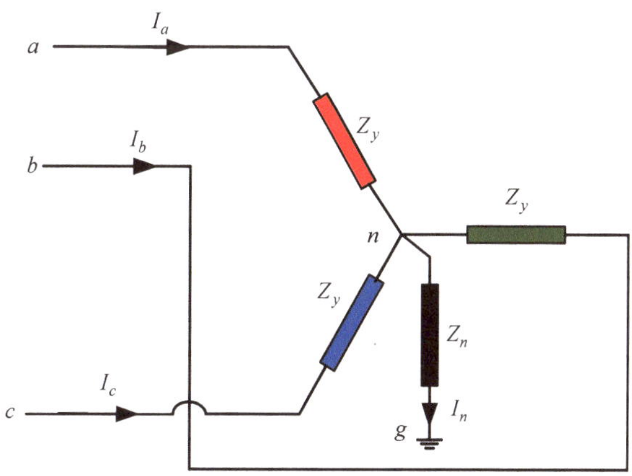

Substituting Eq. (6.147) into Eq. (6.149) yields,

$$V_{ag} = I_a Z_y + Z_n(I_a + I_b + I_c) \tag{6.150}$$

$$V_{ag} = I_a(Z_y + Z_n) + Z_n I_b + Z_n I_c \tag{6.151}$$

Similarly, the voltages of phase b and phase c to the ground point are,

$$V_{bg} = Z_n I_a + (Z_n + Z_y)I_b + Z_n I_c \tag{6.152}$$

$$V_{cg} = Z_n I_a + Z_n I_b + (Z_n + Z_y)I_c \tag{6.153}$$

Equations (6.151), (6.71), and (6.153) can be expressed in the matrix format as,

$$\begin{bmatrix} V_{ag} \\ V_{bg} \\ V_{cg} \end{bmatrix} = \begin{bmatrix} Z_n + Z_y & Z_n & Z_n \\ Z_n & Z_n + Z_y & Z_n \\ Z_n & Z_n & Z_n + Z_y \end{bmatrix} \begin{bmatrix} I_a \\ I_b \\ I_c \end{bmatrix} \tag{6.154}$$

Substituting the expression of unsymmetrical voltages into Eq. (6.154) yields,

$$\begin{bmatrix} 1 & 1 & 1 \\ 1 & a^2 & a \\ 1 & a & a^2 \end{bmatrix} \begin{bmatrix} V_{a0} \\ V_{a1} \\ V_{a2} \end{bmatrix} = \begin{bmatrix} Z_n + Z_y & Z_n & Z_n \\ Z_n & Z_n + Z_y & Z_n \\ Z_n & Z_n & Z_n + Z_y \end{bmatrix} \begin{bmatrix} I_a \\ I_b \\ I_c \end{bmatrix} \tag{6.155}$$

Again, substituting the expression of unsymmetrical currents into Eq. (6.155) yields,

$$\begin{bmatrix} V_{a0} \\ V_{a1} \\ V_{a2} \end{bmatrix} = \frac{1}{3} \begin{bmatrix} 1 & 1 & 1 \\ 1 & a & a^2 \\ 1 & a^2 & a \end{bmatrix} \begin{bmatrix} Z_n + Z_y & Z_n & Z_n \\ Z_n & Z_n + Z_y & Z_n \\ Z_n & Z_n & Z_n + Z_y \end{bmatrix} \begin{bmatrix} 1 & 1 & 1 \\ 1 & a^2 & a \\ 1 & a & a^2 \end{bmatrix} \begin{bmatrix} I_{a0} \\ I_{a1} \\ I_{a2} \end{bmatrix} \tag{6.156}$$

$$\begin{bmatrix} V_{a0} \\ V_{a1} \\ V_{a2} \end{bmatrix} = \frac{1}{3} \begin{bmatrix} 1 & 1 & 1 \\ 1 & a & a^2 \\ 1 & a^2 & a \end{bmatrix} \begin{bmatrix} 3Z_n + Z_y & Z_n + Z_y + Z_n(a^2 + a) & Z_n + Z_y + Z_n(a^2 + a) \\ 3Z_n + Z_y & Z_n + a^2(Z_y + Z_n) + aZ_n & Z_n + a(Z_y + Z_n) + a^2 Z_n \\ 3Z_n + Z_y & Z_n + a^2 Z_n + a(Z_n + Z_y) & Z_n + aZ_n + a^2(Z_y + Z_n) \end{bmatrix} \begin{bmatrix} I_{a0} \\ I_{a1} \\ I_{a2} \end{bmatrix} \tag{6.157}$$

$$\begin{bmatrix} V_{a0} \\ V_{a1} \\ V_{a2} \end{bmatrix} = \frac{1}{3} \begin{bmatrix} 1 & 1 & 1 \\ 1 & a & a^2 \\ 1 & a^2 & a \end{bmatrix} \begin{bmatrix} 3Z_n + Z_y & Z_y & Z_y \\ 3Z_n + Z_y & a^2 Z_y & aZ_y \\ 3Z_n + Z_y & aZ_y & a^2 Z_y \end{bmatrix} \begin{bmatrix} I_{a0} \\ I_{a1} \\ I_{a2} \end{bmatrix} \tag{6.158}$$

$$\begin{bmatrix} V_{a0} \\ V_{a1} \\ V_{a2} \end{bmatrix} = \frac{1}{3} \begin{bmatrix} 9Z_n + 3Z_y & Z_y(a^2 + a + 1) & Z_y(a^2 + a + 1) \\ (3Z_n + Z_y)(a^2 + a + 1) & Z_y + a^3 Z_y + a^3 Z_y & Z_y + a^2 Z_y + a^4 Z_y \\ (3Z_n + Z_y)(a^2 + a + 1) & Z_y + a^4 Z_y + a^2 Z_y & Z_y + a^3 Z_y + a^3 Z_y \end{bmatrix} \begin{bmatrix} I_{a0} \\ I_{a1} \\ I_{a2} \end{bmatrix} \tag{6.159}$$

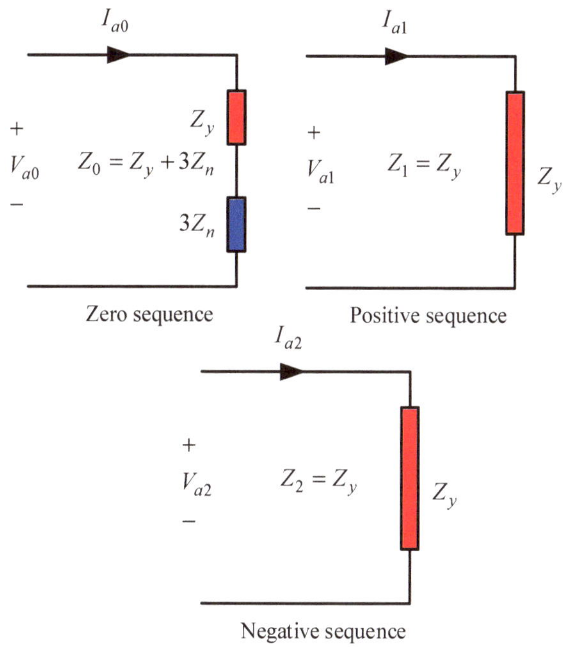

Fig. 6.22 Sequence networks

$$\begin{bmatrix} V_{a0} \\ V_{a1} \\ V_{a2} \end{bmatrix} = \frac{1}{3}\begin{bmatrix} 9Z_n + 3Z_y & 0 & 0 \\ 0 & 3Z_y & 0 \\ 0 & 0 & 3Z_y \end{bmatrix}\begin{bmatrix} I_{a0} \\ I_{a1} \\ I_{a2} \end{bmatrix} \tag{6.160}$$

$$\begin{bmatrix} V_{a0} \\ V_{a1} \\ V_{a2} \end{bmatrix} = \begin{bmatrix} 3Z_n + Z_y & 0 & 0 \\ 0 & Z_y & 0 \\ 0 & 0 & Z_y \end{bmatrix}\begin{bmatrix} I_{a0} \\ I_{a1} \\ I_{a2} \end{bmatrix} \tag{6.161}$$

From Eq. (6.161), the expressions of symmetrical components of voltage for the phase a can be written as,

$$V_{a0} = (3Z_n + Z_y)I_{a0} \tag{6.162}$$

$$V_{a1} = Z_y I_{a1} \tag{6.163}$$

$$V_{a2} = Z_y I_{a2} \tag{6.164}$$

The sequence circuits based on Eqs. (6.162), (6.163), and (6.164) are shown in Fig. 6.22. The neutral impedance of the Y-connection would be zero if it is solidly grounded. Equation (6.162) can be modified as,

$$V_{a0} = Z_y I_{a0} \tag{6.165}$$

The neutral impedance of the Y-connection will provide an infinite quantity if it is not grounded. That means zero sequence circuit will provide an open circuit and nonzero sequence current will flow through an ungrounded Y-connection.

6.9 Zero Sequence Models

In transmission lines, there is no effect on the impedance due to positive and native sequence components of voltages and currents. In this case, positive sequence and negative sequence impedances are equal to each other, that is,

$$Z_1 = Z_2 \tag{6.166}$$

The zero sequence impedance is much higher than the positive or negative sequence impedance due to its ground return path. In this case, the expression can be written as,

$$Z_0 = Z_1 + 3Z_n \tag{6.167}$$

Similarly, the zero sequence reactance can be written as,

$$X_0 = X_1 + 3X_n \tag{6.168}$$

where the neutral reactance (per mile) can be expressed as,

$$X_n = 2.02 \times 10^{-3} f \ln \frac{D_m}{D_s} \; \Omega/\text{mile} \tag{6.169}$$

where

D_m is the geometric mean distance (GMD),
D_s is the geometric mean radius,
f is the frequency.

The sequence networks for transmission lines are shown in Fig. 6.23. There are three impedances, namely subtransient, transient, and direct axis reactance. The positive and negative sequence impedances are equal to the subtransient reactance during a fault condition.

The generator offers a very small reactance due to the leakage flux. Therefore, the zero sequence impedance is smaller than the others. The following equations for a generator can be written as,

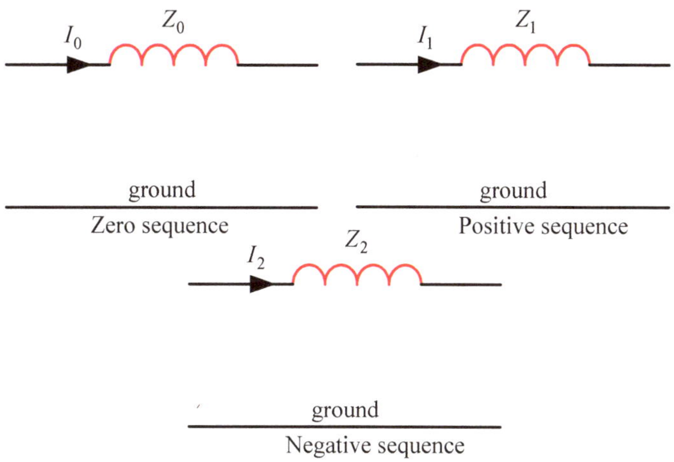

Fig. 6.23 Sequence networks for the transmission line

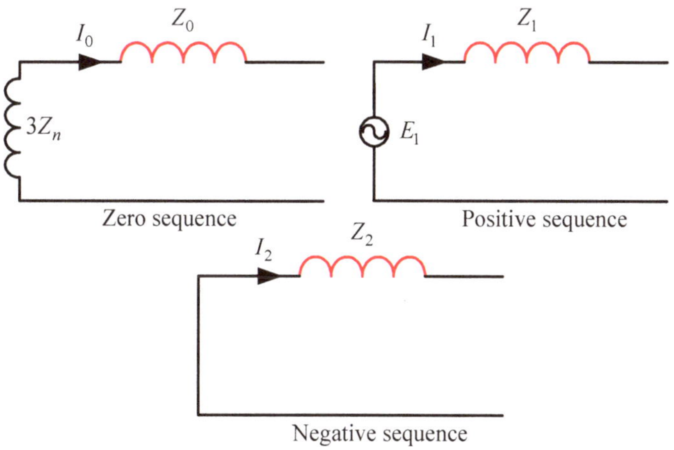

Fig. 6.24 Sequence networks for generator

$$Z_1 = Z_2 = Z_d'' \tag{6.170}$$

$$Z_0 = Z_l \tag{6.171}$$

The sequence networks for the generator are shown in Fig. 6.24. In the transformer, the zero sequence current flows if the neutral is grounded. In this case, the positive and the negative sequence impedances are equal to the zero sequence impedance, that is,

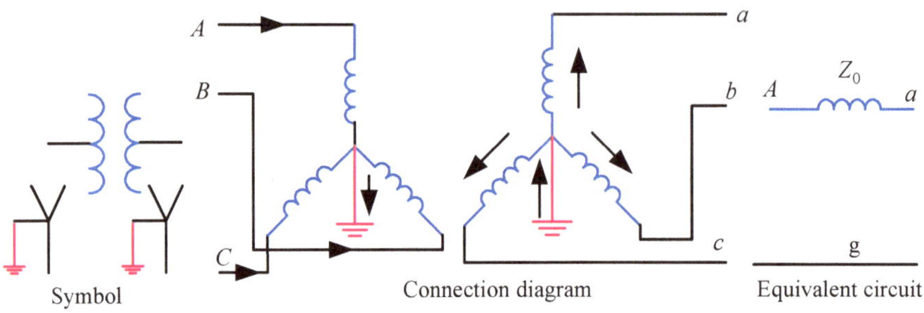

Fig. 6.25 Sequence networks for Y-Y-connection with both sides grounded

$$Z_1 = Z_2 = Z_0 \tag{6.172}$$

Y-Y-Connection with both neutral grounded: In this connection, both neutrals are connected to the ground. Therefore, zero sequence current will flow in the primary and the secondary windings through the two grounded neutrals. As a result, the zero sequence impedance connects the high voltage and the low voltage terminals as shown in Fig. 6.25.

Y-Y-connection with only one grounded: In this arrangement, one side of the Y-connection is not grounded. If anyone of the Y-Y-connection is not grounded, then the zero sequence current will not flow through the ungrounded wye. Therefore, an open circuit will appear between the high voltage and low voltage sides, as shown in Fig. 6.26.

Y-Y-connection with no grounded: In this arrangement, the sum of the phase current in both cases is zero. As a result, no zero sequence current will flow in any of the windings. Hence, there is an open circuit between the high voltage and the low voltage sides, as shown in Fig. 6.27.

Grounded Y- and Δ connections: In this connection, zero sequence current will flow in the Y-connection as its neutral is connected to the ground. The balanced zero sequence current will flow within the phases of the Δ connection, but this

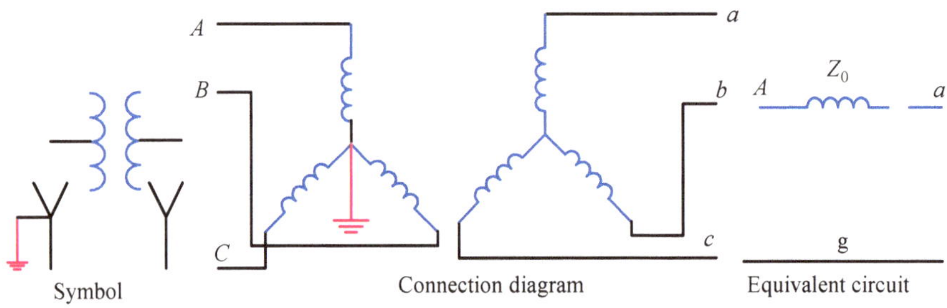

Fig. 6.26 Sequence networks for Y-Y-connection for one side grounded

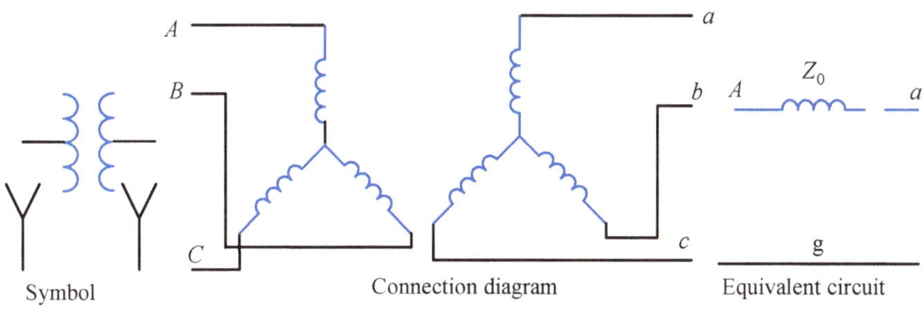

Fig. 6.27 Sequence networks for Y-Y-connection with no ground

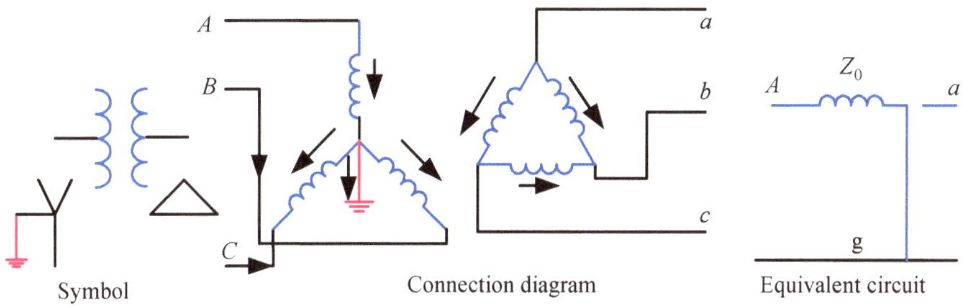

Fig. 6.28 Sequence networks for grounded Y- and delta connection

current will not flow out of the terminal. Therefore, no zero sequence current will flow in the line as shown in Fig. 6.28.

Ungrounded Y- and Δ connections: In this case, there will be no connection between the neutral and the ground. Therefore, the zero sequence current will not flow in the windings of both transformers. As a result, an open circuit will exist between the high and low voltage sides as shown in Fig. 6.29.

Δ-Δ connection: In this connection, no zero sequence current will leave or enter the terminals. However, it is possible for the current components to circulate within the windings. Therefore, there is an open circuit between the high voltage and the low voltage windings. The zero sequence impedance will form a closed path with grounding terminals as shown in Fig. 6.30.

Example 6.7

Figure 6.31 shows a single-line diagram of a three-phase power system. Draw the positive sequence, negative sequence, and zero sequence networks.

Solution

The generator is represented with a voltage source and a series reactance in the positive sequence network. The transformer and transmission lines are also represented by the respective reactance quantities as shown in Fig. 6.32.

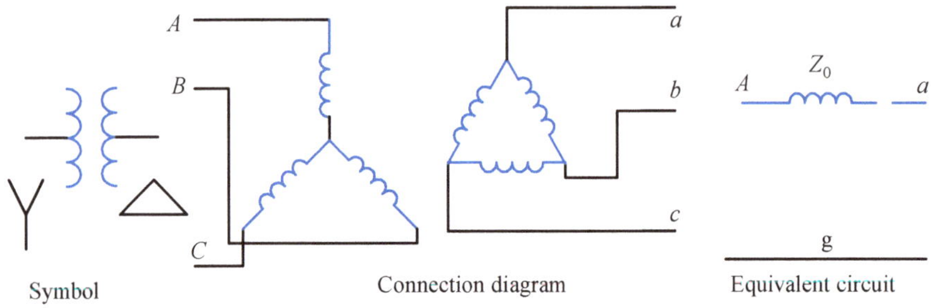

Fig. 6.29 Sequence networks for ungrounded Y- and delta connection

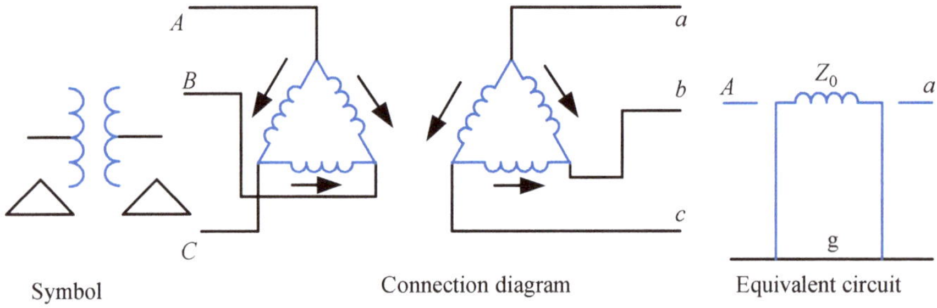

Fig. 6.30 Sequence networks for delta-delta connection

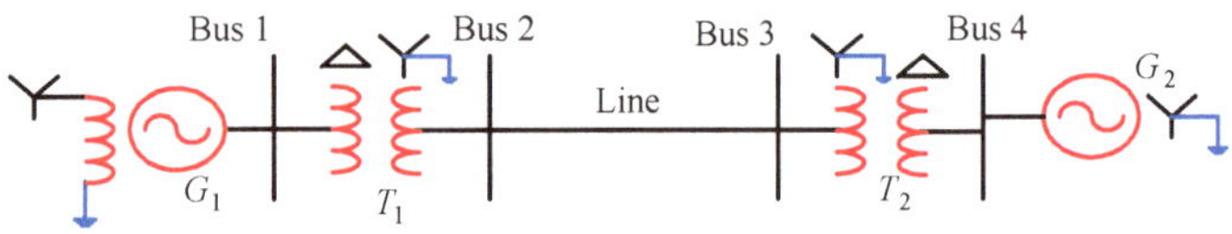

Fig. 6.31 A single-line diagram for Example 6.7

By omitting the voltage sources from the positive sequence network and by replacing the generator reactance components with negative sequence reactance components as shown in Fig. 6.33 has been derived.

The equivalent reactance of the generator G_1 is $X_{g0} + 3X_n$ as it is grounded through a reactance. The total equivalent reactance of the generator G_2 is zero as it is solidly grounded. The primary sides of both the transformers T_1 and T_2 are delta connected. The zero sequence network is, therefore, in an open circuit near the bus 1 and bus 4. The secondary sides of both the transformers are wye-connected and solidly grounded. The zero sequence network is shown in Fig. 6.34.

Practice Problem 6.7

Figure 6.35 shows a single-line diagram of a three-phase power system. Draw the positive sequence, negative sequence, and zero sequence networks.

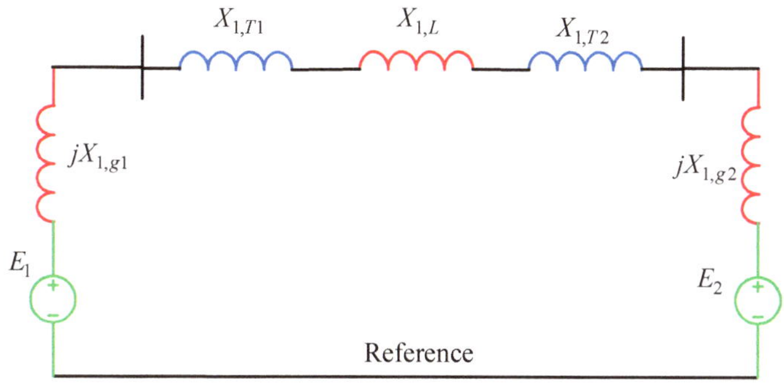

Fig. 6.32 A positive sequence network for Example 6.7

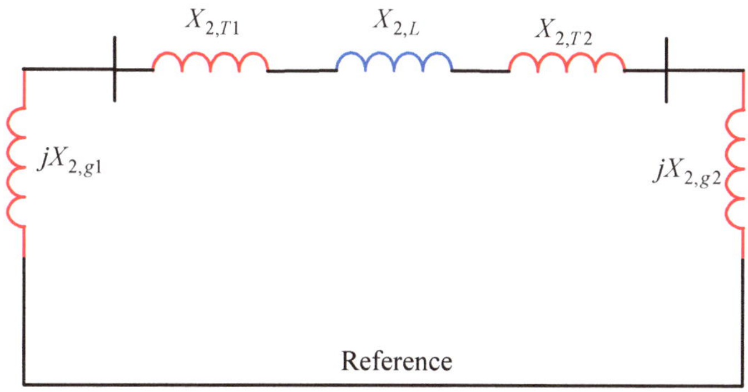

Fig. 6.33 A negative sequence network for Example 6.7

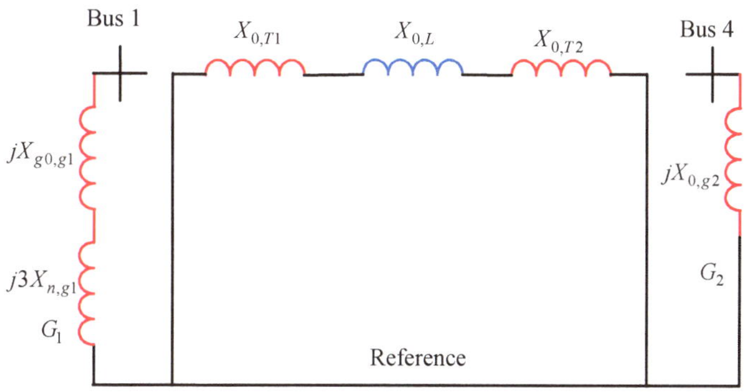

Fig. 6.34 A zero sequence network for Example 6.7

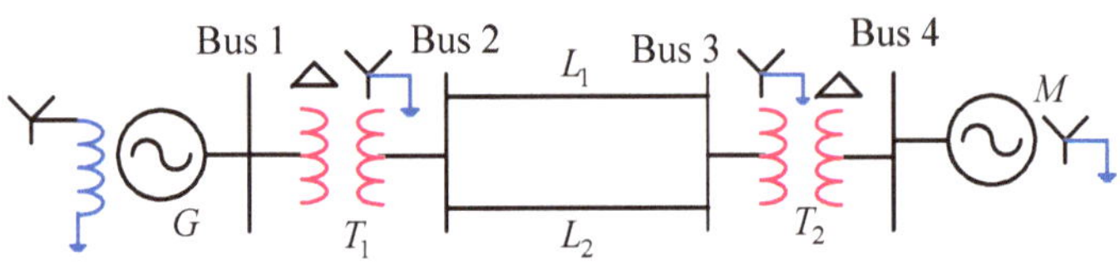

Fig. 6.35 A single-line diagram for Practice Problem 6.7

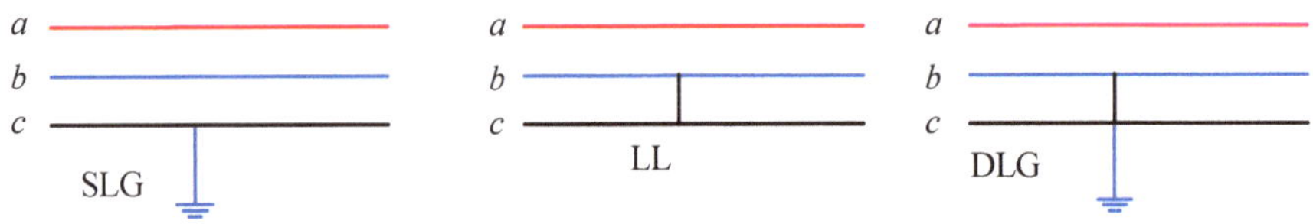

Fig. 6.36 Different types of unsymmetrical faults

6.10 Classification of Unsymmetrical Faults

Unsymmetrical faults are the most common faults that occur in the power system. Due to this fault, the magnitudes of the line currents become unequal, and also these current components observed a phase displacement among them.

In this case, symmetrical components are required to analyze the current and voltage quantities during the unsymmetrical faults. These unsymmetrical faults can be classified into three categories, namely single-line-to-ground fault (SLG), line-to-line fault (LL), and double-line-to-ground fault (DLG). The unsymmetrical faults are shown in Fig. 6.36.

6.11 Sequence Network of an Unloaded Synchronous Generator

A three-phase unloaded synchronous generator is having a synchronous impedance of Z_y per phase, and its neutral is grounded by an impedance Z_n as shown in Fig. 6.37.

In balanced condition, the negative sequence and zero sequence voltages are zero [1, 2]. The expression of neutral current is,

$$I_n = I_a + I_b + I_c \tag{6.173}$$

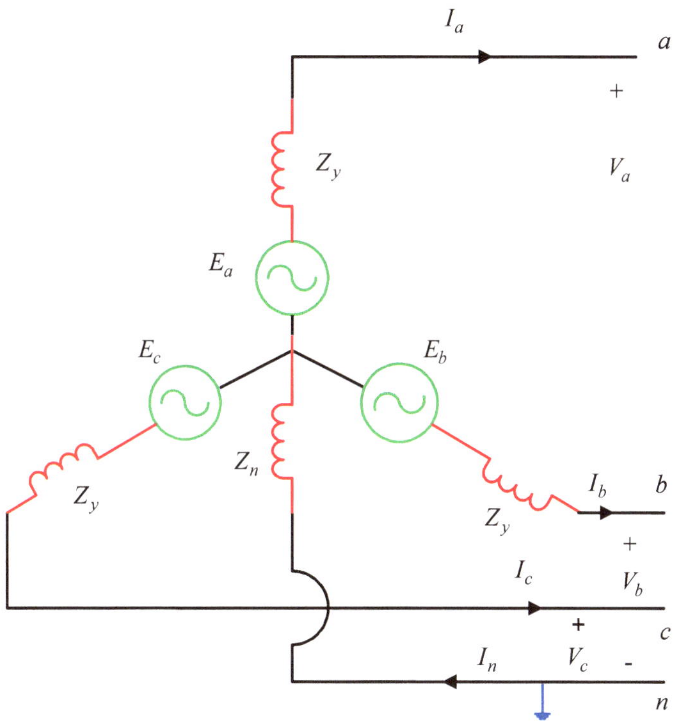

Fig. 6.37 Unloaded three-phase synchronous generator

Applying KVL to the circuit shown in Fig. 6.37, the following equation can be found:

$$V_a = E_a - Z_y I_a - Z_n I_n \tag{6.174}$$

Substituting Eq. (6.173) into Eq. (6.174) yields,

$$E_a = V_a + Z_y I_a + Z_n (I_a + I_b + I_c) \tag{6.175}$$

$$E_a = V_a + (Z_y + Z_n) I_a + Z_n I_b + Z_n I_c \tag{6.176}$$

Similarly, the expression of other voltages can be found as,

$$E_b = V_b + Z_n I_a + (Z_y + Z_n) I_b + Z_n I_c \tag{6.177}$$

$$E_c = V_c + Z_n I_a + Z_n I_b + (Z_y + Z_n) I_c \tag{6.178}$$

Equations (6.176), (6.177), and (6.178) can be written in the matrix form as,

$$\begin{bmatrix} E_a \\ E_b \\ E_c \end{bmatrix} = \begin{bmatrix} V_a \\ V_b \\ V_c \end{bmatrix} + \begin{bmatrix} Z_y + Z_n & Z_n & Z_n \\ Z_n & Z_y + Z_n & Z_n \\ Z_n & Z_n & Z_y + Z_n \end{bmatrix} \begin{bmatrix} I_a \\ I_b \\ I_c \end{bmatrix} \tag{6.179}$$

More concisely, Eq. (6.179) can be written as,

$$[E]_{abc} = [V]_{abc} + [Z]_{abc}[I]_{abc} \tag{6.180}$$

where the following terms can be expressed as,

$$[E]_{abc} = \begin{bmatrix} E_a & E_b & E_c \end{bmatrix}^T \tag{6.181}$$

$$[V]_{abc} = \begin{bmatrix} V_a & V_b & V_c \end{bmatrix}^T \tag{6.182}$$

$$[I]_{abc} = \begin{bmatrix} I_a & I_b & I_c \end{bmatrix}^T \tag{6.183}$$

Multiplying Eq. (6.180) by $[A]^{-1}$ yields,

$$[A]^{-1}[E]_{abc} = [A]^{-1}[V]_{abc} + [A]^{-1}[Z]_{abc}[I]_{abc} \tag{6.184}$$

Let us consider the generator voltage,

$$E_a = E \tag{6.185}$$

According to the a operator, the following relationships can be affirmed,

$$E_b = a^2 E \tag{6.186}$$

$$E_c = aE \tag{6.187}$$

The left-hand side of Eq. (6.184) can be modified as,

$$[A]^{-1}[E]_{abc} = \frac{1}{3}\begin{bmatrix} 1 & 1 & 1 \\ 1 & a & a^2 \\ 1 & a^2 & a \end{bmatrix} \begin{bmatrix} E \\ a^2 E \\ aE \end{bmatrix} \tag{6.188}$$

$$[A]^{-1}[E]_{abc} = \frac{E}{3}\begin{bmatrix} 1 + a + a^2 \\ 1 + a^3 + a^3 \\ 1 + a^4 + a^2 \end{bmatrix} = \frac{E}{3}\begin{bmatrix} 0 \\ 3 \\ 0 \end{bmatrix} = \begin{bmatrix} 0 \\ E \\ 0 \end{bmatrix} \tag{6.189}$$

The first part of the right-hand side of Eq. (6.184) becomes,

$$[A]^{-1}[V]_{abc} = [V]_{012} \tag{6.190}$$

The second part of the right-hand side of Eq. (6.184) can be written as,

$$[A]^{-1}[Z]_{abc}[I]_{abc} = [A]^{-1}[Z]_{abc}[A][I]_{012} \tag{6.191}$$

However, the following relation can be written as,

$$[A]^{-1}[Z]_{abc}[A] = [Z]_{012} \tag{6.192}$$

Substituting Eq. (6.192) into Eq. (6.191) yields,

$$[A]^{-1}[Z]_{abc}[I]_{abc} = [Z]_{012}[I]_{012} \tag{6.193}$$

where

$$[Z]_{012} = \begin{bmatrix} 3Z_n + Z_y & 0 & 0 \\ 0 & Z_y & 0 \\ 0 & 0 & Z_y \end{bmatrix} = \begin{bmatrix} Z_0 & 0 & 0 \\ 0 & Z_1 & 0 \\ 0 & 0 & Z_2 \end{bmatrix} \tag{6.194}$$

In Eq. (6.194),

$Z_0 = Z_y + 3Z_n$ is the zero sequence impedance,
$Z_1 = Z_y$ is the positive sequence impedance,
$Z_2 = Z_y$ is the negative sequence impedance.

Substituting Eqs. (6.189), (6.192), and (6.194) into Eq. (6.184) yields,

$$\begin{bmatrix} 0 \\ E \\ 0 \end{bmatrix} = \begin{bmatrix} V_0 \\ V_1 \\ V_2 \end{bmatrix} + \begin{bmatrix} Z_0 & 0 & 0 \\ 0 & Z_1 & 0 \\ 0 & 0 & Z_2 \end{bmatrix} \begin{bmatrix} I_0 \\ I_1 \\ I_2 \end{bmatrix} \tag{6.195}$$

Equation (6.195) can be modified as,

$$\begin{bmatrix} V_0 \\ V_1 \\ V_2 \end{bmatrix} = \begin{bmatrix} 0 \\ E \\ 0 \end{bmatrix} - \begin{bmatrix} Z_0 & 0 & 0 \\ 0 & Z_1 & 0 \\ 0 & 0 & Z_2 \end{bmatrix} \begin{bmatrix} I_0 \\ I_1 \\ I_2 \end{bmatrix} \tag{6.196}$$

From Eq. (6.196), the voltages are expressed as,

$$V_0 = 0 - I_0 Z_0 \tag{6.197}$$

$$V_1 = E - I_1 Z_1 \tag{6.198}$$

$$V_2 = 0 - I_2 Z_2 \tag{6.199}$$

The sequence networks are shown in Fig. 6.24.

6.12 Single-Line-to-Ground Fault

A three-phase Y-connected unload generator is shown in Fig. 6.38. Initially, the neutral of the generator is grounded with a solid wire. In this case, consider that a single-line-to-ground fault occurs in phase a of the unloaded generator which disturbs the balance of the power system network [3]. For this scenario, the boundary conditions are,

$$V_a = 0 \tag{6.200}$$

$$I_b = 0 \tag{6.201}$$

$$I_c = 0 \tag{6.202}$$

The expressions of symmetrical components of current of phase a are,

$$\begin{bmatrix} I_{a0} \\ I_{a1} \\ I_{a2} \end{bmatrix} = \frac{1}{3}\begin{bmatrix} 1 & 1 & 1 \\ 1 & a & a^2 \\ 1 & a^2 & a \end{bmatrix}\begin{bmatrix} I_a \\ I_b \\ I_c \end{bmatrix} \tag{6.203}$$

Substituting Eqs. (6.201) and (6.202) into Eq. (6.203) yields,

$$\begin{bmatrix} I_{a0} \\ I_{a1} \\ I_{a2} \end{bmatrix} = \frac{1}{3}\begin{bmatrix} 1 & 1 & 1 \\ 1 & a & a^2 \\ 1 & a^2 & a \end{bmatrix}\begin{bmatrix} I_a \\ 0 \\ 0 \end{bmatrix} \tag{6.204}$$

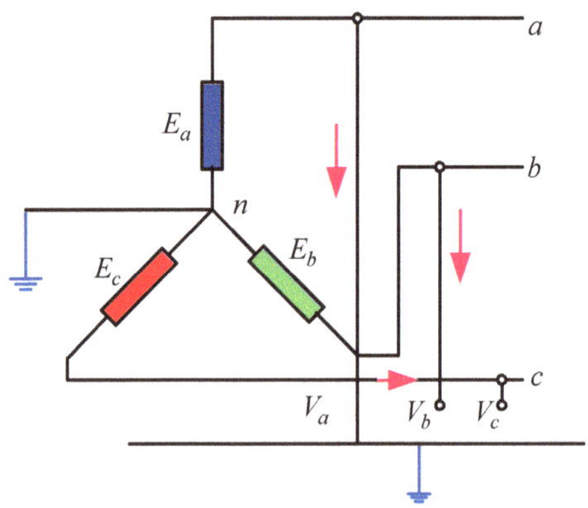

Fig. 6.38 Unloaded generator

From Eq. (6.204), the symmetrical components of the current of phase a can be written as,

$$I_{a0} = I_{a1} = I_{a2} = \frac{I_a}{3} \tag{6.205}$$

From Eq. (6.205), it is observed that the symmetrical components of current are equal in a single-line-to-ground fault.

Substituting the values of symmetrical components of current in Eq. (6.195) yields,

$$\begin{bmatrix} V_{a0} \\ V_{a1} \\ V_{a2} \end{bmatrix} = \begin{bmatrix} 0 \\ E_a \\ 0 \end{bmatrix} - \begin{bmatrix} Z_0 & 0 & 0 \\ 0 & Z_1 & 0 \\ 0 & 0 & Z_2 \end{bmatrix} \begin{bmatrix} I_{a1} \\ I_{a1} \\ I_{a1} \end{bmatrix} \tag{6.206}$$

From Eq. (6.206), the symmetrical components of voltage can be written as,

$$V_{a0} = -Z_0 I_{a1} \tag{6.207}$$

$$V_{a1} = E_a - Z_1 I_{a1} \tag{6.208}$$

$$V_{a2} = -Z_2 I_{a1} \tag{6.209}$$

However, we know that the unbalanced voltage for phase a is,

$$V_a = V_{a0} + V_{a1} + V_{a2} = 0 \tag{6.210}$$

Substituting Eqs. (6.207), (6.208), and (6.209) into Eq. (6.210) yields,

$$-Z_0 I_{a1} + E_a - Z_1 I_{a1} - Z_2 I_{a1} = 0 \tag{6.211}$$

$$I_{a1} = \frac{E_a}{Z_0 + Z_1 + Z_2} \tag{6.212}$$

For a single-line-to-ground fault, the positive sequence current, negative sequence current, and the zero sequence current are equal and can be determined from Eq. (6.212).

From Eq. (6.205), the expression of fault current in phase a can be determined as,

$$I_a = 3I_{a1} \tag{6.213}$$

Substituting Eq. (6.212) into Eq. (6.213) yields,

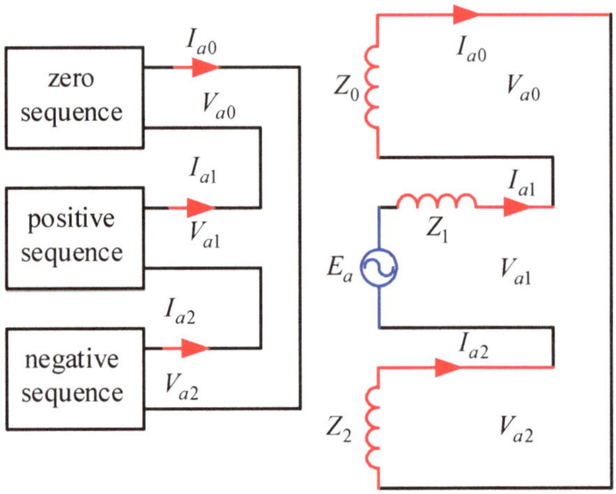

Fig. 6.39 Sequence networks for single-line-to-ground fault without fault impedance

$$I_a = \frac{3E_a}{Z_0 + Z_1 + Z_2} \tag{6.214}$$

The sequence network connection for the single-line-to-ground fault without a fault impedance is shown in Fig. 6.39.

If a single-line-to-ground fault occurs in phase a through the impedance Z_f, then the expression of fault current from Eq. (6.210) can be derived as,

$$V_a = V_{a0} + V_{a1} + V_{a2} = I_a Z_f \tag{6.215}$$

Substituting Eqs. (6.187), (6.188), (6.189), and (6.193) into Eq. (6.195) yields,

$$-Z_0 I_{a1} + E_a - Z_1 I_{a1} - Z_2 I_{a1} = 3Z_f I_{a1} \tag{6.216}$$

$$I_{a1} = \frac{E_a}{Z_0 + Z_1 + Z_2 + 3Z_f} \tag{6.217}$$

Again, substituting Eq. (6.217) into Eq. (6.213), the fault current in phase a can be determined as,

$$I_a = \frac{3E_a}{Z_0 + Z_1 + Z_2 + 3Z_f} \tag{6.218}$$

The sequence network connection for the single-line-to-ground fault with a fault impedance is shown in Fig. 6.40.

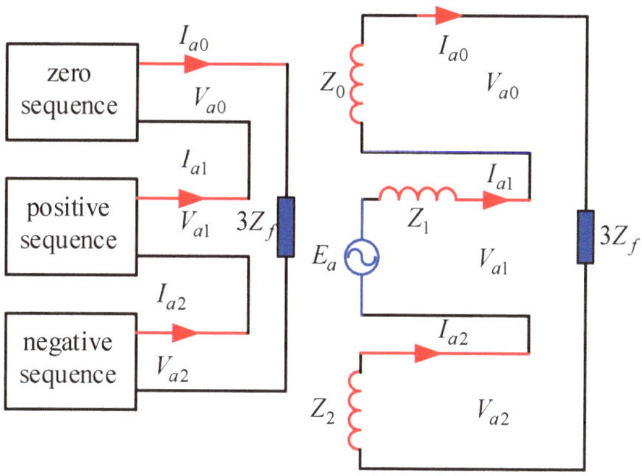

Fig. 6.40 Sequence networks for single-line-to-ground fault with fault impedance

Example 6.8

A three-phase 15 MVA, Y-connected, 11 kV synchronous generator is solidly grounded. The positive, negative, and zero sequence impedances are $j1.5\ \Omega$, $j0.8\ \Omega$, and $j0.3\ \Omega$, respectively. Determine the fault current in phase a if the single-line-to-ground fault occurs in that phase.

Solution

The value of the generated voltage per phase is calculated as,

$$E_a = \frac{11000}{\sqrt{3}} = 6350.8 \text{ V} \tag{6.219}$$

The value of the symmetrical component of the current is determined as,

$$I_{a1} = \frac{E_a}{Z_0 + Z_1 + Z_2} = \frac{6350.8}{j1.5 + j0.8 + j0.3} = -j2442.6 \text{ A} \tag{6.220}$$

The value of the fault current in phase a is calculated as,

$$I_a = 3I_{a1} = 3 \times -j2442.6 = -j7327.8 \text{ A} \tag{6.221}$$

Example 6.9

Figure 6.41 shows a Y-connected, three-phase synchronous generator whose neutral is earthed with solid wire. A single-line-to-ground fault occurs in phase a, and the current in this phase is found to be 100 A. Find the positive sequence, negative sequence, and zero sequence currents for all three phases.

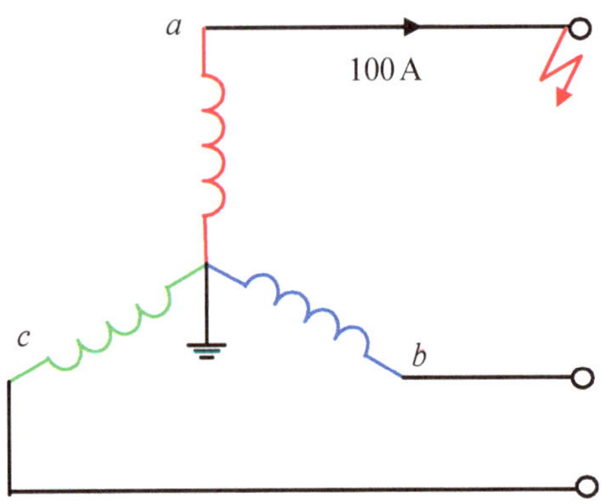

Fig. 6.41 A Y-connected synchronous generator for Example 6.9

Solution
During fault, the currents in different phases are,

$$I_a = 100\ \text{A} \tag{6.222}$$

$$I_b = 0\ \text{A} \tag{6.223}$$

$$I_c = 0\ \text{A} \tag{6.224}$$

The zero sequence components of the current are determined as,

$$I_{a0} = I_{b0} = I_{c0} = \frac{1}{3}(I_a + I_b + I_c) = \frac{1}{3} \times 100 = 33.33\ \text{A} \tag{6.225}$$

The positive sequence components of the current can be determined as,

$$I_{a1} = \frac{1}{3}(I_a + aI_b + a^2 I_c) = \frac{1}{3} \times 100 = 33.33\ \text{A} \tag{6.226}$$

$$I_{b1} = a^2 I_{a1} = 33.33\ \underline{|240^\circ}\ \text{A} \tag{6.227}$$

$$I_{c1} = aI_{a1} = 33.33\ \underline{|120^\circ}\ \text{A} \tag{6.228}$$

The negative sequence components of the current can be determined as,

$$I_{a2} = \frac{1}{3}(I_a + a^2 I_b + aI_c) = \frac{1}{3} \times 100 = 33.33\ \text{A} \tag{6.229}$$

$$I_{b2} = aI_{a2} = 33.33\ \underline{|120^\circ}\ \text{A} \tag{6.230}$$

$$I_{c2} = a^2 I_{a2} = 33.33\ \underline{|240^\circ}\ \text{A} \tag{6.231}$$

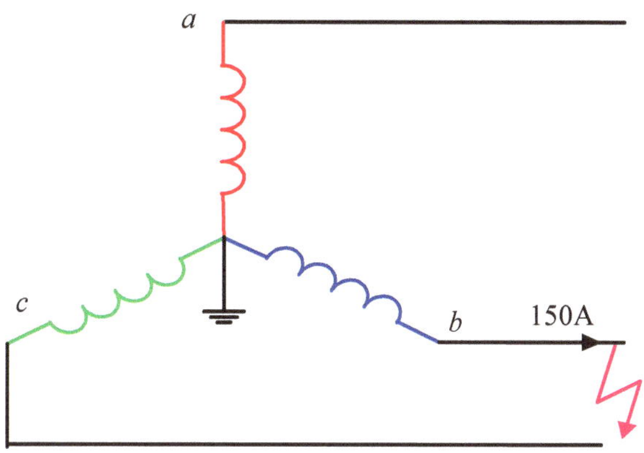

Fig. 6.42 A circuit for Practice Problem 6.5

Practice Problem 6.8
A three-phase 20 MVA, 11 kV synchronous generator is having a subtransient reactance of 0.20 pu, and its neutral is solidly grounded. The negative and zero sequence reactance are 0.30 pu and 0.15 pu, respectively. Determine the fault current if a single-line-to-ground fault occurs in phase *a*.

Practice Problem 6.9
The neutral of a three-phase Y-connected synchronous generator is solidly grounded as shown in Fig. 6.42. A single-line-to-ground fault occurs in phase *b*, and the current in this phase is found to be 150 A. Calculate the positive sequence, negative sequence, and zero sequence currents for all three phases.

6.13 Line-to-Line Fault

A three-phase synchronous generator whose line-to-line fault occurs between phase *b* and phase *c* as shown in Fig. 6.43. In this condition, the voltages at phases *b* and *c* must be the same, whereas the currents in phase *b* and phase *c* must be equal but in opposite direction to each other [4].

In the line-to-line fault, the boundary conditions are,

$$I_a = 0 \tag{6.232}$$

$$I_b = -I_c \tag{6.233}$$

$$V_b = V_c \tag{6.234}$$

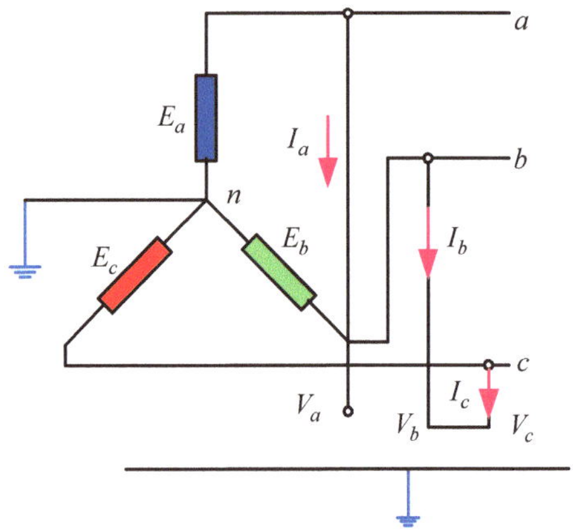

Fig. 6.43 Line-to-line fault on an unloaded synchronous generator

Substituting Eqs. (6.232) and (6.233) into Eq. (6.203) yields,

$$\begin{bmatrix} I_{a0} \\ I_{a1} \\ I_{a2} \end{bmatrix} = \frac{1}{3}\begin{bmatrix} 1 & 1 & 1 \\ 1 & a & a^2 \\ 1 & a^2 & a \end{bmatrix}\begin{bmatrix} 0 \\ I_b \\ -I_b \end{bmatrix} \tag{6.235}$$

$$\begin{bmatrix} I_{a0} \\ I_{a1} \\ I_{a2} \end{bmatrix} = \frac{1}{3}\begin{bmatrix} 0 + I_b - I_b \\ 0 + (a - a^2)I_b \\ 0 + (a^2 - a)I_b \end{bmatrix} \tag{6.236}$$

From Eq. (6.236), the symmetrical components of current in phase a can be written as,

$$I_{a0} = 0 \tag{6.237}$$

$$I_{a1} = \frac{1}{3}(a - a^2)I_b \tag{6.238}$$

$$I_{a2} = \frac{1}{3}(a^2 - a)I_b \tag{6.239}$$

From Eq. (6.237), it is seen that the zero sequence component of current is zero. Hence, the value of the zero sequence voltage can be written as,

$$V_{a0} = I_{a0}Z_0 = 0 \tag{6.240}$$

From Eqs. (6.238) and (6.239), it is also seen that the positive sequence component of current is equal to the negative sequence component of current but in opposite direction.

The symmetrical components of voltage can be determined as,

$$V_b = V_{a0} + a^2 V_{a1} + aV_{a2} \tag{6.241}$$

$$V_c = V_{a0} + aV_{a1} + a^2 V_{a2} \tag{6.242}$$

Substituting Eqs. (6.241) and (6.242) into Eq. (6.234) yields,

$$V_{a0} + a^2 V_{a1} + aV_{a2} = V_{a0} + aV_{a1} + a^2 V_{a2} \tag{6.243}$$

$$(a^2 - a)V_{a1} = (a^2 - a)V_{a2} \tag{6.244}$$

$$V_{a1} = V_{a2} \tag{6.245}$$

Substituting Eqs. (6.208) and (6.209) into Eq. (6.245) yields,

$$E_a - I_{a1}Z_1 = -I_{a2}Z_2 \tag{6.246}$$

$$E_a - I_{a1}Z_1 = I_{a1}Z_2 \tag{6.247}$$

$$I_{a1} = \frac{E_a}{Z_1 + Z_2} \tag{6.248}$$

If there is a fault impedance Z_f in between lines b and c, then the following expression can be written as,

$$V_b = V_c + I_b Z_f \tag{6.249}$$

Substituting Eqs. (6.98), (6.241), and (6.242) into Eq. (6.249) yields,

$$V_{a0} + a^2 V_{a1} + aV_{a2} = V_{a0} + aV_{a1} + a^2 V_{a2} + (I_{a0} + a^2 I_{a1} + aI_{a2})Z_f \tag{6.250}$$

Again, substituting Eqs. (6.237), (6.238), and (6.239) into Eq. (6.250) provides,

$$(a^2 - a)V_{a1} = (a^2 - a)V_{a2} + (a^2 - a)I_{a1}Z_f \tag{6.251}$$

$$V_{a1} = V_{a2} + I_{a1}Z_f \tag{6.252}$$

The interconnection of the sequence network without a fault impedance is shown in Fig. 6.44.

The interconnection of the sequence network with a fault impedance is shown in Fig. 6.45.

Fig. 6.44 Sequence network for line-to-line fault without fault impedance

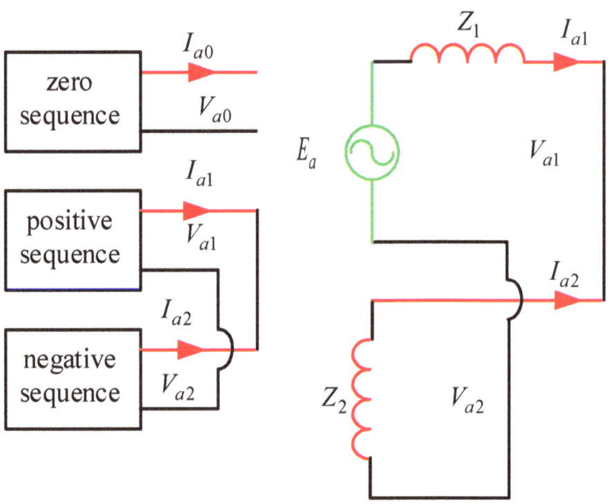

Fig. 6.45 Sequence network for line-to-line fault with a fault impedance

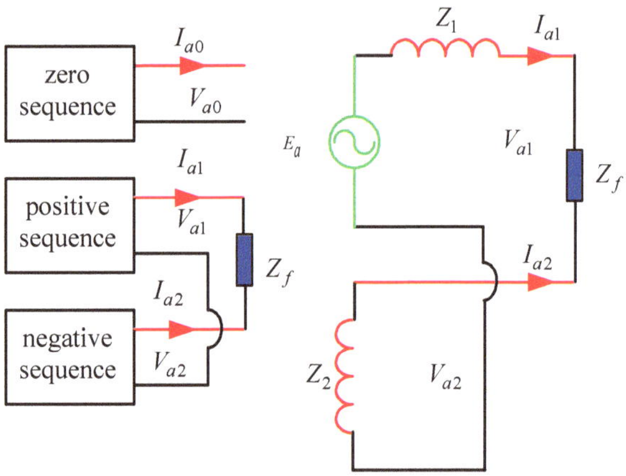

Example 6.10

The positive sequence, negative sequence, and zero sequence reactance of a 15 MVA, 13 kV three-phase Y-connected synchronous generator are 0.4 pu, 0.3 pu, and 0.1 pu, respectively. The neutral point of the generator is solidly grounded and is not supplying current to the load. Calculate the fault current and the actual line-to-line voltages if a line-to-line fault occurs between phase *b* and phase *c*.

Solution

Let us consider that the base values are 15 MVA and 13 kV. The per unit generator voltage can be determined as,

$$E = \frac{13}{13} = 1\underline{|0^\circ}\text{ pu} \tag{6.253}$$

The zero sequence component of current is,

$$I_{a0} = 0 \tag{6.254}$$

The values of the positive and negative sequence components of current are,

$$I_{a1} = -I_{a2} = \frac{E}{Z_1 + Z_2} = \frac{1}{j0.4 + j0.3} = -j1.42\,\text{pu} \tag{6.255}$$

The value of the fault current can be determined as,

$$I_b = I_{a0} + a^2 I_{a1} + a I_{a2} = 0 + 1\,\angle 240^\circ \times 1.42\,\angle -90^\circ + 1\,\angle 120^\circ \times 1.42\,\angle 90^\circ \tag{6.256}$$

$$I_b = 0 + 1.42\,\angle 150^\circ + 1.42\,\angle 210^\circ = -2.44\,\text{pu} \tag{6.257}$$

The value of the base current can be calculated as,

$$I_{\text{base}} = \frac{15 \times 1000000}{\sqrt{3} \times 13 \times 1000} = 666.17\,\text{A} \tag{6.258}$$

The actual value of the fault current is,

$$I_b = 2.44 \times 666.17 = 1625.45\,\text{A} \tag{6.259}$$

The sequence voltages for phase a can be determined as,

$$V_{a0} = -I_{a0} Z_0 = 0\,\angle 0^\circ\,\text{pu} \tag{6.260}$$

$$V_{a1} = E - I_{a1} Z_1 = 1\,\angle 0^\circ + j1.42 \times j0.4 = 0.432\,\angle 0^\circ\,\text{pu} \tag{6.261}$$

$$V_{a2} = -I_{a2} Z_2 = -j1.42 \times j0.3 = 0.436\,\angle 0^\circ\,\text{pu} \tag{6.262}$$

The phase voltages of the generator are calculated as,

$$V_a = V_{a0} + V_{a1} + V_{a2} = 0 + 0.432 + 0.436 = 0.87\,\angle 0^\circ\,\text{pu} \tag{6.263}$$

$$V_b = V_{b0} + a^2 V_{a1} + a V_{a2} = 0 + 0.432\,\angle 240^\circ + 0.436\,\angle 120^\circ = \angle 0^\circ\,\text{pu} \tag{6.264}$$

$$V_c = V_{c0} + a V_{a1} + a^2 V_{a2} = 0 + 0.432\,\angle 120^\circ + 0.436\,\angle 240^\circ = \angle 0^\circ\,\text{pu} \tag{6.265}$$

The line-to-line voltages of the generator are determined as,

$$V_{\text{ab}} = V_a - V_b = 0.87\,\angle 0^\circ - \angle 0^\circ = 0.87\,\angle 0^\circ\,\text{pu} \tag{6.266}$$

$$V_{bc} = V_b - V_c = 0\angle 0^\circ - 0\angle 0^\circ = 0\angle 0^\circ \text{ pu} \tag{6.267}$$

$$V_{ca} = V_c - V_a = 0\angle 0^\circ - 0.87\angle 0^\circ = 0.87\angle -90^\circ \text{ pu} \tag{6.268}$$

The value of the line-to-neutral voltages is calculated as,

$$V_{Ln} = \frac{13}{\sqrt{3}} = 7.51\,\text{kV} \tag{6.269}$$

The actual line-to-line voltages are calculated as,

$$V_{ab} = 7.51 \times 0.87\angle 0^\circ = 6.53\angle 0^\circ \text{ kV} \tag{6.270}$$

$$V_{bc} = 7.51 \times 0\angle 0^\circ = 0\angle 0^\circ \text{ kV} \tag{6.271}$$

$$V_{ca} = 7.51 \times 0.87\angle -90^\circ = 6.53\angle -90^\circ \text{ kV} \tag{6.272}$$

Practice Problem 6.10

The positive sequence, negative sequence, and zero sequence components of the reactance of a 20 MVA, 13.8 kV synchronous generator are 0.3 pu, 0.2 pu, and 0.1 pu, respectively. The neutral point of the generator is solidly grounded, and the generator is not supplying current to load. Find the line-to-line voltages of the generator if the fault occurs between phases *b* and *c*.

6.14 Double-Line-to-Ground Fault

The connection diagram of an unloaded synchronous generator is shown in Fig. 6.46. Consider that a double-line-to-ground fault occurs between phases *b* and *c*. The voltages at phase *b* and phase *c* should be equal to zero, and the current in the phase *a* is equal to zero [5, 6]. In this case, the boundary conditions are,

$$I_a = 0 \tag{6.273}$$

$$V_b = V_c = 0 \tag{6.274}$$

According to Eq. (6.105), the symmetrical components of voltage for phase *a* can be expressed as,

$$\begin{bmatrix} V_{a0} \\ V_{a1} \\ V_{a2} \end{bmatrix} = \frac{1}{3}\begin{bmatrix} 1 & 1 & 1 \\ 1 & a & a^2 \\ 1 & a^2 & a \end{bmatrix}\begin{bmatrix} V_a \\ V_b \\ V_c \end{bmatrix} \tag{6.275}$$

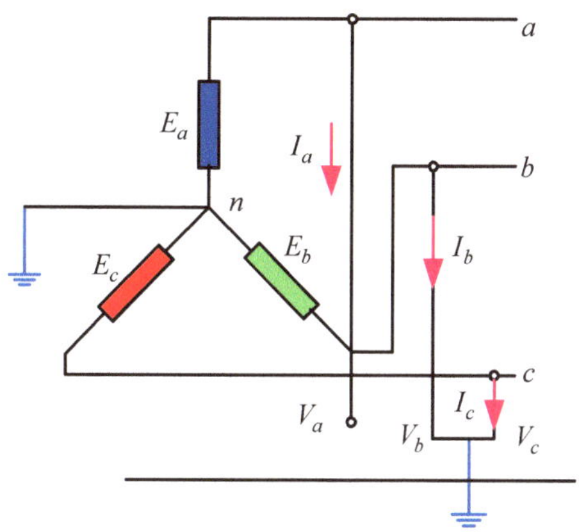

Fig. 6.46 Double-line-to-line fault on an unloaded generator

Substituting Eq. (6.274) into Eq. (6.275) yields,

$$\begin{bmatrix} V_{a0} \\ V_{a1} \\ V_{a2} \end{bmatrix} = \frac{1}{3}\begin{bmatrix} 1 & 1 & 1 \\ 1 & a & a^2 \\ 1 & a^2 & a \end{bmatrix}\begin{bmatrix} V_a \\ 0 \\ 0 \end{bmatrix} \tag{6.276}$$

$$V_{a0} = V_{a1} = V_{a2} = \frac{V_a}{3} \tag{6.277}$$

Initially, the following relation can be written as,

$$V_{a0} = V_{a1} \tag{6.278}$$

Substituting Eqs. (6.207) and (6.208) into Eq. (6.277) yields,

$$-I_{a0}Z_0 = E_a - I_{a1}Z_1 \tag{6.279}$$

$$I_{a0} = -\frac{E_a - I_{a1}Z_1}{Z_0} \tag{6.280}$$

Finally, the following relation can be written as,

$$V_{a2} = V_{a1} \tag{6.281}$$

Substituting Eqs. (6.208) and (6.209) into Eq. (6.280) yields,

$$E_a - I_{a1}Z_1 = -I_{a2}Z_2 \tag{6.282}$$

$$I_{a2} = -\frac{E_a - I_{a1}Z_1}{Z_2} \tag{6.283}$$

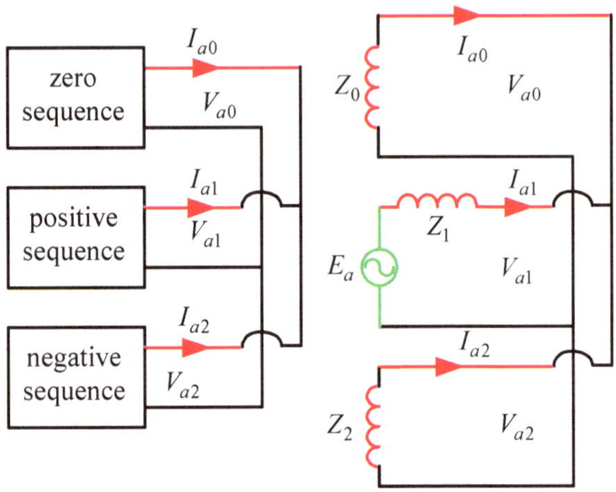

Fig. 6.47 Sequence network for double-line-to-line fault without a fault impedance

The unsymmetrical current for phase a is,

$$I_a = I_{a0} + I_{a1} + I_{a2} \tag{6.284}$$

The sequence network for double-line-to-line fault without a fault impedance is shown in Fig. 6.47.

Substituting Eqs. (6.273), (6.280), and (6.283) into Eq. (6.284) yields,

$$0 = -\frac{E_a - Z_1 I_{a1}}{Z_0} + I_{a1} - \frac{E_a - Z_1 I_{a1}}{Z_2} \tag{6.285}$$

$$I_{a1}\left(1 + \frac{Z_1}{Z_0} + \frac{Z_1}{Z_2}\right) = E_a\left(\frac{1}{Z_0} + \frac{1}{Z_2}\right) \tag{6.286}$$

$$I_{a1} = \frac{E_a\left(\frac{Z_2 + Z_0}{Z_0 Z_2}\right)}{\left(1 + \frac{Z_1 Z_2 + Z_1 Z_0}{Z_0 Z_2}\right)} \tag{6.287}$$

$$I_{a1} = \frac{E_a\left(\frac{Z_2 + Z_0}{Z_0 Z_2}\right)}{\left(\frac{Z_0 Z_2 + Z_1 Z_2 + Z_1 Z_0}{Z_0 Z_2}\right)} \tag{6.288}$$

$$I_{a1} = \frac{E_a(Z_2 + Z_0)}{Z_0 Z_2 + Z_1 Z_2 + Z_1 Z_0} \tag{6.289}$$

$$I_{a1} = \frac{E_a(Z_2 + Z_0)}{Z_0 Z_2 + Z_1(Z_2 + Z_0)} \tag{6.290}$$

$$I_{a1} = \frac{E_a}{Z_1 + \frac{Z_0 Z_2}{Z_2 + Z_0}} \tag{6.291}$$

From Eq. (6.291), it is observed that the zero sequence impedance and negative sequence impedance are connected in parallel, and then it is connected in series with the positive sequence impedance as shown in Fig. 6.47. By applying current divider rule to the circuit shown in Fig. 6.47, the negative sequence and positive sequence currents can be found as,

$$I_{a2} = -I_{a1}\frac{Z_0}{Z_2 + Z_0} \tag{6.292}$$

$$I_{a0} = -I_{a1}\frac{Z_2}{Z_2 + Z_0} \tag{6.293}$$

Again, consider that the double-line-to-ground fault occurs between phases b and c through the ground impedance Z_f as shown in Fig. 6.48. The voltage between the fault terminal and the ground is,

$$V_b = V_c = (I_b + I_c)Z_f \tag{6.294}$$

According to Eqs. (6.98) and (6.99), the following equations can be derived,

$$V_b = V_{a0} + a^2 V_{a1} + aV_{a2} \tag{6.295}$$

$$V_c = V_{a0} + aV_{a1} + a^2 V_{a2} \tag{6.296}$$

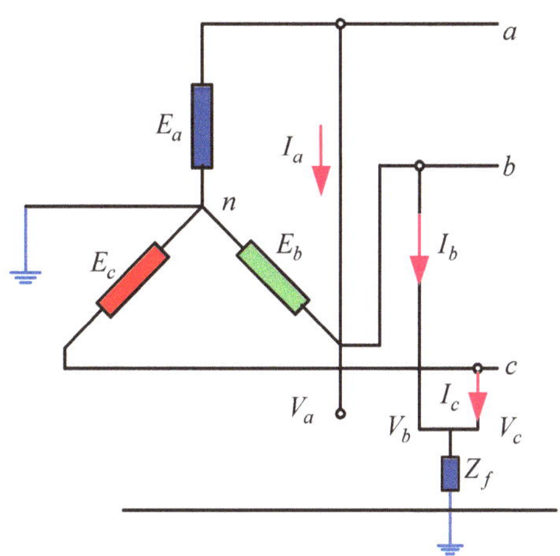

Fig. 6.48 Double-line-to-line fault on an unloaded generator with a fault impedance

Substituting Eqs. (6.98), (6.99), and (6.295) into Eq. (6.294) yields,

$$V_{a0} + a^2 V_{a1} + aV_{a2} = (I_{a0} + a^2 I_{a1} + aI_{a2} + I_{a0} + aI_{a1} + a^2 I_{a2})Z_f \quad (6.297)$$

$$V_{a0} + a^2 V_{a1} + aV_{a2} = 2I_{a0}Z_f + (a^2 + a)I_{a1}Z_f + (a + a^2)I_{a2}Z_f \quad (6.298)$$

$$V_{a0} + a^2 V_{a1} + aV_{a2} = 2I_{a0}Z_f - I_{a1}Z_f - I_{a2}Z_f \quad (6.299)$$

Again, substituting Eq. (6.281) into Eq. (6.299) yields,

$$V_{a0} + (a^2 + a)V_{a1} = 2I_{a0}Z_f - (I_{a1} + I_{a2})Z_f \quad (6.300)$$

$$V_{a0} - V_{a1} = 2I_{a0}Z_f - (I_{a1} + I_{a2})Z_f \quad (6.301)$$

Substituting Eq. (6.273) into Eq. (6.284) yields,

$$0 = I_{a0} + I_{a1} + I_{a2} \quad (6.302)$$

$$I_{a1} + I_{a2} = -I_{a0} \quad (6.303)$$

Again, substituting Eq. (6.303) into Eq. (6.301) yields,

$$V_{a0} - V_{a1} = 3I_{a0}Z_f \quad (6.304)$$

The sequence network with the fault impedance is shown in Fig. 6.49.

Example 6.11
The positive sequence, negative sequence, and zero sequence reactance of a 20 MVA, 11 kV three-phase Y-connected synchronous generator are 0.24 pu, 0.19 pu, and 0.18 pu, respectively. The generator's neutral point is solidly grounded. A double-line-to-line fault occurs between phases b and c. Find the currents in each

Fig. 6.49 Sequence network for double-line-to-line fault with a fault impedance

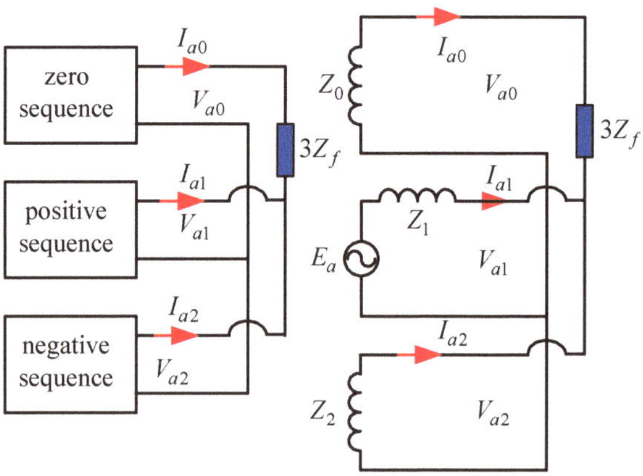

phase during the subtransient period immediately after the fault occurs, and line-to-line voltages.

Solution
Consider that the base values are 20 MVA and 11 kV. Then, the per unit generator voltage is,

$$E = \frac{11}{11} = 1\ \underline{|0^\circ}\ \text{pu} \tag{6.305}$$

The value of the positive sequence component of the current is determined as,

$$I_{a1} = \frac{E}{\left(Z_1 + \frac{Z_0 Z_2}{Z_0 + Z_2}\right)} = \frac{1}{j0.24 + \frac{j0.18 \times j0.19}{j(0.18\ +\ 0.19)}} = -j3.01\ \text{pu} \tag{6.306}$$

The value of the negative sequence component of the current is calculated as,

$$I_{a2} = -I_{a1}\frac{Z_0}{Z_2 + Z_0} = j3.01\frac{j0.18}{j(0.19 + 0.18)} = j1.46\ \text{pu} \tag{6.307}$$

The value of the zero sequence component of the current can be determined as,

$$I_{a0} = -I_{a1}\frac{Z_2}{Z_2 + Z_0} = j3.01\frac{j0.19}{j(0.19 + 0.18)} = j1.55\ \text{pu} \tag{6.308}$$

The value of the base current is calculated as,

$$I_b = \frac{20 \times 1000}{\sqrt{3} \times 11} = 1049.73\ \text{A} \tag{6.309}$$

The per unit values of the phase currents can be calculated as,

$$I_a = I_{a0} + I_{a1} + I_{a2} = j1.55 - j3.01 + j1.46 = 0\ \text{pu} \tag{6.310}$$

$$\begin{aligned} I_b &= I_{b0} + I_{b1} + I_{b2} = I_{a0} + a^2 I_{a1} + a I_{a2} = j1.55 - 3.01\ \underline{|90^\circ + 240^\circ} + 1.46\ \underline{|90^\circ + 120^\circ} \\ &= 4.52\ \underline{|149.01^\circ}\ \text{pu} \end{aligned} \tag{6.311}$$

$$\begin{aligned} I_c &= I_{c0} + I_{c1} + I_{c2} = I_{a0} + a I_{a1} + a^2 I_{a2} = j1.55 - 3.01\ \underline{|90^\circ + 120^\circ} + 1.46\ \underline{|90^\circ + 240^\circ} \\ &= 4.52\ \underline{|30.99^\circ}\ \text{pu} \end{aligned} \tag{6.312}$$

The value of the actual fault current during the subtransient period is,

$$I_b = I_c = 4.52 \times 1049.73 = 4744.78\,\text{A} \tag{6.313}$$

The value of the sequence voltages of phase a can be determined as,

$$V_{a1} = E - I_{a1}Z_1 = 1 + j3.01 \times j0.24 = 0.28\,\text{pu} \tag{6.314}$$

In the double-line-to-ground fault, the symmetrical components of voltage are the same, and it can be written as,

$$V_{a0} = V_{a1} = V_{a2} = 0.28\,\text{pu} \tag{6.315}$$

The values of the phase voltages of the generator can be determined as,

$$V_a = V_{a0} + V_{a1} + V_{a2} = 0.28 + 0.28 + 0.28 = 0.84\,\underline{|0^\circ}\,\text{pu} \tag{6.316}$$

$$V_b = V_{b0} + V_{b1} + V_{b2} = V_{a0} + a^2V_{a1} + aV_{a2} = 0.28 + 0.28\,\underline{|240^\circ} + 0.28\,\underline{|120^\circ} \\ = 0\,\underline{|0^\circ}\,\text{pu} \tag{6.317}$$

$$V_c = V_{c0} + V_{c1} + V_{c2} = V_{a0} + aV_{a1} + a^2V_{a2} = 0.28 + 0.28\,\underline{|120^\circ} + 0.28\,\underline{|240^\circ} \\ = 0\,\underline{|0^\circ}\,\text{pu} \tag{6.318}$$

The per unit line-to-line voltages are determined as,

$$V_{\text{ab}} = V_a - V_b = 0.84\,\underline{|0^\circ}\,\text{pu} \tag{6.319}$$

$$V_{\text{bc}} = V_b - V_c = 0\,\underline{|0^\circ}\,\text{pu} \tag{6.320}$$

$$V_{ca} = V_c - V_a = -0.84\,\underline{|0^\circ} = 0.84\,\underline{|180^\circ}\,\text{pu} \tag{6.321}$$

The value of the phase voltage is determined as,

$$V_p = \frac{11}{\sqrt{3}} = 6.35\,\text{kV} \tag{6.322}$$

The actual line-to-line voltages of the generator can be determined as,

$$V_{\text{ab}} = 0.84 \times 6.35 = 5.33\,\text{kV} \tag{6.323}$$

$$V_{\text{bc}} = 0 \times 6.35 = 0\,\text{kV} \tag{6.324}$$

$$V_{ca} = 0.84 \times 6.35 = 5.33\,\text{kV} \tag{6.325}$$

Practice Problem 6.11

The positive sequence, negative sequence, and zero sequence reactance of a 50 MVA, 16.8 kV, three-phase Y-connected synchronous generator are 0.22 pu, 0.17 pu, and 0.15 pu, respectively. The neutral of the generator is solidly grounded. A double-line-to-line fault occurs between phases *b* and *c*. Calculate the currents in each phase during the subtransient period immediately after the fault occurs.

Example 6.12

Figure 6.50 shows a single-line diagram of a three-phase power system. The ratings of the equipment are as follows.

Generator G_1	100 MVA, 11 kV, $X_1 = X_2 = 0.25\,\text{pu}$, $X_0 = 0.05\,\text{pu}$
Generator G_2	80 MVA, 11 kV, $X_1 = X_2 = 0.15\,\text{pu}$, $X_0 = 0.07\,\text{pu}$
Transformer T_1	100 MVA, 11/66 kV, $X_1 = X_2 = X_0 = 0.09\,\text{pu}$
Transformer T_2	80 MVA, 11/66 kV, $X_1 = X_2 = X_0 = 0.09\,\text{pu}$
Line	$X_1 = X_2 = 15\ \Omega$, $X_0 = 30\ \Omega$

Answer the following questions by considering the power system is initially unloaded.

(i) Draw the positive sequence, negative sequence, and zero sequence networks. Also, find the equivalent sequence impedances.
(ii) A single-line-to-ground fault occurs in line *a* at bus 3. Find the subtransient fault current.
(iii) A line-to-line fault occurs in lines *b* and *c* at bus 3. Calculate the subtransient fault current.
(iv) A double-line-to-ground fault occurs in lines *b* and *c* at bus 3. Determine the subtransient fault current.

Solution

Consider that the base values are 100 MVA and 11 kV for the low voltage side and 66 kV for the high voltage side of the transformer. Based on the common base, the new reactance for the generators and transformers can be determined in the following ways.

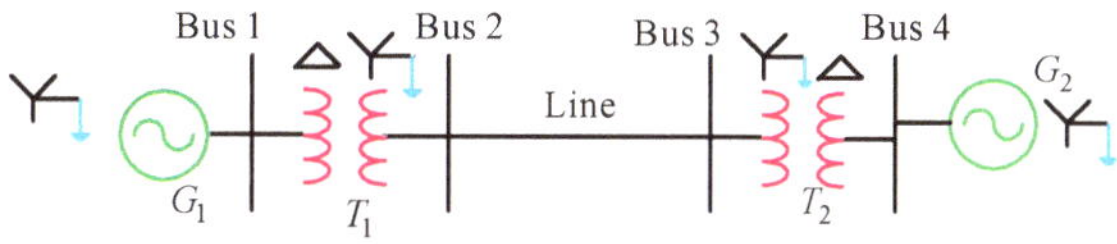

Fig. 6.50 Single-line diagram for Example 6.12

The new reactance for the generator G_1 is,

$$X_1 = X_2 = \frac{100}{100} \times 0.25 = 0.25\,\text{pu} \tag{6.326}$$

$$X_0 = \frac{100}{100} \times 0.05 = 0.05\,\text{pu} \tag{6.327}$$

The new reactance for the generator G_2 is,

$$X_1 = X_2 = \frac{100}{80} \times 0.15 = 0.1875\,\text{pu} \tag{6.328}$$

$$X_0 = \frac{100}{80} \times 0.07 = 0.0875\,\text{pu} \tag{6.329}$$

The new reactance for the transformer T_1 is,

$$X_1 = X_2 = X_0 = \frac{100}{100} \times 0.09 = 0.09\,\text{pu} \tag{6.330}$$

The new reactance for the transformer T_2 is,

$$X_1 = X_2 = X_0 = \frac{100}{80} \times 0.09 = 0.1125\,\text{pu} \tag{6.331}$$

The base voltage for the line can be calculated as,

$$V_b = \frac{11}{a} = 11 \times \frac{66}{11} = 66\,\text{kV} \tag{6.332}$$

The value of the base impedance can be determined as,

$$Z_b = \frac{V_b^2}{S_b} = \frac{66^2}{100} = 43.56\ \Omega \tag{6.333}$$

The values of the new line reactance are determined as,

$$X_1 = X_2 = \frac{15}{43.56} = 0.3443\,\text{pu} \tag{6.334}$$

$$X_0 = \frac{30}{43.56} = 0.6887\,\text{pu} \tag{6.335}$$

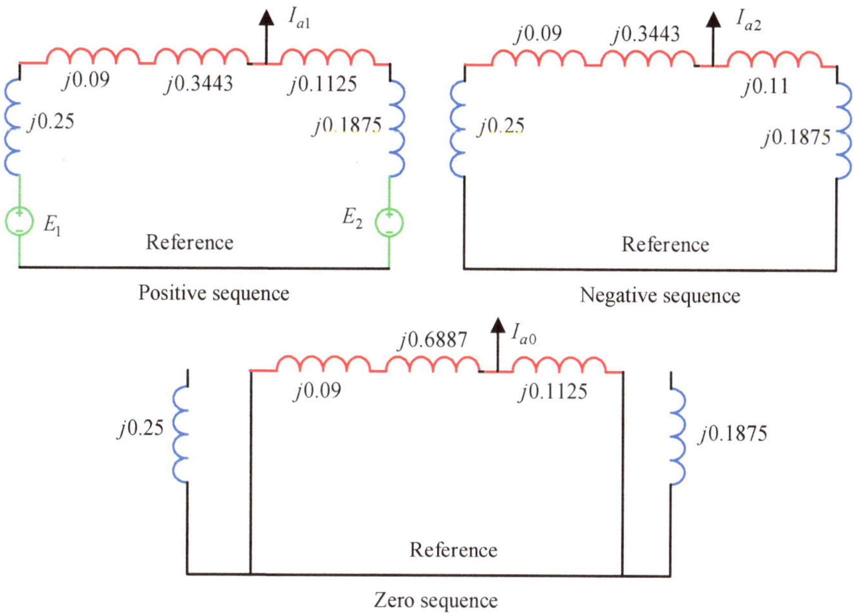

Fig. 6.51 Sequence networks for Example 6.12

(i) Figure 6.51 shows the sequence networks. The equivalent positive sequence and negative sequence impedances can be determined as,

$$Z_1 = Z_2 = \frac{j(0.25 + 0.09 + 0.3443) \times j(0.1125 + 0.1875)}{j(0.6843 + 0.3)} = j0.2085 \text{ pu} \quad (6.336)$$

The value of the equivalent zero sequence impedance can be calculated as,

$$Z_0 = \frac{j(0.09 + 0.6887) \times j(0.1125)}{j(0.7787 + 0.1125)} = j0.0982 \text{ pu} \quad (6.337)$$

(ii) Figure 6.52 shows the sequence network for a single-line-to-ground fault. The sequence components of the current are determined as,

$$I_{a1} = I_{a2} = I_{a0} = \frac{E}{Z_1 + Z_2 + Z_0} = \frac{1\,\angle 0^\circ}{j0.2085 + j0.2085 + j0.0982} = 1.94\,\angle -90^\circ \text{ pu} \quad (6.338)$$

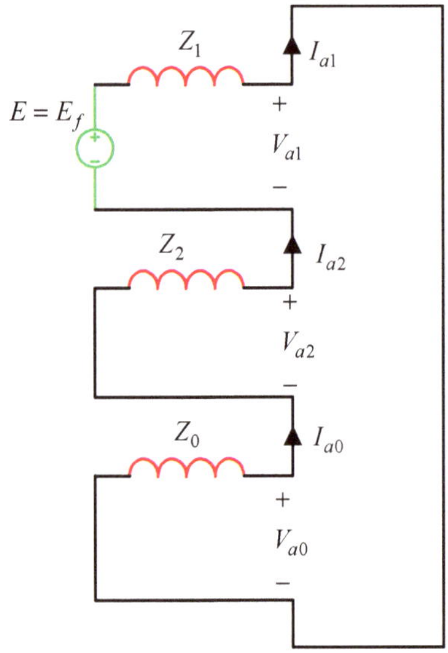

Fig. 6.52 Positive sequence networks for Example 6.12

The value of the fault current in the phase a can be determined as,

$$I_a = I_{a1} + I_{a2} + I_{a0} = 1.94\,\underline{\lfloor -90^\circ} + 1.94\,\underline{\lfloor -90^\circ} + 1.94\,\underline{\lfloor -90^\circ} = 5.82\,\underline{\lfloor -90^\circ}\ \text{pu} \tag{6.339}$$

The value of the base current can be calculated as,

$$I_{\text{base}} = \frac{100 \times 1000}{\sqrt{3} \times 66} = 874.77\ \text{A} \tag{6.340}$$

For a single-line-to-ground fault, the actual value of the fault current can be determined as,

$$I_{f(s-l-g)} = 5.82 \times 874.77 = 5091.16\ \text{A} \tag{6.341}$$

Figure 6.53 shows the simulation result by IPSA software, and the value of the fault current in the busbar 3 is found to be 5119.86 A.

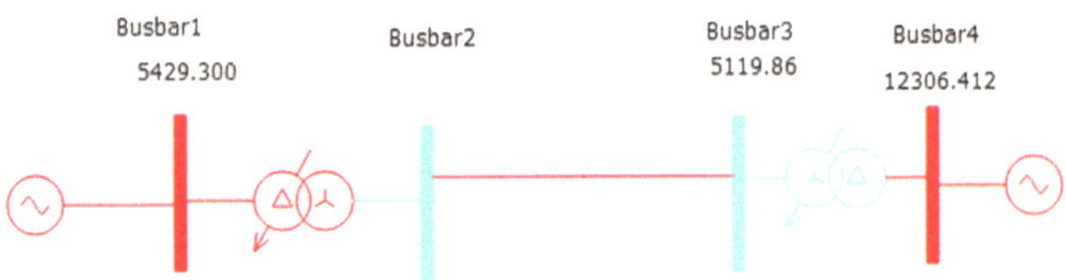

Fig. 6.53 Simulation circuit by IPSA software for Example 6.12

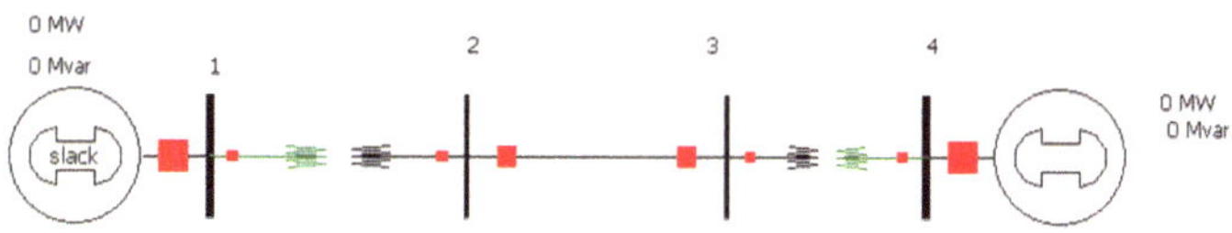

Fig. 6.54 Simulation circuit by PowerWorld software for Example 6.12

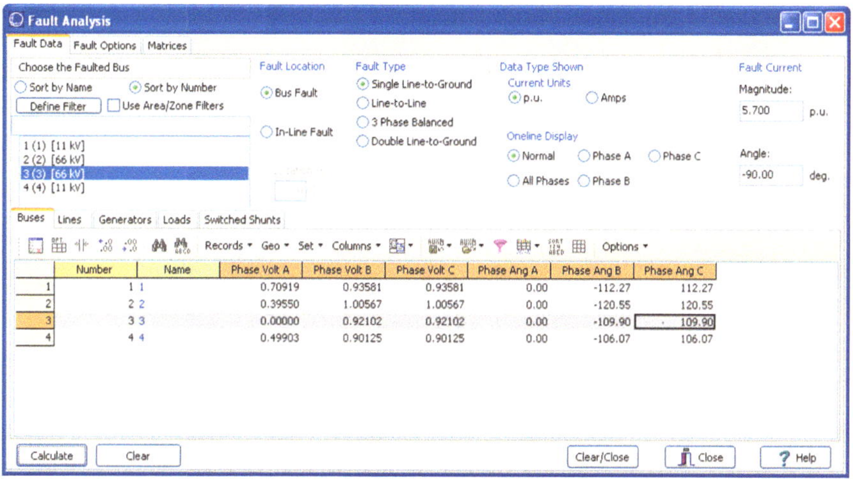

Fig. 6.55 Simulation results by PowerWorld software for Example 6.12

Fig. 6.56 Sequence network for line-to-line fault

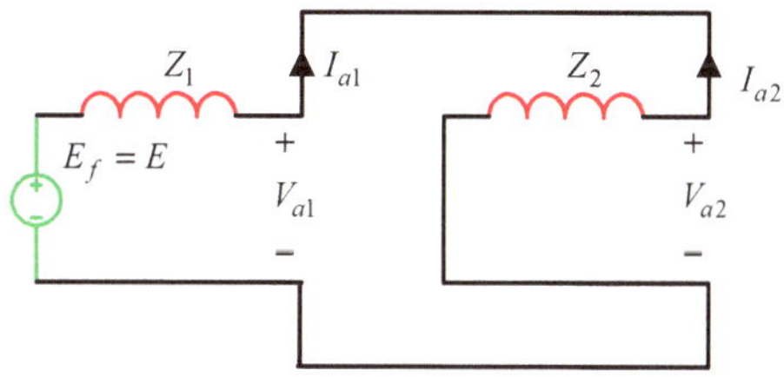

The “PowerWorld” software is also used to calculate the fault current as shown in Fig. 6.54. The values of the sequence reactance of the generators, transformers, and transmission lines are placed in the respective places. Figures 6.54 and 6.55 shows the simulation result for the single-line-to-ground fault, and its value is found to be 6.68 pu. The results are found to be 6.76 pu, which is approximately the same as that of the simulation result.

(iii) Figure 6.56 shows the sequence network for the line-to-line fault. In this case, the value of the positive sequence current can be determined as,

$$I_{a1} = \frac{E}{Z_1 + Z_2} = \frac{1\,\angle 0^\circ}{j0.2085 + j0.2085} = 2.398\,\angle -90^\circ \text{ pu} \tag{6.342}$$

The value of the negative sequence current is determined as,

$$I_{a2} = -I_{a1} = -2.398\,\angle -90^\circ \text{ pu} = 2.398\,\angle 90^\circ \text{ pu} \tag{6.343}$$

The value of the zero sequence current is,

$$I_{a0} = 0\,\angle 0^\circ \text{ pu} \tag{6.344}$$

The value of the fault current in phase b is calculated as,

$$I_b = I_{b0} + I_{b1} + I_{b2} = I_{a0} + a^2 I_{a1} + aI_{a2} = 2.398\,\angle 240^\circ - 90^\circ + 2.398\,\angle 120^\circ + 90^\circ = 4.153\,\angle 180^\circ \text{ pu} \tag{6.345}$$

The value of the fault current in phase c can be calculated as,

$$I_c = -I_b = -4.153\,\angle 180^\circ = 4.153\,\angle 180^\circ - 180^\circ = 4.153\,\angle 0^\circ \text{ pu} \tag{6.346}$$

The actual value of the fault current for line-to-line fault can be determined as,

$$I_{fl-l} = 4.153 \times 874.77 = 3632.919\ \text{A} \tag{6.347}$$

Figure 6.57 shows the IPSA simulation result for line-to-line fault, and the value of the fault current is found to be,

$$I_{f(l-l)IPSA} = 3639.33\ \text{A} \tag{6.348}$$

Figure 6.58 shows the PowerWorld simulation result for line-to-line fault and per unit magnitude of the fault current is the same.

Fig. 6.57 IPSA simulation result for the line-to-line fault

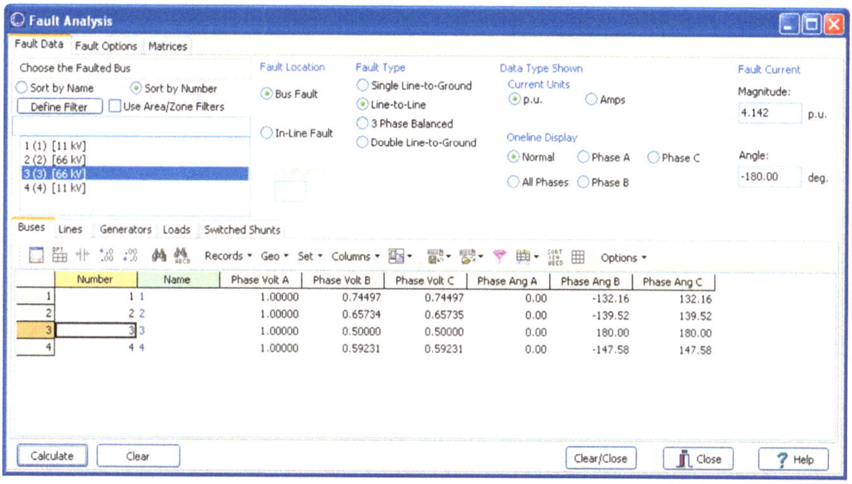

Fig. 6.58 PowerWorld simulation result for the line-to-line fault

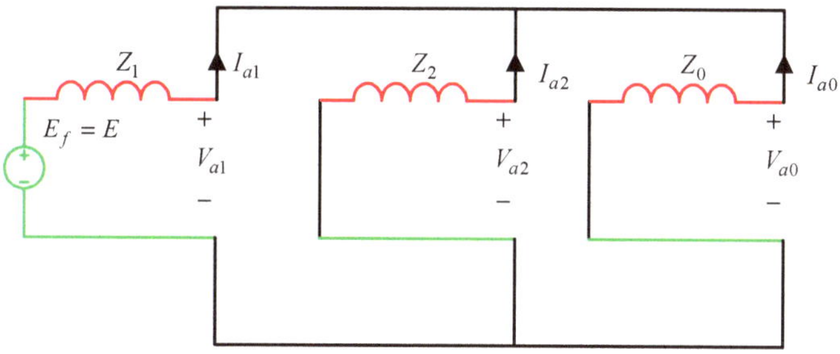

Fig. 6.59 Sequence network for the double-line-to-ground fault

(iii). Figure 6.59 shows the sequence network for the double-line-to-ground fault, and the value of the positive sequence current can be determined as,

$$I_{a1} = \frac{E}{Z_1 + \frac{Z_2 Z_0}{Z_2 + Z_0}} = \frac{1\ \underline{|0^\circ}}{j0.2085 + j\frac{0.2085 \times 0.0982}{0.3067}} = 3.633\ \underline{|-90^\circ}\ \text{pu} \tag{6.349}$$

The value of the negative sequence current can be determined as,

$$I_{a2} = -I_{a1}\frac{Z_0}{Z_0 + Z_2} = -3.633\ \underline{|-90^\circ} \times \frac{j0.0982}{j0.3067} = 1.163\ \underline{|90^\circ}\ \text{pu} \tag{6.350}$$

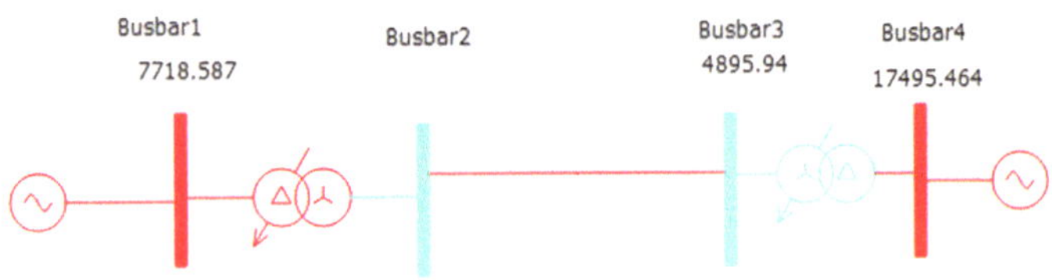

Fig. 6.60 IPSA simulation circuit for the double-line-to-ground fault

The value of the zero sequence current can be calculated as,

$$I_{a0} = -I_{a1}\frac{Z_2}{Z_0 + Z_2} = -3.633\,\underline{|-90^\circ} \times \frac{j0.2085}{j0.3067} = 2.469\,\underline{|90^\circ}\ \text{pu} \tag{6.351}$$

The value of the fault current in phase b is determined as,

$$\begin{aligned} I_b &= I_{b0} + I_{b1} + I_{b2} = I_{a0} + a^2 I_{a1} + aI_{a2} \\ &= 2.469\,\underline{|90^\circ} + 3.633\,\underline{|-90^\circ + 240^\circ} + 1.163\,\underline{|90^\circ + 120^\circ} = 5.565\,\underline{|138.27^\circ}\ \text{pu} \end{aligned} \tag{6.352}$$

The value of the fault current in phase c can be determined as,

$$\begin{aligned} I_c &= I_{c0} + I_{c1} + I_{c2} = I_{a0} + aI_{a1} + a^2 I_{a2} \\ &= 2.469\,\underline{|90^\circ} + 3.633\,\underline{|-90^\circ + 120^\circ} + 1.163\,\underline{|90^\circ + 240^\circ} = 5.565\,\underline{|41.73^\circ}\ \text{pu} \end{aligned} \tag{6.353}$$

The actual value of the fault current can be determined as,

$$I_{f(d-l-g)} = 5.565 \times 874.77 = 4868.095\ \text{A} \tag{6.354}$$

Figure 6.60 shows the IPSA simulation result for the double-line-to-ground fault, and the value of the fault current is found to be,

$$I_{f(dl-g)IPSA} = 4866.46\ \text{A} \tag{6.355}$$

References

1. Duncan Glover J, Overbye T, Sarma M (2017) Power system analysis and design, Sixth edn. Cengage Learning, USA, pp 1–942
2. Nagsarkar TK, Sukhija MS (2014) Power system analysis, Second Edn. Oxford University Press, Oxford, pp 1–726
3. Wildi T (2014) Electrical machines, drives and power systems, Sixth edn. Pearson Education Ltd, USA, pp 1–920

4. Sadat H (2010) Power system analysis, Third edn. PSA Publisher, USA, pp 1–772
5. Wildi T (2006) Electrical machines, drives and power systems, 6th edn. Pearson Education, USA, pp 1–934
6. Guile AE, Paterson W (1977) Electrical power systems, 2nd edn. Pergamon Press, Oxford

Exercise Problems

6.1 Calculate the quantities for (i) a^8, (ii) $a^{10} + a - 3$, and (iii) $a^{12} + 3a - 2$ by considering $a = 1\angle 120^\circ$ and $a^2 = 1\angle 240^\circ$.

6.2 A three-phase system is having the currents of $I_a = 10\angle 30^\circ$ A, $I_b = 15\angle 40^\circ$ A and $I_c = 20\angle 35^\circ$ A. Calculate the symmetrical components of current in phases a and b.

6.3 A three-phase system has the phase voltages of $V_a = 100\angle 35^\circ$ V, $V_b = 200\angle 45^\circ$ V and $V_c = 280\angle 55^\circ$ V. Find the zero sequence, positive sequence and negative sequence components of the voltage for phase a.

6.4 A three-phase system is having the symmetrical components of the voltage of $V_{a0} = 75\angle 45^\circ$ V, $V_{a1} = 155\angle 64^\circ$ V and $V_{a2} = 325\angle 85^\circ$ V for phase a. Find the phase voltages V_a, V_b and V_c.

6.5 The unbalanced currents of a three-phase system are $I_a = 50$ A, $I_b = 30 + j50$ A, $I_c = 40 + j70$ A. Calculate the zero, positive and negative sequence components of the current in phase b.

6.6 A three-phase system is having the symmetrical components of the current of $I_{a0} = 4.54 + j3.5$ A $I_{a1} = 5.34 + j1.45$ A and $I_{a2} = 1.67 - j1.85$ A for phase a. Calculate the unbalanced currents I_a, I_b and I_c if the total neutral current of this system is zero.

6.7 Figure 6.61 shows a three-phase wye-connected unloaded synchronous generator. A single line to ground fault occurs in phase a, and the current in this phase is found to be 1500 A. Calculate the symmetrical components of current in phase b.

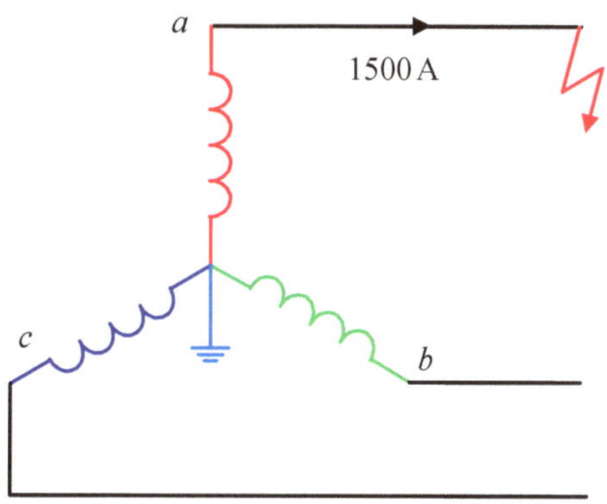

Fig. P6.1 Circuit for Problem 6.7

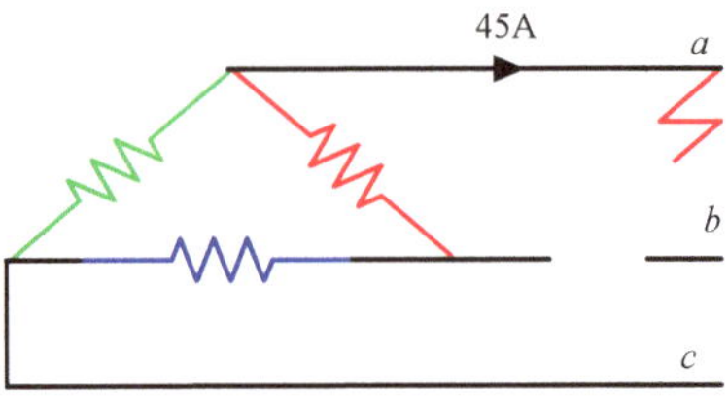

Fig. P6.2 Circuit for Problem 6.8

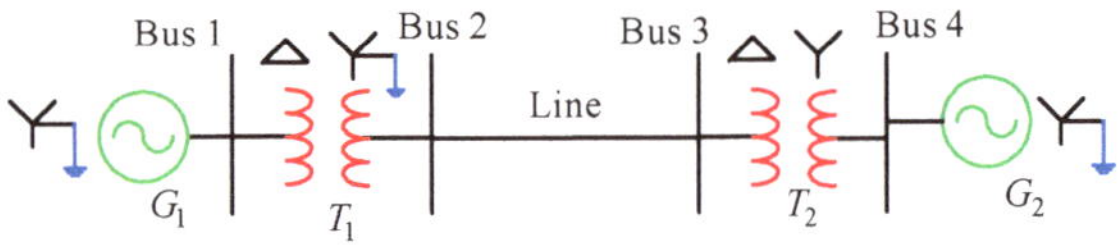

Fig. P6.3 Single-line diagram for Problem 6.12

6.8 A source delivers power to a delta-connected load as shown in Fig. P6.2. The current in phase a is found to be 45 A and phase b is open circuited. Calculate the symmetrical components of the currents in all three phases.

6.9 The positive sequence, negative sequence and zero sequence reactance of a 30 MVA, 11 kV three-phase synchronous generator are measured to be 0.5 pu, 0.4 pu and 0.22 pu, respectively. The generator's neutral is solidly grounded. A single line-to-ground fault occurs in phase a. Find the fault current.

6.10 A 15 MVA, 13.8 kV three-phase synchronous generator is having the positive sequence, negative sequence and zero sequence reactance of 0.3 pu, 0.2 pu and 0.1 pu, respectively. The generator's neutral is solidly grounded and line-to-line fault occurs in phases b and c. Calculate the fault current, sequence voltages for phase a, and phase voltages of the generator.

6.11 A 25 MVA, 13.8 kV three-phase Y-connected synchronous generator is having the positive sequence, negative sequence and zero sequence reactance of 0.34 pu, 0.22 pu and 0.15 pu, respectively. The generator's neutral is solidly grounded and the double line-to-line fault occurs between phases b and c. Calculate the currents in each phase during the sub-transient period immediately after the fault occurs.

6.12 Figure P6.3 shows a single-line diagram of a three-phase power system and the ratings of the equipment are shown below.

Generators G_1, G_2	100 MVA, 20 kV, $X_1 = X_2 = 0.20\,\text{pu}$, $X_0 = 0.06\,\text{pu}$
Transformers T_1, T_2	100 MVA, 20/138 kV, $X_1 = X_2 = X_0 = 0.08\,\text{pu}$
Line	100 MVA, $X_1 = X_2 = 0.11\,\text{pu}$, $X_0 = 0.55\,\text{pu}$

A fault occurs at bus 4. Determine the fault currents in the faulted bus by any simulation software (IPSA/Powerworld) for the single line-to-ground, line-to-line, and double line-to-ground faults.

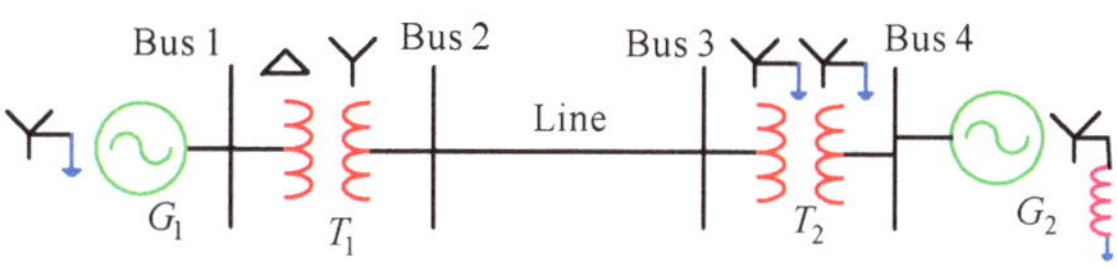

Fig. P6.4 Single-line diagram for Problem 6.13

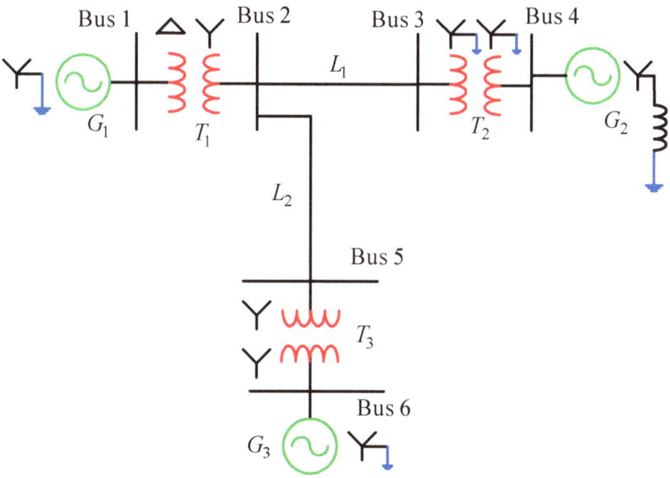

Fig. P6.5 Single-line diagram for Problem 6.14

6.13 A single-line diagram of a three-phase power system is shown in Fig. P6.4. The ratings of the equipment are shown below.

Generator G_1	100 MVA, 11 kV, $X_1 = X_2 = 0.20$ pu, $X_0 = 0.05$ pu
Generator G_2	100 MVA, 20 kV, $X_1 = X_2 = 0.25$ pu, $X_0 = 0.03$ pu, $X_n = 0.05$ pu
Transformer T_1	100 MVA, 11/66 kV, $X_1 = X_2 = X_0 = 0.06$ pu
Transformer T_2	100 MVA, 11/66 kV, $X_1 = X_2 = X_0 = 0.06$ pu
Line	100 MVA, $X_1 = X_2 = 0.15$ pu, $X_0 = 0.65$ pu

A single line-to-ground, line-to-line, and double line-to-ground fault occur at bus 6. Find the fault currents in each case.

6.14 Figure P6.5 shows a single-line diagram of a three-phase power system and the ratings of the equipment are shown below.

Generator G_1	100 MVA, 11 kV, $X_1 = X_2 = 0.20$ pu, $X_0 = 0.05$ pu
Generator G_2	100 MVA, 20 kV, $X_1 = X_2 = 0.25$ pu, $X_0 = 0.03$ pu, $X_n = 0.05$ pu
Generator G_3	100 MVA, 20 kV, $X_1 = X_2 = 0.30$ pu, $X_0 = 0.08$ pu
Transformer T_1	100 MVA, 11/66 kV, Δ-Y, $X_1 = X_2 = X_0 = 0.06$ pu

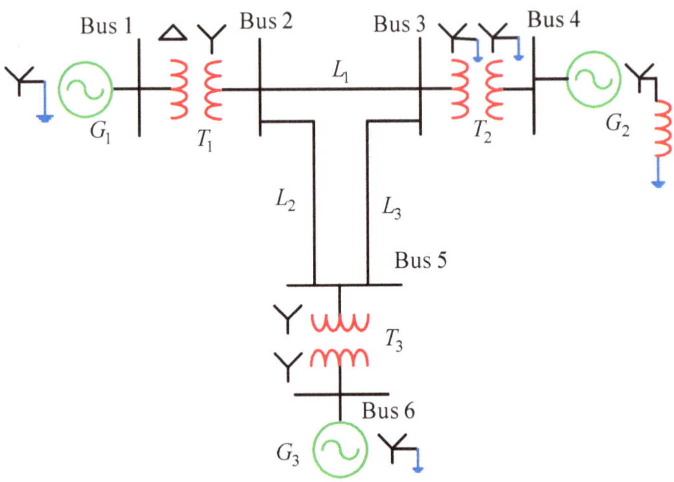

Fig. P6.6 Single-line diagram for Problem 6.15

Transformer T_2	100 MVA, 20/66 kV, both sides earthed Y-Y, $X_1 = X_2 = X_0 = 0.06$ pu
Transformer T_3	100 MVA, 20/66 kV, Y-Y, $X_1 = X_2 = X_0 = 0.04$ pu
Line 1	100 MVA,$X_1 = X_2 = 0.15$ pu, $X_0 = 0.65$ pu
Line 2	100 MVA,$X_1 = X_2 = 0.10$ pu, $X_0 = 0.45$ pu

Calculate the fault currents in case of single line-to-ground, line-line and double line-to-ground faults occurring at bus 6.

6.15 A single-line diagram of a three-phase power station is shown in Fig. P6.6. The ratings of the equipment are shown below.

Generator G_1	100 MVA, 11 kV, $X_1 = X_2 = 0.24$ pu, $X_0 = 0.06$ pu
Generator G_2	100 MVA, 20 kV, $X_1 = X_2 = 0.22$ pu, $X_0 = 0.03$ pu, $X_n = 0.05$ pu
Generator G_3	100 MVA, 20 kV, $X_1 = X_2 = 0.35$ pu, $X_0 = 0.07$ pu
Transformer T_1	100 MVA, 11/66 kV, Δ-Y, $X_1 = X_2 = X_0 = 0.06$ pu
Transformer T_2	100 MVA, 20/66 kV, both sides earthed Y-Y, $X_1 = X_2 = X_0 = 0.06$ pu
Transformer T_3	100 MVA, 20/66 kV, Y-Y, $X_1 = X_2 = X_0 = 0.04$ pu
Line 1	100 MVA, $X_1 = X_2 = 0.15$ pu, $X_0 = 0.65$ pu
Line 2	100 MVA, $X_1 = X_2 = 0.10$ pu, $X_0 = 0.45$ pu
Line 3	100 MVA, $X_1 = X_2 = 0.16$ pu, $X_0 = 0.65$ pu

Find the fault currents in each of single line-to-ground, line-line and double line-to-ground faults occurring at bus 6.

Chapter 7
Load Flow Analysis

7.1 Introduction

Load flow or power flow study is one of the important aspects that is used for power system planning, operation, maintenance, and control. In the planning stage, load flow studies are used to determine if and when the specific power system elements become underloaded and overloaded. In operating studies, load flow studies are used to ensure that each generator runs at their maximum operating point. Based on the load flow study, major investment decisions begin.

In load flow studies, power flows from sending end to the receiving end through transmission lines. The equations in terms of power are known as power flow equations. These power flow equations are normally nonlinear and must be solved by some iterative techniques. Load flow studies are performed to determine the voltage drop on each feeder, voltage magnitude and phase angle at each bus, real and reactive powers flowing in all branches. The total power losses in the system, as well as the power losses in each branch, are also calculated by the load flow study. Also, load flow studies are done before transient stability and contingency studies. Sometimes, a load flow study shows an overloaded connection or transformer; then, preventive actions are taken in the real network to stop this situation. In this case, a large number of load flow analysis is carried out that is called contingency study. Simulation software such as ETAP, CYME, IPSA, and PowerWorld are often used for the studies. In this chapter, the basics of load flow study, some numerical methods are discussed.

Md. A. Salam, *Fundamentals of Electrical Power Systems Analysis*,
https://doi.org/10.1007/978-981-15-3212-2_7

7.2 Classification of Buses

An arrangement of conductor that serves as a common connection for the conductors of two or more circuits is known as bus or busbar. In a bus, voltage and frequency are the same. In a load flow analysis, the four parameters, such as the real power, reactive power, bus voltage, and phase angle, are considered. According to these parameters, busses are classified as load bus, generator bus, and slack bus. A brief description of each bus is given below.

Load bus: The real power (*P*) and the reactive power (*Q*) are known in the load bus. It is often known as *PQ* bus. The magnitude of the bus voltage (*V*) and its phase angle (δ) are calculated from the load flow study. The real and reactive powers supplied to the power system are considered positive, while the real power and reactive power consumed from the power system are considered negative.

Generator bus: The generator bus is also known as a voltage-controlled or regulated bus. The voltage-controlled bus of a power system in any bus is that the magnitude of the voltage can be controlled. The generated voltage (*V*) and power (*P*) are known, and it is often known as the *PV* bus. The reactive power Q and the phase angle δ are not known and need to be computed.

Slack bus: The voltage magnitude and its phase angle are specified. In the slack bus, the magnitude and phase angle of the voltage are normally considered $1\angle 0^\circ$ pu. This bus is often known as a swing bus or reference bus. The real power (*P*) and reactive power (*Q*) are not known and need to be computed from the load flow study. The known and unknown parameters of different types of buses are shown in Table 7.1

7.3 Power Flow in Two-Bus System

A short transmission line is connected between two buses, as shown in Fig. 7.1. Bus 1 represents the sending end, and bus 2 represents the receiving end [1, 2].

Applying KVL to the circuit in Fig. 7.1 yields,

$$V_s = V_r + I(R + jX_L) \tag{7.1}$$

Table 7.1 Known and unknown parameters of buses

Types of buses	Known parameters	Unknown parameters
Slack bus	|V|, δ	*P*, *Q*
Load bus	*P*, *Q*	|V|, δ
Generator bus	*P*, |V|	*Q*, δ

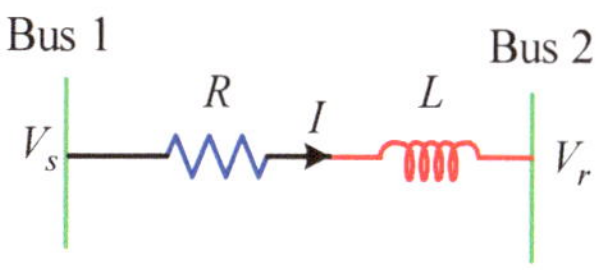

Fig. 7.1 Short transmission line with the two buses

The value of the resistance of the transmission line is very small, as compared to inductive reactance. Therefore, the resistance is negligible and Eq. (7.1) is modified as,

$$I = \frac{V_s - V_r}{jX_L} \tag{7.2}$$

Taking conjugate of Eq. (7.2) yields,

$$I_s^* = I^* = I_r^* = \left(\frac{V_s - V_r}{jX_L}\right)^* \tag{7.3}$$

$$I_s^* = I^* = I_r^* = \frac{V_s^* - V_r^*}{-jX_L} \tag{7.4}$$

Let the receiving end and the sending end voltages are written as,

$$V_s = V_s \angle\delta \tag{7.5}$$

$$V_r = V_r \angle 0^\circ \tag{7.6}$$

Substituting Eqs. (7.5) and (7.6) into Eq. (7.4) yields,

$$I_s^* = I^* = I_r^* = \frac{(V_s \angle\delta)^* - (V_r \angle 0^\circ)^*}{-jX_L} \tag{7.7}$$

$$I_s^* = I^* = I_r^* = \frac{V_s \angle -\delta - V_r \angle 0^\circ}{-jX_L} \tag{7.8}$$

$$I_s^* = I^* = I_r^* = \frac{V_s \angle 90^\circ - \delta - V_r \angle 90^\circ}{X_L} \tag{7.8}$$

The sending end complex power is defined as,

$$S_s = P_s + jQ_s = V_s I_s^* \tag{7.9}$$

Substituting Eq. (7.8) into Eq. (7.9) yields,

$$S_s = P_s + jQ_s = V_s \underline{|\delta} \times \frac{V_s \underline{|90^\circ - \delta} - V_r \underline{|90^\circ}}{X_L} \tag{7.10}$$

$$S_s = P_s + jQ_s = \frac{V_s^2 \underline{|90^\circ - \delta + \delta}}{X_L} - \frac{V_s V_r \underline{|\delta + 90^\circ}}{X_L} \tag{7.11}$$

$$S_s = P_s + jQ_s = \frac{V_s^2}{X_L}(\cos 90^\circ + j \sin 90^\circ) - \frac{V_s V_r}{X_L}[\cos(90^\circ + \delta) + j \sin(90^\circ + \delta)] \tag{7.12}$$

$$S_s = P_s + jQ_s = \frac{V_s^2}{X_L}(0 + j1) - \frac{V_s V_r}{X_L}(-\sin \delta + j \cos \delta) \tag{7.13}$$

$$S_s = P_s + jQ_s = \frac{V_s V_r}{X_L} \sin \delta + j\left(\frac{V_s^2}{X_L} - \frac{V_s V_r}{X_L} \cos \delta\right) \tag{7.14}$$

Equating real and imaginary parts of Eq. (7.14) yields,

$$P_s = \frac{V_s V_r}{X_L} \sin \delta \tag{7.15}$$

$$Q_s = \left(\frac{V_s^2}{X_L} - \frac{V_s V_r}{X_L} \cos \delta\right) \tag{7.16}$$

Similarly, the receiving end complex power is expressed as,

$$S_r = V_r I_r^* \tag{7.17}$$

Substituting Eq. (7.7) into Eq. (7.17) yields,

$$S_r = V_r \times \frac{V_s \underline{|90^\circ - \delta} - V_r \underline{|90^\circ}}{X_L} \tag{7.18}$$

$$S_r = \frac{V_r V_s}{X_L}[\cos(90^\circ - \delta) + j \sin(90^\circ - \delta)] - \frac{V_r^2}{X_L}[\cos 90^\circ + j \sin 90^\circ] \tag{7.19}$$

$$S_r = \frac{V_r V_s}{X_L}[\sin \delta + j \cos \delta] - \frac{V_r^2}{X_L}[0 + j1] \tag{7.20}$$

$$P_r + Q_r = \frac{V_r V_s}{X_L} \sin \delta + j\left[\frac{V_r V_s}{X_L} \cos \delta - \frac{V_r^2}{X_L}\right] \tag{7.21}$$

Again, equating real and imaginary parts of Eq. (7.21) yields,

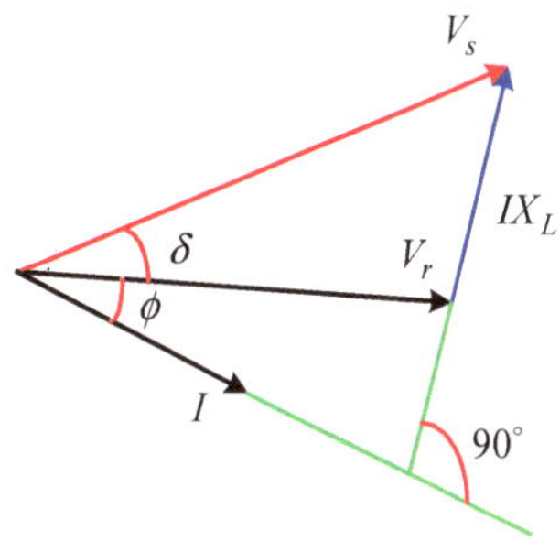

Fig. 7.2 Phasor diagram for a two-bus short transmission line

$$P_r = \frac{V_r V_s}{X_L} \sin\delta \tag{7.22}$$

$$Q_r = \frac{V_r V_s}{X_L} \cos\delta - \frac{V_r^2}{X_L} \tag{7.23}$$

The phasor diagram is drawn by considering negligible resistance and receiving end voltage as a reference phasor as shown in Fig. 7.2.

The real power at both ends will be maximum when the torque angle (δ) is equal to 90°. The expression of maximum real power at both ends is,

$$P_{\text{sm}} = P_{\text{rm}} = \frac{V_r V_s}{X_L} \tag{7.24}$$

The average reactive power flow through the line is expressed as,

$$Q_{\text{av}} = \frac{Q_s + Q_r}{2} \tag{7.25}$$

If the torque angle is zero, the expression of sending end reactive power is modified as,

$$Q_s = \frac{V_s^2}{X_L} - \frac{V_s V_r}{X_L} \times 0 \tag{7.26}$$

$$Q_s = \frac{V_s^2}{X_L} \tag{7.27}$$

Again, at $\delta = 0$, Eq. (7.16) is modified as,

$$Q_r = \frac{V_r V_s}{X_L} \times 0 - \frac{V_r^2}{X_L} \tag{7.28}$$

$$Q_r = -\frac{V_r^2}{X_L} \tag{7.29}$$

Substituting Eqs. (7.27) and (7.29) into Eq. (7.25) yields,

$$Q_{\text{av}} = \frac{1}{2X_L}\left(V_s^2 - V_r^2\right) \tag{7.30}$$

The line loss of the short transmission line is expressed as,

$$P_{\text{line}} = I^2 R \tag{7.31}$$

The general expression of the complex power is,

$$S = P + jQ = VI^* \tag{7.32}$$

$$I^* = \frac{P + jQ}{V} \tag{7.33}$$

$$I = \frac{P - jQ}{V} \tag{7.34}$$

Multiplying Eqs. (7.33) and (7.34) yields,

$$I \times I^* = \frac{P - jQ}{V} \times \frac{P + jQ}{V} \tag{7.35}$$

$$I^2 = \frac{P^2 + Q^2}{V^2} \tag{7.36}$$

Substituting Eq. (7.36) into Eq. (7.31) yields,

$$P_{\text{line}} = R\frac{P^2 + Q^2}{V^2} \tag{7.37}$$

From Eq. (7.37), it is concluded that both real power and reactive power play an important role for line loss.

Example 7.1
A short transmission line is having the sending end voltage 95 V and receiving end voltage 75 V. Calculate the average reactive power of the line if the reactance of the line is 15 Ω.

Solution
The average reactive power is calculated as,

$$Q_{av} = \frac{1}{2X_L}\left(V_s^2 - V_r^2\right) = \frac{1}{2 \times 15}\left(95^2 - 75^2\right) = 113.33\,\text{Var} \tag{7.38}$$

Practice Problem 7.1
A short transmission line is having the sending end voltage 80 V and receiving end voltage 50 V. Calculate the line reactance if the average reactive power of the line is 180 Kvar.

7.4 Load Flow Equations for Two-Bus

Consider a transmission line which is connected between bus i and bus j, as shown in Fig. 7.3. Let the voltage at bus i is $V_i = |V_i|\,\underline{\lfloor \delta_i}$, and the voltage at bus j is $V_j = |V_j|\,\underline{\lfloor \delta_j}$. The voltage and admittance of these two buses can be written as [3],

$$V_i = |V_i|\,\underline{\lfloor \delta_i} = |V_i|(\cos\delta_i + j\sin\delta_i) \tag{7.39}$$

$$V_j = |V_j|\,\underline{\lfloor \delta_j} = |V_j|(\cos\delta_j + j\sin\delta_j) \tag{7.40}$$

The self and mutual admittances between bus i and bus j are written as,

$$Y_{ii} = |Y_{ii}|\,\underline{\lfloor \theta_{ii}} = |Y_{ii}|(\cos\theta_{ii} + j\sin\theta_{ii}) \tag{7.41}$$

$$Y_{ij} = |Y_{ij}|\,\underline{\lfloor \theta_{ij}} = |Y_{ij}|(\cos\theta_{ij} + j\sin\theta_{ij}) \tag{7.42}$$

Again, consider the three-bus single-line where the currents are injected at bus 1, bus 2, and bus 3, as shown in Fig. 7.4.

Applying KCL at bus 1, bus 2, and bus 3 of Fig. 7.4 yields,

$$Y_{11}V_1 + Y_{12}V_2 + Y_{13}V_3 = I_1 \tag{7.43}$$

$$Y_{21}V_1 + Y_{22}V_2 + Y_{23}V_3 = I_2 \tag{7.44}$$

$$Y_{31}V_1 + Y_{32}V_2 + Y_{33}V_3 = I_3 \tag{7.45}$$

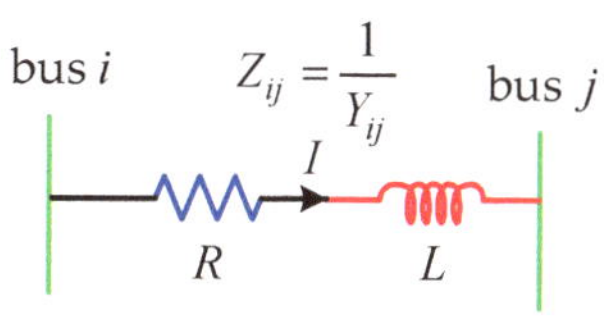

Fig. 7.3 Two-bus single-line diagram

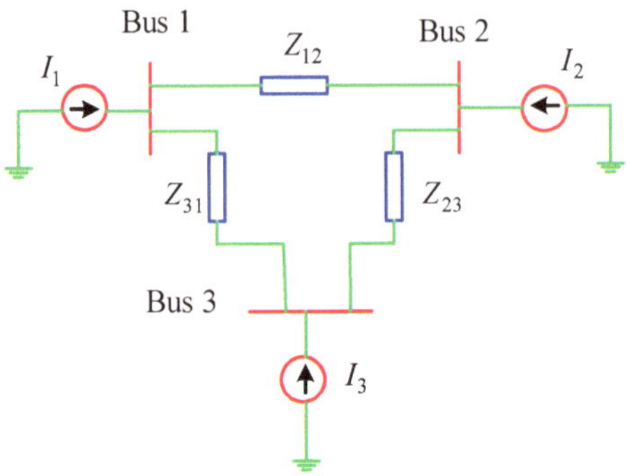

Fig. 7.4 Three-bus single-line diagram

Equations (7.43), (7.44), and (7.45) can be arranged in the matrix form as,

$$\begin{bmatrix} Y_{11} & Y_{12} & Y_{13} \\ Y_{21} & Y_{22} & Y_{23} \\ Y_{31} & Y_{32} & Y_{33} \end{bmatrix} \begin{bmatrix} V_1 \\ V_2 \\ V_3 \end{bmatrix} = \begin{bmatrix} I_1 \\ I_2 \\ I_3 \end{bmatrix} \tag{7.46}$$

In general, Eq. (7.46) can be written as,

$$I_i = \sum_{j=1}^{N} Y_{ij} V_j \tag{7.47}$$

General expression of the complex power is,

$$S = P + jQ = VI^* \tag{7.48}$$

The complex power at bus i is written as,

$$P_i + jQ_i = V_i I_i^* \tag{7.49}$$

$$P_i - jQ_i = V_i^* I_i \tag{7.50}$$

Substituting Eqs. (7.39) and (7.47) into Eq. (7.50) yields,

$$P_i - jQ_i = [|V_i|(\cos\delta_i + j\sin\delta_i)]^* \sum_{j=1}^{N} Y_{ij} V_j \tag{7.51}$$

Again, substituting Eqs. (7.40) and (7.42) into Eq. (7.51) yields,

$$\begin{aligned} P_i - jQ_i = {} & [|V_i|(\cos\delta_i + j\sin\delta_i)]^* \\ & \sum_{j=1}^{N} [|Y_{ij}|(\cos\theta_{ij} + j\sin\theta_{ij})]\,[|V_j|(\cos\delta_j + j\sin\delta_j)] \end{aligned} \tag{7.52}$$

$$P_i - jQ_i = \sum_{j=1}^{N} |V_i||V_j||Y_{ij}|(\cos\delta_i - j\sin\delta_i)\left[\cos(\theta_{ij} + \delta_j) + j\sin(\theta_{ij} + \delta_j)\right] \quad (7.53)$$

$$P_i - jQ_i = \sum_{j=1}^{N} |V_i||V_j||Y_{ij}|\left[\cos(\theta_{ij} + \delta_j - \delta_i) + j\sin(\theta_{ij} + \delta_j - \delta_i)\right] \quad (7.54)$$

From Eq. (7.54), equating real and imaginary parts yields,

$$P_i = \sum_{j=1}^{N} |V_i||V_j||Y_{ij}|\cos(\theta_{ij} + \delta_j - \delta_i) \quad (7.55)$$

$$Q_i = -\sum_{j=1}^{N} |V_i||V_j||Y_{ij}|\sin(\theta_{ij} + \delta_j - \delta_i) \quad (7.56)$$

Based on Eq. (7.48), for bus 2, the following relation can be modified as,

$$P_2 + jQ_2 = V_2 I_2^* \quad (7.57)$$

$$I_2^* = \frac{P_2 + jQ_2}{V_2} \quad (7.58)$$

$$I_2 = \frac{P_2 - jQ_2}{V_2^*} \quad (7.59)$$

Substituting Eq. (7.59) into (7.44) yields,

$$Y_{21}V_1 + Y_{22}V_2 + Y_{23}V_3 = \frac{P_2 - jQ_2}{V_2^*} \quad (7.60)$$

$$Y_{22}V_2 = \frac{P_2 - jQ_2}{V_2^*} - (Y_{21}V_1 + Y_{23}V_3) \quad (7.61)$$

$$V_2 = \frac{1}{Y_{22}}\left[\frac{P_2 - jQ_2}{V_2^*} - (Y_{21}V_1 + Y_{23}V_3)\right] \quad (7.62)$$

In general, the expression of voltage at bus i can be written as,

$$V_i = \frac{1}{Y_{ii}}\left[\frac{P_i - jQ_i}{V_i^*} - \sum_{\substack{j=1 \\ j \neq i}}^{N} Y_{ij}V_j\right] \tag{7.52}$$

7.5 Load Flow Equations for Three-Bus

Consider the three-bus single-line diagram of a power system where the first generator (G_1) is connected to bus 1, and the second generator (G_2) is connected to bus 2 as shown in Fig. 7.5. Here, the load is connected to bus 3.

The equivalent circuit is shown in Fig. 7.6. Applying KCL at node 1 yields,

$$I_1 = V_1 y_{10} + (V_1 - V_2)y_{12} + (V_1 - V_3)y_{13} \tag{7.53}$$

$$(y_{10} + y_{12} + y_{13})V_1 - y_{12}V_2 - y_{13}V_3 = I_1 \tag{7.54}$$

Applying KCL at node 2 yields,

$$I_2 = V_2 y_{20} + (V_2 - V_3)y_{23} + (V_2 - V_1)y_{12} \tag{7.55}$$

$$-y_{12}V_1 + (y_{20} + y_{12} + y_{23})V_2 - y_{23}V_3 = I_2 \tag{7.56}$$

Again, applying KCL at node 3 yields,

$$V_3 y_{30} = (V_1 - V_3)y_{31} + (V_2 - V_3)y_{23} \tag{7.57}$$

$$-y_{13}V_1 - y_{23}V_2 + (y_{30} + y_{31} + y_{32})V_3 = 0 \tag{7.58}$$

Equations (7.54), (7.56), and (7.58) can be arranged in the matrix form as,

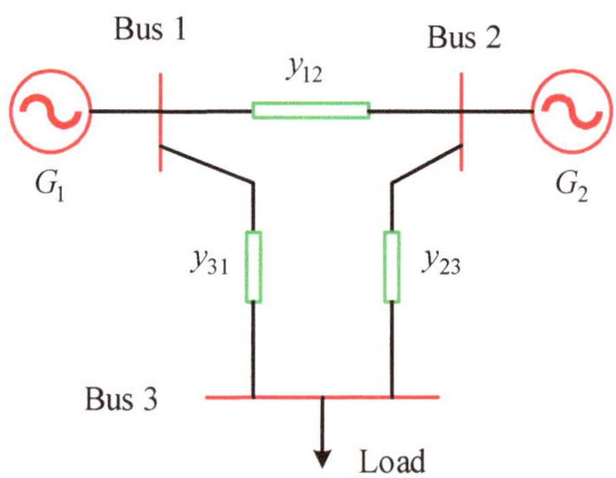

Fig. 7.5 Single-line diagram of a three-bus system

Fig. 7.6 Equivalent circuit

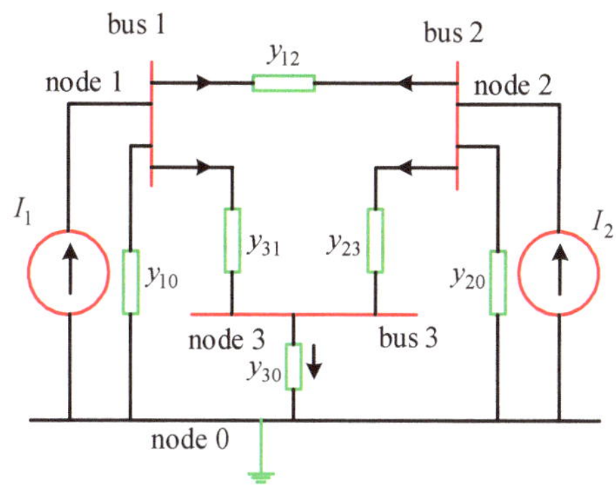

$$\begin{bmatrix} y_{10}+y_{12}+y_{13} & -y_{12} & -y_{13} \\ -y_{12} & y_{20}+y_{12}+y_{23} & -y_{23} \\ -y_{13} & -y_{23} & y_{30}+y_{13}+y_{23} \end{bmatrix} \begin{bmatrix} V_1 \\ V_2 \\ V_3 \end{bmatrix} = \begin{bmatrix} I_1 \\ I_2 \\ 0 \end{bmatrix} \tag{7.59}$$

$$\begin{bmatrix} Y_{11} & Y_{12} & Y_{13} \\ Y_{21} & Y_{22} & Y_{23} \\ Y_{31} & Y_{32} & Y_{33} \end{bmatrix} \begin{bmatrix} V_1 \\ V_2 \\ V_3 \end{bmatrix} = \begin{bmatrix} I_1 \\ I_2 \\ 0 \end{bmatrix} \tag{7.60}$$

where, the following relations are,

$$Y_{11} = y_{10} + y_{12} + y_{13} \tag{7.61}$$

$$Y_{22} = y_{20} + y_{12} + y_{23} \tag{7.62}$$

$$Y_{33} = y_{30} + y_{13} + y_{23} \tag{7.63}$$

$$Y_{12} = Y_{21} = -y_{12} \tag{7.64}$$

$$Y_{13} = Y_{31} = \quad y_{13} \tag{7.65}$$

$$Y_{23} = Y_{32} = -y_{23} \tag{7.66}$$

Equation (7.60) can be expressed in a generalized form as,

$$Y_{\text{bus}} V_{\text{bus}} = I \tag{7.67}$$

where,

Y_{bus} is the bus admittance matrix of any specific power system,
V_{bus} is the bus voltage measured from the reference node.

In Eq. (7.59), the diagonal elements are equal to the sum of all elements that are connected to the different nodes in Fig. 7.6. This element is known as self-admittance or driving point admittance, and it can be expressed as,

$$Y_{ii} = \sum_{\substack{j=0 \\ j \neq i}}^{n} y_{ij} \tag{7.68}$$

The off-diagonal elements are equal to the negative of the admittances, which are connected between the nodes. This element is known as the mutual or transfer admittance, and it can be expressed as,

$$Y_{ij} = Y_{ji} = -y_{ij} \tag{7.69}$$

Example 7.2
The per unit line impedances of a three-bus system is shown in Fig. 7.7. Calculate the Y_{bus} matrix.

Solution
The corresponding line admittances are calculated as,

$$y_{12} = \frac{1}{0.15 + j0.53} = 1.82 \angle{-74.20^\circ} \tag{7.70}$$

$$y_{23} = \frac{1}{0.22 + j0.75} = 1.28 \angle{-73.65^\circ} \tag{7.71}$$

$$y_{31} = \frac{1}{0.25 + j0.69} = 1.36 \angle{-70.08^\circ} \tag{7.72}$$

The values of self-admittances are calculated as,

$$Y_{11} = y_{12} + y_{13} = 1.82 \angle{-74.20^\circ} + 1.36 \angle{-70.08^\circ} = 3.18 \angle{-72.43^\circ} \tag{7.73}$$

$$Y_{22} = y_{12} + y_{23} = 1.82 \angle{-74.20^\circ} + 1.28 \angle{-73.65^\circ} = 3.10 \angle{-73.97^\circ} \tag{7.74}$$

$$Y_{33} = y_{31} + y_{23} = 1.36 \angle{-70.08^\circ} + 1.28 \angle{-73.65^\circ} = 2.64 \angle{-71.81^\circ} \tag{7.75}$$

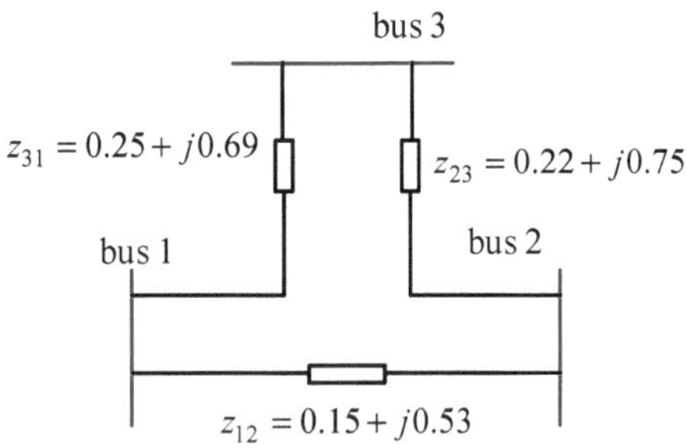

Fig. 7.7 Single-line diagram for Example 7.2

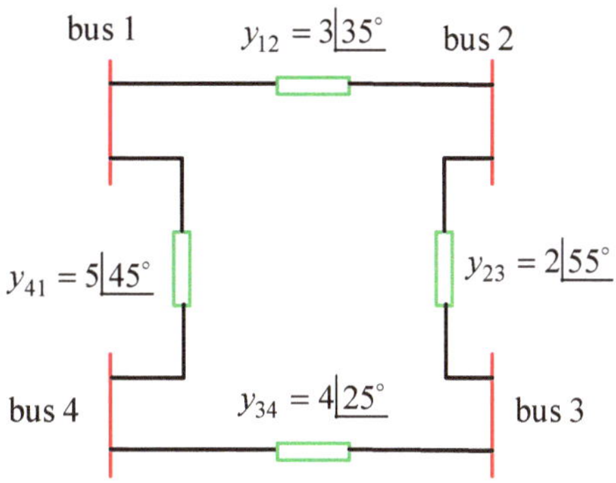

Fig. 7.8 Single-line diagram for Practice Problem 7.2

The values of mutual admittances are calculated as,

$$Y_{12} = Y_{21} = -y_{12} = -1.82\angle -74.20° \tag{7.76}$$

$$Y_{23} = Y_{32} = -y_{23} = -1.28\angle -73.65° \tag{7.77}$$

$$Y_{31} = Y_{13} = -y_{13} = -1.36\angle -70.08° \tag{7.78}$$

The bus admittance matrix can be written as,

$$Y_{bus} = \begin{bmatrix} Y_{11} & Y_{12} & Y_{13} \\ Y_{21} & Y_{22} & Y_{23} \\ Y_{31} & Y_{32} & Y_{33} \end{bmatrix} = \begin{bmatrix} 3.18\angle -72.43° & -1.82\angle -74.20° & -1.36\angle -70.08° \\ -1.82\angle -74.20° & 3.10\angle -73.97° & -1.28\angle -73.65° \\ -1.36\angle -70.08° & -1.28\angle -73.65° & 2.64\angle -71.81° \end{bmatrix} \tag{7.79}$$

Practice Problem 7.2

A four-bus single-line diagram is shown in Fig. 7.8. The units of the admittances are in per unit (pu). Find the Y_{bus} matrix.

7.6 Gauss–Seidel Method

Gauss–Seidel is one of the simplest iterative methods that is widely used to solve the linear equation in power systems. This method works perfectly when the diagonal elements are dominating. Consider a three-bus system to derive the Gauss–Seidel algorithm for power flow equations [4].

$$\begin{bmatrix} Y_{11} & Y_{12} & Y_{13} \\ Y_{21} & Y_{22} & Y_{23} \\ Y_{31} & Y_{32} & Y_{33} \end{bmatrix} \begin{bmatrix} V_1 \\ V_2 \\ V_3 \end{bmatrix} = \begin{bmatrix} I_1 \\ I_2 \\ I_3 \end{bmatrix} \tag{7.80}$$

The current equation for bus one (bus 1) can be written as,

$$Y_{11}V_1 + Y_{12}V_2 + Y_{13}V_3 = I_1 \tag{7.81}$$

The power injection at any bus is defined as,

$$P_1 + jQ_1 = V_1 I_1^* \tag{7.82}$$

Taking conjugate of Eq. (7.82) yields,

$$P_1 - jQ_1 = I_1 V_1^* \tag{7.83}$$

Substituting Eq. (7.83) into Eq. (7.81) yields,

$$Y_{11}V_1 + Y_{12}V_2 + Y_{13}V_3 = \frac{P_1 - jQ_1}{V_1^*} \tag{7.84}$$

The voltage at bus 1 is calculated as,

$$Y_{11}V_1 = \frac{P_1 - jQ_1}{V_1^*} - Y_{12}V_2 - Y_{13}V_3 \tag{7.85}$$

$$V_1 = \frac{1}{Y_{11}}\left[\frac{P_1 - jQ_1}{V_1^*} - Y_{12}V_2 - Y_{13}V_3\right] \tag{7.86}$$

Using $(k + 1)$th iteration, the voltage at bus 1 is calculated as,

$$V_1^{k+1} = \frac{1}{Y_{11}}\left[\frac{P_1 - jQ_1}{(V_1^*)^k} - Y_{12}V_2^k - Y_{13}V_3^k\right] \tag{7.87}$$

The current equation for bus 2 can be written as,

$$Y_{21}V_1 + Y_{22}V_2 + Y_{23}V_3 = I_2 \tag{7.88}$$

The voltage at bus 2 is calculated as,

$$V_2 = \frac{1}{Y_{22}}\left[\frac{P_2 - jQ_2}{V_2^*} - Y_{21}V_1 - Y_{23}V_3\right] \tag{7.89}$$

Using $(k + 1)$th iteration, the voltage at bus 2 is calculated as,

$$V_2^{k+1} = \frac{1}{Y_{22}}\left[\frac{P_2 - jQ_2}{(V_2^*)^k} - Y_{21}V_1^{k+1} - Y_{23}V_3^k\right] \tag{7.90}$$

The current equation for bus 3 can be written as,

$$Y_{31}V_1 + Y_{32}V_2 + Y_{33}V_3 = I_3 \tag{7.91}$$

Using (*k* + 1)th iteration, the voltage at bus 3 is calculated as,

$$V_3^{k+1} = \frac{1}{Y_{33}}\left[\frac{P_3 - jQ_3}{(V_3^*)^k} - Y_{31}V_1^{k+1} - Y_{32}V_2^{k+1}\right] \tag{7.92}$$

In general, the voltage at bus *i* can be calculated as,

$$V_i^{k+1} = \frac{1}{Y_{ii}}\left[\frac{P_i - jQ_i}{(V_i^*)^k} - \sum_{\substack{j=1 \\ j \neq i}}^{n} Y_{ij}V_j^k\right] \tag{7.93}$$

Example 7.3
The two linear equations are given by $8V_1 + 3V_2 = 12$ and $6V_1 + 5V_2 = 18$. Calculate the unknown voltages by the Gauss–Seidel method.

Solution
The voltage equations can be rearranged as,

$$V_1 = \frac{12 - 3V_2}{8} \tag{7.94}$$

$$V_2 = \frac{18 - 6V_1}{5} \tag{7.95}$$

Considering the initial guess value of the V_1 is,

$$V_1 = 0 \tag{7.96}$$

The first iteration values of the voltages are calculated as,

$$V_2^1 = \frac{18 - 0}{5} = 3.6 \tag{7.97}$$

$$V_1^1 = \frac{12 - 3 \times 3.6}{5} = 0.24 \tag{7.98}$$

The second iteration values are calculated as,

$$V_2^2 = \frac{18 - 6 \times 0.24}{5} = 3.312 \tag{7.99}$$

$$V_1^2 = \frac{12 - 3 \times 3.12}{5} = 0.528 \tag{7.100}$$

The third iteration values are calculated as,

$$V_2^3 = \frac{18 - 6 \times 0.528}{5} = 2.9664 \tag{7.101}$$

$$V_1^3 = \frac{12 - 3 \times 2.9664}{5} = 0.62016 \tag{7.102}$$

The fourth iteration values are calculated as,

$$V_2^4 = \frac{18 - 6 \times 0.62016}{5} = 2.855808 \tag{7.103}$$

$$V_1^4 = \frac{12 - 3 \times 2.855808}{5} = 0.6865152 \tag{7.104}$$

The fifth iteration values are calculated as,

$$V_2^5 = \frac{18 - 6 \times 0.6865152}{5} = 2.77618176 \tag{7.105}$$

$$V_1^5 = \frac{12 - 3 \times 2.77618176}{5} = 0.73429 \tag{7.106}$$

The sixth iteration values are calculated as,

$$V_2^6 = \frac{18 - 6 \times 0.73429}{5} = 2.718852 \tag{7.107}$$

$$V_1^6 = \frac{12 - 3 \times 2.718852}{5} = 0.76868 \tag{7.108}$$

The seventh iteration values are calculated as,

$$V_2^7 = \frac{18 - 6 \times 0.7686}{5} = 2.6776 \tag{7.109}$$

$$V_1^7 = \frac{12 - 3 \times 2.6776}{5} = 0.7934 \tag{7.110}$$

The eighth iteration values are calculated as,

$$V_2^8 = \frac{18 - 6 \times 0.7934}{5} = 2.6479 \quad (7.111)$$

$$V_1^8 = \frac{12 - 3 \times 2.6479}{5} = 0.8112 \quad (7.112)$$

The ninth iteration values are calculated as,

$$V_2^9 = \frac{18 - 6 \times 0.8112}{5} = 2.6265 \quad (7.113)$$

$$V_1^9 = \frac{12 - 3 \times 2.6265}{5} = 0.8241 \quad (7.114)$$

The tenth iteration values are calculated as,

$$V_2^{10} = \frac{18 - 6 \times 0.8241}{5} = 2.61108 \quad (7.115)$$

$$V_1^{10} = \frac{12 - 3 \times 2.61108}{5} = 0.8333 \quad (7.116)$$

The eleventh iteration values are calculated as,

$$V_2^{11} = \frac{18 - 6 \times 0.8333}{5} = 2.6 \quad (7.117)$$

$$V_1^{11} = \frac{12 - 3 \times 2.6}{5} = 0.84 \quad (7.118)$$

The twelfth iteration values are calculated as,

$$V_2^{12} = \frac{18 - 6 \times 0.84}{5} = 2.592 \quad (7.119)$$

$$V_1^{12} = \frac{12 - 3 \times 2.592}{5} = 0.844 \quad (7.120)$$

From the eleventh and twelfth iteration values, it is seen that the values of V_1 and V_2 are not changing. Therefore, the real solutions are V_1 = 0.84 V and V_2 = 2.6 V.

Practice Problem 7.3

The three sets of linear equations are given by $12V_1 + 3V_2 + 4V_3 = 10$, $2V_1 + 8V_2 + 3V_3 = 12$, and $4V_1 + 2V_2 + 10V_3 = 18$. Determine the voltages by the Gauss–Seidel method.

Example 7.4

A 12.2 MW, 11 kV (line voltage), 2-pole synchronous generator with 9 MW active generation, 1 Mvar maximum reactive power, and 0.85 power factor is connected to bus 1. The length of the line is 1 mile, and the per unit impedance of the line is $0.12 + j0.23$. A load of 50 MW and 30 Mvar is connected to bus 2, as shown in Fig. 7.9. Consider bus 1 is a slack bus and use the Gauss–Seidel method to calculate the voltage at bus 2, slack bus real and reactive powers, line flows, and line loss.

Solution

The node current equation of Fig. 7.9 is,

$$\begin{bmatrix} Y_{11} & Y_{12} \\ Y_{21} & Y_{22} \end{bmatrix} \begin{bmatrix} V_1 \\ V_2 \end{bmatrix} = \begin{bmatrix} I_1 \\ I_2 \end{bmatrix} \tag{7.121}$$

The self-admittances are calculated as,

$$Y_{11} = Y_{22} = 1.78 - j3.42 \tag{7.122}$$

The mutual admittances are calculated as,

$$Y_{12} = Y_{21} = -y_{12} = -1.78 + j3.42 \tag{7.123}$$

From bus 2, the real power and reactive power are supplied to the load. Therefore, these powers will be negative as $P_2 = -0.5$ pu and $Q_2 = -0.3$ pu. The voltage at bus 2 is calculated as,

$$V_2 = \frac{1}{Y_{22}}\left[\frac{P_2 - jQ_2}{V_2^*} - Y_{21}V_1\right] = \frac{1}{1.78 - j3.42}\left[\frac{-0.5 + j0.3}{V_2^*} - (-1.78 + j3.42)V_1\right] \tag{7.124}$$

$$V_2 = \frac{0.1512\ \underline{|-148.46^\circ}}{V_2^*} + V_1 \tag{7.125}$$

For a slack bus, $V_1 = 1\ \underline{|0^\circ}$, and let the initial guess $V_2^0 = 1\ \underline{|0^\circ}$. The first iteration is calculated as,

$$V_2^1 = \frac{0.1512\ \underline{|-148.46^\circ}}{1\ \underline{|0^\circ}} + 1 = 0.8747\ \underline{|-5.18^\circ} \tag{7.126}$$

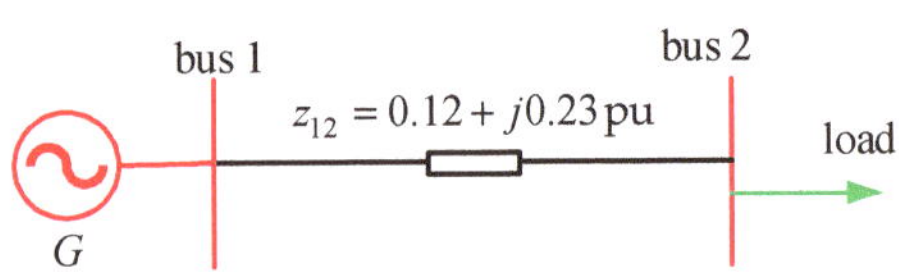

Fig. 7.9 Single-line diagram for Example 7.4

The second iteration is calculated as,

$$V_2^2 = \frac{0.1512\angle{-148.46^\circ}}{0.8747\angle{5.18^\circ}} + 1 = 0.8485\angle{-5.18^\circ} \quad (7.127)$$

The third iteration is calculated as,

$$V_2^3 = \frac{0.1512\angle{-148.46^\circ}}{0.8485\angle{5.18^\circ}} + 1 = 0.8440\angle{-5.37^\circ} \quad (7.128)$$

The fourth iteration is calculated as,

$$V_2^4 = \frac{0.1512\angle{-148.46^\circ}}{0.8440\angle{5.37^\circ}} + 1 = 0.8429\angle{-5.37^\circ} \quad (7.129)$$

The fifth iteration is calculated as,

$$V_2^5 = \frac{0.1512\angle{-148.46^\circ}}{0.8429\angle{5.37^\circ}} + 1 = 0.8427\angle{-5.38^\circ} \quad (7.130)$$

The sixth iteration is calculated as,

$$V_2^6 = \frac{0.1512\angle{-148.46^\circ}}{0.8427\angle{5.38^\circ}} + 1 = 0.8427\angle{-5.38^\circ} \quad (7.131)$$

The voltage at the fifth and sixth iteration is the same. Therefore, the voltage at bus 2 is,

$$V_2 = 0.8427\angle{-5.38^\circ}\ \text{pu} \quad (7.132)$$

The slack bus power is calculated as,

$$P_1 - jQ_1 = V_1^* I_1 = V_1^*(V_1 y_{12} - V_2 y_{21}) = 1[1 \times (1.78 - j3.42) - 0.8427\angle{-5.38^\circ} \times (1.78 - j3.42)] \quad (7.133)$$

$$P_1 - jQ_1 = 0.5568 - j0.4100 \quad (7.134)$$

The value of the real power and the reactive power are calculated as,

$$P_1 = 0.5568\,\text{pu} = 55.68\,\text{MW} \quad (7.135)$$

$$Q_1 = 0.4100\,\text{pu} = 41\,\text{Mvar} \quad (7.136)$$

The line currents are calculated as,

$$I_{12} = y_{12}(V_1 - V_2) = (1.78 - j3.42)(1 - 0.8427 \angle -5.38^\circ) = 0.6914 \angle -36.36^\circ \tag{7.137}$$

$$I_{21} = -I_{12} = -0.6914 \angle -36.36^\circ \tag{7.138}$$

The line flows calculated as,

$$S_{12} = V_1 I_{12}^* = 1 \times 0.6914 \angle 36.36^\circ = 0.5567 + j0.4099\,\text{pu} \tag{7.139}$$

$$S_{12} = 55.67\,\text{MW} + j40.99\,\text{Mvar} \tag{7.140}$$

$$S_{21} = V_2 I_{21}^* = 0.8427 \angle -5.38^\circ \times (-0.6914 \angle 36.36^\circ) = -0.4995 - j0.2999\,\text{pu} \tag{7.141}$$

$$S_{21} = -49.95\,\text{MW} - j29.99\,\text{Mvar}$$

The line loss is determined as,

$$S_{L-12} = S_{12} + S_{21} = 55.67\ \text{MW} + j40.99\ \text{MVar} - 49.95\ \text{MW} - j29.99\ \text{Mvar} \tag{7.142}$$

$$S_{L-12} = 5.72\,\text{MW} + j10.96\,\text{Mvar} \tag{7.143}$$

A two-bus system along with synchronous generator and load is drawn in the CYME software. Then the relevant data are entered, and simulation is carried out using the Gauss–Seidel technique. The CYME simulation results are in Fig. 7.10.

From Fig. 7.10, the voltage at bus 2, line loss, and slack bus real and reactive powers are calculated as,

$$V_{2CYME} = 0.843 \angle -5.37^\circ\ \text{pu} \tag{7.144}$$

$$S_{\text{L12CYME}} = 5.74\ \text{MW} + j11\ \text{Mvar} \tag{7.145}$$

$$P_{\text{1CYME}} = 55.74\ \text{MW} \tag{7.146}$$

$$Q_{\text{1CYME}} = 41\ \text{Mvar} \tag{7.147}$$

Again, a two-bus system single-line diagram is drawn by Interactive Power System Analysis (IPSA) software and is simulated by entering all necessary data. The simulation results are shown in Fig. 7.11.

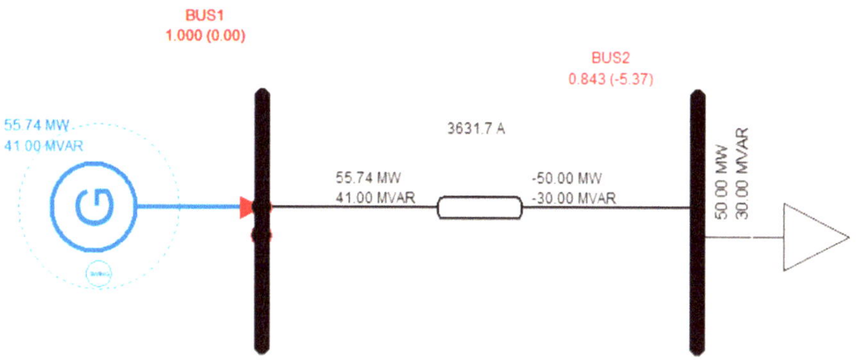

Fig. 7.10 CYME simulation results for Example 7.4

Fig. 7.11 IPSA simulation results for Example 7.4

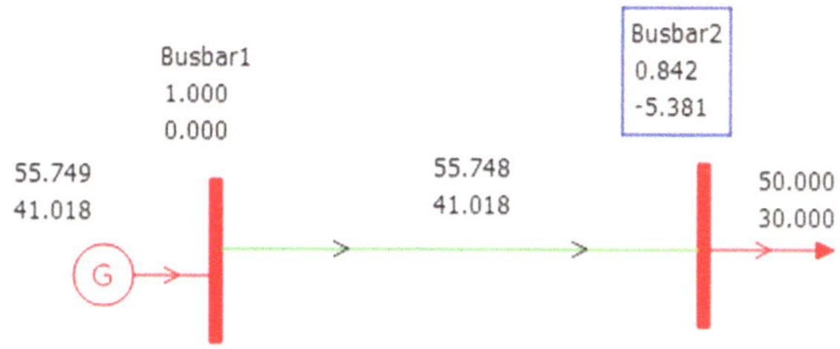

From Fig. 7.11, the voltage at bus 2, line loss, and slack bus real and reactive powers are calculated as,

$$V_{2IPSA} = 0.842 \angle -5.4^\circ \text{ pu} \tag{7.148}$$

$$S_{\text{L12IPSA}} = 5.75\,\text{MW} + j11.02\,\text{Mvar} \tag{7.149}$$

$$P_{\text{1IPSA}} = 55.75\,\text{MW} \tag{7.150}$$

$$Q_{\text{1IPSA}} = 41.02\,\text{Mvar} \tag{7.151}$$

Example 7.5

A 15 MW, 11 kV (line voltage), 2-pole synchronous generator with 11 MW active generation, 2 MW maximum reactive power, and 0.90 power factor is connected to bus 1. The length of the line is 1mile, and the per unit impedance of the lines is given as shown in Fig. 7.12. The load 1 of 30 MW and 10 Mvar is connected to bus 2, and the load 2 of 25 MW and 9 Mvar is connected to bus 3. Consider bus 1 is a slack bus and use the Gauss–Seidel method to calculate the voltage at bus 2, bus 3, and slack bus real power and reactive power.

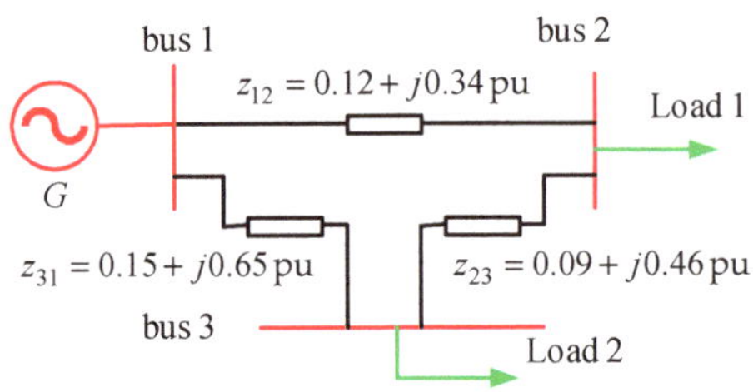

Fig. 7.12 Single-line diagram for Example 7.5

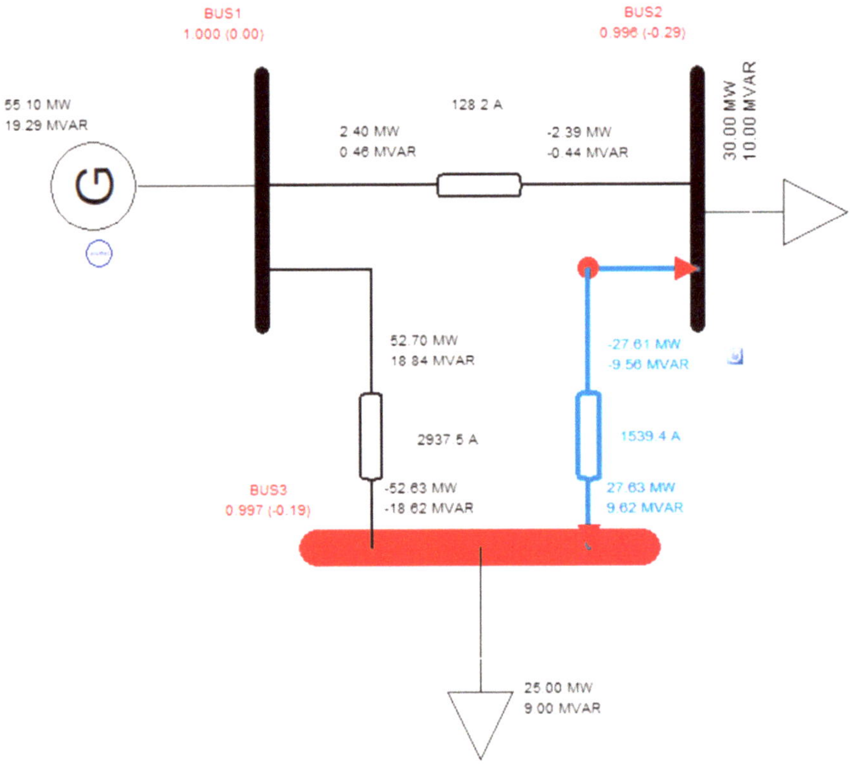

Fig. 7.13 CYME simulation results for Example 7.5

Solution

The three-bus system, along with synchronous generator and loads, is drawn in the CYME software. Then the relevant data are given input and carried out simulation using the Gauss–Seidel technique. The CYME simulation results are in Fig. 7.13.

From Fig. 7.13, the voltage at bus 2, bus 3, line loss, and slack bus real and reactive powers are calculated as,

$$V_{2CYME} = 0.996\ \underline{|-0.29^\circ}\ \text{pu} \tag{7.152}$$

$$V_{3CYME} = 0.997\ \underline{|-0.19^\circ}\ \text{pu} \tag{7.153}$$

$$S_{\text{L12CYME}} = 0.01\,\text{MW} + j0.02\,\text{Mvar} \tag{7.144}$$

$$S_{\text{L23CYME}} = 0.01\,\text{MW} + j0.06\,\text{Mvar} \tag{7.145}$$

$$S_{\text{L31CYME}} = 0.07\,\text{MW} + j0.22\,\text{Mvar} \tag{7.146}$$

$$P_{\text{1CYME}} = 55.10\,\text{MW} \tag{7.147}$$

$$Q_{\text{1CYME}} = 19.29\,\text{Mvar} \tag{7.148}$$

Practice Problem 7.4

A 20 MVA, 11 kV (line voltage), 2-pole synchronous generator with 14 MW active generation, 3 Mvar maximum reactive power, and 0.85 power factor is connected to bus 1. A load of 90 MW and 30 Mvar is connected to bus 2, and the generator is connected to the slack bus 1. The line impedance is $0.02 + j0.06$ pu on a 100 MVA base, and the length of the line is 1 mile. A three-phase 11 kV, 4 MVar shunt capacitor is connected to bus 1 and a three-phase 20 Mvar shunt capacitor is connected to the bus 2 as shown in Fig. 7.14. Use the Gauss–Seidel method to determine the voltage at bus 2 and slack bus real and reactive powers.

Practice Problem 7.5

A 25 MVA hydro generator 11 kV (line voltage) and 0.95 power factor are connected to bus 1. The first load of 250 MW and 135 Mvar is connected to bus 2 and the second load of 115 MW and 90 Mvar is connected to bus 3 as shown in Fig. 7.15. Consider that the generator is connected to the slack bus 1. The lines impedance is in pu on a 100 MVA base, and the length of the line is 1 mile. Use the Gauss–Seidel method to determine the voltage at bus 2, bus 3, and slack bus real and reactive powers.

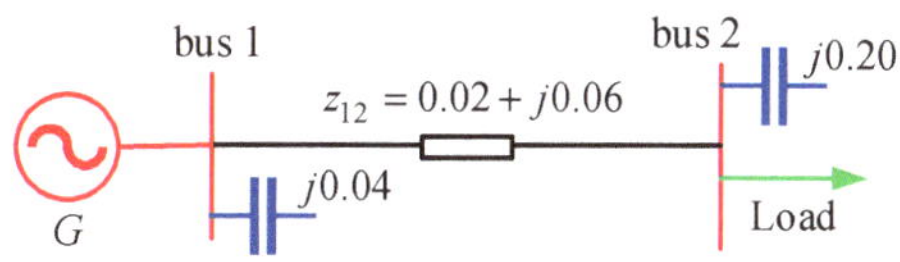

Fig. 7.14 Single-line diagram for Practice Problem 7.4

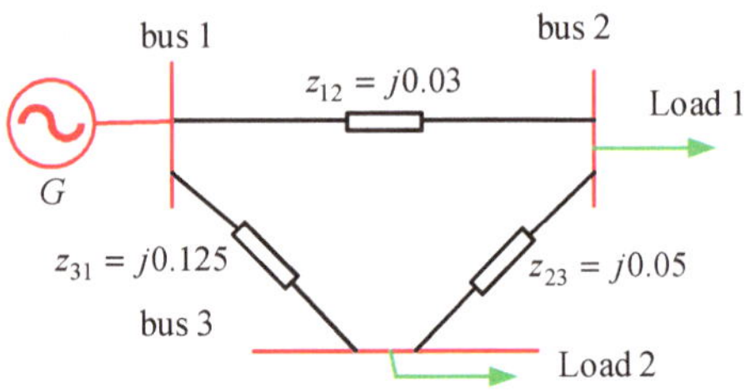

Fig. 7.15 Single-line diagram for Practice Problem 7.5

7.7 Newton–Raphson Method

Newton–Raphson (N-R) method is one of the well-known and powerful methods used for finding solution of a given nonlinear equation. Considering a general nonlinear equation as [5],

$$f(x) = 0 \tag{7.149}$$

Considering a geometry of N-R method for Eq. (7.149) is shown in Fig. 7.16. Let x_0 be the initial guess of the real solution r. Let $f'(x_0)$ be the slope of the line L and it can be determined as,

$$f'(x_0) = \frac{f(x_0) - 0}{x_0 - x_1} \tag{7.150}$$

$$x_0 - x_1 = \frac{f(x_0)}{f'(x_0)} \tag{7.151}$$

$$x_1 = x_0 - \frac{f(x_0)}{f'(x_0)} \tag{7.152}$$

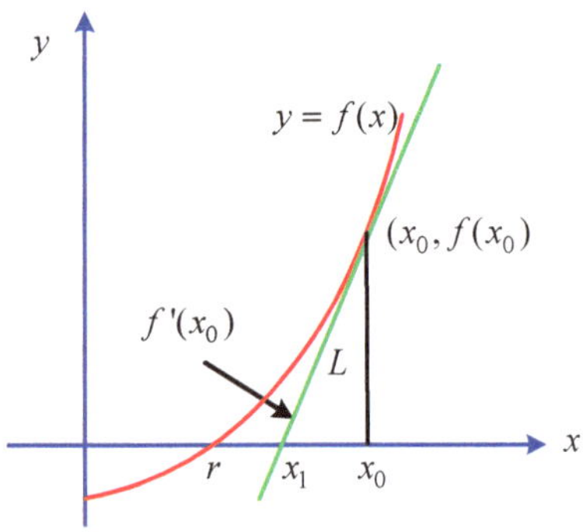

Fig. 7.16 Representation of nonlinear equation with a tangent line for the N-R method

The second iteration of the real solution is written as,

$$x_2 = x_1 - \frac{f(x_1)}{f'(x_1)} \tag{7.153}$$

The third iteration of the real solution is written as,

$$x_3 = x_2 - \frac{f(x_2)}{f'(x_2)} \tag{7.154}$$

Similarly, the *nth* iteration can be written as,

$$x_n = x_{n-1} - \frac{f(x_{n-1})}{f'(x_{n-1})} \tag{7.155}$$

Taking Taylor expansion of Eq. (7.149) yields,

$$f(x) = f(x_0) + \frac{1}{1!}f'(x_0)(x - x_0) + \frac{1}{2!}f''(x_0)(x - x_0)^2 + \ldots\ldots = 0 \tag{7.156}$$

Equation (7.156) is again modified by neglecting al higher terms of $(x - x_0)$ as,

$$f(x) = f(x_0) + \frac{1}{1!}f'(x_0)(x - x_0) = 0 \tag{7.157}$$

$$f(x) = f(x_0) + f'(x_0)(x - x_0) = 0 \tag{7.158}$$

$$-f'(x_0)(x - x_0) = f(x_0) \tag{7.159}$$

$$x - x_0 = -\frac{f(x_0)}{f'(x_0)} \tag{7.160}$$

$$x = x_0 - \frac{f(x_0)}{f'(x_0)} \tag{7.161}$$

Therefore, the first and the second iterations can be written as,

$$x_1 = x_0 - \frac{f(x_0)}{f'(x_0)} \tag{7.162}$$

$$x_2 = x_1 - \frac{f(x_1)}{f'(x_1)} \tag{7.163}$$

In general, the expression for nth iteration can be expressed as,

$$x_n = x_{n-1} - \frac{f(x_{n-1})}{f'(x_{n-1})} \tag{7.164}$$

Example 7.6
Calculate the solution of the nonlinear equation $f(x) = x \sin x + 5$ by the Newton–Raphson method.

Solution
Let the initial guess of the true solution is,

$$x_0 = 5 \tag{7.165}$$

The value of x in the first iteration is calculated as,

$$f(x_0) = 5 \sin 5 + 5 = 0.205 \tag{7.166}$$

$$f'(x) = x \cos x + \sin x \tag{7.167}$$

$$f'(x_0) = 5 \cos 5 + \sin 5 = 0.459 \tag{7.168}$$

$$x_1 = x_0 - \frac{f(x_0)}{f'(x_0)} = 5 - \frac{0.205}{0.459} = 4.5533 \tag{7.169}$$

The value of x in the second iteration is calculated as,

$$f(x_1) = 4.5533 \sin 4.5533 + 5 = 0.504 \tag{7.170}$$

$$f'(x_1) = 4.5533 \cos(4.5533) + \sin 4.5533 = -1.708 \tag{7.171}$$

$$x_2 = x_1 - \frac{f(x_1)}{f'(x_1)} = 4.5533 + \frac{0.504}{1.708} = 4.8483 \tag{7.172}$$

The value of x in the third iteration is calculated as,

$$f(x_2) = 4.8483 \sin 4.8483 + 5 = 0.1963 \tag{7.173}$$

$$f'(x_2) = 4.8483 \cos(4.8483) + \sin 4.8483 = -0.3338 \tag{7.174}$$

$$x_3 = x_2 - \frac{f(x_2)}{f'(x_2)} = 4.8483 + \frac{0.1963}{0.3338} = 5.4363 \tag{7.175}$$

The value of x in the fourth iteration is calculated as,

$$f(x_3) = 5.4363 \sin 5.4363 + 5 = 0.927 \tag{7.176}$$

$$f'(x_3) = 5.4363 \cos(5.4363) + \sin 5.4363 = 2.8513 \tag{7.177}$$

$$x_4 = x_3 - \frac{f(x_3)}{f'(x_3)} = 5.4363 - \frac{0.927}{2.8513} = 5.1111 \tag{7.178}$$

The value of x in the fifth iteration is calculated as,

$$f(x_4) = 5.1111 \sin 5.1111 + 5 = 0.2898 \tag{7.179}$$

$$f'(x_4) = 5.1111 \cos(5.1111) + \sin 5.1111 = 1.0627 \tag{7.180}$$

$$x_5 = x_4 - \frac{f(x_4)}{f'(x_4)} = 5.1111 - \frac{0.2898}{1.0627} = 4.8383 \tag{7.181}$$

The value of x in the sixth iteration is calculated as,

$$f(x_5) = 4.8383 \sin 4.8383 + 5 = 0.2 \tag{7.182}$$

$$f'(x_5) = 4.8383 \cos(4.8383) + \sin 4.8383 = -0.3844 \tag{7.183}$$

$$x_6 = x_5 - \frac{f(x_5)}{f'(x_5)} = 4.8383 + \frac{0.2}{0.3844} = 5.3565 \tag{7.184}$$

The value of x in the seventh iteration is calculated as,

$$f(x_6) = 5.3565 \sin 5.3565 + 5 = 0.7167 \tag{7.185}$$

$$f'(x_6) - 5.3565 \cos(5.3565) + \sin 5.3565 - 2.4168 \tag{7.186}$$

$$x_7 = x_6 - \frac{f(x_6)}{f'(x_6)} = 5.3565 - \frac{0.7167}{2.4168} = 5.0599 \tag{7.187}$$

The difference between the seventh iteration and the sixth iteration is very less. Therefore, the true solution is $x = 5.06$.

Practice Problem 7.6

Use Newton–Raphson method to determine the solution of the equation $5x^2 + 11x - 17 = 0$ correct to four decimal places.

7.8 Newton–Raphson Method for Two Nonlinear Equations

Newton–Raphson (N-R) method is a powerful tool for solving nonlinear equations as its convergence rate is very high compared to the Gauss–Seidel method. Considering two nonlinear equations as [5],

$$P = p(x, y) \tag{7.188}$$

$$Q = q(x, y) \tag{7.189}$$

Let x_0 and y_0 be the initial values of the functions $p(x, y)$ and $q(x, y)$, respectively. Taking Taylor expansion of Eqs. (7.188) and (7.189) with neglecting all higher terms yields,

$$P = p(x_0, y_0) + \frac{1}{1!}\left(\frac{\partial p}{\partial x}\right)_0 (x - x_0) + \frac{1}{1!}\left(\frac{\partial p}{\partial y}\right)_0 (y - y_0) \tag{7.190}$$

$$Q = q(x_0, y_0) + \frac{1}{1!}\left(\frac{\partial q}{\partial x}\right)_0 (x - x_0) + \frac{1}{1!}\left(\frac{\partial q}{\partial y}\right)_0 (y - y_0) \tag{7.191}$$

For further simplification, again, considering the following relations as,

$$\Delta x_0 = x - x_0 \tag{7.192}$$

$$x = x_0 + \Delta x_0 \tag{7.193}$$

$$\Delta y_0 = y - y_0 \tag{7.194}$$

$$y = y_0 + \Delta y_0 \tag{7.195}$$

Substituting Eqs. (7.192) and (7.194) into Eqs. (7.190) and (7.191) yields,

$$P - p(x_0, y_0) = \left(\frac{\partial p}{\partial x}\right)_0 \Delta x_0 + \left(\frac{\partial p}{\partial y}\right)_0 \Delta y_0 \tag{7.196}$$

$$Q - q(x_0, y_0) = \left(\frac{\partial q}{\partial x}\right)_0 \Delta x_0 + \left(\frac{\partial q}{\partial y}\right)_0 \Delta y_0 \tag{7.197}$$

Again, for further simplification, considering the following relations,

$$P - p(x_0, y_0) = \Delta p_0 \tag{7.198}$$

$$Q - q(x_0, y_0) = \Delta q_0 \tag{7.199}$$

Substituting Eqs. (7.198) and (7.199) into Eqs. (7.196) and (7.197) yields,

$$\Delta p_0 = \left(\frac{\partial p}{\partial x}\right)_0 \Delta x_0 + \left(\frac{\partial p}{\partial y}\right)_0 \Delta y_0 \tag{7.200}$$

$$\Delta q_0 = \left(\frac{\partial q}{\partial x}\right)_0 \Delta x_0 + \left(\frac{\partial q}{\partial y}\right)_0 \Delta y_0 \tag{7.201}$$

Equations (7.200) and (7.201) can be expressed in the matrix form as,

$$\begin{bmatrix} \Delta p_0 \\ \Delta q_0 \end{bmatrix} = \begin{bmatrix} \left(\frac{\partial p}{\partial x}\right)_0 & \left(\frac{\partial p}{\partial y}\right)_0 \\ \left(\frac{\partial q}{\partial x}\right)_0 & \left(\frac{\partial q}{\partial y}\right)_0 \end{bmatrix} \begin{bmatrix} \Delta x_0 \\ \Delta y_0 \end{bmatrix} \tag{7.202}$$

$$\begin{bmatrix} \Delta p_0 \\ \Delta q_0 \end{bmatrix} = \begin{bmatrix} \left(\frac{\partial p}{\partial x}\right) & \left(\frac{\partial p}{\partial y}\right) \\ \left(\frac{\partial q}{\partial x}\right) & \left(\frac{\partial q}{\partial y}\right) \end{bmatrix}_0 \begin{bmatrix} \Delta x_0 \\ \Delta y_0 \end{bmatrix} \tag{7.203}$$

$$\begin{bmatrix} \Delta p_0 \\ \Delta q_0 \end{bmatrix} = [J_0] \begin{bmatrix} \Delta x_0 \\ \Delta y_0 \end{bmatrix} \tag{7.204}$$

$$\begin{bmatrix} \Delta x_0 \\ \Delta y_0 \end{bmatrix} = [J_0]^{-1} \begin{bmatrix} \Delta p_0 \\ \Delta q_0 \end{bmatrix} \tag{7.205}$$

where, $[J_0]$ is called the Jacobian matrix.

The better estimation of the solution can be written as,

$$x_1 = x_0 + \Delta x_0 \tag{7.206}$$

$$y_1 = y_0 + \Delta y_0 \tag{7.207}$$

The iteration will be continued until to get required solutions.

Example 7.7

Calculate the solution of the two nonlinear equations given by $2x^2 + 3y^2 - 7 = 0$ and $x^2 + 4y^2 - 10 = 0$ by the Newton–Raphson method.

Solution

Let the initial guess of the true solution are,

$$x_0 = y_0 = 1 \tag{7.208}$$

$$p(x_0, y_0) = 2 \times 1^2 + 3 \times 1^2 = 5 \tag{7.209}$$

$$q(x_0, y_0) = 1^2 + 4 \times 1^2 = 5 \tag{7.210}$$

$$\Delta p_0 = P - p(x_0, y_0) = 7 - 5 = 2 \tag{7.211}$$

$$\Delta q_0 = Q - q(x_0, y_0) = 10 - 5 = 5 \tag{7.212}$$

The Jacobean matrix can be written as,

$$J = \begin{bmatrix} \frac{\partial p}{\partial x} & \frac{\partial p}{\partial y} \\ \frac{\partial q}{\partial x} & \frac{\partial q}{\partial y} \end{bmatrix} = \begin{bmatrix} 4x & 6y \\ 2x & 8y \end{bmatrix} \tag{7.213}$$

$$J_0 = \begin{bmatrix} 4x_0 & 6y_0 \\ 2x_0 & 8y_0 \end{bmatrix} = \begin{bmatrix} 4 & 6 \\ 2 & 8 \end{bmatrix} \tag{7.214}$$

$$\begin{bmatrix} \Delta x_0 \\ \Delta y_0 \end{bmatrix} = [J_0]^{-1} \begin{bmatrix} \Delta p_0 \\ \Delta q_0 \end{bmatrix} \tag{7.215}$$

$$\begin{bmatrix} \Delta x_0 \\ \Delta y_0 \end{bmatrix} = \begin{bmatrix} 4 & 6 \\ 2 & 8 \end{bmatrix}^{-1} \begin{bmatrix} 2 \\ 5 \end{bmatrix} = \begin{bmatrix} 0.4 & -0.3 \\ -0.1 & 0.2 \end{bmatrix} \begin{bmatrix} 2 \\ 5 \end{bmatrix} \tag{7.216}$$

$$\Delta x_0 = 0.4 \times 2 - 0.3 \times 5 = 0.7 \tag{7.217}$$

$$\Delta y_0 = -0.1 \times 2 + 0.2 \times 5 = 0.8 \tag{7.218}$$

First updated values are calculated as,

$$x_1 = x_0 + \Delta x_0 = 1 + 0.7 = 1.7 \tag{7.219}$$

$$y_1 = y_0 + \Delta y_0 = 1 + 0.8 = 1.8 \tag{7.220}$$

$$\Delta p_1 = P - p(x_1, y_1) = 7 - [2(1.7)^2 + 3(1.8)^2] = -8.5 \tag{7.221}$$

$$\Delta q_1 = Q - q(x_1, y_1) = 10 - [(1.7)^2 + 4(1.8)^2] = -5.85 \tag{7.222}$$

$$J_1 = \begin{bmatrix} 4x_1 & 6y_1 \\ 2x_1 & 8y_1 \end{bmatrix} = \begin{bmatrix} 6.8 & 10.8 \\ 3.4 & 14.4 \end{bmatrix} \tag{7.223}$$

$$\begin{bmatrix} \Delta x_1 \\ \Delta y_1 \end{bmatrix} = \begin{bmatrix} 6.8 & 10.8 \\ 3.4 & 14.4 \end{bmatrix}^{-1} \begin{bmatrix} -8.5 \\ -5.85 \end{bmatrix} = \begin{bmatrix} 0.234 & -0.176 \\ -0.055 & 0.111 \end{bmatrix} \begin{bmatrix} -8.5 \\ -5.85 \end{bmatrix} \tag{7.234}$$

$$\Delta x_1 = 0.234(-8.5) + (-0.176)(-5.85) = -0.959 \tag{7.235}$$

$$\Delta y_1 = -0.055(-8.5) + 0.111(-5.85) = -0.181 \tag{7.236}$$

Second updated values are calculated as,

$$x_2 = x_1 + \Delta x_1 = 1.7 - 0.959 = 0.741 \tag{7.237}$$

$$y_2 = y_1 + \Delta y_1 = 1.8 - 0.181 = 1.619 \tag{7.238}$$

$$\Delta p_2 = P - p(x_2, y_2) = 7 - \left[2(0.741)^2 + 3(1.619)^2\right] = -1.96 \tag{7.239}$$

$$\Delta q_2 = Q - q(x_2, y_2) = 10 - \left[(0.741)^2 + 4(1.619)^2\right] = -1.03 \tag{7.240}$$

$$J_2 = \begin{bmatrix} 4x_2 & 6y_2 \\ 2x_2 & 8y_2 \end{bmatrix} = \begin{bmatrix} 2.96 & 9.71 \\ 1.48 & 12.95 \end{bmatrix} \tag{7.241}$$

$$\begin{bmatrix} \Delta x_2 \\ \Delta y_2 \end{bmatrix} = \begin{bmatrix} 2.96 & 9.71 \\ 1.48 & 12.95 \end{bmatrix}^{-1} \begin{bmatrix} -1.96 \\ -1.03 \end{bmatrix} = \begin{bmatrix} 0.54 & -0.40 \\ -0.06 & 0.12 \end{bmatrix} \begin{bmatrix} -1.96 \\ -1.03 \end{bmatrix} \tag{7.242}$$

$$\Delta x_2 = 0.54(-1.96) + (-0.40)(-1.03) = -0.646 \tag{7.243}$$

$$\Delta y_2 = -0.06(-1.96) + 0.12(-1.03) = -0.006 \tag{7.244}$$

Third updated values are calculated as,

$$x_3 = x_2 + \Delta x_2 = 0.741 - 0.646 = 0.095 \tag{7.245}$$

$$y_3 = y_2 + \Delta y_2 = 1.619 - 0.006 = 1.613 \tag{7.246}$$

The values of the third iteration are close to the second iteration values. Therefore, the solutions are,

$$x = 0.095 \tag{7.247}$$

$$y = 1.613 \tag{7.248}$$

Practice Problem 7.7

The two nonlinear equations are given by $2x^2 + 5y^2 - 8 = 0$, and $x^2 - 4y^2 - 7 = 0$. Use the Newton–Raphson method to find the solution.

7.9 Newton–Raphson Method for Power Flow Cases

Newton–Raphson (N-R) method can also be applied to calculate the power flow in different buses, lines, and loads. Consider bus i and bus j as shown in Fig. 7.3. For these buses, the following relations can be written as [4],

$$V_i = |V_i| \underline{|\delta_i} \tag{7.249}$$

$$V_j = |V_j| \underline{|\delta_j} \tag{7.250}$$

$$Y_{ij} = |Y_{ij}| \underline{|\theta_{ij}} \tag{7.251}$$

$$I_i = \sum_{j=1}^{n} Y_{ij} V_j \tag{7.252}$$

The expression of complex power is,

$$P_i - jQ_i = V_i^* I_i \tag{7.253}$$

Substituting Eqs. (7.249) and (7.252) into (7.253) yields,

$$P_i - jQ_i = \sum_{j=1}^{n} |V_i V_j Y_{ij}| \underline{|\theta_{ij} + \delta_j - \delta_i} \tag{7.254}$$

$$P_i - jQ_i = \sum_{j=1}^{n} |V_i V_j Y_{ij}| [\cos(\theta_{ij} + \delta_j - \delta_i) + j\sin(\theta_{ij} + \delta_j - \delta_i)] \tag{7.255}$$

After equating real and imaginary parts of Eq. (7.255) yields,

$$P_i = \sum_{j=1}^{n} |V_i V_j Y_{ij}| \cos(\theta_{ij} + \delta_j - \delta_i) \tag{7.256}$$

$$Q_i = -\sum_{j=1}^{n} |V_i V_j Y_{ij}| \sin(\theta_{ij} + \delta_j - \delta_i) \tag{7.257}$$

Expanding Eqs. (7.256) and (7.257) in Taylor series and neglecting all higher terms and the set of linear equations can be arranged in matrix form as,

$$\begin{bmatrix} \Delta P_2 \\ \vdots \\ \vdots \\ \Delta P_n \\ \hline \Delta Q_2 \\ \vdots \\ \vdots \\ \Delta Q_n \end{bmatrix} = \left[\begin{array}{c|c} \begin{bmatrix} \frac{\partial P_2}{\partial \delta_2} \cdots \frac{\partial P_2}{\partial \delta_n} \\ \vdots \qquad \vdots \\ \frac{\partial P_n}{\partial \delta_2} \cdots \frac{\partial P_n}{\partial \delta_n} \end{bmatrix} & \begin{bmatrix} \frac{\partial P_2}{\partial |V_2|} \cdots \frac{\partial P_2}{\partial |V_n|} \\ \vdots \qquad \vdots \\ \frac{\partial P_n}{\partial |V_2|} \cdots \frac{\partial P_n}{\partial |V_n|} \end{bmatrix} \\ \hline \begin{bmatrix} \frac{\partial Q_2}{\partial \delta_2} \cdots \frac{\partial Q_2}{\partial \delta_n} \\ \vdots \qquad \vdots \\ \frac{\partial Q_n}{\partial \delta_2} \cdots \frac{\partial Q_n}{\partial \delta_n} \end{bmatrix} & \begin{bmatrix} \frac{\partial Q_2}{\partial |V_2|} \cdots \frac{\partial Q_2}{\partial |V_n|} \\ \vdots \qquad \vdots \\ \frac{\partial Q_n}{\partial |V_2|} \cdots \frac{\partial Q_n}{\partial |V_n|} \end{bmatrix} \end{array}\right] \begin{bmatrix} \Delta \delta_2 \\ \vdots \\ \Delta \delta_n \\ \hline \Delta |V_2| \\ \vdots \\ \Delta |V_n| \end{bmatrix} \tag{7.258}$$

Equation (7.258) can be reduced as,

$$\begin{bmatrix} \Delta P \\ \Delta Q \end{bmatrix} = \begin{bmatrix} J_1 & J_2 \\ J_3 & J_4 \end{bmatrix} \begin{bmatrix} \Delta\delta \\ \Delta|V| \end{bmatrix} \quad (7.259)$$

where,

ΔP and ΔQ are the mismatch vectors whose values are known,
$\Delta\delta$ and $\Delta|V|$ are the correction parameters whose values are known.

The Jacobian submatrix can be defined as,

$$J_1 = \begin{bmatrix} \frac{\partial P_2}{\partial \delta_2} \cdots \frac{\partial P_2}{\partial \delta_n} \\ \vdots \qquad \vdots \\ \frac{\partial P_n}{\partial \delta_2} \cdots \frac{\partial P_n}{\partial \delta_n} \end{bmatrix} \approx \frac{\partial P}{\partial \delta} \quad (7.260)$$

$$J_2 = \begin{bmatrix} \frac{\partial P_2}{\partial |V_2|} \cdots \frac{\partial P_2}{\partial |V_n|} \\ \vdots \qquad \vdots \\ \frac{\partial P_n}{\partial |V_2|} \cdots \frac{\partial P_n}{\partial |V_n|} \end{bmatrix} \approx \frac{\partial P}{\partial |V|} \quad (7.261)$$

$$J_3 = \begin{bmatrix} \frac{\partial Q_2}{\partial \delta_2} \cdots \frac{\partial Q_2}{\partial \delta_n} \\ \vdots \qquad \vdots \\ \frac{\partial Q_n}{\partial \delta_2} \cdots \frac{\partial Q_n}{\partial \delta_n} \end{bmatrix} \approx \frac{\partial Q}{\partial \delta} \quad (7.262)$$

$$J_4 = \begin{bmatrix} \frac{\partial Q_2}{\partial |V_2|} \cdots \frac{\partial Q_2}{\partial |V_n|} \\ \vdots \qquad \vdots \\ \frac{\partial Q_n}{\partial |V_2|} \cdots \frac{\partial Q_n}{\partial |V_n|} \end{bmatrix} \approx \frac{\partial Q}{\partial |V|} \quad (7.263)$$

The diagonal and off-diagonal elements of Jacobian submatrix consist of the partial derivatives of P and Q with respect to each of the variables in the Eqs. (7.256) and (7.257). Equation (7.258) can be reduced for the three-bus system as,

$$\begin{bmatrix} \Delta P_2 \\ \Delta P_3 \\ \Delta Q_2 \\ \Delta Q_2 \end{bmatrix} = \begin{bmatrix} \frac{\partial P_2}{\partial \delta_2} & \frac{\partial P_2}{\partial \delta_3} & \frac{\partial P_2}{\partial |V_2|} & \frac{\partial P_2}{\partial |V_3|} \\ \frac{\partial P_3}{\partial \delta_2} & \frac{\partial P_3}{\partial \delta_3} & \frac{\partial P_3}{\partial |V_2|} & \frac{\partial P_3}{\partial |V_3|} \\ \frac{\partial Q_2}{\partial \delta_2} & \frac{\partial Q_2}{\partial \delta_3} & \frac{\partial Q_2}{\partial |V_2|} & \frac{\partial Q_2}{\partial |V_3|} \\ \frac{\partial Q_3}{\partial \delta_2} & \frac{\partial Q_3}{\partial \delta_3} & \frac{\partial Q_3}{\partial |V_2|} & \frac{\partial Q_3}{\partial |V_3|} \end{bmatrix} \begin{bmatrix} \Delta\delta_2 \\ \Delta\delta_3 \\ \Delta|V_2| \\ \Delta|V_3| \end{bmatrix} \quad (7.264)$$

The following procedures need to be considered for N-R method.

- Read the data and form Ybus
- Choose the initial values of $|V|^0$ for all buses and δ^0 of the voltages of all buses except the slack bus
- Calculate the total injected real power P_{cal}^0 and real power mismatch ΔP^0 as

$$\Delta P_i = P_{i,\,specified} - P_{i,\,calculated} \tag{7.265}$$

- Calculate the total injected reactive power Q_{cal}^0 and reactive power mismatch ΔQ^0

$$\Delta Q_i = Q_{i,\,\text{specified}} - Q_{i,\,\text{calculated}} \tag{7.266}$$

- Check if $\Delta P_i < \varepsilon_1$, $\Delta Q_i < \varepsilon_2$, $\Delta|V_i| = \left|V_i^{k+1} - V_i^k\right| \leq \varepsilon_3$. Then, the latest calculated values of voltages are the real solution
- Estimate the Jacobian matrix and solve for ΔV_i and $\Delta\delta$ for all PQ buses, and $\Delta\delta$ all PV buses
- Update voltage and angles at each bus as

$$\delta_i^{k+1} = \delta_i^k + \Delta\delta_i^k \tag{7.267}$$

$$\left|V_i^{k+1}\right| = \left|V_i^k\right| + \Delta\left|V_i^k\right| \tag{7.268}$$

Example 7.8
A single-line diagram of the three-bus power system is shown in Fig. 7.17. All the admittances in the single-line diagram are in per unit values with a base voltage of 11 kV and base 100 MVA. The first generator is connected to the slack bus 1, and the second generator is considered as a fixed generation.

Solution
The values of self-admittances are calculated as,

$$Y_{11} = y_{12} + y_{31} = -j30 = 30\,\underline{|-90^\circ}\tag{7.269}$$

$$Y_{22} = y_{12} + y_{23} = -j26 = 26\,\underline{|-90^\circ}\tag{7.270}$$

$$Y_{33} = y_{31} + y_{23} = -j36 = 36\,\underline{|-90^\circ}\tag{7.271}$$

The values of the mutual admittances are calculated as,

$$Y_{12} = Y_{21} = -y_{12} = j10 = 10\,\underline{|90^\circ}\tag{7.272}$$

$$Y_{23} = Y_{32} = -y_{23} = j16 = 16\,\underline{|90^\circ}\tag{7.273}$$

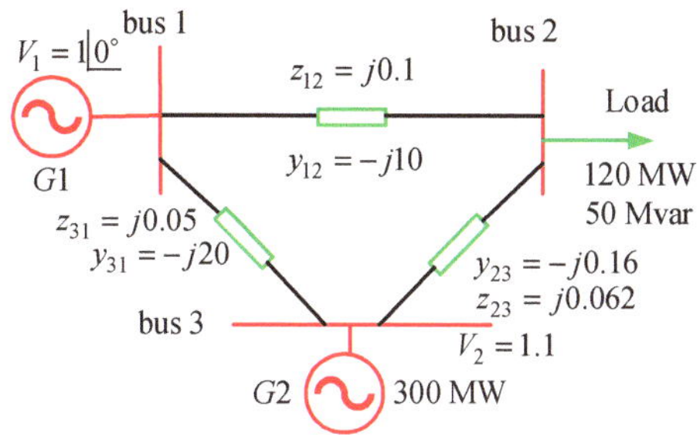

Fig. 7.17 A single-line diagram for Example 7.8

$$Y_{31} = Y_{13} = -y_{13} = j20 = 20\angle 90° \tag{7.274}$$

According to Eqs. (7.256) and (7.257), the expression of real power and reactive power at bus 2 can be written as,

$$p_2 = |V_2||V_1||Y_{21}|\cos(\theta_{21} + \delta_1 - \delta_2) + |V_2|^2|Y_{22}|\cos\theta_{21} + |V_2||V_3||Y_{23}|\cos(\theta_{23} + \delta_3 - \delta_2) \tag{7.275}$$

$$p_2 = 1 \times 10V_2\cos(90° - \delta_2) + 26V_2^2\cos 90° + (1.1)(16)V_2\cos(90° + \delta_3 - \delta_2) \tag{7.276}$$

$$p_2 = 10V_2\cos(90° - \delta_2) + 17.6V_2\cos(90° + \delta_3 - \delta_2) \tag{7.277}$$

$$q_2 = -|V_2||V_1||Y_{21}|\sin(\theta_{21} + \delta_1 - \delta_2) - |V_2|^2|Y_{22}|\sin\theta_{21} - |V_2||V_3||Y_{23}|\sin(\theta_{23} + \delta_3 - \delta_2) \tag{7.278}$$

$$q_2 = -10V_2\sin(90° - \delta_2) - 26V_2^2\sin(-90°) - (1.1)(16)V_2\sin(90° + \delta_3 - \delta_2) \tag{7.279}$$

$$q_2 = -10V_2\sin(90° - \delta_2) + 26V_2^2 - 17.6V_2\sin(90° + \delta_3 - \delta_2) \tag{7.280}$$

The expression of real power at bus 3 can be written as,

$$p_3 = |V_3||V_1||Y_{31}|\cos(\theta_{31} + \delta_1 - \delta_3) + |V_3||V_2||Y_{32}|\cos(\theta_{32} + \delta_2 - \delta_3) + |V_3|^2|Y_{33}|\cos\theta_{33} \tag{7.281}$$

$$p_3 = (1.1)(1)(20)\cos(90° - \delta_3) + (1.1)(16)V_2\cos(90° + \delta_2 - \delta_3) + 36(1.2)^2\cos(-90°) \tag{7.282}$$

$$p_3 = 22\cos(90° - \delta_3) + 17.6V_2\cos(90° + \delta_2 - \delta_3) \tag{7.283}$$

The partial derivatives of Eqs. (7.277) and (7.283) with respect to related angle and voltage are calculated as,

$$\frac{\partial p_2}{\partial \delta_2} = 10V_2 \sin(90^\circ - \delta_2) + 17.6V_2 \sin(90^\circ + \delta_3 - \delta_2) \tag{7.284}$$

$$\frac{\partial p_2}{\partial \delta_3} = -17.6V_2 \sin(90^\circ + \delta_3 - \delta_2) \tag{7.285}$$

$$\frac{\partial p_2}{\partial V_2} = 10 \cos(90^\circ - \delta_2) + 17.6V_2 \cos(90^\circ + \delta_3 - \delta_2) \tag{7.286}$$

$$\frac{\partial p_3}{\partial \delta_2} = -17.6V_2 \sin(90^\circ + \delta_2 - \delta_3) \tag{7.287}$$

$$\frac{\partial p_3}{\partial \delta_3} = 22 \sin(90^\circ - \delta_3) + 17.6V_2 \sin(90^\circ + \delta_2 - \delta_3) \tag{7.288}$$

$$\frac{\partial p_3}{\partial V_2} = 17.6 \sin(90^\circ + \delta_2 - \delta_3) \tag{7.289}$$

$$\frac{\partial q_2}{\partial \delta_2} = 10V_2 \cos(90^\circ - \delta_2) + 17.6V_2 \cos(90^\circ + \delta_3 - \delta_2) \tag{7.290}$$

$$\frac{\partial q_2}{\partial \delta_3} = -17.6V_2 \cos(90^\circ + \delta_3 - \delta_2) \tag{7.291}$$

$$\frac{\partial q_2}{\partial V_2} = -10 \sin(90^\circ - \delta_2) + 52V_2 - 17.6 \sin(90^\circ + \delta_3 - \delta_2) \tag{7.292}$$

Considering the following initial guesses as,

$$\begin{bmatrix} \delta_2 \\ \delta_3 \\ V_2 \end{bmatrix}^0 = \begin{bmatrix} 0 \\ 0 \\ 1 \end{bmatrix}^0 \tag{7.293}$$

The initial updated values are calculated as,

$$p_2^0 = 10 \times 1 \cos(90^\circ) + 17.6 \times 1 \cos(90^\circ) = 0 \tag{7.294}$$

$$p_3^0 = 22 \cos(90^\circ) + 17.6 \times 1 \cos(90^\circ) = 0 \tag{7.295}$$

$$q_2^0 = -10 \times 1 \sin(90^\circ) + 26 \times 1^2 - 17.6 \times 1 \sin(90^\circ) = -1.6 \tag{7.296}$$

$$\Delta p_2^0 = P_2 - p_2^0 = -1.2 - 0 = -1.2 \tag{7.297}$$

$$\Delta p_3^0 = P_3 - p_3^0 = 3 - 0 = 3 \tag{7.298}$$

$$\Delta q_2^0 = Q_2 - q_2^0 = -0.5 - (-1.6) = 1.1 \tag{7.299}$$

The initial Jacobian matrix is calculated as,

$$J^0 = \begin{bmatrix} \frac{\partial p_2}{\partial \delta_2} & \frac{\partial p_2}{\partial \delta_3} & \frac{\partial p_2}{\partial V_2} \\ \frac{\partial p_3}{\partial \delta_2} & \frac{\partial p_3}{\partial \delta_3} & \frac{\partial p_3}{\partial V_2} \\ \frac{\partial q_2}{\partial \delta_2} & \frac{\partial q_2}{\partial \delta_3} & \frac{\partial q_2}{\partial V_2} \end{bmatrix}^0 = \begin{bmatrix} 27.6 & -17.6 & 0 \\ -17.6 & 39.6 & 0 \\ 0 & 0 & 24.4 \end{bmatrix} \tag{7.300}$$

The initial correction parameters are calculated as,

$$\begin{bmatrix} \Delta\delta_2 \\ \Delta\delta_3 \\ \Delta V_2 \end{bmatrix}^0 = \begin{bmatrix} 27.6 & -17.6 & 0 \\ -17.6 & 39.6 & 0 \\ 0 & 0 & 24.4 \end{bmatrix}^{-1} \begin{bmatrix} -1.2 \\ 3 \\ 1.1 \end{bmatrix} = \begin{bmatrix} 0.0067 \\ 0.0787 \\ 0.0450 \end{bmatrix} = \begin{bmatrix} 0.3838^\circ \\ 4.5091^\circ \\ 0.0450 \end{bmatrix} \tag{7.301}$$

The first updated values are calculated as,

$$\delta_2^1 = \delta_2^0 + \Delta\delta_2^0 = 0 + 0.3838^\circ = 0.3838^\circ \tag{7.302}$$

$$\delta_3^1 = \delta_3^0 + \Delta\delta_3^0 = 0 + 4.5091^\circ = 4.5091^\circ \tag{7.303}$$

$$V_2^1 = V_2^0 + \Delta V_2^0 = 1 + 0.0450 = 1.0450 \tag{7.304}$$

$$\begin{aligned} p_2^1 &= 10 \times 1.0450\cos(90^\circ - 0.3838^\circ) + 17.6 \\ &\quad \times 1.0450\cos(90^\circ + 4.5091^\circ - 0.3838^\circ) \\ &= -1.253 \end{aligned} \tag{7.305}$$

$$\begin{aligned} p_3^1 &= 22\cos(90^\circ - 5.5091^\circ) + 17.6 \times 1.0450\cos(90^\circ + 0.3838^\circ - 4.5091^\circ) \\ &= 3.0526 \end{aligned} \tag{7.306}$$

$$\begin{aligned} q_2^1 &= -10 \times 1.0450\sin(90^\circ - 0.3838^\circ) + 26 \times 1.0450^2 \\ &\quad - 17.6 \times 1.0450\sin(90^\circ + 4.5091^\circ - 0.3838^\circ) = -0.4014 \end{aligned} \tag{7.307}$$

The first updated mismatch parameters are calculated as,

$$\Delta p_2^1 = P_2 - p_2^1 = -1.2 - (-1.253) = 0.053 \tag{7.308}$$

$$\Delta p_3^1 = P_3 - p_3^1 = 3 - 3.0526 = -0.0526 \tag{7.309}$$

$$\Delta q_2^2 = Q_2 - q_2^2 = -0.5 - (-0.4014) = -0.0986 \tag{7.310}$$

The first Jacobian matrix is calculated as,

$$J^1 = \begin{bmatrix} \frac{\partial p_2}{\partial \delta_2} & \frac{\partial p_2}{\partial \delta_3} & \frac{\partial p_2}{\partial V_2} \\ \frac{\partial p_3}{\partial \delta_2} & \frac{\partial p_3}{\partial \delta_3} & \frac{\partial p_3}{\partial V_2} \\ \frac{\partial q_2}{\partial \delta_2} & \frac{\partial q_2}{\partial \delta_3} & \frac{\partial q_2}{\partial V_2} \end{bmatrix}^1 = \begin{bmatrix} 28.7941 & -18.3443 & -1.1991 \\ -18.3443 & 40.2762 & 1.2661 \\ -1.253 & 1.323 & 26.7858 \end{bmatrix} \tag{7.311}$$

The first correction parameters are calculated as,

$$\begin{bmatrix} \Delta\delta_2 \\ \Delta\delta_3 \\ \Delta V_2 \end{bmatrix}^1 = \begin{bmatrix} 28.7941 & -18.3443 & -1.1991 \\ -18.3443 & 40.2762 & 1.2661 \\ -1.253 & 1.323 & 26.7858 \end{bmatrix}^{-1} \begin{bmatrix} 0.053 \\ -0.0526 \\ -0.0986 \end{bmatrix} = \begin{bmatrix} 0.0013 \\ -0.0005 \\ -0.0035 \end{bmatrix} = \begin{bmatrix} 0.0744^\circ \\ -0.0286^\circ \\ -0.0035 \end{bmatrix} \tag{7.312}$$

The second updated values are calculated as,

$$\delta_2^2 = \delta_2^1 + \Delta\delta_2^1 = 0.3838^\circ + 0.0744^\circ = 0.4582^\circ \tag{7.313}$$

$$\delta_3^2 = \delta_3^1 + \Delta\delta_3^1 = 4.5091^\circ - 0.0286^\circ = 4.4805^\circ \tag{7.314}$$

$$V_2^2 = V_2^1 + \Delta V_2^1 = 1.0450 - 0.0035 = 1.0415 \tag{7.315}$$

$$\begin{aligned} p_2^2 &= 10 \times 1.0415\cos(90^\circ - 0.4582^\circ) + 17.6 \times 1.0415\cos(90^\circ + 4.4805^\circ \\ &\quad - 0.4582^\circ) \\ &= -1.2024 \end{aligned} \tag{7.316}$$

$$\begin{aligned} p_3^2 &= 22\cos(90^\circ - 4.4805^\circ) + 17.6 \times 1.0415\cos(90^\circ + 0.4582^\circ - 4.4805^\circ) \\ &= 3.0044 \end{aligned} \tag{7.317}$$

$$q_2^1 = -10 \times 1.0415 \sin(90^\circ - 0.4582^\circ) + 26 \times 1.0415^2 - 17.6 \times 1.0415 \sin(90^\circ + 4.4805^\circ - 0.4582^\circ) = -0.4971 \tag{7.318}$$

The second mismatch parameters are calculated as,

$$\Delta p_2^1 = P_2 - p_2^1 = -1.2 - (-1.2024) = 0.0024 \tag{7.319}$$

$$\Delta p_3^1 = P_3 - p_3^1 = 3 - 3.0044 = -0.0044 \tag{7.320}$$

$$\Delta q_2^2 = Q_2 - q_2^2 = -0.5 - (-0.4971) = -0.0029 \tag{7.321}$$

The second Jacobian matrix is calculated as,

$$J^2 = \begin{bmatrix} \frac{\partial p_2}{\partial \delta_2} & \frac{\partial p_2}{\partial \delta_3} & \frac{\partial p_2}{\partial V_2} \\ \frac{\partial p_3}{\partial \delta_2} & \frac{\partial p_3}{\partial \delta_3} & \frac{\partial p_3}{\partial V_2} \\ \frac{\partial q_2}{\partial \delta_2} & \frac{\partial q_2}{\partial \delta_3} & \frac{\partial q_2}{\partial V_2} \end{bmatrix}^2 = \begin{bmatrix} 28.6999 & -18.2852 & -1.1545 \\ -18.2852 & 40.218 & 1.2345 \\ -1.2024 & 1.2857 & 26.6016 \end{bmatrix} \tag{7.322}$$

The second correction parameters are calculated as,

$$\begin{bmatrix} \Delta\delta_2 \\ \Delta\delta_3 \\ \Delta V_2 \end{bmatrix}^2 = \begin{bmatrix} 28.6999 & -18.2852 & -1.1545 \\ -18.2852 & 40.218 & 1.2345 \\ -1.2024 & 1.2857 & 26.6016 \end{bmatrix}^{-1} \begin{bmatrix} 0.0024 \\ -0.0044 \\ -0.0029 \end{bmatrix} = \begin{bmatrix} 0.0009^\circ \\ -0.0056^\circ \\ -0.000103 \end{bmatrix} \tag{7.323}$$

The third updated values are calculated as,

$$\delta_2^3 = \delta_2^2 + \Delta\delta_2^2 = 0.4582^\circ + 0.0009^\circ = 0.4591^\circ \tag{7.324}$$

$$\delta_3^3 = \delta_3^2 + \Delta\delta_3^2 = 4.4805^\circ - 0.0056^\circ = 4.4749^\circ \tag{7.325}$$

$$V_2^3 = V_2^2 + \Delta V_2^2 = 1.0415 - 0.000103 = 1.0413 \tag{7.326}$$

The third Jacobian matrix is calculated as,

$$J^3 = \begin{bmatrix} \frac{\partial p_2}{\partial \delta_2} & \frac{\partial p_2}{\partial \delta_3} & \frac{\partial p_2}{\partial V_2} \\ \frac{\partial p_3}{\partial \delta_2} & \frac{\partial p_3}{\partial \delta_3} & \frac{\partial p_3}{\partial V_2} \\ \frac{\partial q_2}{\partial \delta_2} & \frac{\partial q_2}{\partial \delta_3} & \frac{\partial q_2}{\partial V_2} \end{bmatrix}^3 = \begin{bmatrix} 28.6945 & -18.2818 & -1.1524 \\ -18.2818 & 40.2148 & 1.2325 \\ -1.2 & 1.2834 & 26.5911 \end{bmatrix} \tag{7.327}$$

$$p_2^3 = 10 \times 1.0413\cos(90^\circ - 0.4591^\circ) + 17.6 \times 1.0413\cos(90^\circ + 4.4749^\circ - 0.4591^\circ) = -1.2 \tag{7.328}$$

$$p_3^3 = 22\cos(90^\circ - 4.4749^\circ) + 17.6 \times 1.0413\cos(90^\circ + 0.4591^\circ - 4.4749^\circ) = 2.999 \tag{7.329}$$

$$q_2^3 = -10 \times 1.0413\sin(90^\circ - 0.4591^\circ) + 26 \times 1.0413^2 - 17.6 \times 1.0413\sin(90^\circ + 4.4749^\circ - 0.4591^\circ) = -0.5026 \tag{7.330}$$

The third mismatch parameters are calculated as,

$$\Delta p_2^3 = P_2 - p_2^3 = -1.2 - (-1.2) = 0 \tag{7.331}$$

$$\Delta p_3^3 = P_3 - p_3^3 = 3 - 2.999 = 0.001 \tag{7.332}$$

$$\Delta q_2^3 = Q_2 - q_2^2 = -0.5 - (-0.5026) = 0.0026 \tag{7.333}$$

The third correction parameters are calculated as,

$$\begin{bmatrix} \Delta\delta_2 \\ \Delta\delta_3 \\ \Delta V_2 \end{bmatrix}^3 = \begin{bmatrix} 28.6945 & -18.2818 & -1.1524 \\ -18.2818 & 40.2148 & 1.2325 \\ -1.2 & 1.2834 & 26.5911 \end{bmatrix}^{-1} \begin{bmatrix} 0 \\ 0.001 \\ 0.0026 \end{bmatrix} = \begin{bmatrix} 0.0011^\circ \\ 0.0017^\circ \\ 0.00009 \end{bmatrix} \tag{7.334}$$

The fourth updated values are calculated as,

$$\delta_2^4 = \delta_2^3 + \Delta\delta_2^3 = 0.4591^\circ + 0.0011^\circ = 0.4602^\circ \tag{7.335}$$

$$\delta_3^4 = \delta_3^3 + \Delta\delta_3^3 = 4.4749^\circ + 0.0017^\circ = 4.4766^\circ \tag{7.336}$$

$$V_2^4 = V_2^3 + \Delta V_2^3 = 1.0413 + 0.00009 = 1.0413 \tag{7.337}$$

$$p_2^4 = 10 \times 1.0413\cos(90^\circ - 0.4602^\circ) + 17.6 \times 1.0413\cos(90^\circ + 4.4766^\circ - 0.4602^\circ) = -1.2 \tag{7.338}$$

$$p_3^4 = 22\cos(90^\circ - 4.4766^\circ) + 17.6 \times 1.0413\cos(90^\circ + 0.4602^\circ - 4.4766^\circ) \\ = 3.0007 \tag{7.339}$$

$$q_2^4 = -10 \times 1.0413\sin(90^\circ - 0.4602^\circ) + 26 \times 1.0413^2 \\ - 17.6 \times 1.0413\sin(90^\circ + 4.4766^\circ - 0.4602^\circ) = -0.5025 \tag{7.340}$$

The fourth mismatch parameters are calculated as,

$$\Delta p_2^4 = P_2 - p_2^4 = -1.2 - (-1.2) = 0 \tag{7.341}$$

$$\Delta p_3^4 = P_3 - p_3^4 = 3 - 3.0007 = -0.0007 \tag{7.342}$$

$$\Delta q_2^4 = Q_2 - q_2^4 = -0.5 - (-0.5025) = 0.0025 \tag{7.343}$$

The fourth Jacobian matrix is calculated as,

$$J^4 = \begin{bmatrix} \frac{\partial p_2}{\partial \delta_2} & \frac{\partial p_2}{\partial \delta_3} & \frac{\partial p_2}{\partial V_2} \\ \frac{\partial p_3}{\partial \delta_2} & \frac{\partial p_3}{\partial \delta_3} & \frac{\partial p_3}{\partial V_2} \\ \frac{\partial q_2}{\partial \delta_2} & \frac{\partial q_2}{\partial \delta_3} & \frac{\partial q_2}{\partial V_2} \end{bmatrix}^4 = \begin{bmatrix} 28.6945 & -18.2818 & -1.1524 \\ -18.2818 & 40.2147 & 1.2327 \\ -1.2 & 1.2836 & 26.5911 \end{bmatrix} \tag{7.344}$$

The fourth correction parameters are calculated as,

$$\begin{bmatrix} \Delta\delta_2 \\ \Delta\delta_3 \\ \Delta V_2 \end{bmatrix}^4 = \begin{bmatrix} 28.6945 & -18.2818 & -1.1524 \\ -18.2818 & 40.2147 & 1.2327 \\ -1.2 & 1.2836 & 26.5911 \end{bmatrix}^{-1} \begin{bmatrix} 0 \\ -0.0007 \\ 0.0025 \end{bmatrix} \\ = \begin{bmatrix} 0.000012 \\ -0.000026 \\ 0.000094 \end{bmatrix} = \begin{bmatrix} -0.0006^\circ \\ -0.0014^\circ \\ 0.000094 \end{bmatrix} \tag{7.345}$$

Fifth updated values are calculated as,

$$\delta_2^5 = \delta_2^4 + \Delta\delta_2^4 = 0.4602^\circ - 0.0006^\circ = 0.4596^\circ \tag{7.346}$$

$$\delta_3^5 = \delta_3^4 + \Delta\delta_3^4 = 4.4766^\circ - 0.0014^\circ = 4.4752^\circ \tag{7.347}$$

$$V_2^5 = V_2^4 + \Delta V_2^4 = 1.0413 - 0.000094 = 1.0412 \tag{7.348}$$

The value of the voltage and the angle at the bus 2 are calculated as,

$$V_2 = 1.0412 = 1.041\,\text{pu} \tag{7.349}$$

$$\delta_2 = 0.46^\circ \tag{7.350}$$

The single-line diagrams drawn using CYME and IPSA power system analysis software. Simulations are carried out after entering necessary data and the simulation results are shown in Figs. 7.18 and 7.19, respectively.

$$V_{2\text{CYME}} = 1.01\,\text{pu} \tag{7.351}$$

$$\delta_{2\text{CYME}} = 0.41^\circ \tag{7.352}$$

$$V_{2\text{IPSA}} = 1.04\,\text{pu} \tag{7.353}$$

$$\delta_{2\text{CYME}} = 0.47^\circ \tag{7.354}$$

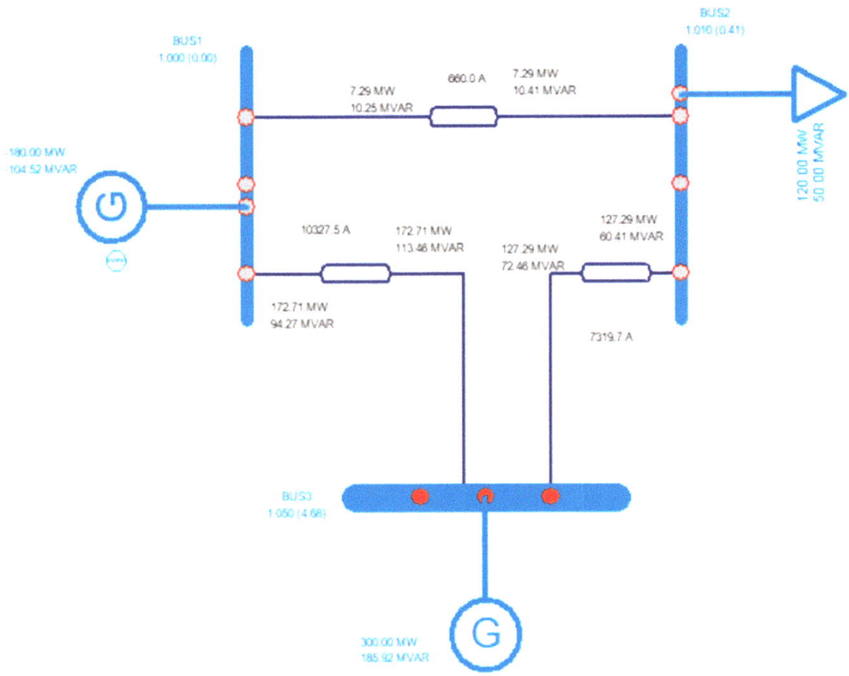

Fig. 7.18 CYME software simulation results for Example 7.8

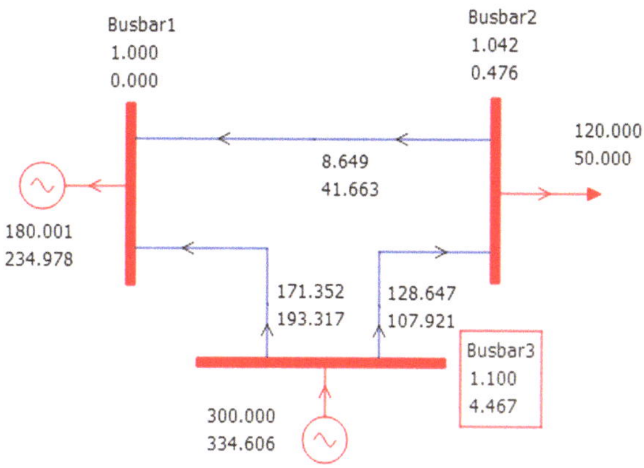

Fig. 7.19 IPSA software simulation results for Example 7.8

Practice Problem 7.8

A 20 MVA, 11 kV (line voltage), 2-pole synchronous generator with 9 MW active generation, 4 Mvar maximum reactive power, and 0.85 power factor is connected to bus 1. A load of 60 MW and 20 Mvar is connected to bus 2, and the generator is connected to the slack bus 1. The line impedances (Ω/mile) and the loads are shown in Fig. 7.20. A three-phase 67 HP, 0.42 kV, 0.91 pf induction motor is connected at the bus 2. Consider $z_{12} = 0.024 + j0.0726\,\Omega/\text{mi}$, $z_{23} = 0.036 + j0.0961\,\Omega/\text{mi}$, $z_{13} = 0.01 + j0.08\,\Omega/\text{mi}$, and the load of 30 MW and 15 Mvar. Use the Newton–Raphson method with any simulation software to calculate the voltages at the bus 2, bus 3, and slack bus real and reactive powers.

7.10 Fast Decoupled Load Flow Method

The transmission lines of a power system have a high ratio of reactance to resistance. This ratio is usually greater than 10, i.e., *X/R* > 10 for transmission lines to reduce the line loss. Whereas, this ratio is in between 2 and 3 for distribution lines to reduce the lines voltage drop. In power system, the changes in the real power is

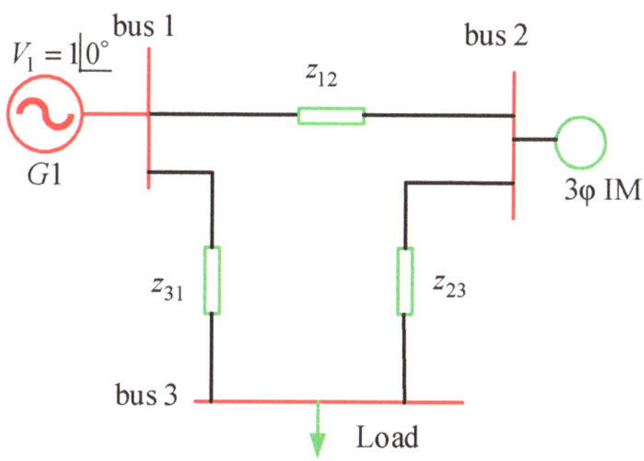

Fig. 7.20 Single-line diagram for Practice Problem 7.8

solely controlled by the changes in the phase angle but not the voltage magnitude. The changes in the reactive power are fully controlled by the changes in the voltage magnitude but not on the phase angle. In the transmission lines, the value of the resistance is very small as compared to the reactance. Therefore, the conductances are quite small as compared the suscentances ($G_{ij} < B_{ij}$). The differences in angles $((\theta_i - \theta_j) \approx 0, \theta_{ij} \approx 0)$ between the two buses are very small under normal steady-state condition and it is considered negligible. Rearrange Eq. (7.256) as [4, 5],

$$P_i = V_i V_i Y_{ii} \cos(\theta_{ii} + \delta_i - \delta_i) + \sum_{\substack{j=1 \\ j \neq i}}^{n} |V_i V_j Y_{ij}| \cos(\theta_{ij} + \delta_j - \delta_i) \tag{7.355}$$

$$P_i = V_i^2 Y_{ii} \cos \theta_{ii} + \sum_{\substack{j=1 \\ j \neq i}}^{n} |V_i V_j Y_{ij}| \cos(-(\delta_i - \delta_j - \theta_{ij})) \tag{7.356}$$

$$P_i = V_i^2 Y_{ii} \cos \theta_{ii} + \sum_{\substack{j=1 \\ j \neq i}}^{n} |V_i V_j Y_{ij}| [\cos(\delta_i - \delta_j) \cos \theta_{ij} + \sin(\delta_i - \delta_j) \sin \theta_{ij}] \tag{7.357}$$

$$P_i = V_i^2 Y_{ii} \cos \theta_{ii} + \sum_{\substack{j=1 \\ j \neq i}}^{n} |V_i V_j| [\cos(\delta_i - \delta_j) |Y_{ij}| \cos \theta_{ij} + \sin(\delta_i - \delta_j) |Y_{ij}| \sin \theta_{ij}] \tag{7.358}$$

$$P_i = V_i^2 G_{ii} + \sum_{\substack{j=1 \\ j \neq i}}^{n} |V_i V_j| [G_{ij} \cos(\delta_i - \delta_j) + B_{ij} \sin(\delta_i - \delta_j)] \tag{7.359}$$

Differentiating Eq. (7.359) with respect to V_i yields,

$$\frac{\partial P_i}{\partial V_i} = 2V_i G_{ii} + \sum_{\substack{j=1 \\ j \neq i}}^{n} |V_j| [G_{ij} \cos(\delta_i - \delta_j) + B_{ij} \sin(\delta_i - \delta_j)] \tag{7.360}$$

Differentiating Eq. (7.359) with respect to V_j yields,

$$\frac{\partial P_i}{\partial V_j} = \sum_{\substack{j=1 \\ j \neq i}}^{n} |V_i|[G_{ij}\cos(\delta_i - \delta_j) + B_{ij}\sin(\delta_i - \delta_j)] \tag{7.361}$$

Setting $G_{ij} = 0$, $\cos(\delta_i - \delta_j) \approx 1$, and $\sin(\delta_i - \delta_j) \approx 0$ in Eqs. (7.360) and (7.361) yields,

$$\frac{\partial P_i}{\partial V_i} = 0 \tag{7.362}$$

$$\frac{\partial P_i}{\partial V_j} = 0 \tag{7.363}$$

According to Eqs. (7.261), (7.362), and (7.363), the Jacobian submatrix can be expressed as,

$$J_2 = 0 \tag{7.364}$$

Again, Eq. (7.257) can be modified as,

$$Q_i = -\sum_{\substack{j=1 \\ j \neq i}}^{n} |V_i V_j Y_{ij}| \sin(-(\delta_i - \delta_j - \theta_{ij})) \tag{7.365}$$

$$Q_i = \sum_{\substack{j=1 \\ j \neq i}}^{n} |V_i V_j Y_{ij}| \sin(\delta_i - \delta_j - \theta_{ij}) \tag{7.366}$$

$$Q_i = \sum_{\substack{j=1 \\ j \neq i}}^{n} |V_i V_j Y_{ij}| [\sin(\delta_i - \delta_j)\cos\theta_{ij} - \cos(\delta_i - \delta_j)\sin\theta_{ij}] \tag{7.367}$$

$$Q_i = \sum_{\substack{j=1 \\ j \neq i}}^{n} |V_i V_j| [G_{ij}\sin(\delta_i - \delta_j) - B_{ij}\cos(\delta_i - \delta_j)] \tag{7.368}$$

Differentiating Eq. (7.368) with respect to δ_i and δ_j yields,

$$\frac{\partial Q_i}{\partial \delta_i} = \sum_{\substack{j=1 \\ j \neq i}}^{n} |V_i V_j| [G_{ij} \cos(\delta_i - \delta_j) + B_{ij} \sin(\delta_i - \delta_j)] \tag{7.369}$$

$$\frac{\partial Q_i}{\partial \delta_j} = \sum_{\substack{j=1 \\ j \neq i}}^{n} |V_i V_j| [-G_{ij} \cos(\delta_i - \delta_j) - B_{ij} \sin(\delta_i - \delta_j)] \tag{7.370}$$

Setting $G_{ij} = 0$, $\cos(\delta_i - \delta_j) \approx 1$, and $\sin(\delta_i - \delta_j) \approx 0$ in Eqs. (7.369) and (7.370) yields,

$$\frac{\partial Q_i}{\partial \delta_i} = 0 \tag{7.371}$$

$$\frac{\partial Q_i}{\partial \delta_j} = 0 \tag{7.372}$$

According to Eqs. (7.262), (7.371), and (7.372), the Jacobian submatrix can be expressed as,

$$J_3 = 0 \tag{7.373}$$

Substituting Eqs. (7.364) and (7.373) into Eq. (7.259) yields,

$$\begin{bmatrix} \Delta P \\ \Delta Q \end{bmatrix} = \begin{bmatrix} J_1 & 0 \\ 0 & J_4 \end{bmatrix} \begin{bmatrix} \Delta\delta \\ \Delta V \end{bmatrix} = \begin{bmatrix} J_{11} & 0 \\ 0 & J_{22} \end{bmatrix} \begin{bmatrix} \Delta\delta \\ \Delta V \end{bmatrix} \tag{7.374}$$

Equation (7.374) can be rearranged as,

$$\Delta P = J_1 \Delta\delta = J_{11} \Delta\delta = \frac{\partial P}{\partial \delta} \Delta\delta \tag{7.375}$$

$$\Delta\delta = \left[\frac{\partial P}{\partial \delta}\right]^{-1} \Delta P \tag{7.376}$$

$$\Delta Q = J_4 \Delta V = J_{22} \Delta V = \frac{\partial Q}{\partial V} \Delta V \tag{7.377}$$

$$\Delta V = \left[\frac{\partial Q}{\partial V}\right]^{-1} \Delta Q \tag{7.378}$$

Again, differentiating Eq. (7.256) with respect to δ_i and δ_j yields,

$$\frac{\partial P_i}{\partial \delta_i} = \sum_{\substack{j=1 \\ j \neq i}}^{n} \left|V_i V_j Y_{ij}\right| \sin(\theta_{ij} + \delta_j - \delta_i) \tag{7.379}$$

$$\frac{\partial P_i}{\partial \delta_j} = -\sum_{\substack{j=1 \\ j \neq i}}^{n} \left|V_i V_j Y_{ij}\right| \sin(\theta_{ij} + \delta_j - \delta_i) \qquad j \neq i \tag{7.380}$$

Equation (7.379) can be reduced as,

$$\frac{\partial P_i}{\partial \delta_i} = \sum_{\substack{j=1 \\ j \neq i}}^{n} \left|V_i V_j Y_{ij}\right| \sin(\theta_{ij} - (\delta_i - \delta_j)) \tag{7.381}$$

$$\frac{\partial P_i}{\partial \delta_i} = \sum_{\substack{j=1 \\ j \neq i}}^{n} \left|V_i V_j Y_{ij}\right| \sin(\theta_{ij} - (\delta_i - \delta_j)) - |V_i V_i Y_{ii}| \sin(\theta_{ij} - (\delta_i - \delta_i)) \tag{7.382}$$

$$\frac{\partial P_i}{\partial \delta_i} = \sum_{\substack{j=1 \\ j \neq i}}^{n} \left|V_i V_j Y_{ij}\right| \sin(\theta_{ij} - \delta_i + \delta_j) - \left|V_i^2\right| |Y_{ii}| \sin\theta_{ij} \tag{7.383}$$

$$\frac{\partial P_i}{\partial \delta_i} = \sum_{\substack{j=1 \\ j \neq i}}^{n} \left|V_i V_j Y_{ij}\right| \sin(\theta_{ij} - \delta_i + \delta_j) - \left|V_i^2\right| B_{ij} \tag{7.384}$$

Replacing the first term of Eq. (7.384) by Q_i yields,

$$\frac{\partial P_i}{\partial \delta_i} = Q_i - \left|V_i^2\right| B_{ij} \tag{7.385}$$

As $Q_i \ll \left|V_i^2\right| B_{ij}$ and assumed $\left|V_i^2\right| = |V_i|$ so that Eq. (7.385) is again simplified as,

$$\frac{\partial P_i}{\partial \delta_i} = -|V_i| B_{ij} \tag{7.386}$$

Equation (7.380) is again modified as,

$$\frac{\partial P_i}{\partial \delta_j} = -\left|V_i V_j Y_{ij}\right| \sin(\theta_{ij} + \delta_j - \delta_i) \qquad j \neq i \tag{7.387}$$

$$\frac{\partial P_i}{\partial \delta_j} = -\left|V_i V_j Y_{ij}\right| \left[\sin\theta_{ij}\cos(\delta_j - \delta_i) + \cos\theta_{ij}\sin(\delta_j - \delta_i)\right] \quad j \neq i \tag{7.388}$$

$$\frac{\partial P_i}{\partial \delta_j} = -\left|V_i V_j\right| \left[B_{ij}\cos(\delta_j - \delta_i) + G_{ij}\sin(\delta_j - \delta_i)\right] \quad j \neq i \tag{7.389}$$

Setting $G_{ij} = 0$, $\cos(\delta_i - \delta_j) \approx 1$, $\sin(\delta_i - \delta_j) \approx 0$, and $\left|V_j\right| \approx 1$ in Eq. (7.389) yields,

$$\frac{\partial P_i}{\partial \delta_j} = -|V_i| B_{ij} \quad j \neq i \tag{7.390}$$

Equation (7.257) can be written as,

$$Q_i = -\left|V_i V_j Y_{ii}\right| \sin(\theta_{ii} + \delta_i - \delta_i) - \sum_{\substack{j=1 \\ j \neq i}}^{n} \left|V_i V_j Y_{ij}\right| \sin(\theta_{ij} + \delta_j - \delta_i) \qquad j = i \tag{7.391}$$

Differentiating Eq. (7.257) with respect to V_i yields,

$$\frac{\partial Q_i}{\partial V_i} = -2|V_i||Y_{ii}| \sin\theta_{ii} - \sum_{\substack{j=1 \\ j \neq i}}^{n} \left|V_j Y_{ij}\right| \sin(\theta_{ij} + \delta_j - \delta_i) \qquad j = i \tag{7.392}$$

$$\frac{\partial Q_i}{\partial V_i} V_i = -2\left|V_i^2\right| B_{ii} - \sum_{\substack{j=1 \\ j \neq i}}^{n} \left|V_i V_j Y_{ij}\right| \sin(\theta_{ij} + \delta_j - \delta_i) \qquad j = i \tag{7.393}$$

$$\frac{\partial Q_i}{\partial V_i} V_i = -2\left|V_i^2\right| B_{ii} + Q_i \qquad j = i \tag{7.394}$$

$$\frac{\partial Q_i}{\partial V_i} V_i = -2\left|V_i^2\right| B_{ii} \qquad j = i \tag{7.395}$$

$$\frac{\partial Q_i}{\partial V_i} = -2|V_i| B_{ii} \qquad j = i \tag{7.396}$$

Again, assuming $2V_i = V_i$ and Eq. (7.396) is modified as,

$$\frac{\partial Q_i}{\partial V_i} = -|V_i|B_{ii} \tag{7.397}$$

Again, differentiating Eq. (7.257) with respect to V_j yields,

$$\frac{\partial Q_i}{\partial V_j} = -\left|V_i Y_{ij}\right| \sin(\theta_{ij} + \delta_j - \delta_i) \qquad j \neq i \tag{7.398}$$

$$\frac{\partial Q_i}{\partial V_j} = -\left|V_i Y_{ij}\right| \left[\sin\theta_{ij}\cos(\delta_j - \delta_i) + \cos\theta_{ij}\sin(\delta_j - \delta_i)\right] \qquad j \neq i \tag{7.399}$$

$$\frac{\partial Q_i}{\partial V_j} = -|V_i| \left[B_{ij}\cos(\delta_j - \delta_i) + G_{ij}\sin(\delta_j - \delta_i)\right] \qquad j \neq i \tag{7.400}$$

Setting $G_{ij} = 0$, $\cos(\delta_i - \delta_j) \approx 1$, and $\sin(\delta_i - \delta_j) \approx 0$ in Eq. (7.400) yields,

$$\frac{\partial Q_i}{\partial V_j} = -|V_i|B_{ij} \qquad j \neq i \tag{7.401}$$

From Eqs. (7.386), (7.390), (7.397), and (7.401), it is seen that values of $\frac{\partial P}{\partial \delta}$ and $\frac{\partial Q}{\partial V}$ are negatives. Hence, Eqs. (7.376) and (7.378) can be modified as,

$$\Delta\delta = -\left[\frac{\partial P}{\partial \delta}\right]^{-1} \Delta P \tag{7.402}$$

$$\Delta V = -\left[\frac{\partial Q}{\partial V}\right]^{-1} \Delta Q \tag{7.403}$$

Substituting Eq. (7.390) into Eq. (7.402) yields,

$$\Delta\delta = [B]^{-1}\frac{\Delta P}{V} \tag{7.404}$$

Again, substituting Eq. (7.397) into Eq. (7.403) yields,

$$\Delta V = [B']^{-1}\frac{\Delta Q}{V} \tag{7.405}$$

where B is the susceptance or imaginary part of the $Y_{\text{bus}} = G + jB$. The elements of B and $B'$are the negative parts of the bus admittance matrix. For three-bus system, the susceptance matrix B can be expressed as,

$$B = \begin{bmatrix} B_{22} & B_{23} \\ B_{32} & B_{33} \end{bmatrix} \tag{7.406}$$

If bus 3 is voltage regulated bus, then the value of B' is,

$$B' = [B_{22}] \tag{7.407}$$

For the three-bus system, Eqs. (7.404) and (7.405) can be extended as,

$$\begin{bmatrix} \Delta\delta_2 \\ \Delta\delta_3 \end{bmatrix} = [B]^{-1} \begin{bmatrix} \frac{\Delta P_2}{V_2} \\ \frac{\Delta P_3}{V_3} \end{bmatrix} \tag{7.408}$$

$$\begin{bmatrix} \Delta V_2 \\ \Delta V_3 \end{bmatrix} = [B']^{-1} \begin{bmatrix} \frac{\Delta Q_2}{V_2} \\ \frac{\Delta Q_3}{V_3} \end{bmatrix} \tag{7.409}$$

Example 7.9
Determine the voltage at bus 2 of the single-line diagram as shown in Fig. 7.17 by Fast Decoupled Load Flow method.

Solution
From Eqs. (7.269)–(7.274), the susceptance matrix B can be expressed as,

$$B = \begin{bmatrix} B_{22} & B_{23} \\ B_{32} & B_{33} \end{bmatrix} = \begin{bmatrix} -26 & 16 \\ 16 & -36 \end{bmatrix} \tag{7.410}$$

Equation (7.408) can be rearranged as,

$$\begin{bmatrix} \Delta\delta_2 \\ \Delta\delta_3 \end{bmatrix} = -\begin{bmatrix} -26 & 16 \\ 16 & -36 \end{bmatrix}^{-1} \begin{bmatrix} \frac{\Delta P_2}{V_2} \\ \frac{\Delta P_3}{1.1} \end{bmatrix} = -\begin{bmatrix} -0.052 & -0.023 \\ -0.023 & -0.038 \end{bmatrix} \begin{bmatrix} \frac{\Delta P_2}{V_2} \\ \frac{\Delta P_3}{1.1} \end{bmatrix} \tag{7.411}$$

The expressions of the angles are written as,

$$\Delta\delta_2 = 0.052\frac{\Delta P_2}{V_2} + 0.023\frac{\Delta P_3}{1.1} = 0.052\frac{\Delta P_2}{V_2} + 0.0209\Delta P_3 \tag{7.412}$$

$$\Delta\delta_3 = 0.023\frac{\Delta P_2}{V_2} + 0.038\frac{\Delta P_3}{1.1} = 0.023\frac{\Delta P_2}{V_2} + 0.0345\Delta P_3 \tag{7.413}$$

The bus 3 is a voltage-controlled bus. In this case, only the bus 2 is considered and other buses are removed. Equation (7.409) can be modified as,

$$\Delta V_2 = -[B']^{-1}\frac{\Delta Q_2}{V_2} = -[B_{22}]^{-1}\frac{\Delta Q_2}{V_2} = \frac{1}{26}\frac{\Delta q_2}{V_2} = 0.0384\frac{\Delta q_2}{V_2} \tag{7.414}$$

Let the initial guesses are $\delta_2^0 = 0$, $\delta_3^0 = 0$, and $V_2^0 = 1$. Then, the initial updated values are calculated as,

$$p_2^0 = 10 \times 1\cos(90^\circ) + 17.6 \times 1\cos(90^\circ) = 0 \tag{7.415}$$

$$p_3^0 = 22\cos(90^\circ) + 17.6 \times 1\cos(90^\circ) = 0 \tag{7.416}$$

$$q_2^0 = -10 \times 1\sin(90^\circ) + 26 \times 1^2 - 17.6 \times 1\sin(90^\circ) = -1.6 \tag{7.417}$$

$$\Delta p_2^0 = P_2 - p_2^0 = -1.2 - 0 = -1.2 \tag{7.418}$$

$$\Delta p_3^0 = P_3 - p_3^0 = 3 - 0 = 3 \tag{7.419}$$

$$\Delta q_2^0 = Q_2 - q_2^0 = -0.5 - (-1.6) = 1.1 \tag{7.420}$$

$$\Delta\delta_2^0 = 0.052\frac{\Delta P_2^0}{V_2^0} + 0.0209\Delta P_3^0 = 0.052\frac{(-1.2)}{1} + 0.0209(3) = 0.0003 = 0.0171^\circ \tag{7.421}$$

$$\Delta\delta_3^0 = 0.023\frac{\Delta P_2^0}{V_2^0} + 0.0345\Delta P_3^0 = 0.023\frac{(-1.2)}{1} + 0.0345(3) = 0.0759 = 4.3487^\circ \tag{7.422}$$

$$\Delta V_2^0 = 0.0384\frac{\Delta q_2^0}{V_2^0} = 0.0384\frac{(1.1)}{1} = 0.0423 \tag{7.423}$$

The first updated values are calculated as,

$$\delta_2^1 = \delta_2^0 + \Delta\delta_2^0 = 0 + 0.0171^\circ = 0.0171^\circ \tag{7.424}$$

$$\delta_3^1 = \delta_3^0 + \Delta\delta_3^0 = 0 + 4.3487^\circ = 4.3487^\circ \tag{7.425}$$

$$V_2^1 = V_2^0 + \Delta V_2^0 = 1 + 0.0423 = 1.0423 \tag{7.426}$$

$$\begin{aligned} p_2^1 &= 10 \times 1.0423\cos(90^\circ - 0.0171^\circ) + 17.6 \times 1.0423\cos(90^\circ + 4.3487^\circ \\ &\quad - 0.0171^\circ) \\ &= -1.3824 \end{aligned} \tag{7.427}$$

$$p_3^1 = 22\cos(90^\circ - 4.3487^\circ) + 17.6 \times 1.0423\cos(90^\circ + 0.0171^\circ - 4.3487^\circ) \\ = 3.0537 \tag{7.428}$$

$$q_2^1 = -10 \times 1.0423\sin(90^\circ - 0.0171^\circ) + 26 \times 1.0423^2 - 17.6 \\ \times 1.0423\sin(90^\circ + 4.3487^\circ - 0.0171^\circ) \tag{7.429}$$

$$q_2^1 = -0.4689 \tag{7.430}$$

$$\Delta p_2^1 = P_2 - p_2^1 = -1.2 - (-1.3824) = 0.1824 \tag{7.431}$$

$$\Delta p_3^1 = P_3 - p_3^1 = 3 - 3.0537 = -0.0537 \tag{7.432}$$

$$\Delta q_2^2 = Q_2 - q_2^2 = -0.5 - (-0.4689) = -0.0311 \tag{7.433}$$

$$\Delta\delta_2^1 = 0.052\frac{\Delta P_2^1}{V_2^1} + 0.0209\Delta P_3^1 = 0.052\frac{(0.1824)}{1.0423} + 0.0209(-0.0537) = 0.0079 \\ = 0.4526^\circ \tag{7.434}$$

$$\Delta\delta_3^1 = 0.023\frac{\Delta P_2^1}{V_2^1} + 0.0345\Delta P_3^1 = 0.023\frac{(0.1824)}{1.0423} + 0.0345(-0.0537) = 0.0021 \\ = 0.1203^\circ \tag{7.435}$$

$$\Delta V_2^1 = 0.0384\frac{\Delta q_2^1}{V_2^1} = 0.0384\frac{(-0.0311)}{1.0423} = -0.0011 \tag{7.436}$$

The second updated values are calculated as,

$$\delta_2^2 = \delta_2^1 + \Delta\delta_2^1 = 0.0171^\circ + 0.4526^\circ = 0.4697^\circ \tag{7.437}$$

$$\delta_3^2 = \delta_3^1 + \Delta\delta_3^1 = 4.3487^\circ + 0.1203^\circ = 4.469^\circ \tag{7.438}$$

$$V_2^2 = V_2^1 + \Delta V_2^1 = 1.0423 - 0.0011 = 1.0412 \tag{7.439}$$

$$p_2^2 = 10 \times 1.0412\cos(90^\circ - 0.4697^\circ) \\ + 17.6 \times 1.0412\cos(90^\circ + 4.469^\circ - 0.4697^\circ) = -1.1927 \tag{7.440}$$

$$p_3^2 = 22\cos(90^\circ - 4.469^\circ) + 17.6 \times 1.0412\cos(90^\circ + 0.4697^\circ - 4.469^\circ)$$
$$= 2.9923 \quad (7.441)$$

$$q_2^2 = -10 \times 1.0412\sin(90^\circ - 0.4697^\circ) + 26 \times 1.0412^2 - 17.6 \times 1.0412\sin(90^\circ + 4.469^\circ - 0.4697^\circ) \quad (7.442)$$

$$q_2^2 = -0.5056 \quad (7.443)$$

$$\Delta p_2^2 = P_2 - p_2^2 = -1.2 - (-1.1927) = -0.0073 \quad (7.444)$$

$$\Delta p_3^2 = P_3 - p_3^2 = 3 - 2.9913 = 0.0077 \quad (7.445)$$

$$\Delta q_2^2 = Q_2 - q_2^2 = -0.5 - (-0.5056) = 0.0056 \quad (7.446)$$

$$\Delta\delta_2^2 = 0.052\frac{\Delta P_2^2}{V_2^2} + 0.0209\Delta P_3^2 = 0.052\frac{(-0.0073)}{1.0412} + 0.0209(0.0077)$$
$$= -0.0002 = -0.0114^\circ \quad (7.447)$$

$$\Delta\delta_3^2 = 0.023\frac{\Delta P_2^2}{V_2^2} + 0.0345\Delta P_3^2 = 0.023\frac{(-0.0073)}{1.0412} + 0.0345(0.0077) = 0.0001$$
$$= 0.0057^\circ \quad (7.448)$$

$$\Delta V_2^2 = 0.0384\frac{\Delta q_2^2}{V_2^2} = 0.0384\frac{(0.0056)}{1.0412} = 0.0002 \quad (7.449)$$

The third updated values are calculated as,

$$\delta_2^3 = \delta_2^2 + \Delta\delta_2^2 = 0.4697^\circ - 0.0114^\circ = 0.4583^\circ \quad (7.450)$$

$$\delta_3^3 = \delta_3^2 + \Delta\delta_3^2 = 4.469^\circ - 0.0057^\circ = 4.4633^\circ \quad (7.451)$$

$$V_2^3 = V_2^2 + \Delta V_2^2 = 1.0412 - 0.0002 = 1.041 \quad (7.452)$$

$$p_2^3 = 10 \times 1.041\cos(90^\circ - 0.4583^\circ) + 17.6 \times 1.041\cos(90^\circ + 4.4633^\circ - 0.4583^\circ)$$
$$= -1.1963 \quad (7.453)$$

$$p_3^3 = 22\cos(90^\circ - 4.4633^\circ) + 17.6 \times 1.041\cos(90^\circ + 0.4583^\circ - 4.4633^\circ)$$
$$= 2.9916 \tag{7.454}$$

$$q_2^3 = -10 \times 1.041\sin(90^\circ - 0.4583^\circ) + 26 \times 1.041^2 - 17.6 \times 1.041\sin(90^\circ + 4.4633^\circ - 0.4583^\circ) \tag{7.455}$$

$$q_2^3 = -0.5108 \tag{7.456}$$

$$\Delta p_2^3 = P_2 - p_2^3 = -1.2 - (-1.1963) = -0.0037 \tag{7.457}$$

$$\Delta p_3^3 = P_3 - p_3^3 = 3 - 2.9916 = 0.0084 \tag{7.458}$$

$$\Delta q_2^3 = Q_2 - q_2^3 = -0.5 - (-0.5108) = 0.0108 \tag{7.459}$$

$$\Delta\delta_2^3 = 0.052\frac{\Delta P_2^3}{V_2^3} + 0.0209\Delta P_3^3 = 0.052\frac{(-0.0037)}{1.041} + 0.0209(0.0084)$$
$$= -0.000009 = -0.00005^\circ \tag{7.460}$$

$$\Delta\delta_3^3 = 0.023\frac{\Delta P_2^3}{V_2^3} + 0.0345\Delta P_3^3 = 0.023\frac{(-0.0037)}{1.041} + 0.0345(0.0084) = 0.0002$$
$$= 0.0114^\circ \tag{7.461}$$

$$\Delta V_2^3 = 0.0384\frac{\Delta q_2^3}{V_2^3} = 0.0384\frac{(0.0108)}{1.041} = 0.0003 \tag{7.462}$$

The fourth updated values are calculated as,

$$\delta_2^4 = \delta_2^3 + \Delta\delta_2^3 = 0.4583^\circ - 0.00005^\circ = 0.4582^\circ \tag{7.463}$$

$$\delta_3^4 = \delta_3^3 + \Delta\delta_3^3 = 4.4633^\circ - 0.0114^\circ = 4.4519^\circ \tag{7.464}$$

$$V_2^4 = V_2^3 + \Delta V_2^3 = 1.041 - 0.0003 = 1.0407 \tag{7.465}$$

$$p_2^4 = 10 \times 1.0407\cos(90^\circ - 0.4582^\circ) + 17.6 \times 1.0407\cos(90^\circ + 4.4519^\circ - 0.4582^\circ) = -1.1924 \tag{7.466}$$

$$p_3^4 = 22\cos(90^\circ - 4.4519^\circ) + 17.6 \times 1.0407\cos(90^\circ + 0.4582^\circ - 4.4519^\circ)$$
$$= 2.9833 \tag{7.467}$$

$$q_2^4 = -10 \times 1.0407 \sin(90^\circ - 0.4582^\circ) + 26 \times 1.0407^2 - 17.6 \times 1.0407 \sin(90^\circ + 4.4519^\circ - 0.4582^\circ) = -0.5190 \quad (7.468)$$

$$\Delta p_2^4 = P_2 - p_2^4 = -1.2 - (-1.1924) = -0.0076 \quad (7.469)$$

$$\Delta p_3^4 = P_3 - p_3^4 = 3 - 2.9833 = 0.0167 \quad (7.470)$$

$$\Delta q_2^4 = Q_2 - q_2^4 = -0.5 - (-0.5190) = 0.019 \quad (7.471)$$

$$\Delta\delta_2^4 = 0.052\frac{\Delta P_2^4}{V_2^4} + 0.0209\Delta P_3^4 = 0.052\frac{(-0.0076)}{1.0407} + 0.0209(0.0167) = -0.00003 = -0.0017^\circ \quad (7.472)$$

$$\Delta\delta_3^4 = 0.023\frac{\Delta P_2^4}{V_2^4} + 0.0345\Delta P_3^4 = 0.023\frac{(-0.0076)}{1.0407} + 0.0345(0.0167) = 0.0004 = 0.0229^\circ \quad (7.473)$$

$$\Delta V_2^4 = 0.0384\frac{\Delta q_2^4}{V_2^4} = 0.0384\frac{(0.019)}{1.0407} = 0.0007 \quad (7.474)$$

The fifth updated values are calculated as,

$$\delta_2^5 = \delta_2^4 + \Delta\delta_2^4 = 0.4582^\circ - 0.0017^\circ = 0.4565^\circ \quad (7.475)$$

$$\delta_3^5 = \delta_3^4 + \Delta\delta_3^4 = 4.4519^\circ - 0.0229^\circ = 4.429^\circ \quad (7.476)$$

$$V_2^5 = V_2^4 + \Delta V_2^4 = 1.0407 - 0.0007 = 1.04 \quad (7.477)$$

Therefore, the final solutions are as,

$$\delta_2 = 0.4565^\circ \quad (7.478)$$

$$\delta_3 = 4.429^\circ \quad (7.479)$$

$$V_2 = 1.04 \quad (7.480)$$

Practice Problem 7.9

A 30 MVA, 11 kV (line voltage), 2-pole synchronous generator with 9 MW active generation, 4 Mvar maximum reactive power, and 0.85 power factor is connected to bus 1. The length of the line is 1mile, and the per unit impedance of the lines are given, as shown in Fig. 7.21. The load 1 of 70 MW and 25 Mvar is connected to bus 2, and the load 2 of 45 MW and 25 Mvar is connected to bus 3. Consider bus 1

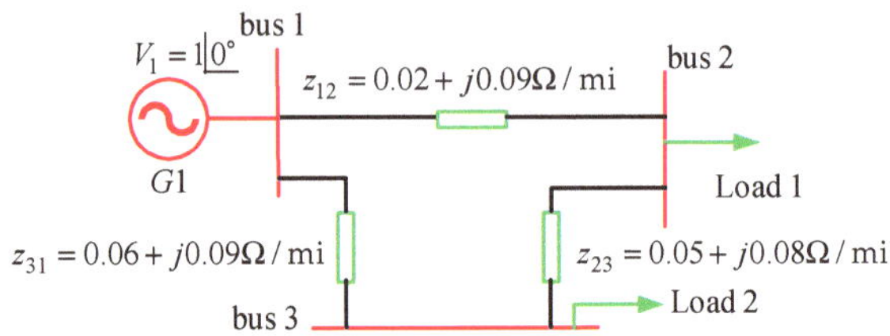

Fig. 7.21 Single-line diagram for Example 7.9

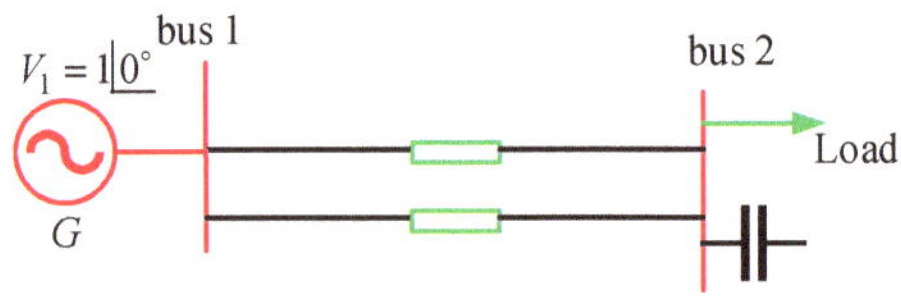

Fig. 7.22 Single-line diagram for Example 7.10

is a slack bus and use the Fast Decoupled Load Flow method to calculate the voltage at bus 2, bus 3, and slack bus real power and reactive power.

Example 7.10
A 15 MVA, 11 kV (line voltage), 2-pole synchronous generator with 10 MW active generation, 5 Mvar maximum reactive power, and 0.85 power factor is connected to bus 1. A load of 50 MW and 25 Mvar, and a three-phase static Var compensator of 25 Mvar is connected to bus 2, as shown in Fig. 7.22. Consider bus 1 is a slack bus and use the Fast Decoupled Load Flow method to calculate the voltage at bus 2, and slack bus real power and reactive power. Consider base values 11 kV and 100 MVA. The line impedances are given as,

Line 1 $z_1 = z_0 = 0.084 + j0.976\ \Omega/\text{mi}$
Line 2 $z_1 = z_0 = 0.0605 + j0.8687\ \Omega/\text{mi}$,

Solution
After drawing the network with CYME power system software, relevant data are given as input. Then, simulate the network as shown in Fig. 7.23.

The voltage and angle at bus 2 are,

$$V_{2\text{CYME}} = 0.965\,\text{pu} \tag{7.481}$$

$$\delta_{2\text{CYME}} = -11.35^\circ \tag{7.482}$$

Slack bus real and reactive powers are,

$$P_{\text{slack}} = 50.79\,\text{MW} \tag{7.483}$$

$$Q_{\text{slack}} = 10.19\,\text{Mvar} \tag{7.484}$$

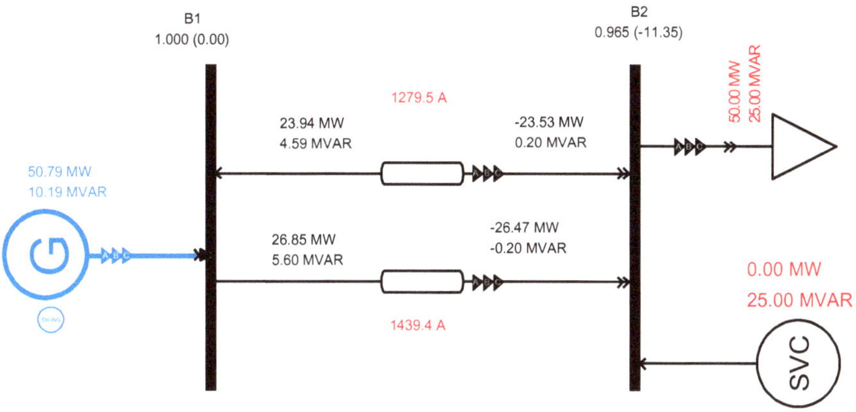

Fig. 7.23 CYME simulation results for Example 7.10

References

1. Duncan JG, Overbye T, Sarma M (2017) Power system analysis and design, 6th edn, Cengage Learning, USA, pp 1–942
2. Wildi T (2014) Electrical machines, drives and power systems, 6th edn. Pearson Education Ltd, USA, pp 1–920
3. Nagsarkar TK, Sukhija MS (2014) Power system analysis. Second Edition, Oxford University Press, pp 1–726
4. Sadat H (2010) Power system analysis, 3rd Edition, PSA Publisher, USA, pp 1–772
5. Wildi T (2006) Electrical machines, drives and power systems, 6th edn, Pearson Education, USA, pp 1–934

Exercise Problems

7.1 A short transmission line has the sending end and receiving end voltages 80 V and 60 V, respectively. Calculate the line average reactive power if the line reactance is $6\,\Omega$.

7.2 A short transmission line has the sending end voltage, receiving end voltages and average reactive power 80 V, 60 V, 80 Var, respectively. Determine the value of the reactance.

7.3 The sending end voltage and average reactive power of a short transmission line are 80 V, and 100 Var. If the line reactance is $10\,\Omega$, calculate the receiving end voltage.

7.4 The per unit impedances of a circuit are shown in Fig. P7.1. Determine the Y_{bus} matrix.

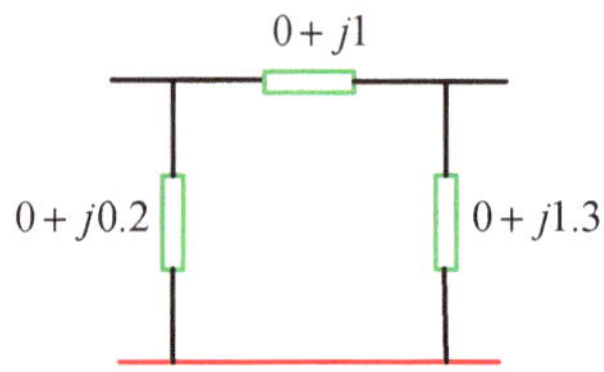

Fig. P7.1 Circuit for Problem 7.4

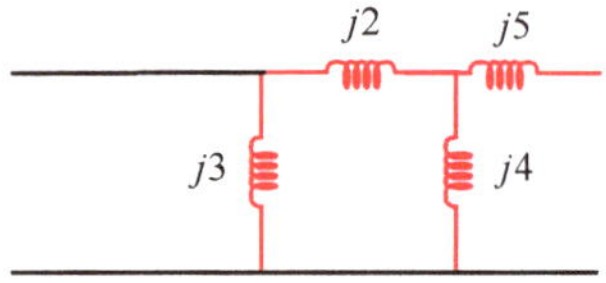

Fig. P7.2 Circuit for Problem 7.5

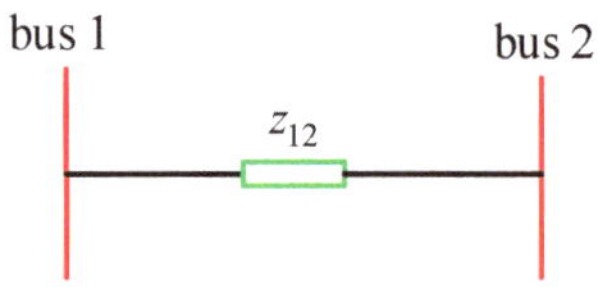

Fig. P7.3 Single-line diagram for Problem 7.6

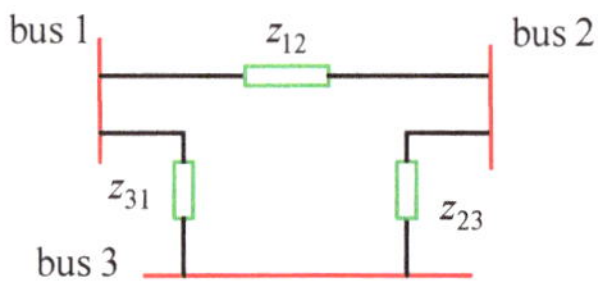

Fig. P7.4 Single-line diagram for Problem 7.7

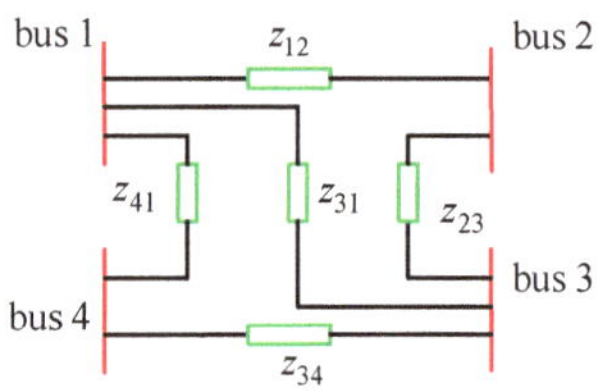

Fig. P7.5 Single-line diagram for Problem 7.8

7.5 Fig. P7.2 shows a circuit where the admittances are in per unit admittances. Calculate the Y_{bus} matrix.

7.6 A two-bus single-line diagram is shown in Fig. P7.3. The per unit impedance is given by $z_{12} = 0.04 + j0.08$. Calculate the Y_{bus} matrix.

7.7 The per unit impedances are given by $z_{12} = 0.04 + j0.08$, $z_{13} = 0.05 + j0.10$, $z_{31} = 0.06 + j0.12$. Find the Y_{bus} matrix of the single-line diagram as shown in Fig. P7.4.

7.8 The per unit impedances are given by $z_{12} = 0.04 + j0.08$, $z_{23} = 0.05 + j0.10$, $z_{34} = 0.06 + j0.12$, $z_{13} = 0.07 + j0.15$, $z_{41} = 0.08 + j0.14$. Calculate the Y_{bus} matrix of the single-line diagram as shown in Fig. P7.5.

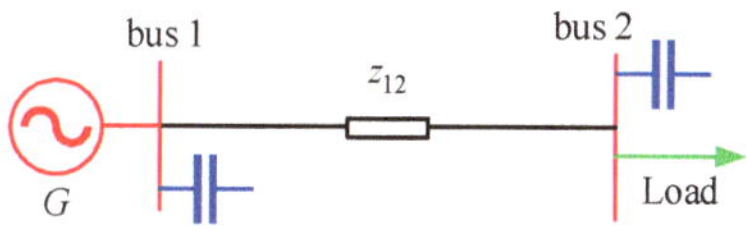

Fig. P7.6 Single-line diagram for Problem 7.11

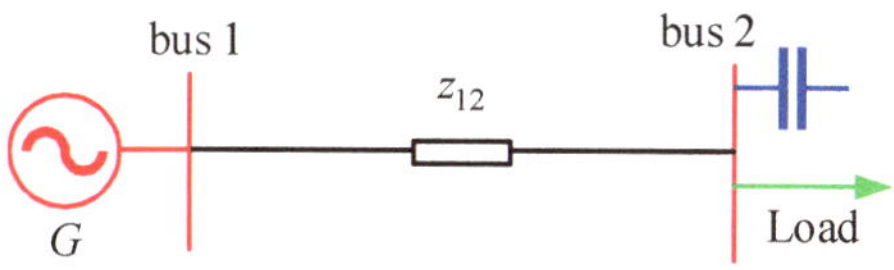

Fig. P7.7 Single-line diagram for Problem 7.12

7.9 Two simultaneous equations of voltages are given $5V_1 + 2V_2 = 11$ and $2V_1 + 6V_2 = 15$. Consider the initial guesses $V_1(0) = 1$ and $V_2(0) = 1$. Calculate the voltages by using the Gauss–Seidel method.

7.10 A voltage equation is given by $V^2 - 5V + 2 = 0$. Use the Gauss–Seidel method to solve the equation by considering the initial guess $V(0) = 1$.

7.11 A load of 125 MW and 55 Mvar is connected to bus 2, and a generator of 15 MVA, 11 kV, 0.85 pf, 2-pole with 8 MW active generation, 4 Mvar reactive power is connected to bus 1 as shown in Fig. P7.6. A transmission line is connected between the two buses whose impedance is $0.04 + j0.08$ pu on a common 100 MVA base and 11 kV base. The transmission line is charging by 10 Mvar shunt capacitor at each end.
Use the Gauss–Seidel method to calculate the voltage at bus 2, and the slack bus real and reactive powers.

7.12 A load of 125 MW and 50 Mvar is connected to the bus 2, and a generator of 20MVA, 11 kV, 0.85 pf, 2-pole with 10 MW active generation, 5 Mvar reactive power is connected to the slack bus 1 as shown in Fig. P7.7. The transmission line impedance is $0.03 + j0.08$ pu on a common 100 MVA and 11 kV base values. A shunt capacitor of 20 Mvar is connected to the bus 2. Use the Gauss–Seidel method to calculate the voltage at bus 2, the slack bus real and reactive powers, line loss, and power supplied by the capacitor.

7.13 Fig. P7.8 shows a three-bus single-line diagram where the impedances are in per unit on a common 100 MVA and 11 kV base values. The first load of 100 MW and 50 Mvar is connected to bus 2, and the second load 150 MW and 60Mvar is connected to bus 3. A generator of 25 MVA, 11 kV, 0.85 pf, 2-pole with 15 MW active generation, 10 Mvar reactive power is connected to the slack bus 1, as shown in Fig. P7.8. Calculate the voltage at bus 2, and bus 3, and slack bus real and reactive powers using the Gauss–Seidel method.

Fig. P7.8 Single-line diagram for Problem 7.13

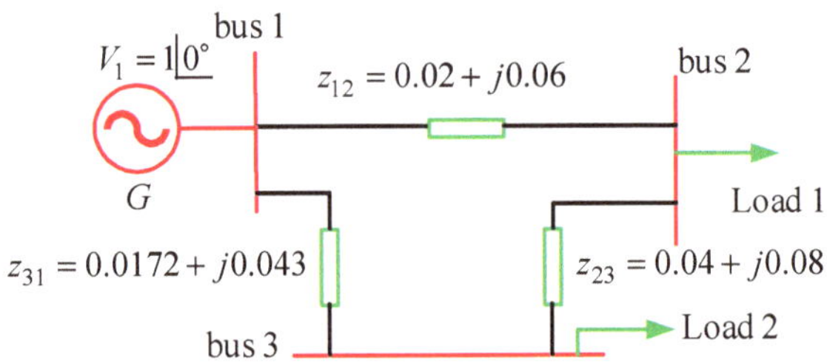

Fig. P7.9 Single-line diagram for Problem 7.16

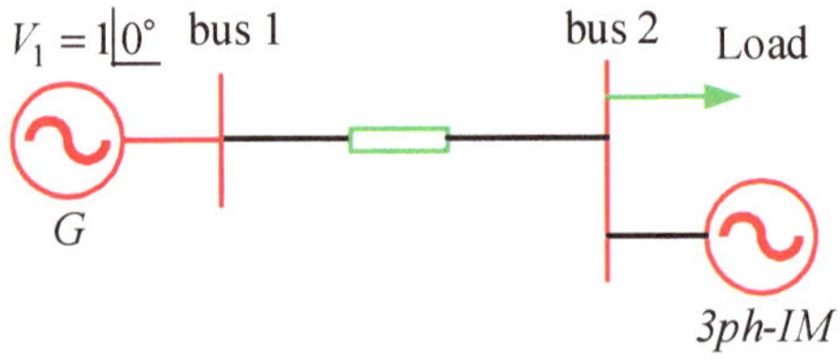

7.14 A voltage function is given by $f(V) = e^{-V} + V - 2$. Use the Newton-Raphson method to calculate the true solution by using the initial guess $V_0 = 1.83$.

7.15 A voltage function is given by $f(V) = 5V^2 + 11V - 17 = 0$. Use the Newton-Raphson method to calculate the true solution by using the initial guess $V_0 = 1$.

7.16 Fig. P7.9 shows a two-bus single-line diagram of a transmission line. The impedance per mile of a line is given by $z_{12} = 0.024 + j0.0726\ \Omega/\text{mi}$ and $B = 4.1\ \mu\text{S/mi}$. A three-phase 11 kV, 10725HP, 93% pf and a load of 90 MW and 50 Mvar are connected in bus 2. Use the Newton-Raphson method to find the magnitude of the voltage and angle at bus 2 at any simulation software.

7.17 A three-bus single-line diagram is shown in Fig. P7.10. A 128MVA, 13.8 kV, 0.85 pf, 2-pole, 15 MW active generation, 5 Mvar reactive generation steam generator is connected at bus 1. A 20 MVA, 13.8 kV delta-11 kVwye, $Z_1 = Z_0 = 8.25\%$, $X_1/R_1 = X_0/R_0 = 22$, single-phase liquid-filled transformer is connected between bus 1 and bus 2. An 11.5 kV shielded hardcore cable is connected between bus 2 and bus 3. A load of 140 MW and 60 Mvar is connected to bus 3. Calculate the slack bus power, voltage and angle at bus 3 using the Newton Raphson method with any simulation software.

Fig. P7.10 Single-line diagram for Problem 7.17

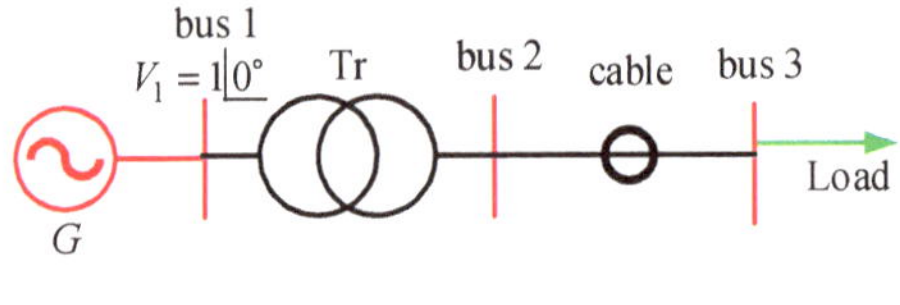

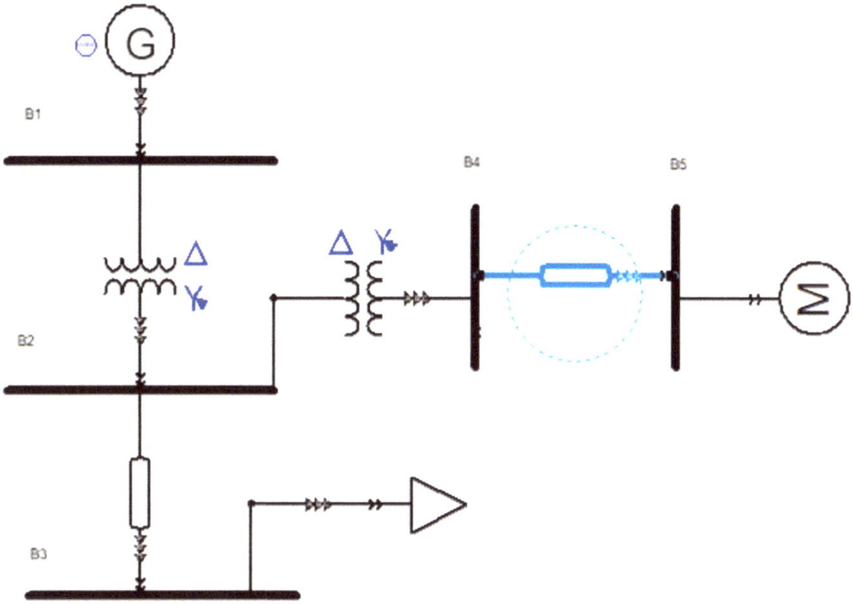

Fig. P7.11 Single-line diagram for Problem 7.18

7.18 A five-bus single-line diagram is shown in Fig. P7.11. The following equipment is connected as below.

Generator (bus 1): 40 MVA, 13.8 kV, 0.85 pf, 2-pole, 20 MW active generation, 10 Mvar reactive generation
Tr1 (bus 1–2): single-phase liquid filled, 20MVA, delta (13.8 kV)–wye (11 kV), $Z_1 = Z_0 = 9\%$, $\frac{X_1}{R_1} = \frac{X_0}{R_0} = 25$
Tr2 (bus 2-4): single-phase liquid filled, 35 MVA, delta (11 kV)–wye (0.42 kV), $Z_1 = Z_0 = 8\%$, $\frac{X_1}{R_1} = \frac{X_0}{R_0} = 15$
Line (bus 2-3): $Z_1 = Z_0 = 0.04 + j1.5\ \Omega/\text{mi}$, $B = 0.66\ \mu\text{S/mi}$
Line (bus 4–5): $Z_1 = Z_0 = 0.036 + j0.098\ \Omega/\text{mi}$, $B = 1.02\ \mu\text{S/mi}$
Load (bus 5): 85 MW, 45 Mvar

Calculate the slack bus power, voltage and angle at bus 3 and bus 5 using the Fast Decoupled method with any simulation software.

Chapter 8
Underground Cables

8.1 Introduction

Electrical power is transferred from one place to another either by overhead transmission or distribution lines or high voltage underground cables. Underground cables have several advantages such as less prone to mechanical damage, elimination of lightning strikes as well as flashover, increase in reliability of supply, less maintenance cost, fewer chances of faults, less voltage drops, and reduction of visual impact. However, the main drawback of the underground cables is that the high initial investment compared to overhead transmission lines. Underground cables are usually used in highly populated areas where overhead transmission lines are difficult to install. High voltage cables are used in the power station, industries, and all other electrical systems when the overhead transmission system is not suitable for use. The primary use of underground cables is in the distribution of electric power in the populated urban areas at low voltages. These cables are laid in ducks or may be buried in the cable trance. Recent improvements in the design and manufacture resulted in an increase in demand for underground cables among the electric utilities and industrial sectors. In this chapter, construction, classification, different types of grading, electrical stress, insulation resistance, etc., will be discussed.

8.2 Construction of Cables

An underground cable consists of three or four conductors, which are surrounded by different types of insulations, as shown in Fig. 8.1. The main components of an underground cable are discussed [1, 2] below.

Md. A. Salam, *Fundamentals of Electrical Power Systems Analysis*,
https://doi.org/10.1007/978-981-15-3212-2_8

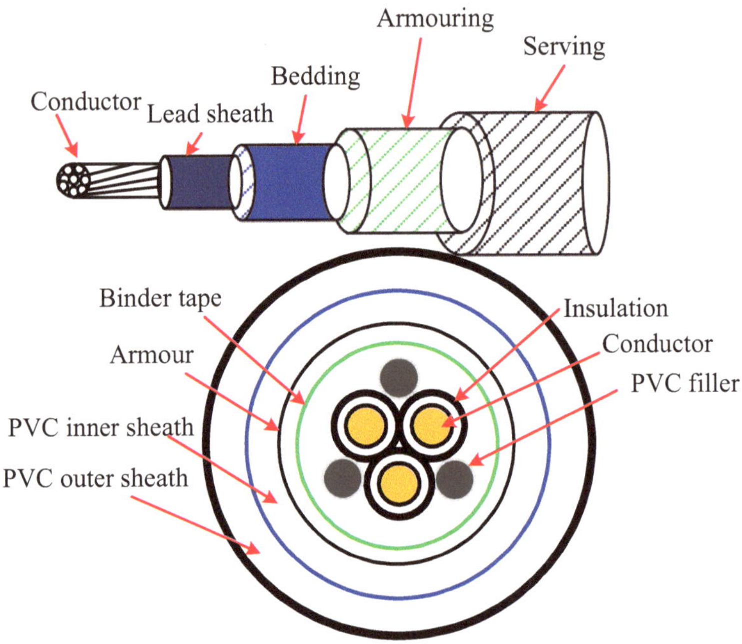

Fig. 8.1 Eight strands cables and cross section of a three-core copper cable

Conductor: The conductor is made of either copper or aluminum. Copper is extensively used for cables. However, the conductor made of aluminum with proper insulation is being used in the overhead transmission line. A cable has one or three cores depending on the type of service voltage. The conductors are usually stranded to get better flexibility of the cable. Electrolytic copper and aluminum ensure a longer life for the conductor.

Insulation: The performance of the cable depends mainly on the insulation, whereas the layer of the insulation depends typically on voltage tolerances. The insulating materials provide grease/oil-resistant property, high dielectric strength, immunity to fire hazards as well as protection from chemicals and microbiological growths. It also provides no damage due to severe bending during laying and safe to fluctuations in temperatures. Different types of rubber and polyvinyl chloride (PVC) are used for cable insulation, which is discussed next.

Vulcanized rubber: Vulcanized rubber has high insulating properties that absorb moisture. When vulcanized rubber is mixed with sulfur and other materials, it does not absorb moisture. As a result, it provides better performance.

Polyvinyl chloride: The polymerization process normally obtains this type of material from the proper mixture of acetylene and plasticizer. The plasticizer is a liquid material that has a high boiling point. PVC has different grades, and it has high insulation resistance, high dielectric strength, and mechanical strength a wide range of temperatures.

Cross-linked polyethylene: It is abbreviated as XLPE. Due to good water resistance and excellent electrical properties, it is used as insulation for many power cables. It has many advantages such as low melting point, high flexibility, lightweight, can carry large current, minimizes the need for armoring, and excellent protection (mechanical, electrical, and chemical) properties.

PVC Filler: The empty spaces between the cores are filled with polyvinyl chloride (PVC) material known as PVC filler. The filler usually is of non-hygroscopic, softer material, and fire-retardant material than cable insulation.

Binder Tape: Separate binder tape is used in the cores to protect conductors from penetration of moisture, chemicals, and jacket compounds. It is also used to maintain cable concentricity and firmness of conductor bundles.

Armoring: Steel wire is used around the binder taped of the cables to provide additional protection for preventing it from any damage. It is used to protect the cables from mechanical damage during direct laying in water-filled ducks and termites attack. Steel tape and steel wire armoring are the two types of armoring that are usually used in the cables. In steel tape armoring, steel tape is wound around the cable in two layers with opposite directions, whereas in steel wire armoring, strong steel wires are used in one or two layers. Cables with aluminum sheath are not armored.

Bedding: Bedding is provided by jute or hessian tape to protect the metallic sheath from mechanical damage due to armoring.

PVC Outer Sheath: PVC sheath is the last stage of the cable, which protects from moisture, gases, and termite in the soil as well as all kinds of external stresses.

Serving: Serving is a kind of protection of armoring from atmospheric conditions. A layer of fibrous material is provided over the armoring, which is known as serving.

8.3 Classification of Cable

Cables are classified according to the number of cores, voltage rating, construction, type, and thickness of insulation and installation types. According to the type and thickness of insulation, cables are classified as polyethylene (PE), polyvinyl chloride (PVC), cross-linked polyethylene (XLPE), low smoke zero halogen (LSZH) insulation materials, and armored instrumentation cables.

Depending on the core, cables are classed as single-core and multi-core (three-core, four-core), whereas according to the voltage rating, underground cables are classified as,

- Low voltage cables up to 1 kV
- High tension cables up to 11 kV
- Supper tension cables from 22 kV to 33 kV
- Extra high tension cables from 33 kV to 66 kV
- Extra super tension cables above 132 kV.

8.4 Insulation Resistance of Single-Core Cable

Consider a single-core cable whose length is l, conductor radius is r, and the sheath radius is R. Again, consider small thickness dx of insulation layer at a distance x from the center of the conductor as shown in Fig. 8.2.

The circumference area of the cable at a distance x is written as,

$$A = 2\pi xl \tag{8.1}$$

The insulation resistance of considered small layer is,

$$dR_{ins} = \rho \frac{dx}{2\pi xl} \tag{8.2}$$

The insulation resistance of the whole cable is calculated as,

$$R_{ins} = \int_r^R dR_{ins} \tag{8.3}$$

Substituting Eq. (8.2) into Eq. (8.3) and integrating yields,

$$R_{ins} = \int_r^R \rho \frac{dx}{2\pi xl} \tag{8.4}$$

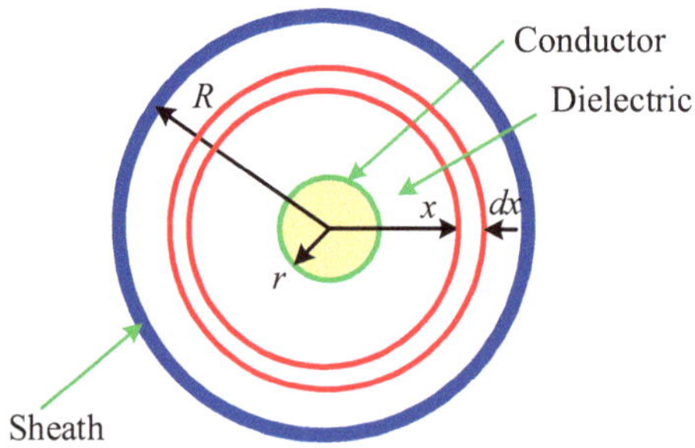

Fig. 8.2 Cable cross section with a small thickness

$$R_{ins} = \frac{\rho}{2\pi l} \ln \frac{R}{r} \tag{8.5}$$

The value of the insulation resistance can be calculated from Eq. (8.5) if the other parameters are given.

Example 8.1
The insulation thickness and radius of a 5 km cable are 8 mm and 14 mm, respectively. Calculate the insulation resistance of a cable with $\rho = 4 \times 10^{10}\,\Omega - \text{m}$.

Solution
The insulation radius of a cable is calculated as,

$$R = 8 + 14 = 22\,\text{mm} = 0.022\,\text{m} \tag{8.6}$$

The insulation resistance of a cable is calculated as,

$$R_{ins} = \frac{\rho}{2\pi l} \ln \frac{R}{r} = \frac{4 \times 10^{10}}{2\pi \times 5000} \ln \frac{0.022}{0.014} = 575.48\ \text{M}\Omega \tag{8.7}$$

Practice Problem 8.1
The insulation thickness diameter and the core diameter of 10 km cable are 3 cm and 7 cm, respectively. Calculate the cable resistivity if the insulation resistance is 500 MΩ.

8.5 Electric Stress of a Single-Core Cable

Figure 8.3 shows a cross section and stress distribution of a single-core cable. Consider r is the radius of the conductor and R is the inner radius of the sheath. Again, consider a small cross section at a distance x from the center of the conductor, whose thickness is dx. If the induced charge on the conductor q, the electric flux density (D) at a distance x is defined as [3, 4],

$$D = \frac{q}{2\pi x} \quad \text{C/m}^2 \tag{8.8}$$

The electric flux density is related to the electric field as,

$$D = \varepsilon E \tag{8.9}$$

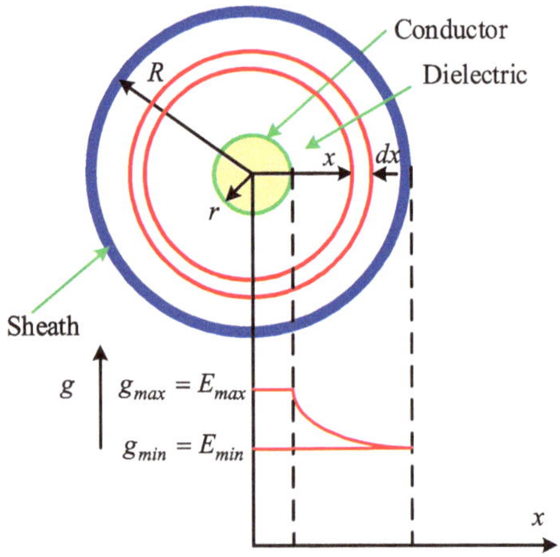

Fig. 8.3 Cable cross section with stress distribution

Substituting Eq. (8.8) into Eq. (8.9) yields,

$$E = \frac{q}{2\pi\varepsilon x} \quad \text{V/m} \tag{8.10}$$

The voltage difference between the conductor and outer sheath can be determined as,

$$V = \int_r^R E\,dx \tag{8.11}$$

Substituting Eq. (8.10) into Eq. (8.11) yields,

$$V = \int_r^R \frac{q}{2\pi\varepsilon x}\,dx \tag{8.12}$$

$$V = \frac{q}{2\pi\varepsilon}\ln\frac{R}{r} \tag{8.13}$$

$$\frac{V}{\ln\frac{R}{r}} = \frac{q}{2\pi\varepsilon} \tag{8.14}$$

Substituting Eq. (8.14) into Eq. (8.10) yields,

$$E = \frac{V}{x\ln\frac{R}{r}} \tag{8.15}$$

From Fig. 8.3, it is seen that the stress distribution over the dielectric is not the same. The maximum stress occurs at the surface of the conductor. It will decrease slowly toward the sheath of the cable, whereas the minimum stress occurs at the outer surface of the sheath. At a distance $x = r$, the expression of maximum electric field or stress is calculated as,

$$E_{max} = \frac{V}{r \ln \frac{R}{r}} \tag{8.16}$$

At $x = R$, the expression of minimum electric field or stress is

$$E_{min} = \frac{V}{R \ln \frac{R}{r}} \tag{8.17}$$

Dividing Eq. (8.16) by Eq. (8.17) yields,

$$\frac{E_{max}}{E_{min}} = \frac{R}{r} \tag{8.18}$$

The maximum stress is usually considered during the design of the high voltage cables.

8.6 Economical Size of Conductor

It is already seen that the maximum stress occurs at the surface of the conductor. For reliable and safe operation, the dielectric strength of the cable should be more than the maximum stress. For designing, working voltage and sheath radius need to be considered constant up to certain values, whereas the core radius needs to vary. Therefore, for a given voltage and sheath radius, the most economical size of the conductor can be determined if the denominator of Eq. (8.16) is minimum. The minimum value of the denominator can be obtained if differentiation of the denominator of Eq. (8.16) with respect to r is equal to zero and it can be obtained as,

$$\frac{d}{dr}\left(r \ln \frac{R}{r}\right) = 0 \tag{8.19}$$

$$\ln \frac{R}{r} + r \frac{1}{\frac{R}{r}}\left(-\frac{R}{r^2}\right) = 0 \tag{8.20}$$

$$\ln \frac{R}{r} - 1 = 0 \tag{8.21}$$

$$\ln\frac{R}{r} = 1 \tag{8.22}$$

$$\frac{R}{r} = e^1 = 2.718 \tag{8.23}$$

From Eq. (8.23), the most economical conductor radius is written as,

$$r = \frac{R}{2.718} \tag{8.24}$$

Substituting Eq. (8.22) into Eq. (8.16) as

$$E_{max} = \frac{V}{r} \tag{8.25}$$

The insulation thickness can be determined as,

$$t_{insu} = R - r \tag{8.26}$$

Substituting Eq. (8.24) into Eq. (8.26) yields,

$$t_{insu} = 2.718r - r = 1.718r \tag{8.27}$$

Example 8.2
The line-to-neutral voltage, conductor radius, and the sheath radius of a single-core cable are 35 kV (rms), 0.85 cm, and 2.5 cm, respectively. Calculate the maximum stress, minimum stress, most economical conductor size, and insulation thickness.

Solution
The value of the maximum stress can be determined as,

$$E_{max} = \frac{V}{r\ln\frac{R}{r}} = \frac{35}{0.85\ln\frac{2.5}{0.85}} = 38.16\,\text{kV/cm} \tag{8.28}$$

The value of minimum stress can be determined as,

$$E_{min} = \frac{V}{R\ln\frac{R}{r}} = \frac{35}{2.5\ln\frac{2.5}{0.85}} = 12.97\,\text{kV/cm} \tag{8.29}$$

The value of the most economical size of the conductor can be determined as,

$$r = \frac{R}{2.718} = \frac{2.5}{2.718} = 0.92\,\text{cm} \tag{8.30}$$

The value of the insulation thickness can be determined as,

$$t_{insu} = 1.718r = 1.718 \times 0.85 = 1.46\,\text{cm} \tag{8.31}$$

Practice Problem 8.2
The core diameter, line-to-neutral voltage, and maximum stress of a single-core cable are 2.5 cm, 15 kV (rms), and 20 kV/cm, respectively. Calculate the sheath radius, minimum stress, and the most economical conductor size.

8.7 Grading of Cables

Dielectric strength and its homogeneous distribution in the cable are very important for safe and reliable operation. The dielectric materials are usually used in the cables to optimize the difference between the maximum and minimum stresses. Unequal distributions of stresses may trigger insulation breakdown as well as an increase in the insulation thickness, which in turn increases the cable size. Therefore, it is necessary to distribute the stresses uniformly in the dielectric throughout the cables. The process of achieving uniform stress distribution in the dielectric of cables is known as grading. There are two methods of cable grading that are normally used for the optimization of the stresses. These are capacitance grading and intersheath grading.

8.7.1 Capacitance Grading

Two or more layers of different dielectrics are used in the capacitance grading. The process of achieving uniformity in the dielectric of cables is known as capacitance grading. Consider a conductor that is surrounded by the three layers of dielectrics and the permeabilities ε_1, ε_2, ε_3 at the radii r_1, r_2, and R, respectively. The core with three layers of dielectrics is shown in Fig. 8.4. The stress at any distance x can be expressed as [5, 6],

Fig. 8.4 Cable core with three layers of insulation

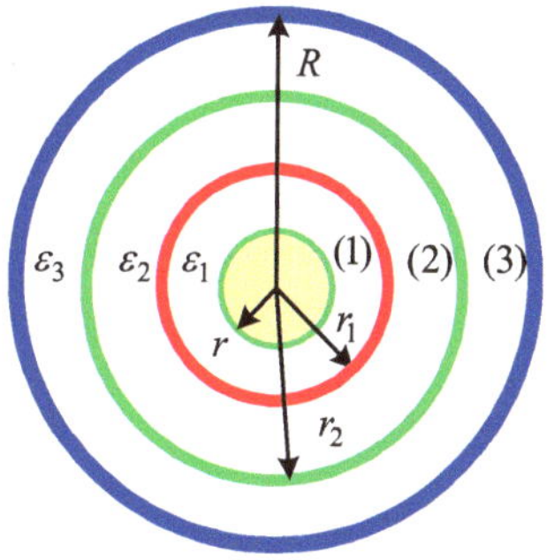

$$E_x = \frac{q}{2\pi\varepsilon_0\varepsilon_x x} \tag{8.32}$$

The stress for insulation layer one which starting from $x = r$ to $x = r_1$ is expressed as,

$$E_{x=r} = \frac{q}{2\pi\varepsilon_0\varepsilon_1 r} \tag{8.33}$$

$$E_{x=r_1} = \frac{q}{2\pi\varepsilon_0\varepsilon_1 r_1} \tag{8.34}$$

The stress for insulation layer two, which starting from $x = r_1$ to $x = r_2$ is expressed as,

$$E_{x=r_1} = \frac{q}{2\pi\varepsilon_0\varepsilon_2 r_1} \tag{8.35}$$

$$E_{x=r_2} = \frac{q}{2\pi\varepsilon_0\varepsilon_2 r_2} \tag{8.36}$$

The stress for insulation layer three which starting from $x = r_2$ to $x = r_3$ is expressed as,

$$E_{x=r_2} = \frac{q}{2\pi\varepsilon_0\varepsilon_3 r_2} \tag{8.37}$$

$$E_{x=R} = \frac{q}{2\pi\varepsilon_0\varepsilon_3 R} \tag{8.38}$$

Assuming all operating at the same electric field, equating Eqs. (8.33), (8.35), and (8.37) yields,

$$\frac{q}{2\pi\varepsilon_0\varepsilon_1 r} = \frac{q}{2\pi\varepsilon_0\varepsilon_2 r_1} = \frac{q}{2\pi\varepsilon_0\varepsilon_3 r_2} \tag{8.39}$$

$$\frac{1}{\varepsilon_1 r} = \frac{1}{\varepsilon_2 r_1} = \frac{1}{\varepsilon_3 r_2} \tag{8.40}$$

$$\varepsilon_1 r = \varepsilon_2 r_1 = \varepsilon_3 r_2 \tag{8.41}$$

The total voltage between the core and the outer sheath is calculated as,

$$V = V_1 + V_2 + V_3 \tag{8.42}$$

$$V = \int_{x=r}^{r_1} E_x dx + \int_{x=r_1}^{r_2} E_x dx + \int_{x=r_2}^{R} E_x dx \tag{8.43}$$

Substituting Eq. (8.32) for three layers into Eq. (8.43) yields,

$$.V = \int_{x=r}^{r_1} \frac{q}{2\pi\varepsilon_0\varepsilon_1 x} dx + \int_{x=r_1}^{r_2} \frac{q}{2\pi\varepsilon_0\varepsilon_2 x} dx + \int_{x=r_2}^{R} \frac{q}{2\pi\varepsilon_0\varepsilon_3 x} dx. \tag{8.44}$$

$$V = \int_{x=r}^{r_1} \frac{q}{2\pi\varepsilon_0\varepsilon_1 x} dx + \int_{x=r_1}^{r_2} \frac{q}{2\pi\varepsilon_0\varepsilon_2 x} dx + \int_{x=r_2}^{R} \frac{q}{2\pi\varepsilon_0\varepsilon_3 x} dx \tag{8.44}$$

$$V = \frac{q}{2\pi\varepsilon_0}\left[\frac{1}{\varepsilon_1}\ln\left(\frac{r_1}{r}\right) + \frac{1}{\varepsilon_2}\ln\left(\frac{r_2}{r_1}\right) + \frac{1}{\varepsilon_3}\ln\left(\frac{R}{r_2}\right)\right] \tag{8.45}$$

$$V = \frac{q}{2\pi\varepsilon_0}\left[\frac{r}{r\varepsilon_1}\ln\left(\frac{r_1}{r}\right) + \frac{r_1}{r_1\varepsilon_2}\ln\left(\frac{r_2}{r_1}\right) + \frac{r_2}{r_2\varepsilon_3}\ln\left(\frac{R}{r_2}\right)\right] \tag{8.46}$$

Substituting Eq. (8.41) into Eq. (8.46) yields,

$$V = \frac{q}{2\pi\varepsilon_0 r\varepsilon_1}\left[r\ln\left(\frac{r_1}{r}\right) + r_1\ln\left(\frac{r_2}{r_1}\right) + r_2\ln\left(\frac{R}{r_2}\right)\right] \tag{8.47}$$

$$V = E_m\left[r\ln\left(\frac{r_1}{r}\right) + r_1\ln\left(\frac{r_2}{r_1}\right) + r_2\ln\left(\frac{R}{r_2}\right)\right] \tag{8.48}$$

Equation (8.49) can be rearranged as,

$$V = E_m\left(r\ln\frac{r_1}{r} + r\ln\frac{r_2}{r_1} - r\ln\frac{r_2}{r_1} + r\ln\frac{R}{r_2} - r\ln\frac{R}{r_2} + r_1\ln\frac{r_2}{r_1} + r_2\ln\frac{R}{r_2}\right) \tag{8.49}$$

$$V = E_m\left(r\ln\frac{r_1}{r}\frac{r_2}{r_1}\frac{R}{r_2} + (r_1 - r)\ln\frac{r_2}{r_1} + (r_2 - r)\ln\frac{R}{r_2}\right) \tag{8.50}$$

Basically, r_1 and r_2 are greater than r. Therefore, the second and the third terms of Eq. (8.50) can be neglected, and Eq. (8.50) can be modified as,

$$V = E_m r\ln\frac{R}{r} \tag{8.51}$$

$$E_m = \frac{q}{2\pi\varepsilon_0 r\varepsilon_1} \tag{8.52}$$

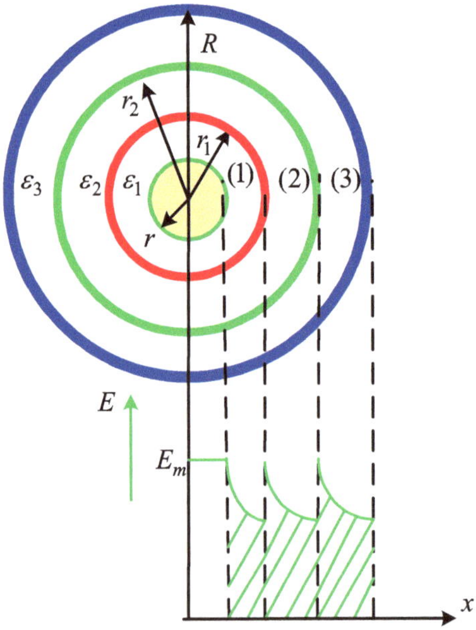

Fig. 8.5 Capacitance grading field distribution

The field distributions of the three-layer dielectrics of single-core cable are shown in Fig. 8.5.

Example 8.3
The permittivities of a three-layer dielectrics capacitance grading are 5, 3, and 2, respectively. The conductor radius is 1 cm, and the overall sheath radius is 7 cm. The three-layer dielectrics are operated at 27 kV/cm. Calculate the radii r_1 and r_2, maximum voltage, and safe working voltage.

Solution
The values of the different radius can be determined as,

$$\varepsilon_1 r = \varepsilon_2 r_1 = \varepsilon_3 r_2 \tag{8.53}$$

$$5 \times 1 = 3r_1 = 2r_2 \tag{8.54}$$

$$5 = 3r_1 \tag{8.55}$$

$$r_1 = 1.67\,\text{cm} \tag{8.56}$$

$$5 = 2r_2 \tag{8.57}$$

$$r_2 = 2.5\,\text{cm} \tag{8.58}$$

The value of the maximum working voltage can be determined as,

$$\begin{aligned} V &= E_m\left(r\ln\frac{r_1}{r} + r_1\ln\frac{r_2}{r_1} + r_2\ln\frac{R}{r_2}\right) \\ &= 27\left(\ln\frac{1.67}{1} + 1.67\ln\frac{2.5}{1.67} + 2.5\ln\frac{7}{2.5}\right) \\ &= 116.6\,\text{kV} \end{aligned} \tag{8.59}$$

The rms value of safe working voltage is calculated as,

$$V_{safe} = \frac{116.6}{\sqrt{2}} = 82.45\,\text{kV} \tag{8.60}$$

Practice Problem 8.3

A single-core cable is graded by three-layer dielectrics whose permittivities are 8, 5, and 4, respectively. The conductor radius is 1.5 cm, and the overall radius is 8 cm. The three-layer dielectrics are worked at 11 kV/cm. Find the radius r_1 and r_2, maximum voltage and safe working voltage.

8.7.2 *Intersheath Grading*

In the intersheath grading, the same insulating material is used throughout the cable. However, it is divided into a few layers by placing metallic intersheaths. These intersheaths are connected to the tapings of a transformer. The potential across each intersheath is maintained at such values that each layer of insulation contributes to the proper share of the total voltage. This arrangement improves the voltage distribution in the dielectric of the cable, which in turn obtains a uniform electric field. The maximum electric field at various intersheaths is the same due to homogenous material. Consider a cable whose core radius is r and the outer sheath radius is R. The two metallic intersheaths of radii r_1 and r_2 are inserted in between the core and the outer sheath as shown in Fig. 8.6.

Let us consider that V_1 is the voltage between the core and the first intersheath, V_2 is the voltage between the first intersheath and the second intersheath, V_3 is the voltage between the second intersheath and the outer sheath. Each sheath can be considered as a homogeneous single-core cable because of a definite voltage difference between the inner and outer radii of each sheath. According to Eq. (8.51), the maximum stress between the core and the first intersheath is expressed as,

$$E_{m1} = \frac{V_1}{r\ln\frac{r_1}{r}} \tag{8.61}$$

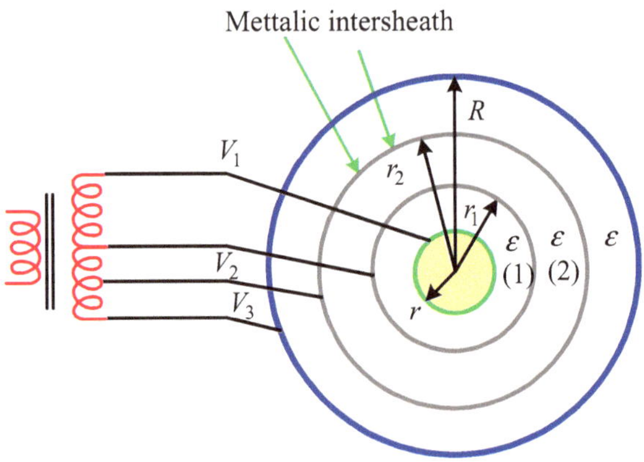

Fig. 8.6 Intersheath tap-changing transformer

The maximum stress between the first intersheath and the second intersheath is expressed as,

$$E_{m2} = \frac{V_2}{r_1 \ln \frac{r_2}{r_1}} \tag{8.62}$$

The maximum stress between the second intersheath and the outer sheath is expressed as,

$$E_{m3} = \frac{V_3}{r_2 \ln \frac{R}{r_2}} \tag{8.63}$$

The dielectric of the cable is homogeneous. Therefore, the maximum stress in each layer is the same, and it can be expressed as,

$$E_{m1} = E_{m2} = E_{m3} = E_m(\text{let}) \tag{8.64}$$

Substituting Eqs. (8.61), (8.62), and (8.63) into Eq. (8.64) yields,

$$E_m = \frac{V_1}{r \ln \frac{r_1}{r}} = \frac{V_2}{r_1 \ln \frac{r_2}{r_1}} = \frac{V_3}{r_2 \ln \frac{R}{r_2}} \tag{8.65}$$

According to behavioral properties, the cable works as like the three capacitors in series. Therefore, all potentials including the outer sheath, metallic sheaths, and the core are in phase, i.e., it can be expressed as,

$$V = V_1 + V_2 + V_3 \tag{8.66}$$

Substituting Eqs. (8.61), (8.62), and (8.63) into Eq. (8.66) yields,

$$V = E_m r \ln\frac{r_1}{r} + E_m r_1 \ln\frac{r_2}{r_1} + E_m r_2 \ln\frac{R}{r_2} \tag{8.67}$$

Again for designing purposes, considering one sheath in between the core and an outer sheath, as shown in Fig. 8.7.

In this case, Eq. (8.67) can be modified as,

$$V = E_m r \ln\frac{r_1}{r} + E_m r_1 \ln\frac{R}{r_1} \tag{8.68}$$

$$E_m = \frac{V}{r \ln\frac{r_1}{r} + r_1 \ln\frac{R}{r_1}} \tag{8.69}$$

In this analysis, the core radius (r) and the outer sheath radius (R) are kept fixed, whereas the intersheath radius (r_1) is only considered the variable parameter. Now, differentiating Eq. (8.69) with respect to r_1 and setting the result equal to zero to get the minimum value for the maximum stress. This can be obtained as,

$$\frac{dE_m}{dr_1} = V\frac{d}{dr_1}\left(r \ln\frac{r_1}{r} + r_1 \ln\frac{R}{r_1}\right)^{-1} = 0 \tag{8.70}$$

$$r_1\frac{r_1}{R}\left(-\frac{R}{r_1^2}\right) + \ln\frac{R}{r_1} + r\frac{r}{r_1}\frac{1}{r} = 0 \tag{8.71}$$

$$-1 + \ln\frac{R}{r_1} + \frac{r}{r_1} = 0 \tag{8.72}$$

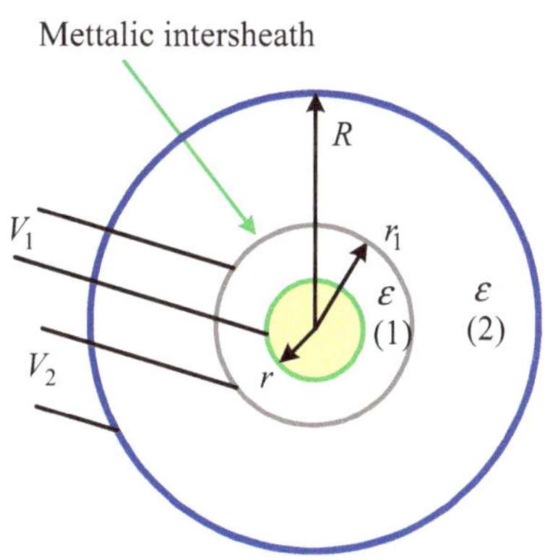

Fig. 8.7 Core with one intersheath

Substituting Eq. (8.23) into Eq. (8.72) yields,

$$-1 + \ln\frac{er}{r_1} + \frac{r}{r_1} = 0 \tag{8.73}$$

$$-1 + \ln e + \ln\frac{r}{r_1} + \frac{r}{r_1} = 0 \tag{8.74}$$

$$-1 + 1 + \ln\frac{r}{r_1} + \frac{r}{r_1} = 0 \tag{8.75}$$

$$\ln\frac{r}{r_1} = -\frac{r}{r_1} \tag{8.76}$$

$$\frac{r}{r_1} = e^{-\frac{r}{r_1}} \tag{8.77}$$

$$\frac{r}{r_1} = 1 + \frac{-\frac{r}{r_1}}{1!} + \frac{\left(-\frac{r}{r_1}\right)^2}{2!} + \frac{\left(-\frac{r}{r_1}\right)^3}{3!} + \cdots \tag{8.78}$$

Neglecting higher terms of Eq. (8.78) yields,

$$\frac{r}{r_1} = 1 + \frac{-\frac{r}{r_1}}{1!} \tag{8.79}$$

$$\frac{r}{r_1} + \frac{r}{r_1} = 1 \tag{8.80}$$

$$r_1 = 2r \tag{8.81}$$

Substituting Eqs. (8.23) and (8.81) into Eq. (8.69) yields,

$$E_m = \frac{V}{r\ln\frac{2r}{r} + 2r\ln\frac{er}{2r}} \tag{8.82}$$

$$E_m = \frac{V}{0.693r + 0.6137r} \tag{8.83}$$

$$E_m = \frac{V}{1.31r} \tag{8.84}$$

From Eq. (8.84), the value of the maximum stress can be calculated is the operating voltage and the core radius are known.

Example 8.4
The core and outer sheath radii of a three-phase 22 kV (rms) cable are found 2 cm and 3.5 cm, respectively. The two metallic intersheaths with radii of 2.5 and 3 cm are inserted in between the core and an outer sheath. The maximum stress in each layer is the same, and calculate the voltages on the first sheath and second sheath.

Solution
The maximum value of the voltage is

$$V_m = \sqrt{2} \times 33 = 31.11\,\text{kV} \tag{8.85}$$

Per phase voltage is calculated as,

$$V_{ph} = \frac{31.11}{\sqrt{3}} = 17.96\,\text{kV} \tag{8.86}$$

The maximum stresses can be determined as,

$$E_{m1} = \frac{V_1}{r \ln \frac{r_1}{r}} = \frac{V_1}{2 \ln \frac{2.5}{2}} = 2.24 V_1 \tag{8.87}$$

$$E_{m2} = \frac{V_2}{r_1 \ln \frac{r_2}{r_1}} = \frac{V_2}{2.5 \ln \frac{3}{2.5}} = 2.19 V_2 \tag{8.88}$$

$$E_{m3} = \frac{V_3}{r_2 \ln \frac{R}{r_2}} = \frac{V_3}{3 \ln \frac{3.5}{3}} = 2.16 V_3 \tag{8.89}$$

The maximum stress in each layer is the same. Then, this relation can be written as,

$$2.24 V_1 = 2.19 V_2 = 2.16 V_3 \tag{8.90}$$

$$V_2 = 1.02 V_1 \tag{8.91}$$

$$2.24 V_1 = 2.16 V_3 \tag{8.92}$$

$$V_3 = 1.03 V_1 \tag{8.93}$$

The total voltage is calculated as,

$$V_1 + V_2 + V_3 = V \tag{8.94}$$

Substituting Eqs. (8.91) and (8.93) into Eq. (8.94) yields,

$$V_1 + 1.02V_1 + 1.03V_1 = 17.96 \tag{8.95}$$

$$V_1 = 5.89\,\text{kV} \tag{8.96}$$

The other values of voltages can be determined as,

$$V_2 = 1.02V_1 = 1.02 \times 5.89 = 6\,\text{kV} \tag{8.97}$$

$$V_3 = 1.03V_1 = 1.03 \times 5.89 = 6.06\,\text{kV} \tag{8.98}$$

The voltage at the first sheath can be determined as,

$$V_{s1} = V - V_1 = 17.96 - 5.89 = 12.07\,\text{kV} \tag{8.99}$$

The voltage at the second sheath can be determined as,

$$V_{s1} = V - V_1 - V_2 = 17.96 - 5.89 - 6 = 6.07\,\text{kV} \tag{8.100}$$

Practice Problem 8.4
The core and the outer sheath radii of a three-phase 66 kV (rms) cable are 1 cm and 2.2 cm, respectively. The two intersheaths with radii of 1.5 cm and 1.9 cm are inserted in between the core and the outer sheath. Calculate the voltages on the first sheath and second sheath. Consider that the maximum stress in each layer is the same.

8.8 Cables Heating

There are three losses which are responsible for heating of the cables. There are copper loss, dielectric loss, and intersheath loss. Copper loss mainly depends on the current passing through the resistance of the conductor. Dielectric loss in the insulation is due to the leakage current, which is originated by the voltage across the insulation. Finally, the intersheath loss is due to the circulating currents in loops, which forms between the sheaths. The copper loss can be expressed as,

$$P_{cu} = I^2 R_c \tag{8.101}$$

where I is the current flows in the cables, and R_c is the resistance of the conductor.

The cable is one kind of capacitor that is formed by the core and the sheath which represents the two plates of a capacitor. These two plates are separated by dielectrics. The equivalent circuit for this is represented by the parallel combination of leakage resistance R and the capacitor C, as shown in Fig. 8.8. The phasor diagram for a cable

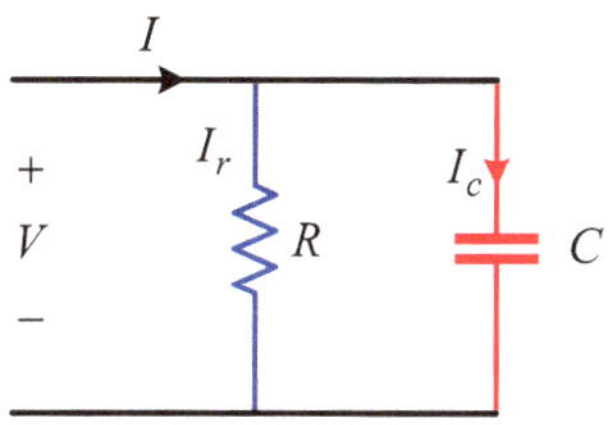

Fig. 8.8 Equivalent circuit of a cable

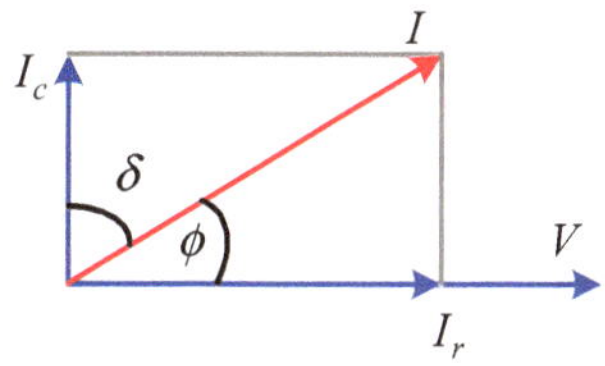

Fig. 8.9 Phasor diagram for cable equivalent circuit

dielectric is shown in Fig. 8.9. The current leads the voltage by an angle less than 90°. There is no loss for a perfect dielectric; however, the cable is not a perfect dielectric. Therefore, there is a loss in the cable due to the dielectric. The loss in the dielectric due to the leakage current flows in the insulation resistance is,

$$P_{lc} = \frac{V^2}{R} \tag{8.102}$$

From Fig. 8.9, the expression of $\tan\delta$ can be written as,

$$\tan\delta = \frac{I_r}{I_c} \tag{8.103}$$

From Fig. 8.8, the current in the leakage resistance (I_r) can be written as,

$$I_r = \frac{V}{R} \tag{8.104}$$

Similarly, the current in the capacitor (I_c) can be written as,

$$I_c = \frac{V}{\frac{1}{\omega C}} = V\omega C \tag{8.105}$$

Substituting Eqs. (8.104) and (8.105) into Eq. (8.103) yields,

$$\tan\delta = \frac{\frac{V}{R}}{V\omega C} \tag{8.106}$$

$$\frac{V}{R} = V\omega C \tan \delta \tag{8.107}$$

Multiplying both sides of Eq. (8.107) by the voltage (V) yields,

$$\frac{V^2}{R} = V^2\omega C \tan \delta \tag{8.108}$$

From Eq. (8.108), the final expression of the total dielectric power loss can be written as,

$$P_{dl} = V^2\omega C \tan \delta \tag{8.109}$$

Here, δ is the dielectric loss angle, and its value is very small. Therefore, $\tan \delta$ is considered as δ. Equation (8.109) can be modified as,

$$P_{dl} = V^2\omega C\delta \tag{8.110}$$

From Eq. (8.110), it is seen that power loss is proportional to the square of the voltage, capacitance, and the dielectric loss angle.

From Fig. 8.9, the power factor angle is expressed as,

$$\phi = 90^\circ - \delta \tag{8.111}$$

$$\cos \phi = \cos(90^\circ - \delta) \tag{8.112}$$

$$\cos \phi = \sin \delta \tag{8.113}$$

The dielectric hysteresis loss (P_{dh}) is the difference between the total dielectric power loss and the power loss due to leakage resistance, i.e.,

$$P_{dh} = P_{dl} - P_{lc} \tag{8.114}$$

Substituting Eqs. (8.102) and (8.110) into Eq. (8.114) yields,

$$P_{dh} = V^2\omega C\delta - \frac{V^2}{R} = V^2\left(\omega C\delta - \frac{1}{R}\right) \tag{8.115}$$

From Eq. (8.115), the values of dielectric hysteresis loss can be determined if others parameters are known.

Example 8.5

A single-core 50 Hz, 6.4 kV (line-to-neutral) cable is having the insulation resistance of 6 MΩ. Calculate the power loss due to leakage of current flows in the insulation, total dielectric loss, and dielectric hysteresis loss. Assume that the dielectric factor is 0.03, and the value of the capacitance of a cable is 0.84 μF.

Solution

The power loss due to leakage current flows in the insulation is calculated as,

$$P_{lc} = \frac{V^2}{R} = \frac{6400^2}{600000} = 6.82\,\text{W} \tag{8.116}$$

The total dielectric loss can be determined as,

$$P_{dl} = V^2\omega C\delta = 6400^2 \times 2\pi \times 50 \times 0.84 \times 10^{-6} \times 0.03 = 324.27\,\text{W} \tag{8.117}$$

The dielectric hysteresis loss can be calculated as,

$$P_{dh} = P_{dl} - P_{lc} = 324.27 - 6.82 = 317.45\,\text{W} \tag{8.118}$$

Practice Problem 8.5

The power loss due to leakage current in the insulation, the capacitance, and the insulation resistance of a single-core 50 Hz cable are 8.5 W, 2 μF, and 4.5 MΩ, respectively. Find the line-to-neutral voltage, total dielectric loss, and dielectric hysteresis loss. Consider the dielectric factor is 0.02.

8.9 Capacitance of a Cable

Capacitance usually forms between the core and the outer sheath due to the voltage difference between the conductor and outer sheath. Let the charge of the capacitor is q, and the capacitance is C. Figure 8.10 shows a cross section of a single-core cable whose core radius is r and the outer sheath radius is R.

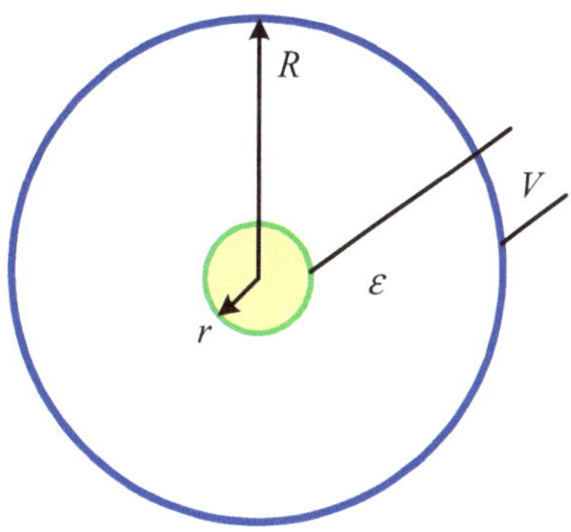

Fig. 8.10 Single-core cable with a sheath

According to Eq. (8.15), the expression of voltage difference between the core and the sheath is

$$V = \frac{q}{2\pi\varepsilon} \ln \frac{R}{r} \tag{8.119}$$

$$q = \frac{2\pi\varepsilon V}{\ln \frac{R}{r}} \tag{8.120}$$

The capacitance per unit length of a single-core cable is usually defined as,

$$C = \frac{q}{V} \tag{8.121}$$

Substituting Eq. (8.120) into Eq. (8.121) yields,

$$C = \frac{\frac{2\pi\varepsilon V}{\ln \frac{R}{r}}}{V} \tag{8.122}$$

$$C = \frac{2\pi\varepsilon}{\ln \frac{R}{r}} \tag{8.123}$$

But, the permittivity of any medium is written as,

$$\varepsilon = \varepsilon_0 \varepsilon_r \tag{8.124}$$

The value of the permittivity for free space is

$$\varepsilon_0 = 8.85 \times 10^{-12}\ \text{F/m} \tag{8.125}$$

Substituting Eqs. (8.124) and (8.125) into Eq. (8.123) yields,

$$C = \frac{2\pi \times 8.85 \times 10^{-12} \varepsilon_r}{\ln \frac{R}{r}} \tag{8.126}$$

$$C = \frac{5.56 \times 10^{-11} \varepsilon_r}{\ln \frac{R}{r}} \quad \text{F/m} \tag{8127}$$

$$C = \frac{0.0556 \varepsilon_r}{\ln \frac{R}{r}} \quad \mu\text{F/km} \tag{8.128}$$

The admittance due to capacitance can be expressed as,

$$Y = \frac{1}{Z_c} = \frac{1}{\frac{1}{\omega C}} \tag{8.129}$$

$$Y = \omega C = 2\pi f C \tag{8.130}$$

The expression of charging current per phase can be written as,

$$I_c = 2\pi f\, C\, V_{L-N} \tag{8.131}$$

Example 8.6
A 2.5 km long single-core cable is used to deliver power in the 6.4 kV, 50 Hz three-phase induction motor in the oil field. The cable core radius is 3 cm, and an outer sheath radius is 11 cm. Calculate the capacitance and the charging current per phase. Assume the relative permittivity is 2.

Solution
The value of the capacitance per km can be determined as,

$$C = \frac{0.0556\varepsilon_r}{\ln\frac{R}{r}} = \frac{0.0556 \times 2}{\ln\frac{11}{3}} = 0.085\ \mu\text{F/km} \tag{8.132}$$

$$C = 0.085 \times 2.5 = 0.21\ \mu\text{F} \qquad (8.132)$$

The value of the voltage per phase is

$$V_p = V_{L-N} = \frac{11 \times 1000}{\sqrt{3}} = 6350.85\ \text{V} \tag{8.133}$$

The value of the charging current per phase can be determined as,

$$I_c = 2\pi f C V_{L-N} = 2\pi \times 50 \times 0.21 \times 10^{-6} \times 6350.85 = 0.42\ \text{A} \tag{8.134}$$

Practice Problem 8.6
A 2 km long single-core cable is having the core and outer sheath radii 1 cm and 2 cm, respectively. The relative permittivity of insulating material is 4. Calculate the capacitance of the cable.

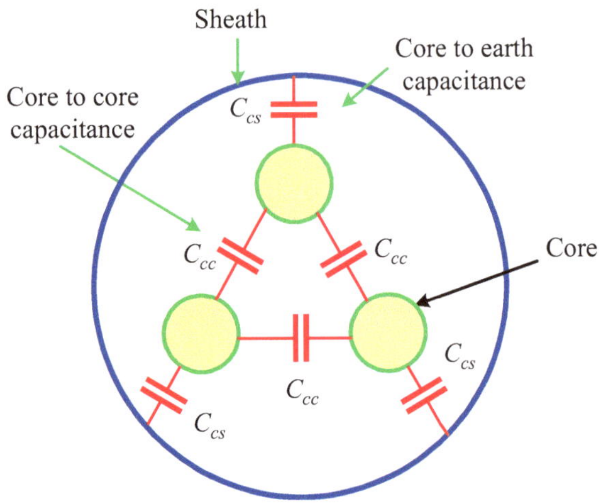

Fig. 8.11 Different types of the capacitance of a cable

8.10 Capacitance of a Three-Core Cable

Consider a three-core cable, as shown in Fig. 8.11. In a cable, the conductors are much closer to each other than to the sheath. In a three-core cable, the voltage energizes three cores, and zero voltage energizes the sheath, which creates electrostatic fields. Due to electrostatic fields around the cable, capacitance forms between core-to-core and core-to-sheath. The calculation of capacitance is simple if the distribution of dielectric is the same. However, it is not possible to get a uniform distribution of dielectrics. Different types of capacitances of a three-core cable are shown in Fig. 8.12.

Consider C_{cc} is the core capacitance formed between the core-to-core, C_{cs} is the sheath capacitance formed between the core-to-sheath, and C_{eq} is the equivalent capacitance.

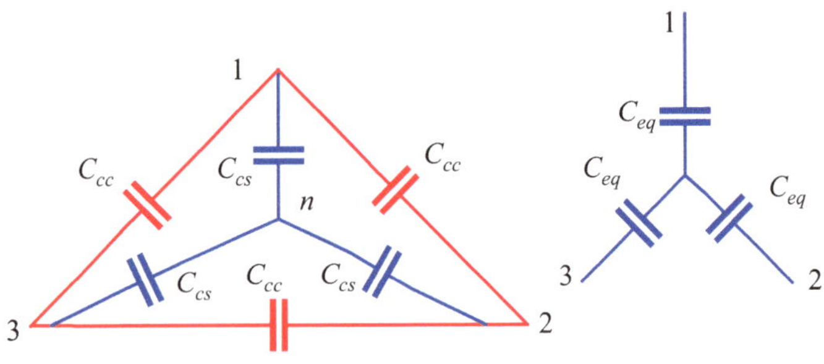

Fig. 8.12 Equivalent capacitance of a cable

According to Fig. 8.12, the three core-to-core capacitances are formed a delta circuit, and it can be converted into a wye circuit. The sheath represents the neutral point, and the core-to-sheath capacitance forms a wye circuit.

The total capacitance between the terminals 1 and 2 for both the delta and the wye connections is the same, and it can be expressed as,

$$C_{12(\Delta)} = C_{cc} + \frac{C_{cc}C_{cc}}{C_{cc} + C_{cc}} = C_{cc} + \frac{C_{cc}}{2} \tag{8.135}$$

$$C_{12(Y)} = \frac{C_{eq}C_{eq}}{C_{eq} + C_{eq}} = \frac{C_{eq}}{2} \tag{8.136}$$

Equations (8.135) and (8.136) are the same, and it becomes,

$$C_{cc} + \frac{C_{cc}}{2} = \frac{C_{eq}}{2} \tag{8.137}$$

$$C_{eq} = 3C_{cc} \tag{8.138}$$

The total capacitance between each conductor to the neutral (sheath represents the neutral) is expressed as,

$$C_n = C_{1n} = C_{2n} = C_{3n} = C_{cs} + 3C_{cc} \tag{8.139}$$

The capacitive reactance can be expressed as,

$$X_n = \frac{1}{\omega(C_{cs} + 3C_{cc})} \tag{8.140}$$

For a given phase voltage, $V_{ph} = V_{L-N}$, the expression of the charging current is expressed as,

$$I_c = 2\pi f V_{L-N}(C_{cs} + 3C_{cc}) \tag{8.141}$$

From Eq. (8.141), the value of the charging can be determined if the other parameters are given.

8.11 Measurement of Capacitance

The dielectric distribution in a cable is not homogeneous. In this situation, measurement of the core-to-core and core-to-sheath capacitances needs to be carried out. Initially, connect the two cores directly to the sheath, as shown in Fig. 8.13. Then, measure the capacitance (C_p) between the remaining core and the sheath. From Fig. 8.13, it is seen that the two core-to-core capacitances are connected in parallel, and the equivalent capacitance is expressed as,

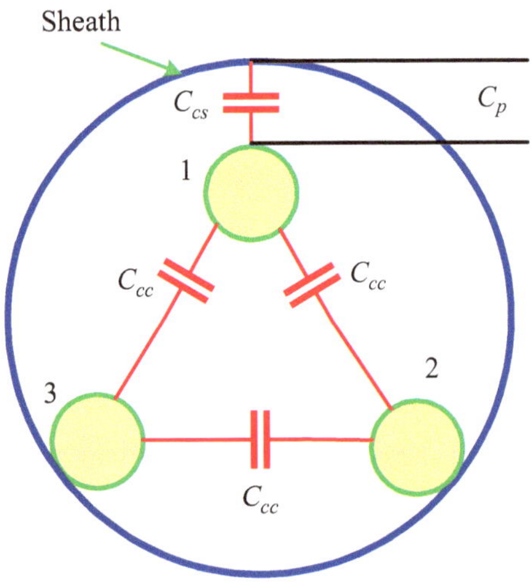

Fig. 8.13 Two cores of a cable connected to sheath

$$C_{eq} = C_{cc} + C_{cc} = 2C_{cc} \tag{8.142}$$

The equivalent capacitance is, therefore, parallel with the core-to-sheath capacitance, and the expression of total capacitance is written as,

$$C_p = C_{cs} + 2C_{cc} \tag{8.143}$$

For the second measurement, three cores are connected in such a way to eliminate core-to-core capacitances, as shown in Fig. 8.14. From Fig. 8.14, it is observed that the three core-to-sheath capacitances are connected in parallel with the common point at the sheath. Therefore, their equivalent capacitance is written as,

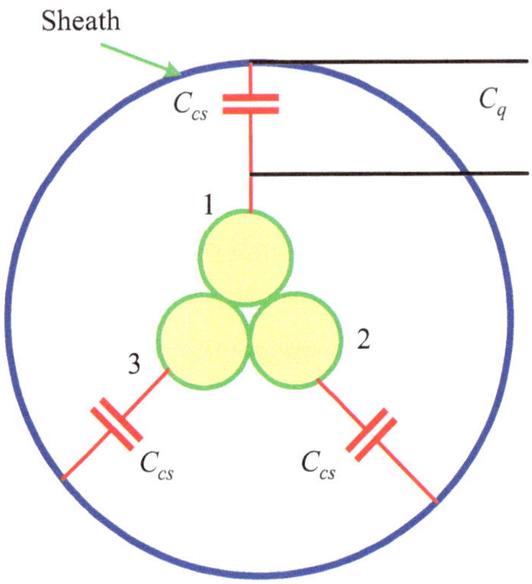

Fig. 8.14 Three cores of a cable close to each other

$$C_q = C_{cs} + C_{cs} + C_{cs} = 3C_{cs} \tag{8.144}$$

Substituting Eq. (8.144) into Eq. (8.143) yields,

$$C_p = \frac{C_q}{3} + 2C_{cc} \tag{8.145}$$

$$C_{cc} = \frac{C_p}{2} - \frac{C_q}{6} \tag{8.146}$$

Substituting Eqs. (8.144) and (8.146) into the Eq. (8.139) yields,

$$C_n = \frac{C_q}{3} + 3\left(\frac{C_p}{2} - \frac{C_q}{6}\right) \tag{8.147}$$

$$C_n = \frac{1}{6}\left(9C_p - C_q\right) \tag{8.148}$$

From Eq. (8.148), it is seen that the neutral capacitance can be calculated if the other parameters are known.

Example 8.7
A 6.4 kV, 50 Hz three-phase system uses 8 km long three-core cable. The capacitance between three bunched cores and sheath is measured and found 0.94 μF/km. The value of the capacitance is again measured 0.65 μF/km between one core and the sheath when the other two cores that are connected to the sheath. Calculate the core-to-sheath capacitance, core-to-core capacitance, line-to-neutral capacitance, and charging current per phase.

Solution
The value of the core-to-sheath capacitance per km can be determined as,

$$C_q = 0.94 = 3C_{cs} \tag{8.149}$$

$$C_{cs} = 0.31\ \mu\text{F/km} \tag{8.150}$$

The actual value of the core-to-sheath capacitance is calculated as,

$$C_{cs} = 0.31 \times 8 = 2.48\ \mu\text{F} \tag{8.151}$$

The value of the core-to-core capacitance per km can be determined as,

$$C_{cc} = \frac{C_p}{2} - \frac{C_q}{6} = \frac{0.65}{2} - \frac{0.94}{6} = 0.17\ \mu\text{F/km} \tag{8.152}$$

The actual value of the core-to-core capacitance is calculated as,

$$C_{cc} = 0.17 \times 8 = 1.35\,\mu\text{F} \tag{8.153}$$

The value of the line-to-neutral capacitance is calculated as,

$$C_n = C_{cs} + 3C_{cc} = 2.48 + 3 \times 1.35 = 6.53\,\mu\text{F} \tag{8.154}$$

The value of the charging current per phase can be determined as,

$$I_c = 2\pi f V_{L-N} C_n = 2\pi \times 50 \times \frac{6400}{\sqrt{3}} \times 6.53 \times 10^{-6} = 7.58\,\text{A} \tag{8.155}$$

Practice Problem 8.7

A 10 km long three-core cable is used on the 11 kV, 60 Hz, three-phase system. The measured capacitance between the three bunched cores and the sheath of a cable is found to be 0.95 μF/km. The value of the capacitance is again measured 0.40 μF/km between one core and the sheath when the other two cores that are connected to the sheath. Determine the line-to-neutral capacitance and charging current per phase.

8.12 Measurement of Insulation Resistance

Electrical equipment and high voltage cables need proper insulation resistance for safe operation. High voltage cables are insulated using different types of materials with high electrical resistance to stop the flow of current outside of the conductor or core. However, the properties of these insulating materials change due to the long time in operation and environmental effects. These changes reduce the electrical resistivity of the insulating materials. As a result, increasing leakage currents in the cable leads to incidents that may seriously affect operator safety as well as increase the operation cost.

Two types of tests are usually carried out to identify the cable performance, namely dielectric strength test (withstand test or destructive test) and non-destructive test [7]. In a destructive test, high voltage and current are applied to expose weak insulation. However, in a non-destructive test, a megohmmeter is used to generate lower currents and DC voltage in different magnitudes, which in turn provides the insulation resistance in kΩ, MΩ, GΩ, and TΩ.

Fluke Corporation manufactured a high voltage insulation tester F1555 (see Fig. 8.15) is widely used to measure the insulation resistance of high voltage cables. This meter can measure the insulation resistance from the test voltages 250 V to 10000 V (50/100 V steps) such as the step patterns are 250 V-500 V-1 kV-2.5 kV-5 kV-10 kV.

Fig. 8.15 Insulation tester F1555 courtesy by Fluke Corporation

A few parameters, such as resistance, leakage current, polarization index (PI), dielectric absorption ratio (DAR), are usually displayed. According to IEEE 43-2013 Standard, the polarization index and DAR are defined in the following ways.

Polarization Index (*PI*): After voltage application, two readings are taken at 1 min and 10 min. The ratio of the insulation resistance measured at 10 min to the insulation resistance measured at 1 min is known as the polarization index. Mathematically, it can be expressed as,

$$PI = \frac{IR_{10\min}}{IR_{1\min}} \tag{8.156}$$

Table 8.1 *PI* values corresponding insulation conditions

PI value	Insulation condition
<1	Bad insulation
1–2	Questionable
2–4	Good
>4	Excellent

where

$IR_{1\min}$ is the insulation resistance reading after the application of voltage for 1 min,
$IR_{10\min}$ is the insulation resistance reading after the application of voltage for 10 min.

The *PI* values with corresponding insulation conditions are shown in Table 8.1.

Dielectric Absorption Ratio (*DAR*): There are some insulation materials in which the absorption current decreases sharply. In this situation, measurements after 30 and 60 s are more than enough to qualify the insulation of the material. Mathematically, DAR is defined as,

$$DAR = \frac{IR_{60s}}{IR_{30s}} \tag{8.157}$$

The *DAR* values with corresponding insulation conditions are shown in Table 8.2.

In addition to *PI* and *DAR*, the dielectric discharge (*DD*) test is also carried out in a high voltage cable. The DD test is an insulation diagnostic test that identifies the cable aging, voids, and deterioration of insulation. This test was originally developed by EDF, France's Power Utility Company. Initially, the high voltage cable needs to charge for 30 min for full absorption of the insulating material. Then, the test voltage *V* (volts) is disconnected, and after one minute, the discharge current (mA) and the capacitance *C* (Farads) need to be measured. Mathematically, *DD* can be expressed as,

$$DD = \frac{I_{1\min}}{V \times C} \tag{8.158}$$

The *DD* values with corresponding insulation conditions are shown in Table 8.3.

Table 8.2 *DAR* values corresponding insulation conditions

DAR value	Insulation condition
<1	Poor
1–1.4	Acceptable
1.4–1.6	Excellent

Table 8.3 *DD* values corresponding insulation conditions

DD value	Insulation condition
> 7	Bad
4–7	Poor
2–4	Questionable
<2	Good
0	Homogeneous

References

1. Duncan Glover J, Overbye T, Sarma M (2017) Power system analysis and design, 6th edn. Cengage Learning, USA, pp 1–942
2. Nagsarkar TK, Sukhija MS (2014) Power system analysis, 2nd edn. Oxford University Press, Oxford, pp 1–726
3. Wildi T (2014) Electrical machines, drives and power systems, 6th edn. Pearson Education Ltd, USA, pp 1–920
4. Sadat H (2010) Power system analysis, 3rd edn. PSA Publisher, USA, pp 1–772
5. Wildi T (2006) Electrical machines, drives and power systems, 6th edn. Pearson Education, USA, pp 1–934
6. El-Hawary ME (1995) Electrical power systems, Revised Edition. Willey-IEEE Press, USA, pp 1–808
7. IEEE 43–2013 (2013) IEEE recommended practice for testing insulation resistance of electric machinery. In: IEEE power and energy society. USA, pp 1–37

Exercise Problems

8.1 The insulation thickness and radius of a 5 km cable are 9 mm and 16 mm, respectively. Calculate the insulation resistance of a cable with $\rho = 3 \times 10^{10}\,\Omega - \mathrm{m}$.

8.2 The insulation thickness diameter and the core diameter of 10 km cable are 4 cm and 9 cm, respectively. Find the cable resistivity if the insulation resistance is 350 MΩ.

8.3 A single-core cable is energized by a phase voltage of 11 kV (rms). The conductor radius and the sheath radius are given as 0.6 cm and 1.9 cm, respectively. Calculate the maximum stress, minimum stress, insulation thickness, and most economical conductor size.

8.4 The core and sheath diameters of a 33 kV (rms) single-core cable are 2.2 cm and 6.4 cm, respectively. Find the maximum stress, minimum stress, and most economical conductor size.

8.5 A single-core cable is having the maximum and minimum stresses are 33 kV/cm and 8 kV/cm, respectively. If the radius of the core is 1.1 cm, calculate the insulation thickness and operating voltage.

8.6 The minimum stress of an 11 kV (rms) single-core cable is 10 kV/cm. The sheath radius of the cable is found to be 3 cm. Determine the core radius and the most economical conductor size.

8.7 A cable used capacitance grading by three dielectrics whose permittivities are 5, 3, and 2.5, respectively. The conductor radius and the overall radius are found to be 0.9 cm and 5.5 cm, respectively. Consider three dielectrics are worked at the 20 kV/cm. Calculate the radii r_1 and r_2, maximum voltage, and safe working voltage.

8.8 The core and sheath radii of a single-core cable are found to 1.5 cm and 6 cm, respectively. The cable is graded by two dielectrics whose permittivities are given 4 and 3. The maximum stresses of these two dielectrics are 35 and 25 kV/cm. Calculate the inner radial thickness, outer radial thickness, and safe working voltage.

8.9 The core and outer sheath radii of a three-phase 11 kV (rms) cable are 1 cm and 3.5 cm, respectively. The 1.5 cm and 2.5 cm radii of intersheaths are inserted in between the core and an outer sheath. If the maximum stress in each layer is the same, find the voltages on the first sheath and the second sheath.

8.10 A single-core 50 Hz, 11 kV (line-to-neutral) cable is having the insulation resistance and the capacitance 3 MΩ and 1.2 μF, respectively. Calculate the power loss due to leakage current flows in the insulation, total dielectric loss, and dielectric hysteresis loss. Consider the dielectric factor is 0.03.

8.11 The insulation resistance of a single-core 50 Hz, 11 kV (line-to-line) cable is 4.4 MΩ. The total dielectric loss of the cable is found to be 1.2 kW. Find the power loss due to leakage current flows in the insulation, capacitance of the cable, and dielectric hysteresis loss. Assume that the dielectric factor is 0.05.

8.12 The core and sheath radii of a 3 km long single-core 11 kV, 60 Hz, three-phase system cable are 3 cm and 8 cm, respectively. Calculate the capacitance and charging current per phase. Assume that the relative permittivity is 4.

8.13 The core radius and charging current of a 1.5 km long 11 kV, 60 Hz, three-phase system single-core cable are given as 1.2 cm and 0.95 A, respectively. The relative permittivity of insulating material is given as 3. Calculate the capacitance and sheath radius.

8.14 A 1.5 km long three-core cable is used in the 0.44 kV, 50 Hz, three-phase system. The per kilometer capacitance measured between three bunched cores and sheath is 0.68 μF. The value of the capacitance is again measured 0.12 μF/km between one core and the sheath when the other two cores that are connected to the sheath. Find the core-to-sheath capacitance, core-to-core capacitance, line-to-neutral capacitance, and charging current per phase.

8.15 The 11 kV, 50 Hz three-phase voltage is connected to an industry by the 5 km long three-core cable. The measured capacitance between the three bunched cores and sheath is found to be 0.88 μF/km. Two cores are placed on the sheath, and the measured capacitance between the third core and sheath is found to be 0.50 μF/km. Calculate the core-to-sheath capacitance, core-to-core capacitance, and line-to-neutral capacitance.

Chapter 9
Power System Stability Analysis

9.1 Introduction

The high demand for electrical power is becoming apparent due to increasing population, commercial buildings, and the industrial sector. As a result, the size and the interconnected networks are increasing to meet this high demand. Power utility companies are giving more importance for maximum power transfer from generating station to the consumer terminals through large interconnected networks. Sometimes, it is difficult to transfer maximum power with the presence of many synchronous machines along with large interconnected networks due to different forms of disturbances. These disturbances are classified into two categories, namely small disturbance (perturbation) and large disturbance (perturbation). Random changes in the load that occur in the system continuously are an example of small perturbation. A power system may be subjected to large perturbations such as the occurrence of faults on the line, loss of large generating units, loss of major transmission facilities, and loss of large loads. Any disturbances in the power system hamper the power system stability. The stability means that any system will remain in stable condition after disturbance. The stability of the power system is defined as the ability of the system to remain in the state of equilibrium or synchronism after disturbances occur on the system. Depending on nature and the magnitude, stability studies are classified into three categories, namely transient stability, steady stability, and dynamic stability.

Transient stability is the ability of a system to respond to a large disturbance within a short duration. Transient stability is normally occurred due to line switching, sudden changes of load, and power transfer. The system response to this disturbance is normally counted within 1 s. In practice, transient stability is determined by computer calculations, whereas manual calculation is only possible in simple cases.

Md. A. Salam, *Fundamentals of Electrical Power Systems Analysis*,
https://doi.org/10.1007/978-981-15-3212-2_9

Steady-state stability is the ability of the system to regain or remain in synchronism when subject to slow and small disturbances.

Dynamic stability is the ability of the system to respond to small disturbances. Due to these small disturbances, oscillations are produced on the system. The system is said to be dynamically stable due to smaller amplitudes of the oscillation and does not acquire more than a certain amplitude and die out quickly. If these oscillations continuously grow in amplitude, the system is dynamically unstable. These types of disturbances normally occur due to an interaction between control systems. The duration of dynamic stability is from 5 to 10 s, and sometimes up to 30 s. The dynamic stability of a given power system can be improved through the use of power system stabilizers.

Single machine to an infinite bus, swing equation, equal area criterion, and different types of stability analysis, etc. will be discussed in this chapter.

9.2 Pattern of Stability Curves

The pattern of the different types of stability curves depends on the nature of the faults and its duration in the power system. In steady-state stability, the synchronous generator regains synchronism after a small disturbance, as shown in Fig. 9.1.

In transient stability, the generator speeds up for a short duration after losing synchronism, as shown in Fig. 9.2.

The dynamic stability is the sustained rotor angle oscillations of a synchronous generator caused by a fault, as shown in Fig. 9.3.

Fig. 9.1 Steady-state stability curve

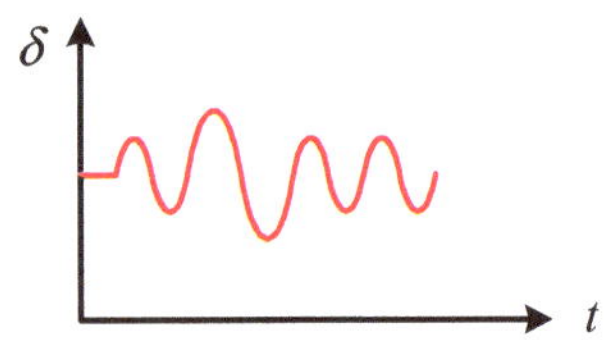

Fig. 9.2 Transient stability curve

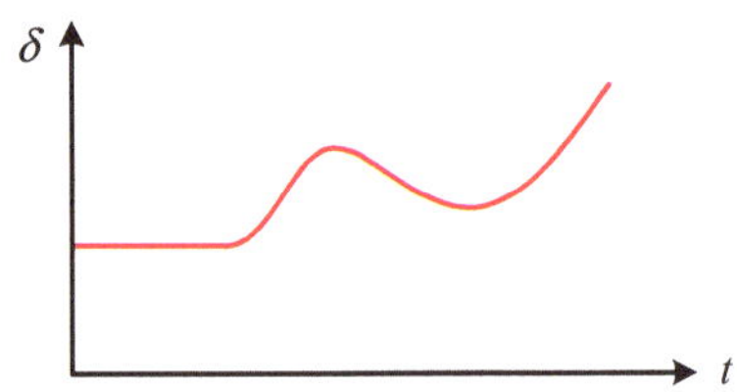

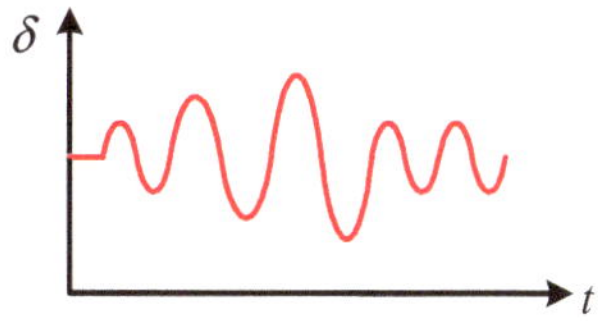

Fig. 9.3 Dynamic stability curve

9.3 Synchronous Machine Dynamics

A synchronous machine or alternator delivers instantaneous power ($p = P + p_i$). The expression of power with a sinusoidal current can be expressed as,

$$p = i_2 R = RI^2 \sin^2 \omega t = I^2 R(1 - \cos 2\omega t) = P + p_i \tag{9.1}$$

Here, the expression of active power due to fundamental voltage and current is,

$$P = I^2 R = \frac{V^2}{R} \tag{9.2}$$

The intrinsic power (p_i) is expressed as,

$$p_i = -P \cos 2\omega t \tag{9.3}$$

In steady-state, this intrinsic power cannot be supplied by the prime mover. Generally, it is supported by the kinetic energy ($KE = J\omega^2/2$) stored in the rotating masses. Here, J is the total momentum (alternator rotor plus prime mover) of inertia and its unit is kg m^2. The kinetic energy of a synchronous machine in Mega-Joules can be written as,

$$K \cdot E = \frac{1}{2} J\omega_{sm}^2 \times 10^{-6} \quad \text{MJ} \tag{9.4}$$

where ϖ_{sm} is the synchronous speed in rad/s.

The relationship between the electrical and mechanical angles can be written as,

$$\theta_e = \frac{P}{2}\theta_m \tag{9.5}$$

According to Eq. (9.5), the synchronous speed in electrical radians is expressed as,

$$\omega_s = \frac{P}{2}\omega_{sm} \tag{9.5}$$

$$\omega_{sm} = \frac{2}{P}\omega_s \tag{9.6}$$

Substituting Eq. (9.6) into Eq. (9.4) yields,

$$K \cdot E = \frac{1}{2}J\left(\frac{2}{P}\omega_s\right)^2 \times 10^{-6} \quad \text{MJ} \tag{9.7}$$

$$K \cdot E = \frac{1}{2}\left(J\left(\frac{2}{P}\right)^2 \omega_s \times 10^{-6}\right)\omega_s \quad \text{MJ} \tag{9.8}$$

$$K \cdot E = \frac{1}{2}M\omega_s \quad \text{MJ} \tag{9.9}$$

The moment of inertia (M) in MJ-s/electrical radians is,

$$M = J\left(\frac{2}{P}\right)^2 \omega_s \times 10^{-6} \tag{9.10}$$

The inertia constant (H) is defined as the ratio of the kinetic energy of a rotor of a synchronous machine to the rating (S) of a machine (in MVA), and it can be expressed as,

$$H = \frac{K \cdot E}{S} \tag{9.11}$$

Substituting Eq. (9.9) into Eq. (9.11) yields,

$$H = \frac{1}{2}\frac{M\omega_s}{S} \tag{9.12}$$

$$SH = K \cdot E = \frac{1}{2}M\omega_s \quad \text{MJ} \tag{9.13}$$

$$M = \frac{2SH}{\omega_s} = \frac{SH}{\pi f_s} \quad \text{MJ s/elec rad} \tag{9.14}$$

$$M = \frac{2SH}{\omega_s} = \frac{SH}{180 f_s} \quad \text{MJ s/elec deg} \tag{9.15}$$

9.4 Single Machine with Infinite Bus

Figure 9.4 shows a single-line diagram where the generator is connected to an infinite bus through a lossless transmission line. An infinite bus is an ideal voltage source that maintains constant values of voltage magnitude, frequency, and phase angle. Here, δ is the angle of the machine with respect to the infinite bus. This angle is known as torque or load angle. The torque angle or load angle is defined as the angular displacement in electrical degrees from the synchronously rotating reference axis. The generator delivers active power to the infinite bus through transmission lines. The synchronous machine is represented by a voltage source $\left(E\,\underline{|\delta}\right)$ behind the direct axis reactance $\left(X'_d\right)$. The equivalent circuit of a synchronous machine is shown in Fig. 9.5.

The voltage at the generator terminal can be written as,

$$V_s = E\,\underline{|\delta} \tag{9.16}$$

The voltage at the infinite bus is,

$$V_r = V\,\underline{|0^\circ} \tag{9.17}$$

The current in the circuit in Fig. 9.5 can be written as [1, 2],

$$I_s = \frac{V_s - V_r}{jX'_d + jX_{\mathrm{TL}}} = \frac{V_s - V_r}{jX} \tag{9.18}$$

Substituting Eqs. (9.16) and (9.17) into Eq. (9.18) yields,

$$I_s = \frac{E\,\underline{|\delta} - V\,\underline{|0^\circ}}{jX} \tag{9.19}$$

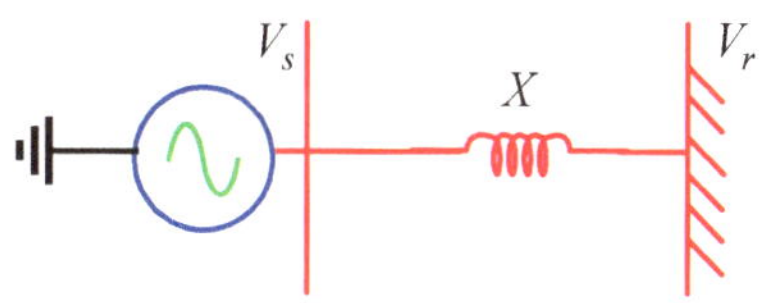

Fig. 9.4 Synchronous machine with an infinite bus

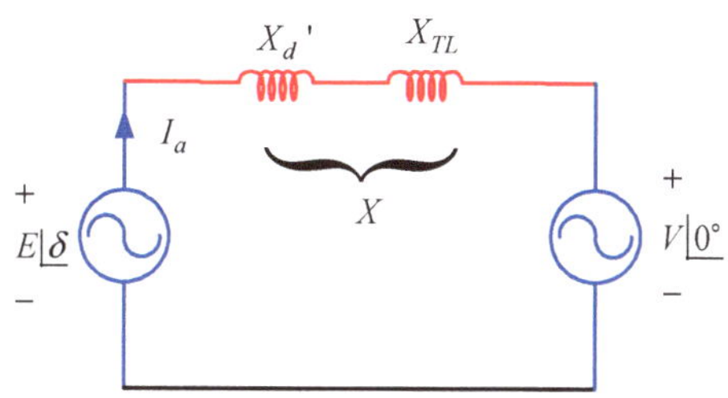

Fig. 9.5 Equivalent circuit of a machine to an infinite bus

The conjugate value of the current is,

$$I_s^* = \frac{E\,\underline{|-\delta} - V\,\underline{|0^\circ}}{X\,\underline{|-90^\circ}} \tag{9.20}$$

The single-phase sending end real power and reactive power can be derived as,

$$S = P_s + jQ_s = V_s I_s^* \tag{9.21}$$

Substituting Eqs. (9.16) and (9.21) into Eq. (9.21) yields,

$$S = P_s + jQ_s = E\,\underline{|\delta}\,\frac{E\,\underline{|-\delta} - V\,\underline{|0^\circ}}{X\,\underline{|-90^\circ}} \tag{9.22}$$

$$P_s + jQ_s = \frac{E^2}{X}\,\underline{|90^\circ} - \frac{EV}{X}\,\underline{|90^\circ + \delta} \tag{9.23}$$

$$P_s + jQ_s = \frac{E^2}{X}(\cos 90^\circ + j\sin 90^\circ) - \frac{EV}{X}(\cos(90^\circ + \delta) + j\sin(90^\circ + \delta)) \tag{9.24}$$

$$P_s + jQ_s = j\frac{E^2}{X} + \frac{EV}{X}\sin\delta - j\frac{EV}{X}\cos\delta \tag{9.25}$$

Equating the real and imaginary parts of Eq. (9.25) yields,

$$P_s = \frac{EV}{X}\sin\delta \tag{9.26}$$

$$Q_s = \frac{E^2}{X} - \frac{EV}{X}\cos\delta \tag{9.27}$$

For lossless transmission lines, the sending power is equal to the receiving end power. In case of three-phase system, this power can be written as,

$$P_e = P_s = P_r = 3\frac{EV}{X}\sin\delta \tag{9.28}$$

At $\delta = 90^\circ$, $P_e = P_{\max}$, Eq. (9.28) becomes,

$$P_{\max} = 3\frac{EV}{X} \tag{9.29}$$

Substituting Eq. (9.29) into Eq. (9.28) yields,

$$P_e = P_{\max} \sin \delta \tag{9.30}$$

At $\delta = \delta_0$, $P_e = P_0$, Eq. (9.30) becomes,

$$P_0 = P_{\max} \sin \delta_0 \tag{9.31}$$

$$\sin \delta_0 = \frac{P_0}{P_{\max}} \tag{9.32}$$

$$\delta_0 = \sin^{-1}\left(\frac{P_0}{P_{\max}}\right) \tag{9.33}$$

The output power of a synchronous machine depends on the torque produced by the mechanical turbine and generator angular speed. In this case, the electrical power output of a synschronous machine can be written as,

$$P_e = T \times \omega_s \tag{9.34}$$

$$T = \frac{P_e}{\omega_s} \tag{9.35}$$

Substituting Eq. (9.30) into Eq. (9.35) yields,

$$T = \frac{P_{\max}}{\omega_s} \sin \delta = T_{\max} \sin \delta \tag{9.36}$$

where the expression of maximum torque is,

$$T_{\max} = \frac{P_{\max}}{\omega_s} \tag{9.37}$$

Substituting Eq. (9.29) into Eq. (9.37) yields,

$$T_{\max} = 3\frac{EV}{X\omega_s} \tag{9.38}$$

The variations of power and torque at different torque angles are shown in Fig. 9.6. From Eqs. (9.30) to (9.36), it is seen that the power and torque vary sinusoidally with a power or torque angle. Torque and power will increase until the value of the torque angle at 90°. If the prime-mover input to the generator increases further, the power output will decrease and will be zero at the torque angle 180°. As a result, the excess power goes into accelerating the generator, thereby increasing its speed and causing it to pull out of synchronism. Hence, the steady-state stability limit is reached when $\delta = 90°$. For normal steady-state operating conditions, the value of the power angle or torque angle is usually less than 90°.

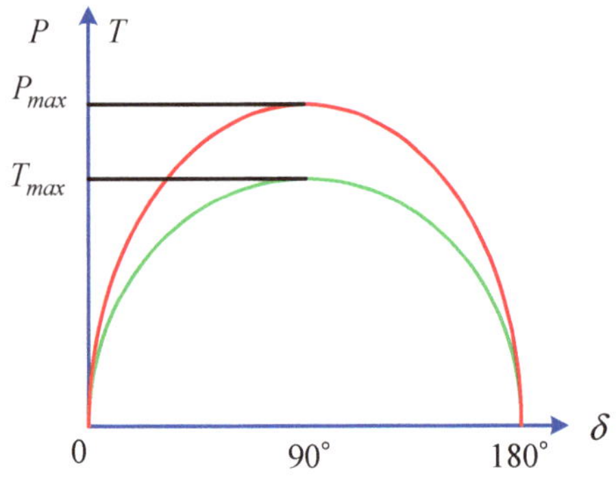

Fig. 9.6 Variation of power and torque with a torque angle

Example 9.1
The terminal voltage of a three-phase wye connected synchronous generator is 33 kV. The generator is connected to an infinite bus through a transmission line. The reactance of the generator and transmission line are 1.5 Ω and 0.6 Ω, respectively. The generator delivers the power of 60 MW to the infinite bus, and the infinite bus voltage is found to be 11 kV. Calculate the torque angle, transmission line current, and three-phase reactive power consumed by the transmission line.

Solution
Per phase synchronous generator voltage is,

$$E_f = \frac{33}{\sqrt{3}} = 19052.55 \text{ V} \tag{9.39}$$

The per phase voltage at the infinite bus is calculated as,

$$V_{\text{ib}} = \frac{11}{\sqrt{3}} = 6350.85 \text{ V} \tag{9.40}$$

The value of the torque angle can be determined as,

$$60 = 3\frac{\frac{11}{\sqrt{3}}\frac{33}{\sqrt{3}}}{(0.6+1.5)}\sin\delta \tag{9.41}$$

$$\delta = 20.31^\circ \tag{9.42}$$

The value of the current in the transmission line can be calculated as,

$$I_L = \frac{19052.55\,\angle 20.31^\circ - 6350.85\,\angle 0^\circ}{2\,\angle 90^\circ} = 6640.38\,\angle -60.14^\circ \text{ A} \tag{9.43}$$

The value of the three-phase reactive power consumed can be calculated as,

$$Q = 3I_L^2 X_L = 3 \times 6640.38^2 \times 0.6 = 79.37\,\text{Mvar} \tag{9.44}$$

Practice Problem 9.1
A three-phase synchronous generator is having the terminal voltage 22 kV, and it is connected to an infinite bus through a transmission line. The reactance of the generator and transmission line are 0.9 Ω and 2.2 Ω, respectively. The torque angle is found to be 35°, and the bus voltage at the infinite bus is found to be 6.6 kV. Find the three-phase real power delivers to the infinite bus and the transmission line current.

9.5 Swing Equation

The swing equation is very important to study the stability analysis of a power system. Under steady-state conditions, the rotor of the synchronous generator rotates smoothly. However, the rotor will accelerate or decelerate with respect to the rotating air gap of the generator and start oscillation based on the disturbance. The equation representing this rotor oscillation is known as the swing equation. Consider a three-phase synchronous machine is connected with a prime mover. The prime mover with the torque of T_m drives a three-phase synchronous generator, which develops an electromagnetic torque T_e as shown in Fig. 9.7. According to Newton's second law, the rotor motion can be expressed as [2],

$$J\alpha_m + T_{f\&d} + T_e = T_m \tag{9.45}$$

Equation (9.45) can be modified by neglecting the friction and damping torque $(T_{f\&d})$ as,

$$J\alpha_m = T_m - T_e = T_a \tag{9.46}$$

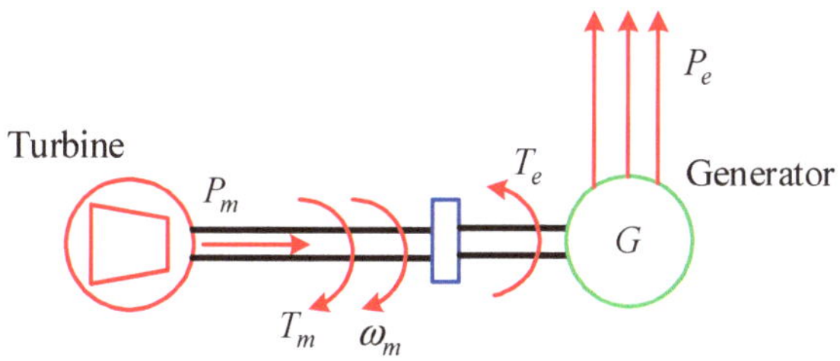

Fig. 9.7 Generator with the turbine

where

J is the combined moment of inertia of the generator and prime mover in kg m^2,
T_a is the accelerating torque in N m,
T_m is the mechanical torque in N m,
T_e is the electromagnetic torque in N m,
α_m is the rotor angular acceleration in rad/s^2.

The rotor angular acceleration is defined as the time rate of change of angular speed, and it can be expressed as,

$$\alpha_m = \frac{d\omega_m}{dt} \tag{9.47}$$

However, the rotor angular velocity (ω_m) is defined as the time rate of change of angular displacement (θ_m), and it can be expressed as,

$$\omega_m = \frac{d\theta_m}{dt} \tag{9.48}$$

Substituting Eq. (9.47) into Eq. (9.48) yields,

$$\alpha_m = \frac{d}{dt}\left(\frac{d\theta_m}{dt}\right) = \frac{d^2\theta_m}{dt^2} \tag{9.49}$$

Substituting Eq. (9.49) into Eq. (9.46) yields,

$$J\frac{d^2\theta_m}{dt^2} = T_m - T_e = T_a \tag{9.50}$$

where

θ_m is the rotor angular position with respect to fixed axis in rad,
ω_m is the rotor angular velocity in rad/s.

The main interest for stability study is the rotor speed with respect to the fixed reference axis on the stator as shown in Fig. 9.8. The rotor angular position with reference to the stator axis is shown in Fig. 9.9, and the rotor angular position with respect to a reference axis which rotates at synchronous speed is given by,

$$\theta_m = \omega_{sm}t + \delta_m \tag{9.51}$$

where the following parameters are,

ω_{sm} is the synchronous angular velocity of the rotor in rad/s,
δ_m is the rotor angular position with respect to synchronously rotating reference axis in rad.

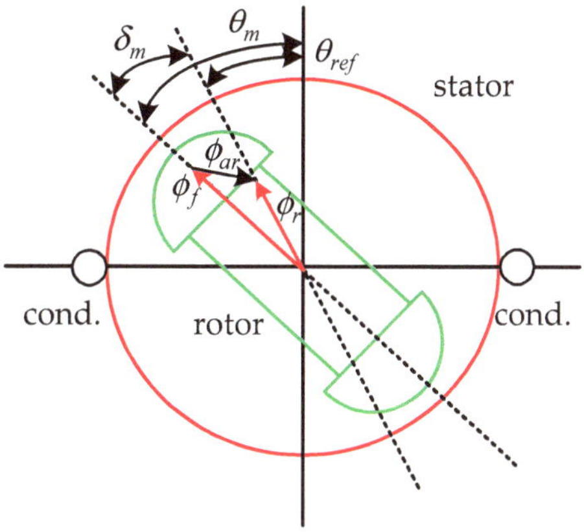

Fig. 9.8 Rotor with the reference axis

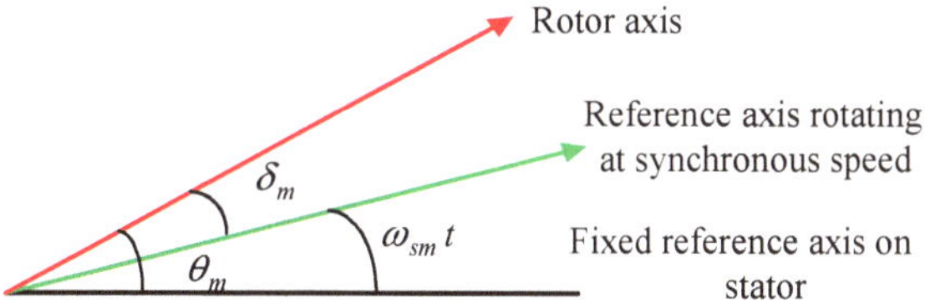

Fig. 9.9 Rotor angular position to the reference axis

Differentiating Eq. (9.51) with respect to the time yields,

$$\frac{d\theta_m}{dt} = \omega_{sm} + \frac{d\delta_m}{dt} \tag{9.52}$$

Again differentiating Eq. (9.52) with respect to time yields,

$$\frac{d^2\theta_m}{dt^2} = \frac{d^2\delta_m}{dt^2} \tag{9.53}$$

Substituting Eq. (9.53) into Eq. (9.50) yields,

$$J\frac{d^2\delta_m}{dt^2} = T_m - T_e = T_a \tag{9.54}$$

Multiplying Eq. (9.54) by ω_m yields,

$$J\omega_m\frac{d^2\delta_m}{dt^2} = \omega_m T_m - \omega_m T_e \tag{9.55}$$

In general, the power is equal to torque times the angular velocity. Therefore, Eq. (9.55) is again modified as,

$$J\omega_m \frac{d^2\delta_m}{dt^2} = P_m - P_e \tag{9.56}$$

The angular momentum (M) is equal to the moment of inertia (J) times angular synchronous velocity ($\omega_m = \omega_s$). Therefore, Eq. (9.56) can be modified as,

$$M\frac{d^2\delta_m}{dt^2} = P_m - P_e \tag{9.57}$$

The angular momentum (M) is constant before the stability of a synchronous machine is disturbed or lost. In this region, the ω_m is constant, and it can be represented by ω_{sm}. Equation (9.14) can be modified as,

$$M = \frac{2HS}{\omega_{sm}} \tag{9.58}$$

$$S = \frac{M\omega_{sm}}{2H} \tag{9.59}$$

Substituting Eq. (9.58) into Eq. (9.57) yields,

$$\frac{2HS}{\omega_{sm}}\frac{d^2\delta_m}{dt^2} = P_m - P_e \tag{9.60}$$

Equation (9.57) is known as swing equation. It is more convenient to express the swing equation in terms of electrical power angle rather mechanical power angle. Let the number of poles of a synchronous generator is P. The relationship between the electrical power angle δ and the mechanical power angle δ_m is,

$$\delta = \frac{P}{2}\delta_m \tag{9.61}$$

According to Eq. (9.61), the electrical radian frequency is expressed as,

$$\omega = \frac{P}{2}\omega_m \tag{9.62}$$

Similarly, the synchronous electrical radian frequency is expressed as,

$$\omega_s = \frac{P}{2}\omega_{\mathrm{sm}} \tag{9.63}$$

Substituting Eq. (9.63) into Eq. (9.58) yields,

$$M = \frac{2HSP}{2\omega_s} \tag{9.64}$$

$$\frac{M}{P} = \frac{HS}{\omega_s} \tag{9.65}$$

$$\frac{2M}{P} = \frac{2HS}{\omega_s} \tag{9.65}$$

Substituting Eq. (9.65) into Eq. (9.60) yields,

$$\frac{2M}{P}\frac{d^2\delta}{dt^2} = P_m - P_e \tag{9.66}$$

Power system analysis is usually done based on the power unit values. Therefore, dividing Eq. (9.66) by the base power S_b yields,

$$\frac{2M}{P}\frac{1}{S_b}\frac{d^2\delta}{dt^2} = \frac{P_m}{S_b} - \frac{P_e}{S_b} \tag{9.67}$$

$$\frac{2M}{P}\frac{1}{S_b}\frac{d^2\delta}{dt^2} = P_{m(\text{pu})} - P_{e(\text{pu})} \tag{9.68}$$

Substituting Eq. (9.59) into Eq. (9.68) yields,

$$\frac{2M}{P}\frac{2H}{M\omega_{\text{sm}}}\frac{d^2\delta}{dt^2} = P_{m(\text{pu})} - P_{e(\text{pu})} \tag{9.69}$$

$$\frac{2}{P}\frac{2H}{\omega_{sm}}\frac{d^2\delta}{dt^2} = P_{m(\text{pu})} - P_{e(\text{pu})} \tag{9.70}$$

Substituting Eq. (9.63) into Eq. (9.70) yields,

$$\frac{2}{P}\frac{2H}{1}\frac{P}{2\omega_s}\frac{d^2\delta}{dt^2} = P_{m(\text{pu})} - P_{e(\text{pu})} \tag{9.71}$$

$$\frac{2H}{\omega_s}\frac{d^2\delta}{dt^2} = P_{m(\text{pu})} - P_{e(\text{pu})} \tag{9.72}$$

Again, expressing ω_s in terms of frequency and the subscript pu is omitted from Eq. (9.72) to express the general swing equation as,

$$\frac{2H}{2\pi f_0}\frac{d^2\delta}{dt^2} = P_m - P_e \tag{9.73}$$

$$\frac{H}{\pi f_0}\frac{d^2\delta}{dt^2} = P_m - P_e \tag{9.74}$$

In Eq. (9.74), δ represents the load angle of the generator internal emf in radians, and it dominates the amount of power that can be transferred. If the load angle express in electrical degrees, then Eq. (9.74) is further modified as,

$$\frac{H}{180 f_0}\frac{d^2\delta}{dt^2} = P_m - P_e \tag{9.75}$$

Alternative form of Swing equation: Equation (9.52) can be written as,

$$\omega_m = \frac{d\theta_m}{dt} = \omega_{sm} + \frac{d\delta_m}{dt} \tag{9.76}$$

Equation (9.45) can be further modified by introducing damping coefficient D_d (Nms) for mechanical rotational loss due to friction and windage as,

$$J\frac{d\omega_m}{dt} + D_d\,\omega_m = T_m - T_e \tag{9.77}$$

Substituting Eq. (9.76) into Eq. (9.77) yields,

$$J\frac{d}{dt}\left(\omega_{sm} + \frac{d\delta_m}{dt}\right) + D_d\left(\omega_{sm} + \frac{d\delta_m}{dt}\right) = T_m - T_e \tag{9.78}$$

$$J\frac{d^2\delta_m}{dt^2} + D_d\frac{d\delta_m}{dt} = T_m - T_e \tag{9.79}$$

Multiplying both sides of Eq. (9.79) by ω_{sm} yields,

$$J\omega_{sm}\frac{d^2\delta_m}{dt^2} + \omega_{sm}D_d\frac{d\delta_m}{dt} = \omega_{sm}(T_m - T_e) \tag{9.80}$$

$$J\omega_{sm}\frac{d^2\delta_m}{dt^2} + \omega_{sm}D_d\frac{d\delta_m}{dt} = \omega_{sm}\left(\frac{P_m}{\omega_m} - \frac{P_e}{\omega_m}\right) \tag{9.81}$$

During a small disturbance, the speed of a synchronous machine is normally close to synchronous speed so that $\omega_m \approx \omega_{sm}$ and Eq. (9.81) becomes,

$$J\omega_{sm}\frac{d^2\delta_m}{dt^2} + \omega_{sm}D_d\frac{d\delta_m}{dt} = P_m - P_e \tag{9.82}$$

Again, substituting $M_m = J\omega_{sm}$ is the angular momentum of the rotor at synchronous speed and $D_m = \omega_{sm}D_d$ is the damping coefficient in Eq. (9.81) as,

$$M_m\frac{d^2\delta_m}{dt^2} + D_m\frac{d\delta_m}{dt} = P_m - P_e \tag{9.83}$$

Equation (9.83) is the fundamental swing equation derived from the rotor dynamics.

Example 9.2
A 4-pole, 60 Hz, 30 MVA synchronous generator is having the inertia constant 8 MJ/MVA. The input power and output power of the generator are found to be 25 MW and 18 MW, respectively. Calculate the kinetic energy stored in the rotor at synchronous speed, accelerating power, acceleration, and torque angle at 10 cycles.

Solution
The value of the kinetic energy stored in the rotor is calculated as,

$$KE = 30 \times 8 = 240\,\text{MJ} \tag{9.84}$$

The value of the accelerating power is calculated as,

$$P_a = P_m - P_e = 25 - 18 = 7\,\text{MW} \tag{9.85}$$

The value of the acceleration can be determined as,

$$M = \frac{2HS_b}{\omega_{sm}} = \frac{HS_b}{\pi f} = \frac{8 \times 30}{\pi \times 60} = 1.27 \tag{9.86}$$

Substituting necessary values in Eq. (9.66) yields,

$$\frac{2 \times 1.27}{4}\frac{d^2\delta}{dt^2} = 7 \tag{9.87}$$

$$\frac{d^2\delta}{dt^2} = 11.02\,\text{rad/s}^2 \tag{9.88}$$

The value of the torque angle at 10 cycles can be determined as,

$$t = \frac{10}{60} = 0.166\,\text{s} \tag{9.89}$$

Multiply both sides of Eq. (9.80) by $2\frac{d\delta}{dt}$ yields,

$$2\frac{d\delta}{dt} \times \frac{d^2\delta}{dt^2} = 11.02 \times 2\frac{d\delta}{dt} \quad (9.90)$$

Integrating Eq. (9.90) yields,

$$\int 2\frac{d\delta}{dt} \times \frac{d^2\delta}{dt^2} dt = 22.04 \int \frac{d\delta}{dt} dt \quad (9.91)$$

$$\left(\frac{d\delta}{dt}\right)^2 = 22.04\delta + C \quad (9.92)$$

At $t = 0$, $\frac{d\delta}{dt} = 0$ and $C = 0$

$$\left(\frac{d\delta}{dt}\right)^2 = 22.04\delta \quad (9.93)$$

$$\frac{d\delta}{dt} = 4.69\delta^{0.5} \quad (9.94)$$

Integrating Eq. (9.94) yields,

$$\int \delta^{-0.5} d\delta = 4.69 \int dt \quad (9.95)$$

$$\frac{\delta^{-0.5+1}}{-0.5+1} = 4.69t \quad (9.96)$$

$$\delta^{0.5} = 2.34t \quad (9.97)$$

$$\delta = (2.34t)^2 = (2.34 \times 0.166)^2 = 0.15 \text{ rad} \quad (9.98)$$

Practice Problem 9.2

The stored energy in the rotor of a 4-pole, 50 Hz, 25 MVA synchronous generator is found to be 180 MJ. The input power and output power of the generator are written as 35 MW and 28 MW, respectively. Calculate the inertia constant, accelerating power, and acceleration.

9.6 Steady-State Stability Analysis

The steady-state stability of a system triggers a small disturbance. However, the synchronous generator will not lose its synchronism due to a small disturbance. In this condition, mechanical power input (P_m) equals the electrical power output (P_e) of the generator. To study the steady-state stability analysis, consider a single synchronous machine is connected to an infinite bus. Substituting Eq. (9.30) into the swing Eq. (9.74) yields [3],

$$\frac{H}{\pi f_0}\frac{d^2\delta}{dt^2} = P_m - P_{\max}\sin\delta \tag{9.99}$$

Due to a small disturbance, there is a small deviation $\Delta\delta$ in the torque angle from the initial torque angle δ_0. In this condition, the expression of torque angle is expressed as,

$$\delta = \delta_0 + \Delta\delta \tag{9.100}$$

Substituting Eq. (9.100) into Eq. (9.99) yields,

$$\frac{H}{\pi f_0}\frac{d^2(\delta_0 + \Delta\delta)}{dt^2} = P_m - P_{\max}\sin(\delta_0 + \Delta\delta) \tag{9.101}$$

$$\frac{H}{\pi f_0}\frac{d^2\delta_0}{dt^2} + \frac{H}{\pi f_0}\frac{d^2\Delta\delta_0}{dt^2} = P_m - P_{\max}(\sin\delta_0\cos\Delta\delta + \cos\delta_0\sin\Delta\delta) \tag{9.102}$$

The value of the rotor deviation angle $(\Delta\delta)$ is very small. Therefore, $\sin\Delta\delta \cong \Delta\delta$ and $\cos\Delta\delta \cong 1$. Equation (9.102) can be modified as,

$$\frac{H}{\pi f_0}\frac{d^2\delta_0}{dt^2} + \frac{H}{\pi f_0}\frac{d^2\Delta\delta_0}{dt^2} = P_m - P_{\max}(\sin\delta_0 + \Delta\delta\cos\delta_0) \tag{9.103}$$

At $\delta = \delta_0$, the swing equation can be expressed as,

$$\frac{H}{\pi f_0}\frac{d^2\delta_0}{dt^2} = P_m - P_{\max}\sin\delta_0 \tag{9.104}$$

Substituting Eq. (9.104) into Eq. (9.103) yields,

$$\frac{H}{\pi f_0}\frac{d^2\Delta\delta_0}{dt^2} + P_m - P_{\max}\sin\delta_0 = P_m - P_{\max}(\sin\delta_0 + \Delta\delta\cos\delta_0) \tag{9.105}$$

$$\frac{H}{\pi f_0}\frac{d^2\Delta\delta_0}{dt^2} + (P_{\max}\cos\delta_0)\Delta\delta = 0 \tag{9.106}$$

The term $P_{\max}\cos\delta_0$ in Eq. (9.106) is the slope of the power angle curve at an angle δ_0. This term is known as the synchronizing power coefficient (S_p). Equation (9.106) can be modified as,

$$\frac{H}{\pi f_0}\frac{d^2\Delta\delta_0}{dt^2} + S_p\Delta\delta = 0 \tag{9.107}$$

where the synchronizing power coefficient is expressed as,

$$S_p = P_{\max}\cos\delta_0 \tag{9.108}$$

Equation (9.106) can be modified as,

$$\frac{d^2\Delta\delta_0}{dt^2} + \frac{\pi f_0 S_p}{H}\Delta\delta = 0 \tag{9.109}$$

The roots of Eq. (9.109) are calculated as,

$$s^2 = \frac{-\pi f_0}{H}S_p \tag{9.110}$$

The system will lose stability for the negative values of synchronizing power. However, for the positive synchronizing power coefficient, the angular frequency of the undamped oscillations is written as,

$$\omega_n = \sqrt{\frac{2\pi f_0}{2H}S_p} \tag{9.111}$$

$$\omega_n = \sqrt{\frac{\omega_s}{2H}S_p} \tag{9.112}$$

The frequency of oscillation is expressed as,

$$f_n = \frac{1}{2\pi}\sqrt{\frac{\omega_s}{2H}S_p} \tag{9.113}$$

The rotor of a synchronous machine will accelerate or decelerate with respect to synchronously rotating air-gap mmf due to a small disturbance. As a result, an induction motor action will resume between them, and torque will establish on the rotor to reduce the difference between the rotor and air-gap mmf angular velocities. This torque is known as damping torque. According to Eqs. (9.83) and (9.107) can be modified as,

$$\frac{H}{\pi f_0}\frac{d^2\Delta\delta_0}{dt^2} + D_m\frac{d\Delta\delta_0}{dt} + S_p\Delta\delta = 0 \tag{9.114}$$

$$\frac{d^2\Delta\delta_0}{dt^2} + \frac{\pi f_0}{H}D_m\frac{d\Delta\delta_0}{dt} + \frac{\pi f_0}{H}S_p\Delta\delta = 0 \tag{9.115}$$

Again substituting dimensionless damping ratio, ξ, and natural frequency of oscillation, ϖ_n, in Eq. (9.115) yields,

$$\frac{d^2\Delta\delta_0}{dt^2} + 2\xi\omega_n\frac{d\Delta\delta_0}{dt} + \omega_n^2\Delta\delta = 0 \tag{9.116}$$

where the following parameters are written as,

$$\xi = \frac{D_m}{2}\sqrt{\frac{\pi f_0}{H S_p}} \tag{9.117}$$

$$\omega_n = \sqrt{\frac{\pi f_0}{H}S_p} \tag{9.118}$$

The characteristics equation is,

$$s^2 + 2\xi\omega_n s + \omega_n^2 = 0 \tag{9.119}$$

For normal operating condition, $\xi < 1$ and the roots are in complex and it can be expressed as,

$$S_{1,2} = -\xi\omega_n \pm j\omega_n\sqrt{1-\xi^2} = -\xi\omega_n \pm j\omega_d \tag{9.120}$$

where the damped natural frequency (ω_d) is,

$$\omega_d = \omega_n\sqrt{1-\xi^2} \tag{9.121}$$

After some state variable manipulations, the solution of Eq. (9.119) can be written as,

$$\Delta\delta(s) = \frac{(s + 2\xi\omega_n)\Delta\delta_0}{s^2 + 2\xi\omega_n s + \omega_n^2} \tag{9.122}$$

$$\Delta\omega(s) = -\frac{\omega_n^2\Delta\delta_0}{s^2 + 2\xi\omega_n s + \omega_n^2} \tag{9.123}$$

Taking inverse Laplace transform of Eqs. (9.122) and (9.123) yields,

$$\Delta\delta = \frac{\Delta\delta_0}{\sqrt{1-\xi^2}} e^{-\xi\omega_n t} \sin(\omega_d t + \theta) \tag{9.124}$$

$$\Delta\omega = -\frac{\omega_n \Delta\delta_0}{\sqrt{1-\xi^2}} e^{-\xi\omega_n t} \sin \omega_d t \tag{9.125}$$

The rotor angle relative to the synchronously revolving field is expressed as,

$$\delta = \delta_0 + \Delta\delta = \delta_0 + \frac{\Delta\delta_0}{\sqrt{1-\xi^2}} e^{-\xi\omega_n t} \sin(\omega_d t + \theta) \tag{9.126}$$

The rotor angular frequency is expressed as,

$$\delta = \omega_0 + \Delta\omega = \omega_0 - \frac{\Delta\delta_0}{\sqrt{1-\xi^2}} e^{-\xi\omega_n t} \sin \omega_d t \tag{9.127}$$

From Eqs. (9.117) and (9.118), it is seen that the natural frequency and the damping ratio will decrease with increasing the inertia constant. As a result, settling time will be longer.

Example 9.3
A synchronous machine is having the inertia constant 6 MJ/MVA. The machine is connected to an infinite bus through a transmission line as shown in Fig. 9.10. The generator delivers a real power of 0.9 per unit at a 0.9 power factor lagging to the infinite bus. A small disturbance occurs in the system, and the deviation of the torque angle is found to be 9°. Calculate the per unit apparent power, line current, generated voltage, synchronizing power coefficient, undamped angular frequency of oscillation, and period of oscillation.

Solution
The reactance between the generator and infinite bus is calculated as,

$$X = j(0.24 + 0.12 + 0.34) = j0.70 \tag{9.128}$$

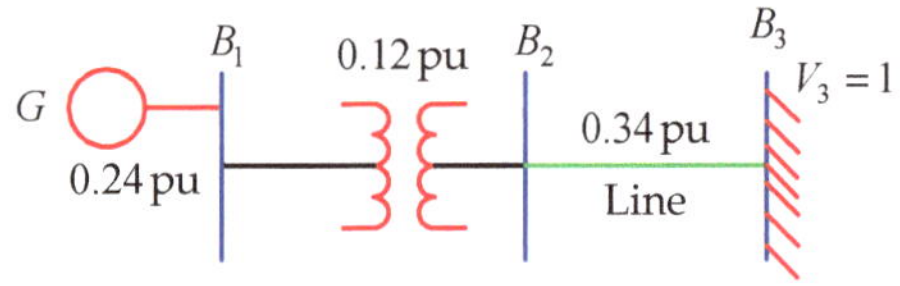

Fig. 9.10 Single-line diagram for Example 9.3

The value of the apparent power is calculated as,

$$S = \frac{0.9}{0.9}\angle \cos^{-1} 0.9 = 1\angle 25.84^\circ \tag{9.129}$$

The value of the line current is calculated as,

$$I_L = \frac{S^*}{V_2^*} = \frac{1\angle -25.84^\circ}{1.0\angle 0^\circ} = 1\angle -25.84^\circ \text{ pu} \tag{9.130}$$

The value of the generated voltage is calculated as,

$$E_g = V_2 + I_L X = 1 + 0.70\angle 90^\circ \times 1\angle -25.84^\circ = 1.45\angle 25.76^\circ \text{ pu} \tag{9.131}$$

The value of the synchronizing power coefficient is calculated as,

$$S_p = P_{\max}\cos\delta_0 = \frac{1.45 \times 1}{0.70}\cos(25.76^\circ - 9^\circ) = 1.98 \tag{9.132}$$

The value of an undamped angular frequency is calculated as,

$$\omega_n = \sqrt{\frac{\omega_s}{2H}S_p} = \sqrt{\frac{\pi \times 50}{6} \times 1.98} = 7.2 \text{ rad/s} \tag{9.133}$$

The frequency of oscillation is determined as,

$$f_n = \frac{7.2}{2\pi} = 1.14\,\text{Hz} \tag{9.134}$$

The period of oscillation is calculated as,

$$T = \frac{1}{f_n} = \frac{1}{1.14} = 0.88 \text{ s} \tag{9.135}$$

Practice Problem 9.3

A 50 Hz synchronous machine is having the inertia constant of 8 MJ/MVA and the excitation voltage $E_f = 1.5\angle 20^\circ$ pu. The generator is connected to an infinite bus through a transmission line, and the infinity bus voltage is $V_{ib} = 1\angle 0^\circ$ pu. The total reactance between the generator and the infinite bus is found to be 0.23 pu. Find the synchronizing power coefficient, undamped angular frequency of oscillation, frequency of oscillation, and period of oscillation.

9.7 Swing Equation for Multimachine

In the case of multimachine, the swing equation needs to be generalized on a common base system. Let the machine base MVA is S_{machine} and the system base MVA is S_{system}. Multiplying both sides of Eq. (9.73) by $S_{\text{machine}}/S_{\text{system}}$ yields [4],

$$\frac{S_{\text{machine}}}{S_{\text{system}}}\frac{H_{\text{machine}}}{\pi f_0}\frac{d^2\delta_1}{dt^2} = \frac{S_{\text{machine}}}{S_{\text{system}}}(P_m - P_e) \tag{9.136}$$

$$\frac{H_{\text{system}}}{\pi f_0}\frac{d^2\delta_1}{dt^2} = P_m - P_e \quad \text{pu on a system base} \tag{9.137}$$

where the expression of system inertia constant is,

$$H_{\text{system}} = H_{\text{machine}}\frac{S_{\text{machine}}}{S_{\text{system}}} \tag{9.138}$$

9.8 Swing Equations of Coherent Machines

Depends on the high demand of energy by the consumer, sometimes more than one machine is connected in the power station to generate more voltage. The machines when swing together it is known as coherent machines. The swing equation for coherent machine can be combined together to form a single-machine swing equation. The swing equations of machine one and machine on a common base system are expressed as [5],

$$\frac{H_1}{\pi f_0}\frac{d^2\delta_1}{dt^2} = P_{m1} - P_{e1} \tag{9.139}$$

$$\frac{H_2}{\pi f_0}\frac{d^2\delta_2}{dt^2} = P_{m2} - P_{e2} \tag{9.140}$$

Subtracting Eq. (9.139) from Eq. (9.140) yields,

$$\frac{d^2\delta_1}{dt^2} - \frac{d^2\delta_2}{dt^2} = \pi f_0\left(\frac{P_{m1} - P_{e1}}{H_1} - \frac{P_{m2} - P_{e2}}{H_2}\right) \tag{9.141}$$

$$\frac{d^2(\delta_1 - \delta_2)}{dt^2} = \pi f_0\left(\frac{P_{m1} - P_{e1}}{H_1} - \frac{P_{m2} - P_{e2}}{H_2}\right) \tag{9.142}$$

$$\frac{2}{2\pi f_0}\frac{d^2\delta_{12}}{dt^2} = \left(\frac{P_{m1} - P_{e1}}{H_1} - \frac{P_{m2} - P_{e2}}{H_2}\right) \tag{9.143}$$

$$\frac{2}{\omega_s}\frac{d^2\delta_{12}}{dt^2} = \left(\frac{P_{m1} - P_{e1}}{H_1} - \frac{P_{m2} - P_{e2}}{H_2}\right) \tag{9.144}$$

where, the rotor angle for the two machines is,

$$\delta_{12} = \delta_1 - \delta_2 \tag{9.145}$$

Multiplying Eq. (9.144) by the term $\frac{H_1 H_2}{H_1 + H_2}$ yields,

$$\frac{2}{\omega_s}\frac{H_1 H_2}{H_1 + H_2}\frac{d^2\delta_{12}}{dt^2} = \frac{H_1 H_2}{H_1 + H_2}\left(\frac{P_{m1} - P_{e1}}{H_1} - \frac{P_{m2} - P_{e2}}{H_2}\right) \tag{9.145}$$

$$\frac{2}{\omega_s}\frac{H_1 H_2}{H_1 + H_2}\frac{d^2\delta_{12}}{dt^2} = \left(\frac{H_2 P_{m1} - H_1 P_{m2}}{H_1 + H_2} - \frac{H_2 P_{e1} - H_1 P_{e1}}{H_1 + H_2}\right) \tag{9.146}$$

$$\frac{2}{\omega_s} H_{12}\frac{d^2\delta_{12}}{dt^2} = P_{m12} - P_{e12} \tag{9.147}$$

where the equivalent parameters are represented as,

$$H_{12} = \frac{H_1 H_2}{H_1 + H_2} \tag{9.148}$$

$$P_{m12} = \frac{H_2 P_{m1} - H_1 P_{m2}}{H_1 + H_2} \tag{9.149}$$

$$P_{e12} = \frac{H_2 P_{e1} - H_1 P_{e1}}{H_1 + H_2} \tag{9.150}$$

Again, the machine one is considered as synchronous generator and the machine two is considered as synchronous motor, the following equations can be written as,

$$P_{m1} = -P_{m2} = P_m \tag{9.151}$$

$$P_{e1} = -P_{e2} = P_e \tag{9.152}$$

Substituting Eqs. (9.151) and (9.152) into Eqs. (9.147), (9.149), and (9.150) yields,

$$\frac{2}{\omega_s} H_{12}\frac{d^2\delta_{12}}{dt^2} = P_m - P_e \tag{9.153}$$

$$P_m = \frac{H_2 P_{m1} - H_1 P_{m2}}{H_1 + H_2} \tag{9.154}$$

$$P_e = \frac{H_2 P_{e1} - H_1 P_{e1}}{H_1 + H_2} \tag{9.155}$$

Again adding Eqs. (9.139) and (9.140) yields,

$$\frac{H_1}{\pi f_0}\frac{d^2\delta_1}{dt^2} + \frac{H_2}{\pi f_0}\frac{d^2\delta_2}{dt^2} = P_{m1} - P_{e1} + P_{m2} - P_{e2} \tag{9.156}$$

Since the rotors of the machines are in unison so that the following relation can be written as,

$$\delta_1 = \delta_2 = \delta \tag{9.157}$$

Substituting Eq. (9.157) into Eq. (9.156) yields,

$$\left(\frac{H_1}{\pi f_0} + \frac{H_2}{\pi f_0}\right)\frac{d^2\delta}{dt^2} = P_m - P_e \tag{9.158}$$

$$\frac{H_{eq}}{\pi f_0}\frac{d^2\delta}{dt^2} = P_m - P_e \tag{9.159}$$

where the following equivalent parameters are expressed as,

$$H_{eq} = H_{1\text{machine}}\frac{S_{1\text{machine}}}{S_{\text{system}}} + H_{2\text{machine}}\frac{S_{2\text{machine}}}{S_{\text{system}}} \tag{9.160}$$

$$P_m = P_{m1} + P_{m2} \tag{9.161}$$

$$P_e = P_{e1} + P_{e2} \tag{9.162}$$

Example 9.4
Two generators are connected in parallel to a busbar. The necessary data for the generators are given below.

Generator 1 300 MVA, 0.8 power factor, 11 kV, 3000 rpm, H_1 = 6 MJ/MVA
Generator 2 500 MVA, 0.8 power factor, 11 kV, 1800 rpm, H_2 = 4 MJ/MVA.

Calculate the equivalent inertia constant and combined total kinetic energy, and rotor acceleration if the input power of generator 1 is 380 MW. Consider a common base of 100 MVA.

Solution
The equivalent inertia constant is calculated as,

$$H_{eq} = \frac{H_1 S_1}{S_b} + \frac{H_2 S_2}{S_b} = \frac{6 \times 300}{100} + \frac{4 \times 500}{100} = 38\,\text{MJ/MVA} \tag{9.163}$$

The value of the combined energy is calculated as,

$$E = S_b H_{eq} = 100 \times 38 = 3800\ \text{MJ} \tag{9.164}$$

The following parameter is calculated as,

$$M = \frac{SH}{\pi f} = \frac{6 \times 300}{180 \times 50} = \frac{1}{5} \tag{9.165}$$

The value of the rotor acceleration is calculated by using Eq. (9.159) as,

$$\frac{1}{5}\frac{d^2\delta}{dt^2} = 380 - 375 \tag{9.166}$$

$$\frac{d^2\delta}{dt^2} = 25\,\text{elec.deg/s}^2 \tag{9.167}$$

Practice Problem 9.4
Two generators are connected in parallel to a busbar. The necessary data for the generators are given below.

Generator 1 60 MVA, 0.9 power factor, 11 kV, $H_1 = 7$ MJ/MVA
Generator 2 50 MVA, 0.9 power factor, 11 kV, $H_2 = 5$ MJ/MVA

Calculate the equivalent inertia constant and combined energy total kinetic energy, and rotor acceleration if the input power of generator 2 is 65 MW. Consider a common base of 80 MVA.

9.9 Equal Area Criterion

The accelerating power of the swing equation contains the sine term. Therefore, the solution of a swing equation will be a nonlinear differential equation. The solution of a nonlinear equation is not simple or direct. A direct method named as equal area criterion is used to determine the stability of the system. This method is based on the graphical representation of stored energy in the rotating machine. It is often known as a graphical method for stability analysis. This method is only suitable for one machine to an infinite bus. For an unstable system, the torque angle (δ) increases indefinitely with time. Whereas for a stable system, torque angle (δ)

undergoes oscillations, and after some time, it will die out. Equation (9.74) can be rearranged as [2],

$$\frac{d^2\delta}{dt^2} = \frac{\pi f_0}{H}(P_m - P_e) \tag{9.168}$$

Multiplying Eq. (9.168) by $2\frac{d\delta}{dt}$ yields,

$$2\frac{d\delta}{dt}\frac{d^2\delta}{dt^2} = \frac{2\pi f_0}{H}(P_m - P_e)\frac{d\delta}{dt} \tag{9.169}$$

$$\frac{d}{dt}\left[\left(\frac{d\delta}{dt}\right)^2\right] = \frac{2\pi f_0}{H}(P_m - P_e)\frac{d\delta}{dt} \tag{9.170}$$

$$d\left[\left(\frac{d\delta}{dt}\right)^2\right] = \frac{2\pi f_0}{H}(P_m - P_e)d\delta \tag{9.171}$$

Integrating Eq. (9.171) from δ_0 to δ yields,

$$\left(\frac{d\delta}{dt}\right)^2 = \frac{2\pi f_0}{H}\int_{\delta_0}^{\delta}(P_m - P_e)d\delta \tag{9.172}$$

$$\frac{d\delta}{dt} = \sqrt{\frac{2\pi f_0}{H}\int_{\delta_0}^{\delta}(P_m - P_e)d\delta} \tag{9.173}$$

Equation (9.173) represents the speed of the machine with respect to the synchronously rotating reference axis. The machine rotor angles are plotted for both stable and unstable conditions as shown in Fig. 9.11. In an unstable condition, the rotor angles increase with time. However, in a stable condition, it oscillates and dies

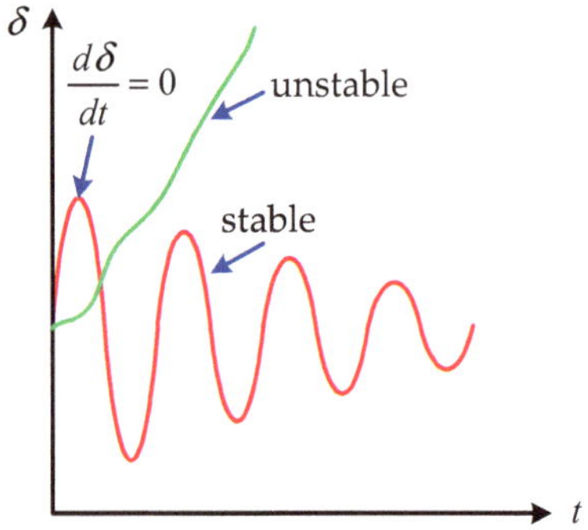

Fig. 9.11 Variation of rotor angles with time

out due to damping. Under the stable condition, the rate of change of the rotor angle will be zero at some instants.

For stable condition of a system setting $\frac{d\delta}{dt} = 0$ and Eq. (9.173) becomes,

$$\frac{d\delta}{dt} = \sqrt{\frac{2\pi f_0}{H}\int_{\delta_0}^{\delta}(P_m - P_e)d\delta} = 0 \tag{9.174}$$

$$\frac{2\pi f_0}{H}\int_{\delta_0}^{\delta}(P_m - P_e)d\delta = 0 \tag{9.175}$$

$$\int_{\delta_0}^{\delta}(P_m - P_e)d\delta = 0 \tag{9.176}$$

From Eq. (9.176), it is seen that the positive (accelerating) area under the graph is equal to the negative (decelerating) area for stability of a system. This condition is known as the equal area criterion.

From Fig. 9.12, it is seen that the point a corresponding δ_0 represents the steady-state operating point or equilibrium point. At this point, the mechanical input power to the machine is equal to the electrical output power, i.e., $P_{m0} = P_{e0}$. Consider a sudden increase in the input power of the machine as represented by the line P_{m1}. As a result, the accelerating power of the machine will increase, which increases the power angle. Therefore, the rotor will move forward toward the point c. Due to increasing energy in the rotor, the power and the rotor angle will continuously increase to reach the point c. As a result, the mechanical input power to the rotor will decrease, and the rotor will decelerate at synchronous speed until to reach the torque angle δ_2 or point b that represents the steady-state operating point. For stability of a system, an equal area criterion requires the following condition,

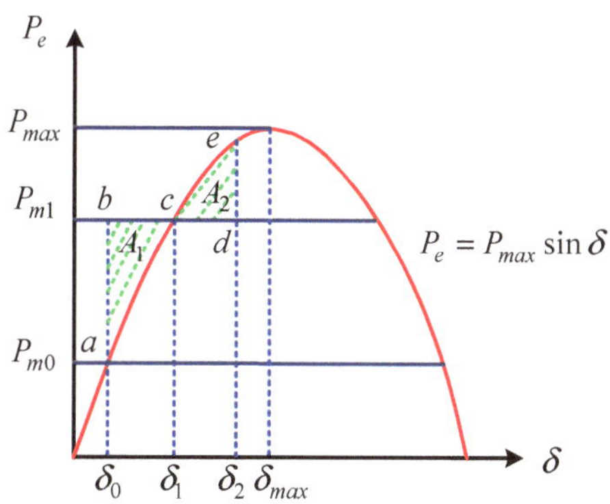

Fig. 9.12 Power curve for an equal area criterion

$$\text{Area } A_1 = \text{Area } A_2 \tag{9.177}$$

$$\int_{\delta_0}^{\delta_1} (P_m - P_{max}\sin\delta)d\delta = \int_{\delta_1}^{\delta_2} (P_{max}\sin\delta - P_m)d\delta \tag{9.178}$$

$$P_m(\delta_1 - \delta_0) + P_{max}|\cos\delta|_{\delta_0}^{\delta_1} = -P_{max}|\cos\delta|_{\delta_1}^{\delta_2} - P_m(\delta_2 - \delta_1) \tag{9.179}$$

$$P_m(\delta_1 - \delta_0 + \delta_2 - \delta_1) + P_{max}(\cos\delta_1 - \cos\delta_0) = P_{max}(\cos\delta_1 - \cos\delta_2) \tag{9.180}$$

$$P_m(\delta_2 - \delta_0) = P_{max}(\cos\delta_0 - \cos\delta_2) \tag{9.181}$$

Substituting the expression $P_m = P_{max}\sin\delta_1$ into Eq. (9.181) yields,

$$P_{max}(\delta_2 - \delta_0)\sin\delta_1 = P_{max}(\cos\delta_0 - \cos\delta_2) \tag{9.182}$$

$$(\delta_2 - \delta_0)\sin\delta_1 = \cos\delta_0 - \cos\delta_2 \tag{9.183}$$

$$(\delta_2 - \delta_0)\sin\delta_1 + \cos\delta_2 = \cos\delta_0 \tag{9.184}$$

From Eq. (9.184), it is seen that the angle δ_2 can be calculated if the angles δ_0 and δ_1 are known.

9.10 Critical Clearing Angle and Time

The torque angle of a synchronous machine usually increases due to a disturbance in the system. Therefore, the system will be unstable if the torque angle increases indefinitely with respect to time as mentioned before in Fig. 9.11. Therefore, there is a critical value of the torque angle for clearing the fault to keep the system remain stable and satisfy the equal area criterion. This value of the torque angle is known as the critical clearing angle, and it is represented by the δ_c.

Consider one machine connected to an infinite bus as shown in Fig. 9.13 to find a critical clearing angle and time. This power system operates at a steady-state point a as shown in Fig. 9.14. Considering a three-phase fault occurs at bus B_2. As a result, terminal voltage, as well as electrical power output, will be zero at bus 2, and

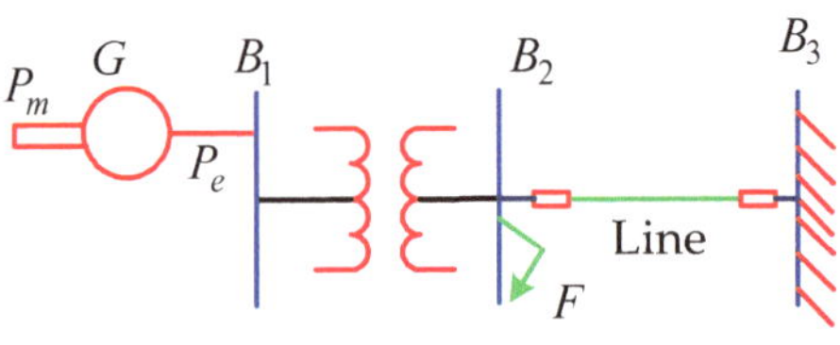

Fig. 9.13 One machine to an infinite bus with a fault

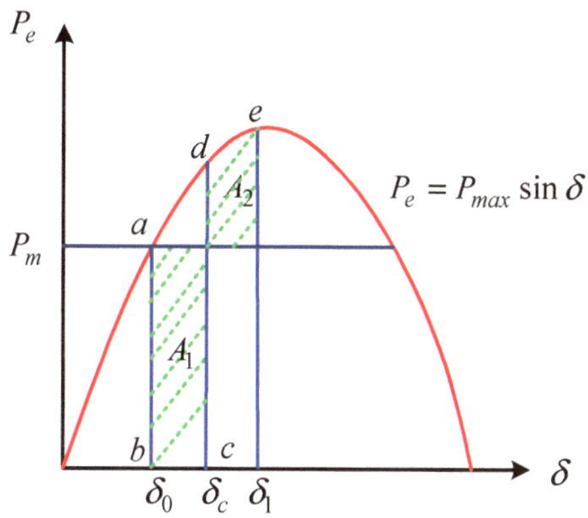

Fig. 9.14 Power curve with clearing time and angle

the state point drops to b. The accelerating area A_1 will increase, while the point moves from b to c. After the fault is cleared by the time known as the critical clearing time t_c corresponding δ_c, the system will become healthy and transmit normal power and reach the d. Afterward, the rotor will decelerate at synchronous speed, and the decelerating area A_2 will increase, while the point moves from d to e. For a stability system, to find the critical clearing angle, both the areas must be the same ($A_1 = A_2$).

The accelerating area can be calculated as,

$$A_1 = \int_{\delta_0}^{\delta_c} P_m \, d\delta = P_m(\delta_c - \delta_0) \tag{9.185}$$

The decelerating area can be calculated as,

$$A_2 = \int_{\delta_c}^{\delta_{max}} (P_{\max} \sin\delta - P_m) \, d\delta \tag{9.186}$$

$$A_2 = -P_{\max}(\cos\delta_{\max} - \cos\delta_c) - P_m(\delta_{\max} - \delta_c) \tag{9.187}$$

$$A_2 = P_{\max}(\cos\delta_c - \cos\delta_{\max}) - P_m(\delta_{\max} - \delta_c) \tag{9.188}$$

Equating Eqs. (9.185) and (9.188) yields,

$$P_m(\delta_c - \delta_0) = P_{\max}(\cos\delta_c - \cos\delta_{\max}) - P_m(\delta_{\max} - \delta_c) \tag{9.189}$$

The limit of stability occurs when the maximum torque angle is at the intersection of P_m line as shown in Fig. 9.15. From Fig. 9.15, the following angle can be determined as,

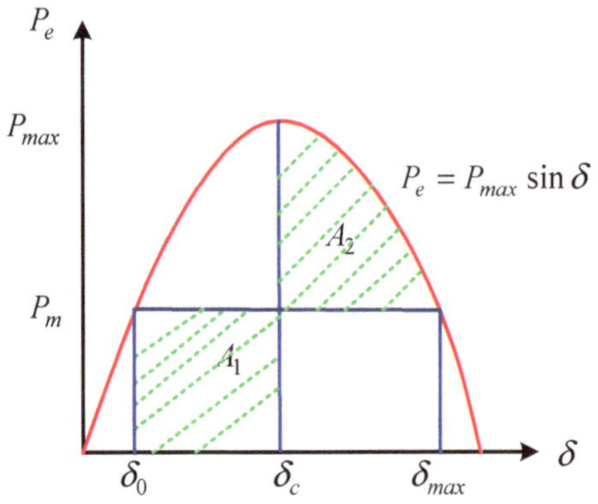

Fig. 9.15 Power curve with critical clearing time

$$\delta_{\max} = \pi - \delta_0 \tag{9.190}$$

Substituting Eq. (9.190) into Eq. (9.189) yields,

$$P_m(\delta_c - \delta_0) = P_{\max}(\cos\delta_c - \cos(\pi - \delta_0)) - P_m(\pi - \delta_0 - \delta_c) \tag{9.191}$$

$$P_m(\delta_c - \delta_0) = P_{max}(\cos\delta_c + \cos\delta_0) - P_m(\pi - \delta_0 - \delta_c) \tag{9.192}$$

$$P_m(\delta_c - \delta_0 + \pi - \delta_0 - \delta_c) = P_{\max}(\cos\delta_c + \cos\delta_0) \tag{9.193}$$

$$P_m(\pi - 2\delta_0) = P_{max}(\cos\delta_c + \cos\delta_0) \tag{9.194}$$

Substituting the expression $P_m = P_{\max}\sin\delta_0$ into Eq. (9.194) yields,

$$P_{\max}(\pi - 2\delta_0)\sin\delta_0 = P_{\max}(\cos\delta_{cr} + \cos\delta_0) \tag{9.195}$$

$$(\pi - 2\delta_0)\sin\delta_0 = (\cos\delta_{cr} + \cos\delta_0) \tag{9.196}$$

$$\cos\delta_{\mathrm{cr}} = (\pi - 2\delta_0)\sin\delta_0 - \cos\delta_0 \tag{9.197}$$

$$\delta_{\mathrm{cr}} = \cos^{-1}[(\pi - 2\delta_0)\sin\delta_0 - \cos\delta_0] \tag{9.198}$$

From Eq. (9.198), it is seen that the critical angle can be determined if the initial torque angle is known.

The electrical power output P_e is zero at the fault condition. In this case, the swing equation in (9.74) can be rewritten as,

$$\frac{H}{\pi f_0}\frac{d^2\delta}{dt^2} = P_m - 0 \tag{9.199}$$

$$\frac{d^2\delta}{dt^2} = \frac{\pi f_0}{H} P_m \tag{9.200}$$

Integrating Eq. (9.200) from 0 to t yields,

$$\frac{d\delta}{dt} = \int_0^t \frac{\pi f_0}{H} P_m \, dt \tag{9.201}$$

$$\frac{d\delta}{dt} = \frac{\pi f_0}{H} P_m t \tag{9.202}$$

Again, integrating Eq. (9.202) yields,

$$\delta = \frac{\pi f_0}{2H} P_m t^2 + C \tag{9.203}$$

Setting the boundary condition at $t = 0$, $\delta = \delta_0$ in Eq. (9.203) yields,

$$C = \delta_0 \tag{9.204}$$

Substituting Eq. (9.204) into Eq. (9.203) yields,

$$\delta = \frac{\pi f_0}{2H} P_m t^2 + \delta_0 \tag{9.205}$$

For a critical clearing angle δ_{cr}, the expression of critical clearing time can be written as,

$$\delta_{cr} = \frac{\pi f_0}{2H} P_m t_{cr}^2 + \delta_0 \tag{9.206}$$

$$t_{cr}^2 = \frac{2H}{\pi f_0 P_m} (\delta_{cr} - \delta_0) \tag{9.207}$$

$$t_{cr} = \sqrt{\frac{2H}{\pi f_0 P_m} (\delta_{cr} - \delta_0)} \tag{9.208}$$

Example 9.5
A synchronous machine with an inertia constant of 5 MJ/MVA is connected to an infinite bus as shown in Fig. 9.16. The generator delivers real power and reactive power to the infinite bus which are 0.75 pu and 0.95, respectively. A three-phase

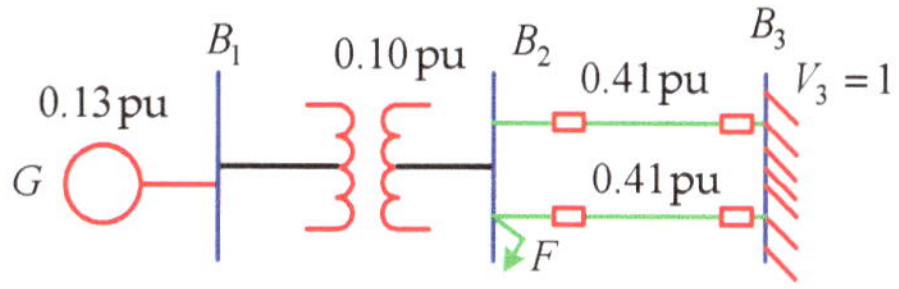

Fig. 9.16 Single-line diagram for Example 9.5

fault occurs on the system at point F and subsequently cleared by the circuit breaker operation. Under prefault condition of the lines, calculate the generator transient voltage, power angle equation, initial torque angle, maximum torque angle, critical clearing angle, and critical clearing time. During fault (impedance 0.11 pu), calculate the fault clearing angle and time.

Solution

The value of the line current is calculated as,

$$I_L = \frac{S^*}{V^*} = \frac{0.75 - j0.95}{1} = 1.21\,\underline{|-51.71^\circ}\ \text{pu} \tag{9.209}$$

The total reactance between the generator and the infinite bus is calculated as,

$$X = j(0.13 + 0.10 + \frac{0.41}{2}) = j0.44 \tag{9.210}$$

The value of the generator transient voltage is calculated as,

$$E_g = V_{ib} + jXI_L = 1 + 1.21\,\underline{|-51.71^\circ} \times 0.44\,\underline{|90^\circ} = 1.46\,\underline{|13.09^\circ}\ \text{pu} \tag{9.211}$$

The power equation is calculated as,

$$P_e = P_{\max} \sin\delta = \frac{1.46 \times 1}{0.44} \sin\delta = 3.32 \sin\delta \tag{9.212}$$

From Eq. (9.212), the value of the initial torque angle can be determined as,

$$P_e = 3.32 \sin\delta_0 = 0.75 \tag{9.213}$$

$$\delta_0 = 13.06^\circ = 0.23\,\text{rad} \tag{9.214}$$

From Eq. (9.190), the value of the maximum torque angle can be determined as,

$$\delta_{\max} = 180^\circ - \delta_0 = 180^\circ - 13.06^\circ = 166.94^\circ \tag{9.215}$$

From Eq. (9.198), the value of the critical clearing angle can be determined as,

$$\delta_{cr} = \cos^{-1}[(\pi - 2 \times 0.23) \sin 13.06^\circ - \cos 13.06^\circ] = 111.6^\circ = 1.95\,\text{rad} \tag{9.216}$$

From Eq. (9.208), the value of the critical clearing time can be determined as,

$$t_c = \sqrt{\frac{2 \times 5}{\pi \times 50 \times 0.75}(1.95 - 0.23)} = 0.38\ \text{s} \tag{9.217}$$

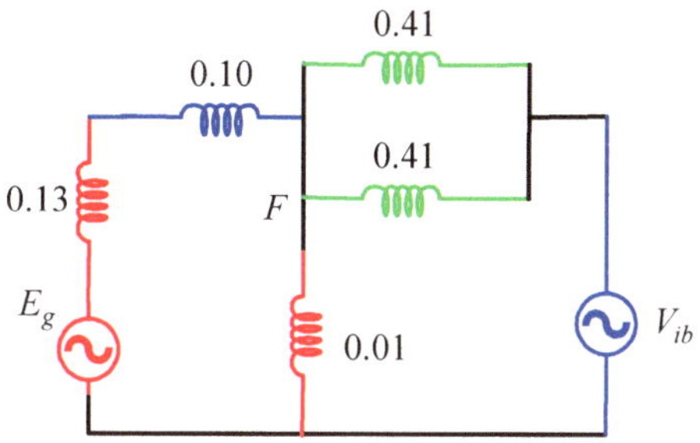

Fig. 9.17 Reactance diagram during fault for Example 9.5

During the fault, the circuit is drawn as shown in Fig. 9.17. The reactances need to be rearranged. The reactances 0.13 pu and 0.10 pu are in series, and its equivalent reactance is calculated as,

$$X_{11} = 0.13 + 10 = 0.23\,\text{pu} \tag{9.218}$$

The reactances 0.41 pu and 0.41 pu are in parallel, and the equivalent reactance is calculated as,

$$X_{12} = \frac{0.41}{2} = 0.20\text{ pu} \tag{9.219}$$

The reactances 0.23 pu, 0.20 pu and the reactances due to fault create the wye circuit and then converted to delta circuit as shown in Fig. 9.18.

The values of the reactances are calculated as,

$$X_1 = \frac{0.23 \times 0.20 + 0.20 \times 0.11 + 0.11 \times 0.23}{0.11} = 0.85\,\text{pu} \tag{9.220}$$

$$X_2 = \frac{0.23 \times 0.20 + 0.20 \times 0.11 + 0.11 \times 23}{0.20} = 0.47\,\text{pu} \tag{9.221}$$

$$X_3 = \frac{0.23 \times 0.20 + 0.20 \times 0.11 + 0.11 \times 23}{0.23} = 0.41\,\text{pu} \tag{9.222}$$

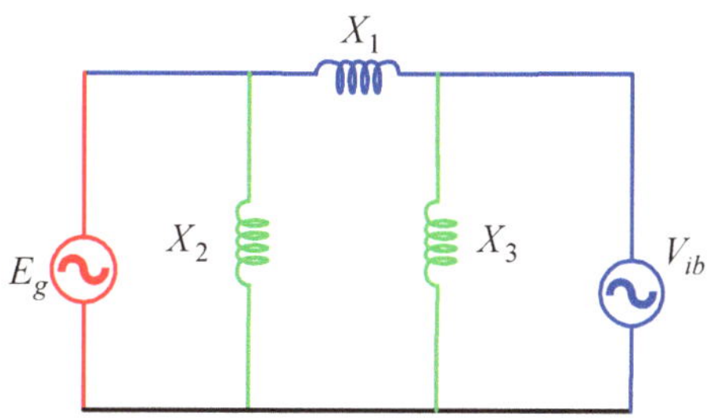

Fig. 9.18 Single-line diagram for Example 9.5

The power equation is calculated as,

$$P_e = P_{\max} \sin\delta = \frac{1.46 \times 1}{0.85} \sin\delta = 1.72 \sin\delta \tag{9.223}$$

From Eq. (9.223), the value of the initial torque angle can be determined as,

$$P_e = 1.72 \sin\delta_0 = 0.75 \tag{9.224}$$

$$\delta_0 = 25.85^\circ = 0.45\,\text{rad} \tag{9.225}$$

From Eq. (9.190), the value of the maximum torque angle can be determined as,

$$\delta_{\max} = 180^\circ - \delta_0 = 180^\circ - 25.85^\circ = 154.15^\circ \tag{9.226}$$

From Eq. (9.198), the value of the critical clearing angle can be determined as,

$$\delta_{cr} = \cos^{-1}[(\pi - 2 \times 0.45)\sin 25.85^\circ - \cos 25.85^\circ] = 85.56^\circ = 1.49\,\text{rad} \tag{9.227}$$

From Eq. (9.208), the value of the critical clearing time can be determined as,

$$t_c = \sqrt{\frac{2 \times 5}{\pi \times 50 \times 0.75}(1.49 - 0.45)} = 0.3\,\text{s} \tag{9.228}$$

Practice Problem 9.5

A 50 Hz generator with a 7 MJ/MVA inertia constant is connected to an infinite bus through a single transmission line as shown in Fig. 9.19. The real power delivers to the infinite bus is found to be 0.65 pu. A temporary three-phase fault occurs on the line, at busbar 6. Calculate the fault clearing angle and fault clearing time at prefault and during fault conditions.

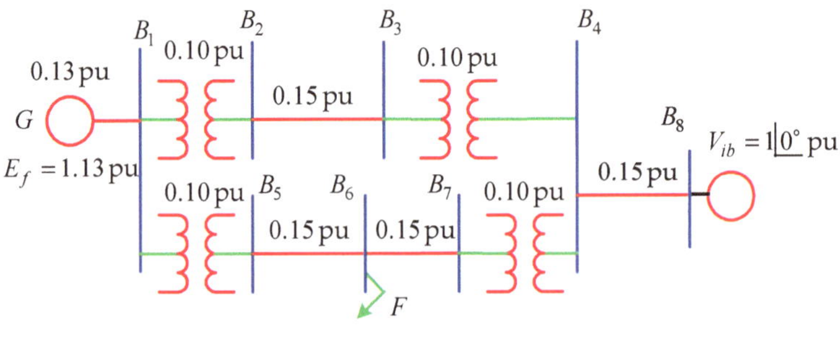

Fig. 9.19 Single-line diagram for Practice Problem 9.5

9.11 Step-by-Step Solution of Swing Equation

The swing equation is a nonlinear equation, and its normal solution is difficult. There is a simple and conventional method to solve the swing equation which is known as step-by-step method. In this method, the accelerating power P_a and the rotor angular velocity are assumed to be constant from the middle of the preceding interval to the middle of the interval being considered. The accelerating power at 0^+ can be determined as,

$$P_{a(0+)} = P_m - P_{e(0+)} \tag{9.229}$$

The swing equation can be modified as,

$$\frac{d\delta^2}{dt^2} = \alpha_{(0+)} = \frac{\pi f_0}{H} P_{a(0+)} \tag{9.230}$$

From Fig. 9.20, the change in rotor angular velocity can be written as,

$$\Delta\omega' = \alpha_{(0+)}\,\Delta t \tag{9.231}$$

For the first interval, the change in angular velocity is expressed as,

$$\Delta\omega_1 = \alpha_{(0+)}\,\Delta t \tag{9.232}$$

The angular velocity in the first interval is expressed as,

$$\omega_1 = \omega_0 + \Delta\omega_1 \tag{9.233}$$

Substituting Eq. (9.232) into Eq. (9.233) yields,

$$\omega_1 = \omega_0 + \alpha_{(0+)}\,\Delta t \tag{9.234}$$

Similarly, for the first interval, the change in power angle is expressed as,

$$\Delta\delta_1 = \Delta\omega_1\,\Delta t \tag{9.235}$$

For the first interval, the power angle is expressed as,

$$\delta_1 = \delta_0 + \Delta\delta_1 \tag{9.236}$$

Substituting Eq. (9.235) into Eq. (9.236) yields,

$$\delta_1 = \delta_0 + \Delta\omega_1\,\Delta t \tag{9.237}$$

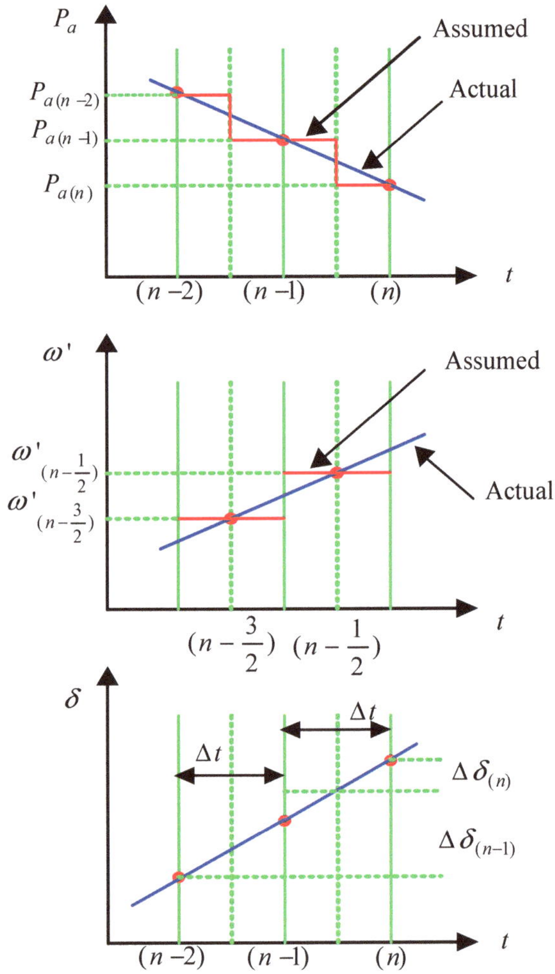

Fig. 9.20 Assumed and actual solution of swing equation

Again, substituting Eq. (9.232) into Eq. (9.237) yields,

$$\delta_1 = \delta_0 + \alpha_{(0+)}\,(\Delta t)^2 \tag{9.238}$$

Considering that the discontinuity occurs at the middle of the interval, then the accelerating power is expressed as,

$$P_a = P_m - \text{output during the fault} \tag{9.239}$$

Finally, considered that the fault is cleared due to circuit breaker operation, and the accelerating power is expressed as,

$$P_a = P_m - \text{output power after the fault cleared} \tag{9.240}$$

The different equations for the $(n - 1)$th interval can be written as,

$$P_{a(n-1)} = P_m - P_{e(n-1)} \tag{9.241}$$

$$P_{e(n-1)} = \frac{E_f V_{\text{inf}}}{X} \sin \delta_{(n-1)} \tag{9.242}$$

$$\alpha_{(n-1)} = \frac{\pi f_0}{H} P_{a(n-1)} \tag{9.243}$$

According to Eq. (9.231), the following relation can be written as,

$$\Delta\omega'_{(n-\frac{1}{2})} = \alpha_{(n-1)} \, \Delta t \tag{9.244}$$

According to Eqs. (9.234) and (9.235), the following relations can be written as,

$$\omega'_{(n-\frac{1}{2})} = \omega'_{(n-\frac{3}{2})} + \Delta\omega'_{(n-\frac{1}{2})} = \omega'_{(n-\frac{3}{2})} + \alpha_{(n-1)} \, \Delta t \tag{9.245}$$

$$\Delta\delta_{(n)} = \omega'_{(n-\frac{1}{2})} \, \Delta t \tag{9.246}$$

The rotor angle δ change during $(n - 1)$th interval is,

$$\Delta\delta_{(n-1)} = \delta_{(n-1)} - \delta_{(n-2)} = \Delta t \omega'_{(n-\frac{3}{2})} \tag{9.247}$$

Substituting Eq. (9.245) into Eq. (9.246) yields,

$$\Delta\delta_n = \left(\omega'_{(n-\frac{3}{2})} + \alpha_{(n-1)} \, \Delta t\right) \Delta t \tag{9.248}$$

$$\Delta\delta_{(n)} = \omega'_{(n-\frac{3}{2})} \Delta t + \alpha_{(n-1)} (\Delta t)^2 \tag{9.249}$$

Substituting Eq. (9.247) into Eq. (9.249) yields,

$$\Delta\delta_{(n)} = \Delta\delta_{(n-1)} + \alpha_{(n-1)} (\Delta t)^2 \tag{9.250}$$

Again, substituting Eq. (9.243) into Eq. (9.250) yields,

$$\Delta\delta_{(n)} = \Delta\delta_{(n-1)} + \frac{\pi f_0}{H} P_{a(n-1)} (\Delta t)^2 \tag{9.251}$$

$$\delta_{(n)} = \delta_{(n-1)} + \Delta\delta_{(n)} \tag{9.252}$$

From Eq. (9.252), it is seen that the rotor angle can be determined if other parameters are known.

9.12 Alternate Solution of Swing Equation

The well-known swing equation is related to a synchronous generator's rotor swing angle δ to its accelerating power P_a. In one machine to an infinite bus (OMIB) system, the power output of the single machine connected to an infinite bus at any instant of time is $P_e = P_{\max} \sin \delta$ when the resistances in the network are considered negligible and $P_{\max}$ is the maximum power output of the machine. In classical modeling, $P_{\max}$ is the product of constant internal voltage behind transient reactance of the machine and the infinite bus voltage divided by the net series reactance between the machine and the infinite bus. Therefore, the maximum power $P_{\max}$ is constant so long the said reactance does not change. Since for a given fault location in the system, the reactance before the fault is different from that after removing the fault $P_{\max}$ will have two constant values P_1 and P_2, respectively for $0 \le t \le t_r$ (during fault, i.e., in between fault occurrence time $t = 0$ and fault removal time $t = t_r$) and for $t \ge t_r$ (i.e., after the fault is removed). In other words, the maximum value of the machine output power, i.e., $P_{\max}$ can be represented as a two-valued step function as shown in Fig. 9.21. From Fig. 9.21, an expression for $P_{\max}$ can be derived as [2, 5–7],

$$P_{\max} = P_1 U(t) + \Delta U(t - t_r) \tag{9.253}$$

where,

$U(t)$ is the unit step function starting at $t = 0$,
$U(t - t_r)$ is the unit step function starting at $t = t_r$,
Δ is the difference between P_2 and P_1.

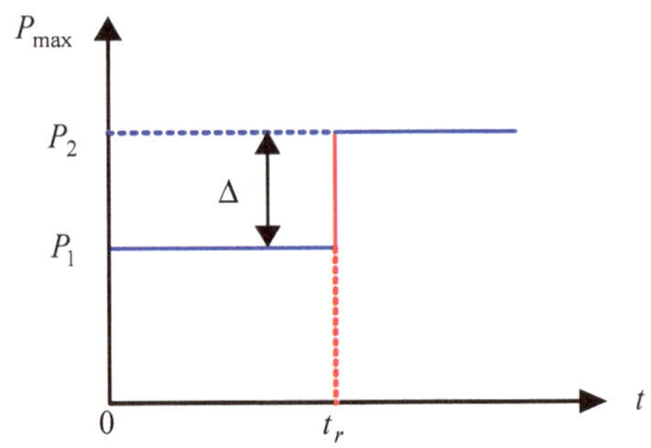

Fig. 9.21 Maximum power output during and after faults

The rotor swing angle, δ, which has been defined relative to a reference axis rotating at the synchronous angular velocity, ω_s, can be related to the actual ω as

$$\delta = (\omega - \omega_s)t = \omega' t \tag{9.254}$$

where $\omega' = \frac{d\delta}{dt} = 2\pi f'$ is a measure of the frequency (f') of rotor oscillation relative to the synchronously rotating reference axis. Combining Eqs. (9.253) and (9.254) and the machines output power P_e can be expressed as,

$$P_e = P_1 \sin \omega' t U(t) + \Delta \sin \omega' t U(t - t_r) \tag{9.255}$$

Equations (9.57), (9.253), (9.254) and (9.255) can now be combined into a model of a closed-loop system for the machine as in Fig. 9.22 in which the reference input is P_m, the controlled variable is δ, the error signal is $P_{a,}$ and the feedback signal is P_e. It should be noted that P_e could be represented as a function of a variable y when the quantity $\sin \delta$, i.e., $\sin \omega' t$ of Eq. (9.153) is denoted by y. The transfer functions of various blocks in Fig. 9.22 have been obtained by taking Laplace transform of Eqs. (9.57), (9.254), (9.255) and $y = \sin \omega' t$ with zero initial conditions. The equations are,

$$Ms^2\delta(s) = P_a(s) = P_m(s) - P_e(s) \tag{9.256}$$

$$\delta(s) = \frac{\omega'}{s^2} \tag{9.257}$$

$$P_e(s) = \frac{P_1 \omega'}{s^2 + \omega'^2} + \frac{\Delta e^{-t_r s}(\omega' \cos \omega' t_r + s \sin \omega' t_r)}{s^2 + \omega'^2} \tag{9.258}$$

$$Y(s) = \frac{\omega'}{s^2 + \omega'^2} \tag{9.259}$$

In taking Laplace transforms, the rotor oscillation angular velocity ω' has been considered as remaining constant at an average value. The closed-loop transfer

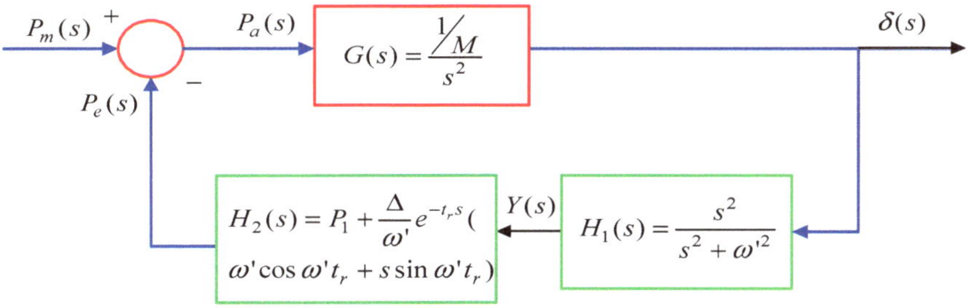

Fig. 9.22 Swing equation into a feedback system model

function of the feedback system of Fig. 9.22 corresponding to a synchronous machine is,

$$\frac{\delta(s)}{P_m(s)} = \frac{G(s)}{1 + G(s)H_1(s)H_2(s)} \tag{9.260}$$

Substituting the expressions for $G(s)$, $H_1(s)$, $H_2(s)$ into Eq. (9.260) yields,

$$\frac{\delta(s)}{P_m(s)} = \frac{(1/M)(s^2 + \omega'^2)}{s^2\left[s^2 + k_1 + k_2 e^{-t_r s} + k_3 s e^{-t_r s}\right]} \tag{9.261}$$

where the following expressions are,

$$k_1 = \omega'^2 + \frac{P_1}{M} \tag{9.262}$$

$$k_2 = \frac{1}{M}\Delta \cos \omega' t_r \tag{9.263}$$

$$k_3 = \frac{\Delta \sin \omega' t_r}{\omega' M} \tag{9.264}$$

9.12.1 Dominant Root

From Eq. (9.261), the characteristic equation can be written as,

$$s^2 + k_1 + k_2 e^{-t_r s} + k_3 s e^{-t_r s} = 0 \tag{9.265}$$

Equation (9.265) can be rewritten as follows,

$$f(s) = f_1(s) - f_2(s) = 0 \tag{9.266}$$

where the following functions are,

$$f_1(s) = s^2 + k_1 \tag{9.267}$$

$$f_2(s) = k_2 e^{-t_r s}\left(-1 - \frac{k_3}{k_2} s\right) \tag{9.268}$$

A root of (9.266) is the point of intersection of the functions $f_1(s)$ and $f_2(s)$. This point has been termed the "dominant" root. Figure 9.23 sketches the forms of the function $f_1(s)$ and $f_2(s)$. It can be seen that function $f_1(s)$ is greater than $f_2(s)$, i.e., $f(s) = f_1(s) - f_2(s)$ is positive for all values of s on the right of and up to a point $s = -k_2/k_3$. At $s = -k_2/k_3$, the function $f_2(s)$ is zero and beyond this point $f_2(s)$ starts increasing, so that $f_2(s)$ intersects $f_1(s)$ at a point having its real part equal to s_{root}.

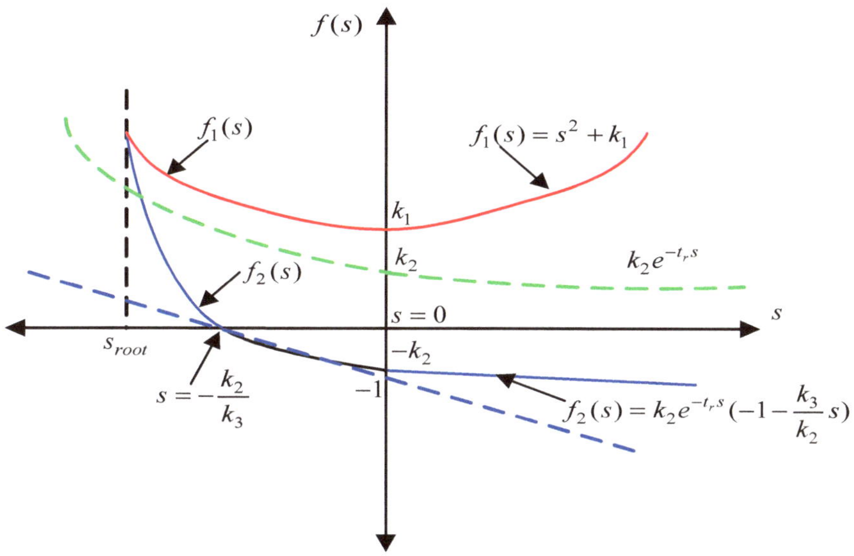

Fig. 9.23 Sketches of functions $f_1(s)$ and $f_2(s)$

This point is the desired dominant root to be searched for. Beyond this point s_{root}, $f_2(s)$ is greater than $f_1(s)$, i.e., the function $f(s)$ is negative. Therefore, a search for the real axis bounds of the region containing the dominant root can be made starting from the point $s = -k_2/k_3$ and continued by increasing the absolute value of s until the function $f(s)$ becomes negative.

9.12.2 Extension for a Multimachine System

Equation (9.57) holds good for ith $(i = 1, 2, 3, \ldots n)$ machine in a system with n number of machines if i is used as a subscript to each of δ, M, P_a, P_m, and P_e. The electrical power output P_{ei} is written as,

$$P_{ei} = E_i^2 G_{ii} + \sum_{j=1, j \neq i}^{n} E_i E_j Y_{ij} \cos\left[\theta_{ij} - \left(\delta_i - \delta_j\right)\right] \tag{9.269}$$

where E_i and E_j are magnitudes (assumed constant in the classical model) of the internal EMFs behind transient reactances at ith and jth machine nodes respectively such that $E_i = E_i \angle \delta_i$ and $E_j = E_j \angle \delta_j$, $Y_{ij} \underline{|\theta_{ij}} = G_{ij} + jB_{ij}$ is the transfer admittance between ith and jth machine nodes, and G_{ij} is the real part, i.e., conductance of the driving point admittance at ith machine node. The quantities G_{ij} and Y_{ij} are obtained from the admittance matrix of the whole system reduced to only the internal nodes of the machine, i.e., to $n \times n$ size. Equation (9.167) gives the electrical power output in any network condition, i.e., prefault, during, and after faults provided the conductance and admittance terms are taken from the corresponding matrix reduced at the machine internal nodes. In addition, it follows that from (9.269) that $\theta_{ij} = \delta_i - \delta_j$ results in maximum power output, which for the ith machine is P_{maxi} as,

$$P_{\text{maxi}} = E_i^2 G_{ii} + \sum_{j=1, j \neq i}^{n} E_i E_j Y_{ij} \tag{9.270}$$

Now, if an OMIB system with a single machine connected to an infinite bus is such that the machine's inertia constant and mechanical power input are the same as those of the ith unit of the multimachine system. But its rotor swing angle is δ_i^r, while the internal EMF behind its transient reactance, the infinite bus voltage, and the reactance between the machine and the infinite bus are of such magnitudes that with the resistances neglected machines electrical power output is $P_{\text{maxi}} \sin \delta_i^r$ and equal to P_{ei} given in Eq. (9.269) when P_{maxi} is equal to that in the Eq. (9.270). Then, this particular OMIB system can be considered as equivalent to the ith machine of the multimachine system. In this way, each machine of the original multimachine system is replaceable by an equivalent OMIB system which can be represented by Eq. (9.57), and by the feedback model of Fig. 9.22 changing the notations δ, M, P_a, P_m, P_e, and $P_{\max}$, respectively, by δ_i^r, M_i, P_{ai}, P_{mi}, P_{ei}, and P_{maxi}. Also, the notation "Δ" in the feedback model will then be $\Delta_i = P_{2i} - P_{1i}$ when P_{2i} and P_{1i} are the maximum electrical power outputs of the ith unit of the multimachine system, i.e., P_{maxi} in the after and the during fault conditions, respectively, both being obtained from Eq. (9.168) but using the reduced admittance matrix for the corresponding condition. The real axis limits of the narrowest possible range in s-plane containing the dominant root of the characteristic equation $f(s) = 0$ of each feedback model is searched in the same way as mentioned in Sect. 9.12.1, starting from the point $s = -k_2/k_3$. It is an advantage that for a given t_r and a chosen average value ω' for rotor oscillation angular velocities of all the machines, the starting point is the same for each machine as may be seen substituting Δ_i and M_i respectively for Δ and M in Eqs. (9.263) and (9.264).

Example 9.6

Figure 9.24 shows a single-line diagram where two generators are connected with a grid source through transformers (delta-earthed wye) and transmission lines. Some data are shown in Tables 9.1, 9.2, and 9.3. A three-phase occurs on line 4–5 near bus 4 at 10 cycles and is cleared at 16 cycles.

Transformer T_1	10 MVA, three-phase shell, 11 kV (delta)/33 kV (wye), $Z_1 = Z_0 = 8.25\%$, $X_1/R_1 = X_0/R_0 = 15$
Transformer T_2	10 MVA, three-phase shell, 11 kV (delta)/33 kV (wye), $Z_1 = Z_0 = 8.25\%$, $X_1/R_1 = X_0/R_0 = 15$
Grid source	100 MVA, 33 kV, $Z_1 = 0.95 + j9.96$ pu, $Z_0 = 0.05 + j1.006$ pu

Calculate the transient stability by CYME software.

Fig. 9.24 Single-line diagram for Example 9.6

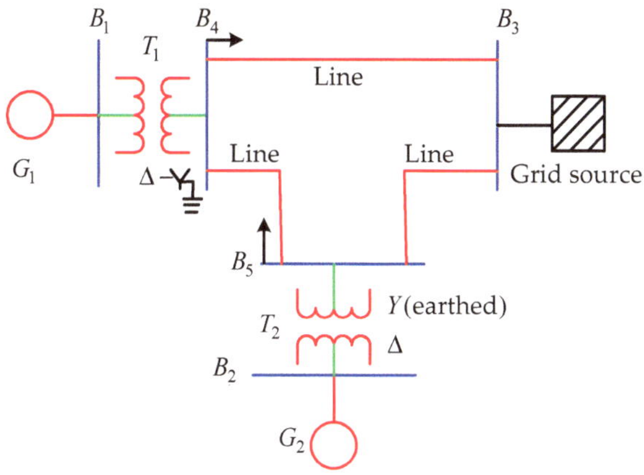

Table 9.1 Machine data

Gen.	Rating	Active and reactive generations	H (MJ/MVA)
1	15 MVA, 11 kV, 2P, 90% pf, round rotor type 5	10 MW, 3 Mvar	11.2
2	30 MVA, 11 kV, 2P, 90% pf, round rotor type 5	20 MW, 4 Mvar	8

Table 9.2 Load data

	Load	
Bus no.	Real power (MW)	Reactive power (Mvar)
4	100	44
5	50	16

Table 9.3 Line data

Bus no.	R (Ω/mi)	X (Ω/mi)	B (μS/mi)
3–4	0.12	2.56	6.34
3–5	0.07	3.12	7.06
4–5	0.64	2.25	8.05

Solution

The circuit for simulation is drawn by CYME software and put all necessary data of the equipment. The circuit under simulation is shown in Fig. 9.25, where the three-phase fault is considered on line 4–5 near bus 4.

A three-phase fault is considered at bus 4 at 10 cycles, and it is cleared at 16 cycles. The simulation is carried out for 300 cycles, and the respective parameters are drawn with respect to simulation time as shown in Figs. 9.26, 9.27, and 9.28.

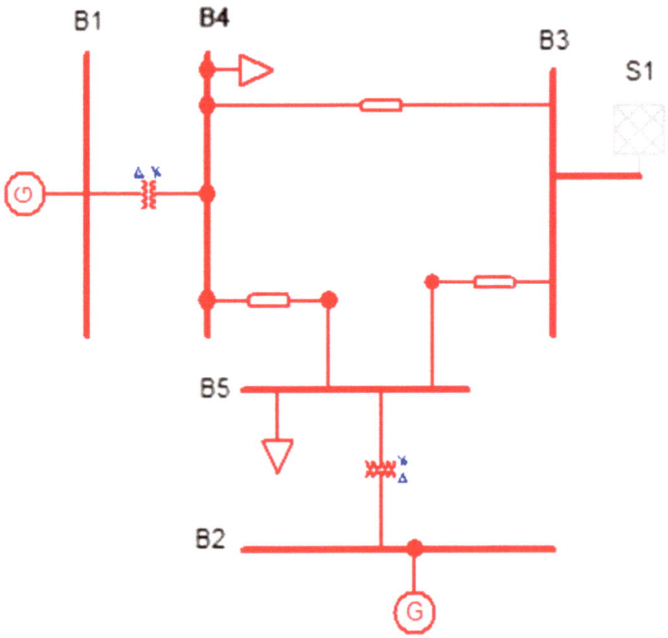

Fig. 9.25 CYME simulation circuit for Example 9.6

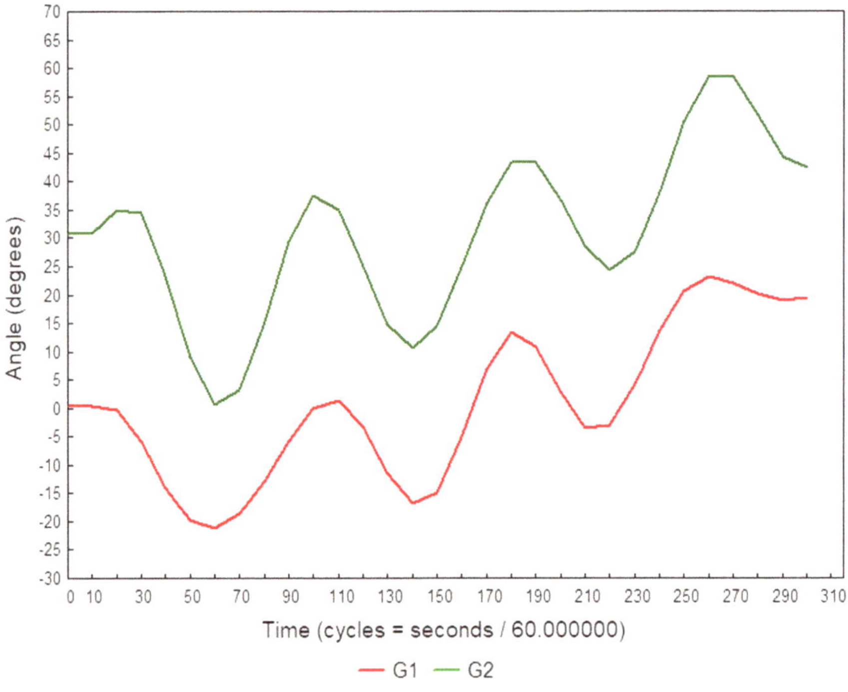

Fig. 9.26 Machines rotor angles versus time for Example 9.6

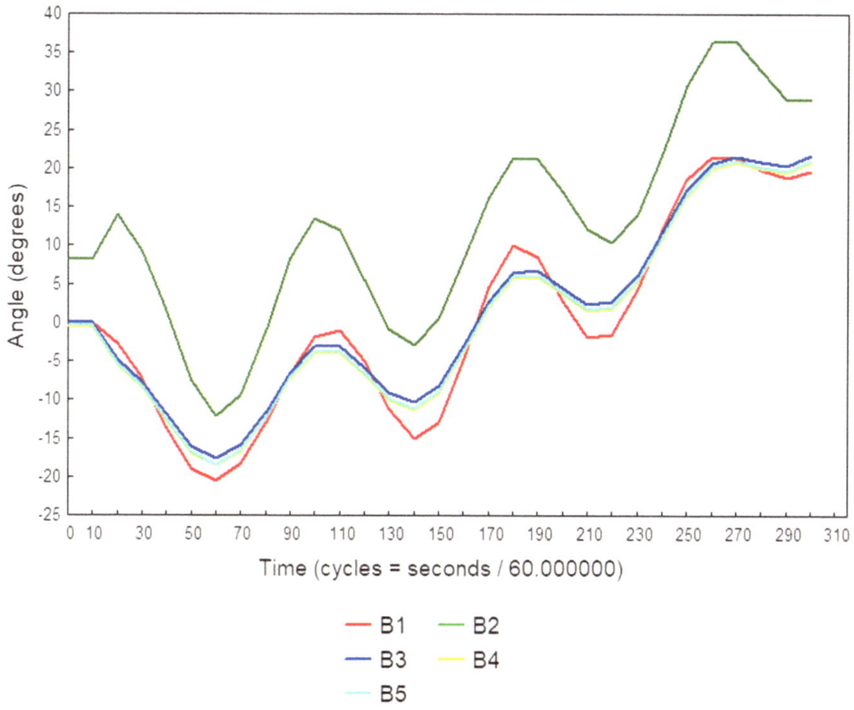

Fig. 9.27 Different bus angles versus time for Example 9.6

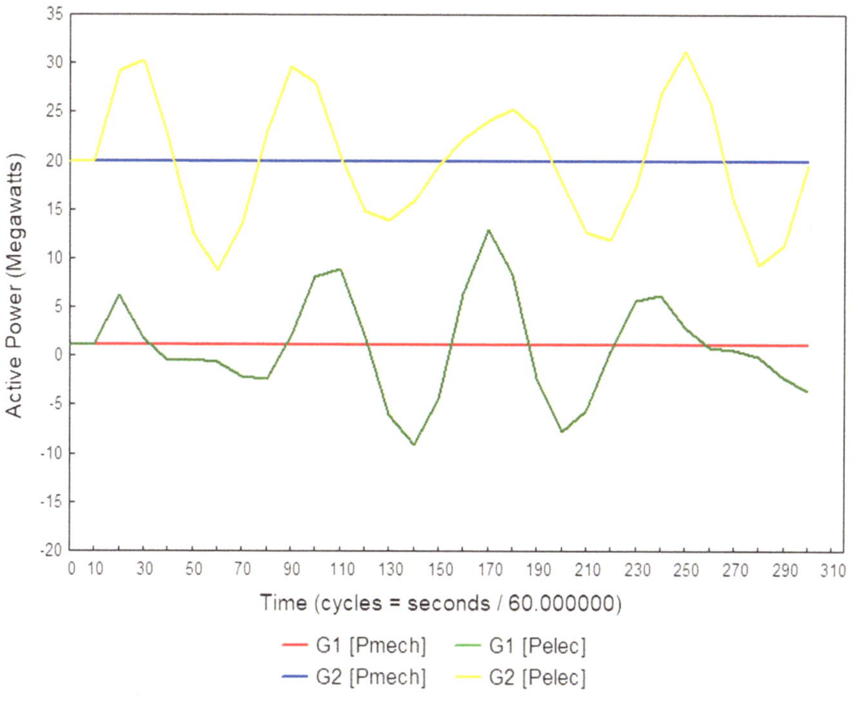

Fig. 9.28 Machines electrical and mechanical powers versus time for Example 9.6

References

1. Glover JD, Overbye T, Sarma M (2017) Power system analysis and design, 6th edn. Cengage Learning, USA, pp 1–942
2. Salam MA (2009) Fundamentals of power systems. Narosa Publishing House Pvt. Ltd., India, pp 1–408
3. Nagsarkar TK, Sukhija MS (2014) Power system analysis, 2nd edn. Oxford University Press, pp 1–726
4. Wildi T (2014) Electrical machines, drives and power systems, 6th edn. Pearson Education Ltd., USA, pp 1–920
5. Sadat H (2010) Power system analysis, 3rd edn. PSA Publisher, USA, pp 1–772
6. Wildi T (2006) Electrical machines, drives and power systems, 6th edn. Pearson Education, USA, pp 1–934
7. El-Hawary ME (1995) Electrical power systems, Revised edn. Willey-IEEE Press, USA, pp 1–808

Exercise Problems

9.1 The terminal voltage of a three-phase wye connected synchronous generator is found to be 16 kV. The generator is connected to an infinite bus through a transmission line. The reactance of the generator and transmission line is given by 1.4 Ω and 1.0 Ω, respectively. The generator delivers the power of 85 MW to the infinite bus, and the infinite bus voltage is found to be 13 kV. Calculate the torque angle, transmission line current, and three-phase reactive power consumed by the transmission line.

9.2 A three-phase synchronous generator is having the terminal voltage 11 kV, and it is connected to an infinite bus through a transmission line. The reactance of the generator and transmission line are 0.12 Ω and 2.54, respectively. The torque angle is found to be 35°, and the bus voltage at the infinite bus is found to be 6.4 kV. Calculate the three-phase real power delivers to the infinite bus and the transmission line current.

9.3 The input and output powers of a 50 Hz, 80 MVA, 8 MJ/MVA, 11 kV, synchronous machine are found to be 20 MW and 12 MW, respectively. Determine the kinetic energy stored in the rotor at synchronous speed, accelerating power, acceleration, and torque angle at 15 cycles.

9.4 The stored energy in the rotor of a 4-pole, 50 Hz, 25 MVA synchronous machine is found to be 175 MJ. Whereas the input and the output powers of the machines are given as 25 MW and 21 MW, respectively. Calculate the inertia constant, accelerating power, and acceleration.

9.5 The inertia constant of a synchronous machine is 5 MJ/MVA, and it is connected to an infinite bus through a transmission line as shown in Fig. P9.1. The generator delivers a real power of 0.7 per unit at a 0.95 power factor lagging to the infinite bus. Find the per unit apparent power, line current, generated voltage, synchronizing power coefficient, undamped angular frequency of oscillation, and period of oscillation.

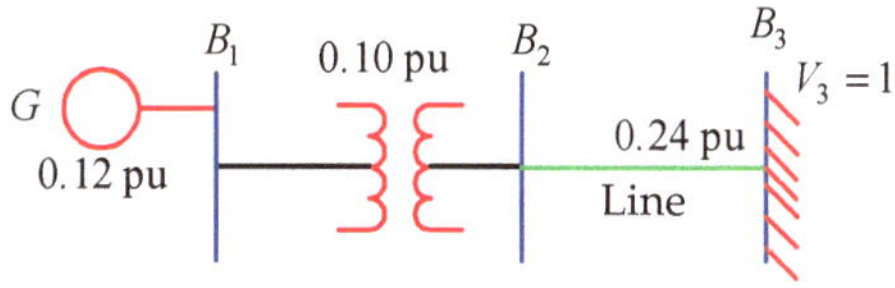

Fig. P9.1 Single-line diagram for Problem 9.5

9.6 The inertia constant of A 50 Hz synchronous machine is 5 MJ/MVA and the excitation voltage is $E_f = 1.1\angle 15^\circ$ pu. The generator is connected to an infinite bus through a transmission line, and the infinity bus voltage is $V_{ib} = 1\angle 0^\circ$ pu. The total reactance between the generator and the infinite bus is found to be 0.42 pu. Determine the synchronizing power coefficient, undamped angular frequency of oscillation, and period of oscillation.

9.7 A substation supplies 120 kV through the transmission lines to a load terminal. The voltage at the load terminals is found 11 kV, and the total reactance is given as 0.24 pu. Calculate the maximum power transferred to the load.

9.8 Two generators are connected in parallel to a busbar. The necessary data for the generators are given below.

Generator 1	100 MVA, 0.85 power factor, 13.8 kV, 3000 rpm, H_1 = 7 MJ/MVA
Generator 2	150 MVA, 0.85 power factor, 13.8 kV, 3000 rpm, H_2 = 5 MJ/MVA

Find the equivalent inertia constant, total kinetic energy, and rotor acceleration if the input power of generator 1 is 90 MW. Consider a common base of 100 MVA.

9.9 Two generators are connected in parallel to a busbar. The necessary data for the generators are given below.

Generator 1	160 MVA, 0.95 power factor, 11.75 kV, H_1 = 8 MJ/MVA
Generator 2	150 MVA, 0.95 power factor, 11.75 kV, H_2 = 6 MJ/MVA

Calculate the equivalent inertia constant, and the total combined energy kinetic energy, and rotor acceleration if the input power of generator 2 is 180 MW. Consider a common base of 100 MVA.

9.10 A synchronous machine with inertia constant of 6 MJ/MVA is connected to an infinite bus as shown in Fig. P9.2. The generator delivers real power and reactive power to the infinite bus are 0.65 pu and 0.85, respectively.

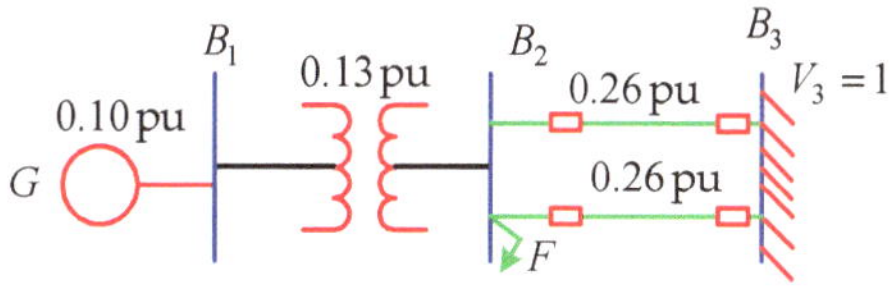

Fig. P9.2 Single-line diagram for Problem 9.10

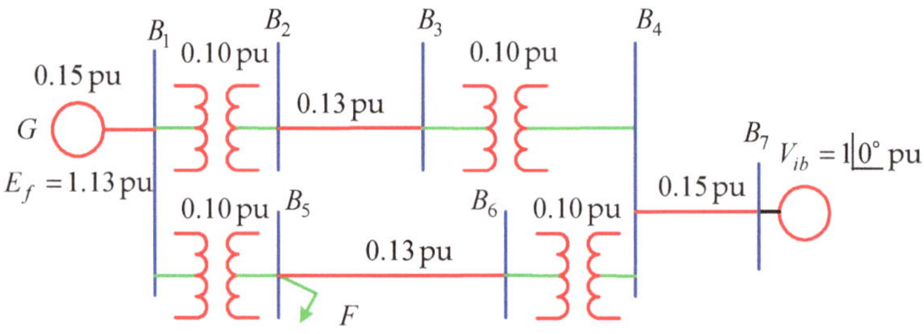

Fig. P9.3 Single-line diagram for Problem 9.11

Fig. P9.4 Single-line diagram for Problem 9.12

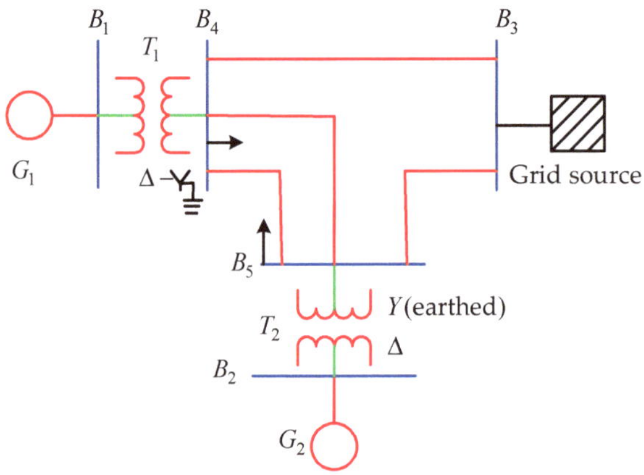

A three-phase fault occurs on the system at point F and subsequently cleared by the circuit breaker operation. Under prefault condition of the lines, calculate the generated voltage, initial torque angle, maximum torque angle, critical clearing angle, and critical clearing time. During fault (impedance 0.09 pu), calculate the fault clearing angle and time.

9.11 A 50 Hz generator with a 6 MJ/MVA inertia constant is connected to an infinite bus through a single transmission line as shown in Fig. P9.3. The real power delivers to the infinite bus is found to be 0.84 pu. A temporary three-phase fault occurs on the line, at busbar 6. Find the fault clearing angle and fault clearing time at pre-fault.

9.12 Fig. P9.4 shows a single-line diagram where two generators are connected with a grid source through transformers (delta-earthed wye) and transmission lines. Some data are shown in Tables P9.1, P9.2, and P9.3. A three-phase occurs on line 3-4 near bus 4 at 12 cycles and is cleared at 16 cycles.

Transformer T_1	25 MVA, three-phase shell, 11 kV (delta)/33 kV (wye), $Z_1 = Z_0 = 10\%$, $X_1/R_1 = X_0/R_0 = 25$
Transformer T_2	40 MVA, three-phase shell, 11 kV (delta)/33 kV (wye), $Z_1 = Z_0 = 12\%$, $X_1/R_1 = X_0/R_0 = 30$

Table P9.1 Machine data

Gen.	Rating	Active and reactive generations	*H* (MJ/MVA)
1	25 MVA, 11 kV, 2P, 85% pf, round rotor type 5	10 MW, 3 Mvar	10
2	35 MVA, 11 kV, 2P, 85% pf, round rotor type 5	20 MW, 4 Mvar	15

Table P9.2 Load data

	Load	
Bus no.	Real power (MW)	Reactive power (Mvar)
4	100	44
5	50	16

Table P9.3 Line data

Bus no.	*R* (Ω/mi)	*X* (Ω/mi)	*B* (μS/mi)
3–4	0.12	2.56	6.34
3–5	0.07	3.12	7.06
4–5	0.64	2.25	8.05
4–5	0.34	2.2	7.67

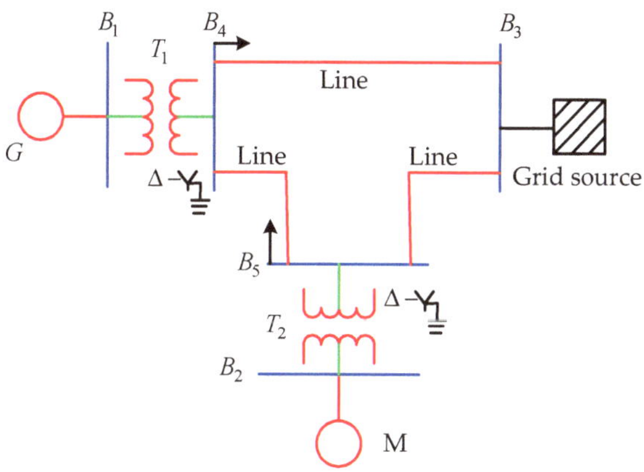

Fig. P9.5 Single-line diagram for Problem 9.13

Grid source 100 MVA, 33 kV, $Z_1 = 0.83 + j10.5$ pu, $Z_0 = 0.07 + j1.2$ pu

Calculate the transient stability by CYME software.

9.13 Fig. P9.5 shows a single-line diagram where two generators are connected with a grid source through transformers (delta-earthed wye) and transmission lines. Some data are shown in Tables P9.4, P9.5, and P9.6. A three-phase occurs on line 3-4 near bus 4 at 12 cycles and is cleared at 16 cycles.

Table P9.4 Machine data

Gen./motor	Rating	Active and reactive generations	*H* (MJ/MVA)
G	25 MVA, 11 kV, 2P, 85% pf, round rotor type 5	10 MW, 3 Mvar	10
M	Three-phase, 1200 kW, 6.4 kV, 1800 rpm, 85% pf, 90% efficiency	–	–

Table P9.5 Load data

	Load	
Bus no.	Real power (MW)	Reactive power (Mvar)
4	120	50
5	76	30

Table P9.6 Line data

Bus no.	*R* (Ω/mi)	*X* (Ω/mi)	*B* (μS/mi)
3–4	0.12	2.56	6.34
3–5	0.07	3.12	7.06
4–5	0.34	2.2	7.67

Transformer T_1 35 MVA, three-phase shell, 11 kV (delta)/22 kV (wye), $Z_1 = Z_0 = 11\%$, $X_1/R_1 = X_0/R_0 = 22$

Transformer T_2 45 MVA, three-phase shell, 22 kV (delta)/0.48 kV (wye), $Z_1 = Z_0 = 13\%$, $X_1/R_1 = X_0/R_0 = 28$

Grid source 100 MVA, 22 kV, $Z_1 = 0.93 + j10.2$ pu, $Z_0 = 0.08 + j1.25$ pu

Calculate the transient stability by CYME software.

Chapter 10
Power System Harmonics

10.1 Introduction

The concept of power system harmonics is not a new phenomenon. In 1916, scientist Steinmetz studied and published the effect of harmonics in three-phase power systems. At the time, the main focus was given on third harmonic currents caused by saturated iron in electrical machines such as transformers and machines. Steinmetz introduced the first delta connections for blocking third harmonic currents.

Oscillation of voltage and current waveforms that follow the mathematical form of a sine or cosine function is known as a clean waveform or undistorted waveform. Voltage waveform distortions are typically created by generators, while current distortion results from loads. These distortions of voltage or current occur in the form of oscillations, which occur more frequently than 50 or 60 Hz and are called harmonics, as shown in Fig. 10.1.

Harmonics affect the quality of AC system delivery to home and other facilities and reduce the performance of electronics equipment that uses the AC system. Harmonics increase energy costs and reduce the lifespan of the electronics hardware system. It can overheat a conductor that can trigger fires. A sine wave or clean wave is a periodic wave that follows symmetrical and repeating patterns. The distorted periodic waveform is defined as the combination of a fundamental wave and one or more harmonic waves. A perfect sinusoid with a constant frequency and amplitude is known as perfect power quality. Less than perfect power quality occurs when a voltage waveform is distorted by transients or harmonics, changes in amplitude, or deviations in frequency. Frequencies that are not integer multiples of fundamental power system frequency are called interharmonics.

Md. A. Salam, *Fundamentals of Electrical Power Systems Analysis*,
https://doi.org/10.1007/978-981-15-3212-2_10

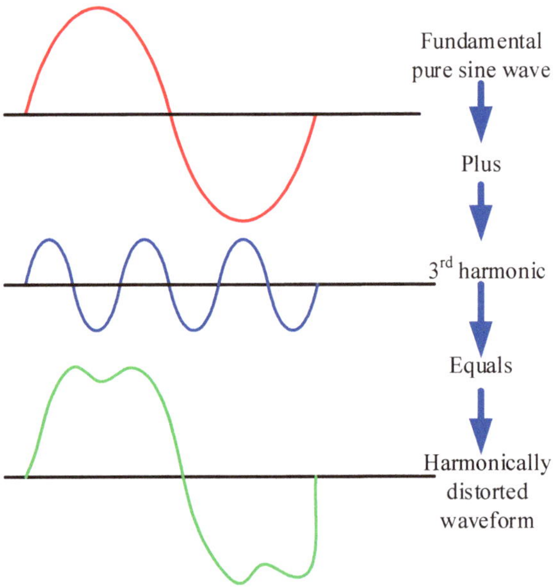

Fig. 10.1 Harmonic patterns

10.2 Generation of Harmonics

The loads are classified as linear load and nonlinear load. The linear load is defined as a load which draws current from the supply that is proportional to the supply voltage. The nonlinear load is defined as a load that its impedance changes with the supply voltage. Due to this changing impedance, the current draw by the nonlinear load is also nonlinear, i.e., non-sinusoidal in nature.

Power semiconductor devices are used during power conversion from AC to DC, DC to DC, DC to AC, and AC to AC. During this power conversion, semiconductor devices constitute the largest nonlinear loads connected to the electric power systems. In the industry, these converters are used for adjustable speed (or variable frequency) drives, uninterruptable power supplies, switch-mode power supplies and draw nonlinear currents. As a result, the supply voltage is distorted at the point of common coupling (PCC).

Harmonics come from commercial and residential single-phase supplies such as fluorescent lighting, adjustable speed drivers for AVAC, and elevators and industrial loads.

Nonlinear industrial loads are generally grouped into three categories. There are three-phase power converters, arcing devices (arc furnaces, arc welder), and saturable devices (transformer, motor).

10.3 Single-Phase Circuit with Linear Load

Consider a single-phase circuit where sinusoidal voltage supplies current to a linear load as shown in Fig. 10.2. In this circuit, the expressions for voltage and current can be written as,

$$v(t) = \sqrt{2}V \sin \omega t \tag{10.1}$$

$$i(t) = \sqrt{2}I \sin(\omega t - \phi) \tag{10.2}$$

Instantaneous power is defined as the product of instantaneous voltage and current, and it is expressed as,

$$p(t) = v(t)i(t) \tag{10.3}$$

Substituting Eqs. (10.1) and (10.2) into Eq. (10.3) yields,

$$p(t) = 2VI \sin \omega t \sin(\omega t - \phi) \tag{10.4}$$

$$p(t) = VI \cos \phi - VI \cos(2\omega t - \phi) \tag{10.5}$$

$$p(t) = VI \cos \phi - VI \cos 2\omega t \cos \phi - VI \sin 2\omega t \sin \phi \tag{10.6}$$

$$p(t) = VI \cos \phi(1 - \cos 2\omega t) - VI \sin \phi \sin 2\omega t = p_a + p_q \tag{10.7}$$

$$p_a = VI \cos \phi(1 - \cos 2\omega t) \tag{10.8}$$

$$p_q = -VI \sin \phi \sin 2\omega t \tag{10.9}$$

The average active power is the average of instantaneous power during time period t_0 to $T + t_0$. Then, the following equation for active power can be written as,

$$P = \frac{1}{T}\int_{t_0}^{T+t_0} p(t)\, dt = VI \cos \phi \tag{10.10}$$

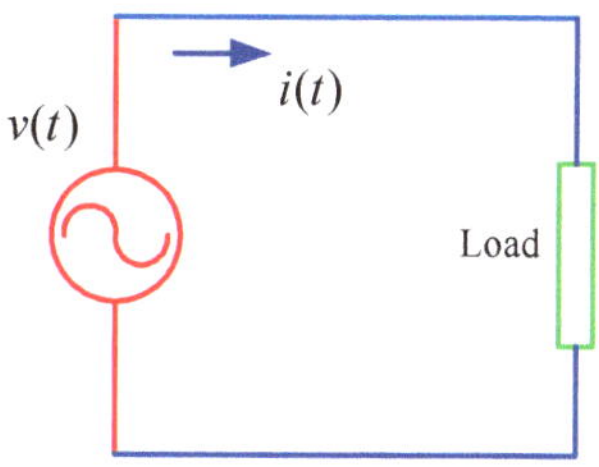

Fig. 10.2 A circuit with a load

The reactive power can be written as,

$$Q \triangleq VI \sin \phi \tag{10.11}$$

Substituting Eqs. (10.10) and (10.11) into Eq. (10.6) yields,

$$p(t) = P(1 - \cos 2\omega t) - Q \sin 2\omega t \tag{10.12}$$

Example 10.1
A linear load of impedance $Z_L = 4 + 12\,\Omega$ is connected in series with a voltage source of $v(t) = \sqrt{2}(220 \sin \omega t)$. Determine the instantaneous active power, instantaneous reactive power, average power, reactive power, apparent power, and power factor.

Solution
The current in the circuit is,

$$i(t) = \frac{v(t)}{Z_L} = \frac{\sqrt{2}(220 \sin \omega t)}{4 + j12} = 24.6 \sin(\omega t - 71.56°)\text{A} \tag{10.13}$$

According to Eq. (10.8), the instantaneous reactive power is calculated as,

$$p_a = \sqrt{2} \times 220 \times 24.6 \cos 71.56°(1 - \cos 2 \times 314t) \tag{10.15}$$

$$p_a = 2420.96(1 - \cos 628t)\text{ W} \tag{10.16}$$

According to Eq. (10.9), the instantaneous reactive power is,

$$p_q = -VI \sin \phi \sin 2\omega t = -\sqrt{2} \times 220 \times 24.6 \sin 71.56° \sin 628t \tag{10.17}$$

$$p_q = -7260.75 \sin 628t\text{ Var} \tag{10.18}$$

Average power is calculated as,

$$P = 2420.96\text{ W} \tag{10.19}$$

The reactive power is determined as,

$$Q = 7260.75\text{ Var} \tag{10.20}$$

The apparent power is calculated as,

$$S = \sqrt{2} \times 220 \times 24.6 = 7653.72\text{VA} \tag{10.21}$$

The power factor is calculated as,

$$pf = \cos 71.56^\circ = 0.31 \tag{10.22}$$

Practice Problem 10.1

A voltage source $v(t) = \sqrt{2}(240 \sin \omega t)$ V is connected in series with a linear load whose impedance is $Z_L = 3 + 8\,\Omega$. Calculate the total instantaneous power, instantaneous active power, instantaneous reactive power, average power, reactive power, and apparent power.

10.4 Single-Phase with Nonlinear Load

Distorted current waveform usually creates heat in the power delivery equipment resolving a nonlinear waveform into a sinusoidal component and is called harmonic analysis. Harmonic analysis will determine heating effect due to nonlinear current flowing to circuit breakers and transformers. Consider the nonlinear load and the expressions for the voltage and current are [1, 2]

$$v(t) = \sqrt{2} V \sin \omega t \tag{10.23}$$

$$i(t) = \sqrt{2} \sum_{n=1}^{\infty} I_n \sin(n\omega t - \phi_n) \tag{10.24}$$

Substituting Eqs. (10.23) and (10.24) into Eq. (10.3) yields,

$$p(t) = \left(\sqrt{2} V \sin \omega t\right) \sqrt{2} \sum_{n=1}^{\infty} I_n \sin(n\omega t - \phi_n) \tag{10.25}$$

$$p(t) = V \sum_{n=1}^{\infty} I_n 2 \sin \omega t \sin(n\omega t - \phi_n) \tag{10.26}$$

$$p(t) = V \sum_{n=1}^{\infty} I_n 2 \sin \omega t \sin(2n\omega t - \phi_n - n\omega t) \tag{10.27}$$

$$p(t) = V \sum_{n=1}^{\infty} I_n [\cos \phi_n - \cos(2n\omega t - \phi_n)] \tag{10.28}$$

$$p(t) = VI_1 2 \sin \omega t \sin(\omega t - \phi_1) + \sum_{n=2}^{\infty} I_n 2 \sin \omega t \sin(n\omega t - \phi_n) \tag{10.29}$$

$$p(t) = VI_1 2 \sin \omega t \sin(\omega t - \phi_1) + V \sum_{n=2}^{\infty} I_n[\cos \phi_n - \cos(2n\omega t - \phi_n)] \tag{10.30}$$

$$\begin{aligned} p(t) &= VI_1 \cos \phi_1 - VI_1 \cos(2\omega t - \phi_1) \\ &\quad + \sum_{n=2}^{\infty} VI_n[\cos \phi_n - \cos 2n\omega t \cos \phi_n - \sin \phi_n \sin 2n\omega t] \end{aligned} \tag{10.31}$$

$$\begin{aligned} p(t) &= VI_1 \cos \phi_1 - VI_1 \cos(2\omega t - \phi_1) - VI_1 \sin \phi_1 \sin 2\omega t \\ &\quad + \sum_{n=2}^{\infty} VI_n[\cos \phi_n - \cos 2n\omega t \cos \phi_n - \sin \phi_n \sin 2n\omega t] \end{aligned} \tag{10.32}$$

From Eq. (10.32), the average active power (P) and reactive power (Q) can be written as,

$$P = P_1 = VI_1 \cos \phi_1 \tag{10.33}$$

$$Q = Q_1 = VI_1 \sin \phi_1 \tag{10.34}$$

Squared Eqs. (10.33) and (10.34) and adding yields,

$$P_1^2 + Q_1^2 = (VI_1 \cos \phi_1)^2 + (VI_1 \sin \phi_1)^2 \tag{10.35}$$

$$P_1^2 + Q_1^2 = (VI_1)^2 (\sin \phi_1^2 + \cos \phi_1^2) \tag{10.36}$$

$$VI_1 = \sqrt{P_1^2 + Q_1^2} \tag{10.37}$$

The apparent power (S) can be written as,

$$S = VI \tag{10.38}$$

Squared Eq. (10.38) yields,

$$S^2 = V^2 I^2 \tag{10.39}$$

The current (I) can be expressed as,

$$I = \sqrt{I_1^2 + I_2^2 + I_3^2 + I_4^2 + ---} \tag{10.40}$$

Substituting Eq. (10.40) into Eq. (10.39) yields,

$$S^2 = V^2(I_1^2 + I_2^2 + I_3^2 + I_4^2 + ---) \quad (10.41)$$

$$S^2 = V^2I_1^2 + V^2(I_2^2 + I_3^2 + I_4^2 + ---) \quad (10.42)$$

The first part of Eq. (10.42) can be modified as,

$$S^2 = (VI_1 \cos\phi_1)^2 + (VI_1 \sin\phi_1)^2 + V^2(I_2^2 + I_3^2 + I_4^2 + ---) \quad (10.43)$$

Substituting Eqs. (10.33) and (10.34) into Eq. (10.43) yields,

$$S^2 = P^2 + Q^2 + H^2 \quad (10.44)$$

$$H^2 = V^2(I_2^2 + I_3^2 + I_4^2 + ---) \quad (10.45)$$

where H is the harmonic power and its unit is volt-ampere (VA). It depends on the harmonic currents whereas P and Q depend on the fundamental component of the current.

To extend the knowledge on harmonics, fundamental power factor and distortion factor need to be defined. Fundamental power factor is defined as the cosine angle between the fundamental voltage and fundamental current, and it can be expressed as,

$$f_{pf} = \cos(\phi_{V_1} - \phi_{I_1}) = \cos\phi_1 \quad (10.46)$$

The ratio of fundamental apparent power to the total apparent power is known as distortion factor (DF), and it can be expressed as,

$$DF = \frac{\sqrt{P_1^2 + Q_1^2}}{VI} \quad (10.47)$$

Substituting Eqs. (10.33) and (10.34) into Eq. (10.47) yields,

$$DF = \frac{\sqrt{(VI_1 \cos\phi_1)^2 + (VI_1 \sin\phi_1)^2}}{VI} \quad (10.48)$$

$$DF = \frac{VI_1}{VI} \quad (10.49)$$

$$DF = \frac{I_1}{I} = \cos\beta \quad (10.50)$$

The ratio of average active power to the total apparent power is known as power factor, and it can be expressed as,

$$pf = \frac{P}{S} \tag{10.51}$$

Substituting Eqs. (10.33) and (10.38) into Eq. (10.51) yields,

$$pf = \frac{VI_1 \cos\phi_1}{VI} \tag{10.52}$$

$$pf = \frac{I_1}{I}\cos\phi_1 \tag{10.53}$$

Substituting Eq. (10.50) into Eq. (10.53) yields,

$$pf = \cos\beta\cos\phi_1 \tag{10.54}$$

From Eq. (10.54), it is seen that the power factor is reduced by a factor of $\cos\beta$ due to the presence of harmonics in the load current.

Example 10.2
The fundamental, second, third harmonic components of the current of a 120 V, 0.9 pf electrical system are found to be 1A, 2A, and 3A, respectively. Calculate the *H*, *P*, *Q*, and distribution factor (*DF*).

Solution
The value of the current (*I*) is calculated as,

$$I = \sqrt{I_1^2 + I_2^2 + I_3^2 + I_4^2 + ---} = \sqrt{1^2 + 2^2 + 3^2} = 3.74\,\text{A} \tag{10.55}$$

The average active power (*P*) is calculated as,

$$P = P_1 = VI_1\cos\phi_1 = 120 \times 1 \times 0.9 = 108\,\text{W} \tag{10.56}$$

The average reactive power (*Q*) is calculated as,

$$Q = Q_1 = VI_1\sin\phi_1 = 120 \times 1 \times \sin(\cos^{-1}0.9) = 52.31\,\text{W} \tag{10.57}$$

The harmonic power (*H*) is calculated as,

$$H = V\sqrt{(I_2^2 + I_3^2 + I_4^2 + ---)} = 120\sqrt{(2^2 + 3^2)} = 432.67\,\text{W} \tag{10.58}$$

The distribution factor (DF) is calculated as,

$$DF = \frac{I_1}{I} = \frac{1}{3.74} = 0.27 \tag{10.59}$$

Practice Problem 10.2
The distribution factor of a 110 V, 0.85 pf electrical system is found to be 0.35. The fundamental and second harmonic components of the current are 2A and 4A, respectively. Calculate the third harmonic component of the current, P, Q, and H.

10.5 Non-sinusoidal Voltage and Nonlinear Load

Consider a non-sinusoidal voltage is connected to nonlinear loads. In this case, the voltage and the current can be expressed as [2],

$$v(t) = V_{dc} + \sum_{n=1}^{\infty} \sqrt{2}V_n \sin(n\omega t - \phi_{vn}) \tag{10.60}$$

$$i(t) = I_{dc} + \sum_{n=1}^{\infty} \sqrt{2}I_n \sin(n\omega t - \phi_{in}) \tag{10.61}$$

Instantaneous power is defined as the power received at any instance of time at its terminals, and it is expressed as the product of instantaneous voltage and current.

From Eqs. (10.60) and (10.61), the instantaneous power $p(t)$ can be expressed as,

$$p(t) = \left[V_{dc} + \sum_{n=1}^{\infty} \sqrt{2}V_n \sin(n\omega t - \phi_{vn})\right]\left[I_{dc} + \sum_{n=1}^{\infty} \sqrt{2}I_n \sin(n\omega t - \phi_{in})\right] \tag{10.62}$$

$$\begin{aligned} p(t) = {} & V_{dc}I_{dc} + I_{dc}\sum_{n=1}^{\infty} \sqrt{2}V_n \sin(n\omega t - \phi_{vn}) + V_{dc}\sum_{n=1}^{\infty} \sqrt{2}I_n \sin(n\omega t - \phi_{in}) \\ & + \sum_{n=1}^{\infty} \sqrt{2}I_n \sin(n\omega t - \phi_{in}) \sum_{n=1}^{\infty} \sqrt{2}V_n \sin(n\omega t - \phi_{vn}) \end{aligned} \tag{10.63}$$

The fourth term of instantaneous power is expressed as,

$$p_{iv}(t) = \sum_{n=1}^{\infty} V_n I_n [\cos(\phi_{in} - \phi_{iv}) - \cos(2n\omega t - \phi_{in} - \phi_{vn})] \tag{10.64}$$

$$p_{iv}(t) = \sum_{n=1}^{\infty} V_n I_n [\cos(\phi_{in} - \phi_{iv}) - \cos(2n\omega t - (\phi_{in} - \phi_{iv}) - 2\phi_{vn})] \quad (10.65)$$

The phase angle between nth harmonic current and voltage is defined as,

$$\phi_n = \phi_{in} - \phi_{vn} \quad (10.66)$$

Substituting Eq. (10.66) into Eq. (10.65) yields,

$$p_{iv}(t) = \sum_{n=1}^{\infty} V_n I_n [\cos\phi_n - \cos(2n\omega t - 2\phi_{vn} - \phi_n)] \quad (10.67)$$

Equation (10.67) can be expanded as,

$$p_{iv}(t) = \sum_{n=1}^{\infty} V_n I_n [\cos\phi_n - \cos(2n\omega t - 2\phi_{vn})\cos\phi_n - \sin(2n\omega t - 2\phi_{vn})\sin\phi_n] \quad (10.68)$$

10.6 Non-sinusoidal Voltage and Nonlinear Loads Active Power

The active power is usually consumed by an active element of a circuit. The active power is always positive. From Eqs. (10.63) and (10.68), the expression for active power is written as [2, 3],

$$p_{active}(t) = V_{dc} I_{dc} + \sum_{n=1}^{\infty} V_n I_n \cos\phi_n [1 - \cos(2n\omega t - 2\phi_{vn})] \quad (10.69)$$

The average active power is defined as,

$$P = \frac{1}{T}\int_0^T p(t)\,dt \quad (10.70)$$

Substituting Eq. (10.69) into Eq. (10.70) yields,

$$P = \frac{1}{T}\int_0^T \left[V_{dc} I_{dc} + \sum_{n=1}^{\infty} V_n I_n \cos\phi_n [1 - \cos(2n\omega t - 2\phi_{vn})] \right] dt \quad (10.71)$$

The average power of $\cos(2n\omega t - 2\phi_{vn})$ is zero; Eq. (10.71) can be modified as,

$$P = V_{dc}I_{dc} + \sum_{n=1}^{\infty} V_n I_n \cos\phi_n \tag{10.72}$$

Expanding Eq. (10.72) yields,

$$P = V_{dc}I_{dc} + V_1 I_1 \cos\phi_1 + V_2 I_2 \cos\phi_2 + V_3 I_3 \cos\phi_3 + \cdots \tag{10.73}$$

$$P = P_{dc} + P_1 + P_H \tag{10.74}$$

where,

$P_{dc} = V_{dc}I_{dc}$ is the average active power due to DC components,

$P_1 = V_1 I_1 \cos\phi_1$ is the average fundamental active power,

$P_H = V_2 I_2 \cos\phi_2 + V_3 I_3 \cos\phi_3 + ---- = \sum_{h\neq 1} V_h I_h \cos\phi_h$ is the average harmonic active power.

10.7 Non-sinusoidal Voltage and Nonlinear Loads Reactive Power

The maximum pulsating power over one cycle is known as reactive power. It can be positive or negative depending on the phase angle between the voltage and the current. The reactive power is positive when the coil consumes it. It is negative when is usually consumed by an active element of a circuit. The reactive power is represented by Q, and its unit of reactive power is Var, which stands for volt-ampere reactive. The reactive power can be written as [2, 3],

$$Q = \sum_{n=1}^{\infty} V_n I_n \sin\phi_n \tag{10.75}$$

Expanding Eq. (10.75) yields,

$$Q = V_1 I_1 \sin\phi_1 + V_2 I_2 \sin\phi_2 + V_3 I_3 \sin\phi_3 + \cdots \tag{10.76}$$

$$Q = Q_1 + Q_H \tag{10.77}$$

where

$Q_1 = V_1 I_1 \sin\phi_1$ is the fundamental reactive power,

$Q_H = V_2 I_2 \sin\phi_2 + V_3 I_3 \sin\phi_3 + \cdots = \sum_{h \neq 1} V_h I_h \sin\phi_h$ is the harmonic reactive power.

Example 10.3
The single-phase current and voltage expressions of an electrical system are given as, $v(t) = 80 + \sqrt{2} \times 110 \sin\omega t + \sqrt{2} \times 65 \sin(2\omega t - 25^\circ) + \sqrt{2} \times 35 \sin(3\omega t - 50^\circ)$ and $i(t) = 10 + \sqrt{2} \times 15 \sin(\omega t - 25^\circ) + \sqrt{2} \times 9 \sin(2\omega t - 50^\circ) + \sqrt{2} \times 4 \sin(3\omega t - 75^\circ)$. Determine the average active power due to dc component, average fundamental and harmonic active and reactive powers.

Solution
The value of the average active power due to the dc component is calculated as,

$$P_{dc} = V_{dc} I_{dc} = 80 \times 10 = 800\text{W} \tag{10.78}$$

The value of the average active and reactive power due to the fundamental components are calculated as,

$$P_1 = V_1 I_1 \cos\phi_1 = 110 \times 15 \times \cos(0^\circ - 25^\circ) = 1.5\,\text{kW} \tag{10.79}$$

$$Q_1 = V_1 I_1 \sin\phi_1 = 110 \times 15 \times \sin(0^\circ - 25^\circ) = 697.32\text{W} \tag{10.80}$$

The value of the harmonic active and reactive powers is calculated as,

$$\begin{aligned} P_H &= V_2 I_2 \cos\phi_2 + V_3 I_3 \cos\phi_3 \\ &= 65 \times 9 \times \cos(25^\circ - 50^\circ) + 35 \times 4 \times \cos(50^\circ - 75^\circ) = 657.07\text{W} \end{aligned} \tag{10.81}$$

$$\begin{aligned} Q_H &= V_2 I_2 \sin\phi_2 + V_3 I_3 \sin\phi_3 \\ &= 65 \times 9 \times \sin(25^\circ - 50^\circ) + 35 \times 4 \times \sin(50^\circ - 75^\circ) = 306.4\text{W} \end{aligned} \tag{10.82}$$

Practice Problem 10.3
The single-phase current and voltage expressions of an electrical system are given as, $v(t) = 40 + \sqrt{2} \times 120 \sin\omega t + \sqrt{2} \times 85 \sin(2\omega t - 30^\circ) + \sqrt{2} \times 40 \sin(3\omega t - 60^\circ)$ and $i(t) = 20 + \sqrt{2} \times 25 \sin(\omega t - 30^\circ) + \sqrt{2} \times 12 \sin(2\omega t - 60^\circ) + \sqrt{2} \times 7 \sin(3\omega t - 90^\circ)$. Calculate the average active power due to dc component, average fundamental and harmonic active and reactive powers.

10.8 Non-sinusoidal Voltage and Nonlinear Loads Apparent Power

The product of root mean square (rms) voltage and rms current is known as apparent power. Apparent power is represented by S and its unit is volt-ampere (VA). The rms values of voltage and current for nonlinear loads can be written as [2, 4, 5],

$$V = V_{dc}^2 + V_1^2 + V_2^2 + V_3^2 + \cdots \tag{10.83}$$

$$I = I_{dc}^2 + I_1^2 + I_2^2 + I_3^2 + \cdots \tag{10.84}$$

The apparent power is expressed as,

$$S = VI \tag{10.85}$$

Substituting Eqs. (10.83) and (10.84) into Eq. (10.85) yields,

$$S = \left(\sqrt{V_{dc}^2 + V_1^2 + V_2^2 + V_3^2 + \cdots}\right)\left(\sqrt{I_{dc}^2 + I_1^2 + I_2^2 + I_3^2 + \cdots}\right) \tag{10.86}$$

$$S = \left(\sqrt{V_{dc}^2 + V_1^2 + V_H^2}\right)\left(\sqrt{I_{dc}^2 + I_1^2 + I_H^2}\right) \tag{10.87}$$

where the expressions for harmonic voltage and current are written as,

$$V_H^2 = V_2^2 + V_3^2 + \cdots = \sum_{h \neq 1} V_h^2 \tag{10.88}$$

$$I_H^2 = I_2^2 + I_3^2 + \cdots = \sum_{h \neq 1} I_h^2 \tag{10.89}$$

Squared Eq. (10.87) yields,

$$S^2 = \left(V_{dc}^2 + V_1^2 + V_H^2\right)\left(I_{dc}^2 + I_1^2 + I_H^2\right) \tag{10.90}$$

$$S^2 = V_{dc}^2 I_{dc}^2 + V_1^2 I_{dc}^2 + V_H^2 I_{dc}^2 + V_{dc}^2 I_1^2 + V_1^2 I_1^2 + V_H^2 I_1^2 + V_{dc}^2 I_H^2 + V_1^2 I_H^2 + V_H^2 I_H^2 \tag{10.91}$$

$$S^2 = V_{dc}^2 I_{dc}^2 + V_1^2 I_1^2 + V_H^2 I_H^2 + V_{dc}^2 (I_1^2 + I_H^2) + I_{dc}^2 (V_1^2 + V_H^2) + V_1^2 I_H^2 + V_H^2 I_1^2 \tag{10.92}$$

$$S^2 = S_{dc}^2 + S_1^2 + S_H^2 + S_D^2 \tag{10.93}$$

where the expressions for the apparent powers due to dc, fundamental, harmonic, and distortion components are written as,

$$S_{dc}^2 = V_{dc}^2 I_{dc}^2 \tag{10.94}$$

$$S_1^2 = V_1^2 I_1^2 \tag{10.95}$$

$$S_H^2 = V_H^2 I_H^2 \tag{10.96}$$

$$S_D^2 = V_{dc}^2 (I_1^2 + I_H^2) + I_{dc}^2 (V_1^2 + V_H^2) + V_1^2 I_H^2 + V_H^2 I_1^2 \tag{10.97}$$

Equation (10.93) again can be modified by introducing apparent power due to non-fundamental term (S_N) as,

$$S^2 = S_1^2 + S_N^2 \tag{10.98}$$

$$S_N^2 = V_{dc}^2 I_{dc}^2 + V_H^2 I_H^2 + V_{dc}^2 (I_1^2 + I_H^2) + I_{dc}^2 (V_1^2 + V_H^2) + V_1^2 I_H^2 + V_H^2 I_1^2 \tag{10.99}$$

In power system, the effect of dc component is usually considered negligible. Then, the terms associated with dc component are considered zero, and the final expression for non-fundamental apparent power is,

$$S_N^2 = V_H^2 I_H^2 + V_1^2 I_H^2 + V_H^2 I_1^2 \tag{10.100}$$

$$S_N^2 = S_H^2 + D_I^2 + D_V^2 \tag{10.101}$$

$$S_N = \sqrt{S_H^2 + D_I^2 + D_V^2} \tag{10.102}$$

where the expression for the apparent power due to distortion of voltage (D_V) is expressed as,

$$D_V = V_H I_1 \tag{10.103}$$

Apparent power due to voltage distortion or simply voltage distortion power is equal to the product of the total harmonic voltage and fundamental component of the current.

Substituting Eq. (10.88) into Eq. (10.103) yields,

$$D_V = I_1 \sum_{h \neq 1} V_h^2 \quad \text{(Var)} \tag{10.104}$$

Similarly, the apparent power due to current distortion or simply current distortion power is equal to the product of the total harmonic current and fundamental component of the voltage.

$$D_I = V_1 I_H \tag{10.105}$$

Substituting Eq. (10.89) into Eq. (10.105) yields,

$$D_I = V_1 \sum_{h \neq 1} I_h^2 \quad \text{(Var)} \tag{10.106}$$

Apparent power due to harmonic or simply harmonic apparent power is expressed as,

$$S_H = V_H I_H \tag{10.107}$$

Again, substituting Eqs. (10.88) and (10.89) into Eq. (10.107) yields,

$$S_H = \sqrt{\sum_{h \neq 1} V_h^2 \sum_{h \neq 1} I_h^2} \tag{10.108}$$

The total harmonic distortion due to voltage and current is defined as the ratio of harmonic voltage or current to the fundamental component of voltage or current. These can be expressed as,

$$THD_V = \frac{V_H}{V_1} = \frac{\sqrt{\sum_{h \neq 1} V_h^2}}{V_1} \tag{10.109}$$

$$THD_I = \frac{I_H}{I_1} = \frac{\sqrt{\sum_{h \neq 1} I_h^2}}{I_1} \tag{10.110}$$

From Eqs. (10.109) and (10.110), harmonic voltage and current can be expressed as,

$$V_H = THD_V V_1 \tag{10.111}$$

$$I_H = I_1 THD_I \tag{10.112}$$

Substituting Eq. (10.111) into Eq. (10.103) and Eq. (10.112) into Eq. (10.105) yields,

$$D_V = THD_V V_1 I_1 \tag{10.113}$$

$$D_I = V_1 I_1 THD_I \tag{10.114}$$

Again, substituting Eqs. (10.111) and (10.112) into Eq. (10.107) yields,

$$S_H = THD_V V_1 I_1 THD_I \tag{10.115}$$

Equations (10.113), (10.114) and (10.115) can be modified by putting the expression of fundamental apparent power ($S_1 = V_1 I_1$) as,

$$D_V = THD_V S_1 \tag{10.116}$$

$$D_I = S_1 THD_I \tag{10.117}$$

$$S_H = THD_V S_1 THD_I \tag{10.118}$$

From Eq. (10.116), the apparent power due to voltage is defined as the product of fundamental apparent power and the total harmonic distortion due to voltage. Similarly, from Eq. (10.117), the apparent power due to current is defined as the product of fundamental apparent power and the total harmonic distortion due to current. Whereas from Eq. (10.118), the apparent power due to harmonic is equivalent to the product of the total harmonic distortion due to current and voltage and the fundamental apparent power.

Substituting Eqs. (10.116), (10.117), and (10.118) into Eq. (10.101) yields,

$$S_N^2 = (THD_V S_1 THD_I)^2 + (THD_V S_1)^2 + (THD_I S_1)^2 \tag{10.119}$$

In a power system, the total harmonic distortion due to current dominates more than the voltage. Therefore, the effect of THD_V on the calculation of non-fundamental apparent power is neglected. In this case, Eq. (10.119) can be modified by neglecting the first and second terms as,

$$S_N^2 \approx (THD_I S_1)^2 \tag{10.120}$$

$$S_N \approx THD_I S_1 \tag{10.121}$$

From Eq. (10.121), the non-fundamental apparent power is defined as the product of the total harmonic distortion due to the current and the apparent power due to the fundamental component. Non-fundamental apparent power will increase due to increasing the total harmonic distortion due to current. Therefore, the loss will be increased during the transfer of electrical power, and the power system network will be less efficient.

Example 10.4

The single-phase current and voltage expressions of an electrical system are given as, $v(t) = 40 + \sqrt{2} \times 110 \sin \omega t + \sqrt{2} \times 75 \sin(3\omega t - 25^\circ) + \sqrt{2} \times 45 \sin(5\omega t - 50^\circ) + \sqrt{2} \times 23 \sin(7\omega t - 75^\circ)$ and $i(t) = 8 + \sqrt{2} \times 12 \sin(\omega t - 25^\circ) + \sqrt{2} \times 7 \sin(3\omega t - 50^\circ) + \sqrt{2} \times 5 \sin(5\omega t - 75^\circ) + \sqrt{2} \times 3 \sin(7\omega t - 100^\circ)$. Calculate the S_H, THD_V, THD_I, D_V, D_I, and S_N.

Solution

The value of the harmonic voltage is calculated as,

$$V_H = \sqrt{V_3^2 + V_5^2 + V_7^2} = \sqrt{75^2 + 45^2 + 23^2} = 90.44\,\mathrm{V} \tag{10.122}$$

The value of the harmonic current is calculated as,

$$I_H = \sqrt{I_3^2 + I_5^2 + I_7^2} = \sqrt{7^2 + 5^2 + 3^2} = 9.11\,A \tag{10.123}$$

The value of the apparent power due to harmonic components of voltage and current is calculated as,

$$S_H = V_H I_H = 90.44 \times 9.11 = 823.90\,\mathrm{VA} \tag{10.124}$$

The total harmonic distortion due to voltage is calculated as,

$$THD_V = \frac{V_H}{V_1} = \frac{90.44}{110} = 0.82 \tag{10.125}$$

The total harmonic distortion due to current is calculated as,

$$THD_I = \frac{I_H}{I_1} = \frac{9.11}{12} = 0.76 \tag{10.126}$$

The value of the apparent power due to distortion of voltage (D_V) is calculated as,

$$D_V = V_H I_1 = 90.44 \times 12 = 1085.28\,\mathrm{VA} \tag{10.127}$$

The value of the apparent power due to the distortion of current (D_I) is calculated as,

$$D_I = V_1 I_H = 110 \times 9.11 = 1002.1\,\mathrm{VA} \tag{10.128}$$

The value of the non-fundamental apparent power is calculated as,

$$S_N = \sqrt{S_H^2 + D_I^2 + D_V^2} = \sqrt{823.90^2 + 1002.1^2 + 1085.28^2} = 1691.40\,\mathrm{VA} \tag{10.129}$$

Practice Problem 10.4
An electrical system has the following single-phase current and voltage.

$$v(t) = 320 + \sqrt{2} \times 120 \sin \omega t + \sqrt{2} \times 67 \sin(2\omega t - 30°) + \sqrt{2} \times 34 \sin(3\omega t - 60°) + \sqrt{2} \times 17 \sin(4\omega t - 90°)$$

$i(t) = 5 + \sqrt{2} \times 16 \sin(\omega t - 30°) + \sqrt{2} \times 9 \sin(2\omega t - 60°) + \sqrt{2} \times 6 \sin(3\omega t - 90°) + \sqrt{2} \times 4 \sin(4\omega t - 120°)$. Find S_N.

10.9 Modeling Concept of Load

The term load represents a single device that consumes active and reactive power connected to a power system. The load is broadly divided into static and dynamic. Static load represents the active and reactive powers at any instant of time as functions of the bus voltage magnitude and the frequency at the same instant. Static loads are counted for passive elements such as resistive, inductive, and capacitive. The dynamic load is a load that represents the active and reactive powers at any instant of time as functions of the bus voltage magnitude and the frequency at the last instant of time. Examples of dynamic loads are the loads driven by the motor. According to the basic definition of power (P), the following relation can be written as,

$$P = VI = \frac{V^2}{R} = I^2 R \tag{10.130}$$

According to Eq. (10.130), the static load model is classified as constant current, constant impedance, and constant power. In a constant current static load model, the power varies directly with the voltage magnitude. In a constant impedance static load model, the power varies directly with the square of the voltage magnitude. It is often known as a constant admittance load model. Whereas in a constant power static load model, the power does not change with changes the voltage magnitude.

Consider a load device whose active and reactive power at the rated voltage (V_0) is P_0 and Q_0, respectively. However, these parameters are considered initial bus values during modeling. Again, consider the load consumes active power P and reactive power Q at voltage V. At constant current static load model, the active and reactive power can be expressed as [6],

$$\frac{P}{P_0} = \left(\frac{V}{V_0}\right)^1 \tag{10.131}$$

$$\frac{Q}{Q_0} = \left(\frac{V}{V_0}\right)^1 \tag{10.132}$$

At constant impedance static load model, the active and reactive power can be expressed as,

$$\frac{P}{P_0} = \left(\frac{V}{V_0}\right)^2 \tag{10.133}$$

$$\frac{Q}{Q_0} = \left(\frac{V}{V_0}\right)^2 \tag{10.134}$$

At constant power static load model, the active and reactive power can be expressed as,

$$\frac{P}{P_0} = \left(\frac{V}{V_0}\right)^0 \tag{10.135}$$

$$\frac{Q}{Q_0} = \left(\frac{V}{V_0}\right)^0 \tag{10.136}$$

The static load models can be further extended as exponential and polynomial. The static load model in an exponential case can be expressed as,

$$P = P_0\left(\frac{V}{V_0}\right)^{n_p} \tag{10.137}$$

$$Q = Q_0\left(\frac{V}{V_0}\right)^{n_p} \tag{10.138}$$

where n_p and n_q are the exponent parameters of the load. By setting these parameters to 0, 1, and 2 represent the constant power, constant current and constant impedance load models. The values of the exponential parameters based on reference [6] are mentioned in Table 10.1.

Table 10.1 Values of exponential parameters for certain loads [6]

Loads	n_p	n_q
Induction motor (half load)	0.2	1.5
Induction motor (full load)	0.1	2.8
Incandescent lamp	1.6	0
Fluorescent lamp	1.2	3
Heating	2	0

The static load model in a polynomial case can be represented by the combination of constant power, constant current, and constant impedance load models. These can be represented as,

$$P = P_0\left[a_0 + a_1\left(\frac{V}{V_0}\right) + a_2\left(\frac{V}{V_0}\right)^2\right] \tag{10.139}$$

$$Q = Q_0\left[b_0 + b_1\left(\frac{V}{V_0}\right) + b_2\left(\frac{V}{V_0}\right)^2\right] \tag{10.140}$$

where a_0, a_1, a_2 and b_0, b_1, b_2 are the constant parameters of the load and the sum of these parameters is equal to one, i.e.,

$$a_0 + a_1 + a_2 = 1 \tag{10.141}$$

$$b_0 + b_1 + b_2 = 1 \tag{10.142}$$

The frequency dependency of a static load model is represented by the frequency deviation and frequency sensitivity parameter as,

$$f_d = 1 + a_f(f - f_0) \tag{10.143}$$

where f is the frequency of the bus voltage, f_0 is the rated frequency of the load, and a_f is the sensitivity parameter.

The static load modeling with the frequency-dependent is obtained by multiplying Eq. (10.143) to any of exponential or polynomial expressions. Now, multiplying Eq. (10.137) by (10.143) yields,

$$P = P_0\left(\frac{V}{V_0}\right)^{n_p}\left[1 + a_f(f - f_0)\right] \tag{10.144}$$

10.10 Resistive Load Modeling

A resistive load radiates heat when energized with rated voltage. Ohm's law represents the relationship between the resistance, voltage, and current as,

$$V = IR \tag{10.145}$$

The following expression defines the resistance as,

$$R = \rho \frac{l}{A} \tag{10.146}$$

where l is the length of the conductor, A is the cross-sectional area of the conductor, and ρ is the resistivity of a conductor. The resistivity is temperature-dependent and is expressed as,

$$\rho = \rho_0[1 + \alpha(T - T_0)] \tag{10.147}$$

where

ρ_0 is the resistivity at 300°K,
α is the temperature coefficient at 300°K,
T_0 is the temperature at 300°K,
T is the normal operating temperature.

In an operating condition, the heating of resistance depends on T. This temperature (T) depends on the current flowing in the resistance. Therefore, T is equivalent to I and α is replaced by a current coefficient, β. Equation (10.147) can be modified as,

$$\rho = \rho_0(1 + \beta I) \tag{10.148}$$

Substituting Eq. (10.148) into Eq. (10.146) yields,

$$R = \rho_0(1 + \beta I)\frac{l}{A} \tag{10.149}$$

$$R = \rho_0 \frac{l}{A} + \beta \frac{l}{A} I \tag{10.150}$$

$$R = R_0 + R_1 I \tag{10.151}$$

where the expression of resistances is written as,

$$R_0 = \rho_0 \frac{l}{A} \tag{10.152}$$

$$R_1 = \beta \frac{l}{A} \tag{10.153}$$

Substituting Eq. (10.151) into Eq. (10.145) yields,

$$V = I(R_0 + R_1 I) \tag{10.154}$$

$$V = R_1 I^2 + R_0 I \tag{10.155}$$

$$I_{1,2} = \frac{-R_0 \pm \sqrt{R_0^2 + 4VR_1}}{2R_1} \tag{10.156}$$

$$I_{1,2} = \frac{-R_0}{2R_1} \pm \sqrt{\frac{R_0^2}{4R_1^2} + \frac{V}{R_1}} \tag{10.157}$$

The power can be expressed as,

$$P = V\left[\frac{-R_0}{2R_1} \pm \sqrt{\frac{R_0^2}{4R_1^2} + \frac{V}{R_1}}\right] = \left[\frac{-R_0}{2R_1} \pm \sqrt{\frac{R_0^2}{4R_1^2} + \frac{V}{R_1}}\right]^2 R \tag{10.158}$$

From Eq. (10.158), it is seen that the power can be calculated if other parameters are known.

10.11 Modeling of Induction Motor Load

Three-phase and single-phase induction motors with different ratings are used in different types of industries. Students can get more knowledge from the course of electrical machines on classifications, equivalent circuits, and applications of induction motors. However, in a power system, most of the stability analysis uses dynamic model based on the equivalent circuit as shown in Fig. 10.3.

The no-load circuit either is used across the source or across the load with any significant error. The no-load circuit consists of a parallel combination of magnetizing resistance and inductance. However, the value of the magnetizing resistance is very small compared to magnetizing inductance, and it can be neglected in the dynamic model. The more simplified dynamic model is shown in Fig. 10.4.

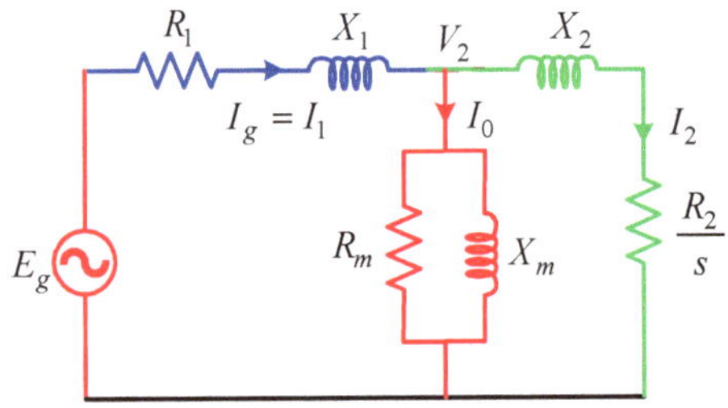

Fig. 10.3 Equivalent circuit of induction motor

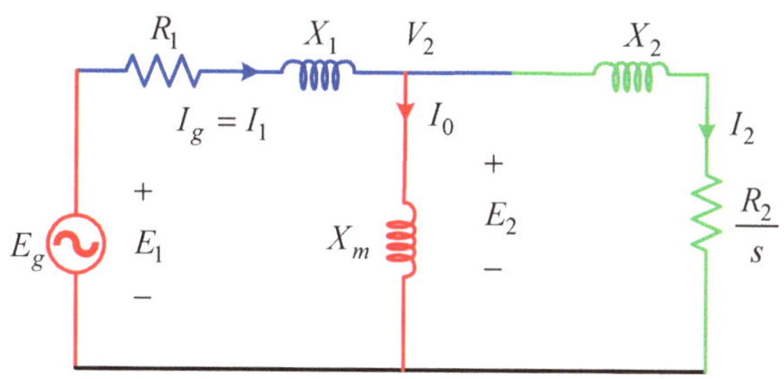

Fig. 10.4 Simplified equivalent circuit

From the circuit in Fig. 10.4, the expression of current I_2 is written as,

$$I_2 = \frac{E_2}{\frac{R_2}{s} + X_2} \tag{10.159}$$

From Fig. 10.3, the active power and the reactive power absorbed by the circuit is expressed as,

$$P = \frac{V_2^2}{R_m} + I_1^2 R_1 + I_2^2 \frac{R_2}{s} \tag{10.160}$$

$$Q = \frac{V_2^2}{X_m} + I_1^2 X_1 + I_2^2 X_2 \tag{10.161}$$

10.12 Harmonic Simulation

Harmonics are generated by a nonlinear load such as battery charger, variable frequency drives, arc furnace, and shunt frequency source. They draw current at the fundamental frequency and inject current to the network at a higher frequency by an odd multiplier (3, 5, 7, 9, etc.) of the fundamental frequency. Harmonic currents create excess heating in the conductor, motor and transformer due to I^2R whereas the harmonic voltages distort the sinusoidal voltage waveforms that create the high peaks. These high peaks damage the insulation and shorten the life of the high voltage equipment. The normal operation of protective devices such as fuse, relay, and circuit breaker is disturbed due to the presence of excessive harmonic current. However, a suitable rating of the capacitor is used in the network to partially and fully remove the harmonic currents as the capacitive reactance decreases with increasing frequency. A five-bus power system is shown in Fig. 10.5. Here, the grid source is connected at bus 1, the generator is connected at bus 5, and a shunt frequency source is connected at bus 3. The line and transformer data are shown in Tables 10.1 and 10.2, respectively.

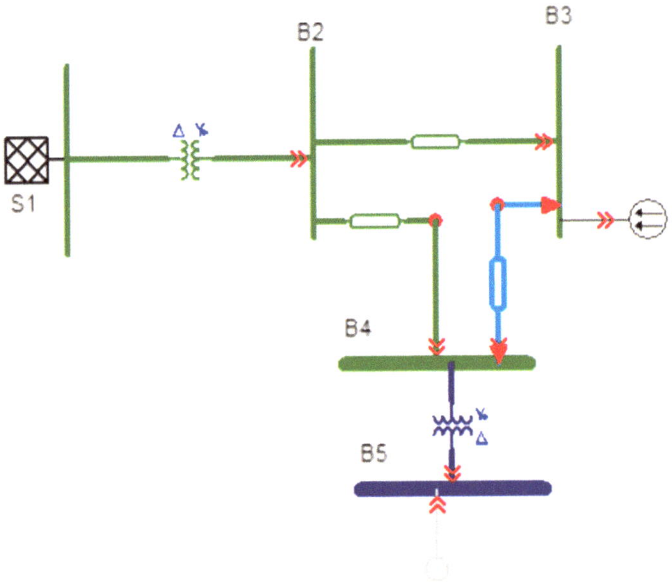

Fig. 10.5 A single-line diagram drawn by CYME

Table 10.2 Line data

Line	*R* (Ω/mi)	*X* (Ω/mi)	*B* (μS/mi)
2–3	0.24	0.85	6.78
2–4	0.14	4.57	10.34
3–4	0.21	2.95	8.68

Source S1 100 MVA, operating and nominal voltage is 12.47 kV, $Z_1 = 0.1 + j0.6$ Ω, $Z_0 = 0.5 + j2.0$ Ω.

Generator 17.5 MVA hydro, 7.33 kV, 0.80 pf.

The data of each equipment such as source, transformer, generator, load, and balanced overhead line need to be input through the equipment link. In addition, the harmonic order along with magnitudes and phase angles needs to be assigned through the equipment link as shown in Table 10.6. Once all equipment data are entered, select the harmonic analysis and keep the default global setting and run the software. Finally, different graphs such as voltage and current distortions and impedance scans are plotted (Tables 10.3 and 10.4).

CYME simulation results are plotted as shown in Figs. 10.6, 10.7, 10.8, 10.9, 10.10, 10.11 and 10.12.

Table 10.3 Transformer data

Transformer	Impedance (%)	Connection (D-Y earthed)	$X_1/R_1 = X_0/R_0$
T1: 3ϕ shell, 15MVA, liquid-filled	$Z_1 = Z_0 = 8.25$	12.47 kV/33 kV	18
T2: 3ϕ shell, 5MVA, liquid-filled	8.25	7.33 kV/33 kV	12

Table 10.4 Line data

Harmonic order	Magnitude (A)	Angles (deg.)
3	4	30
5	8	30
7	12	30
9	6	30
11	19	30

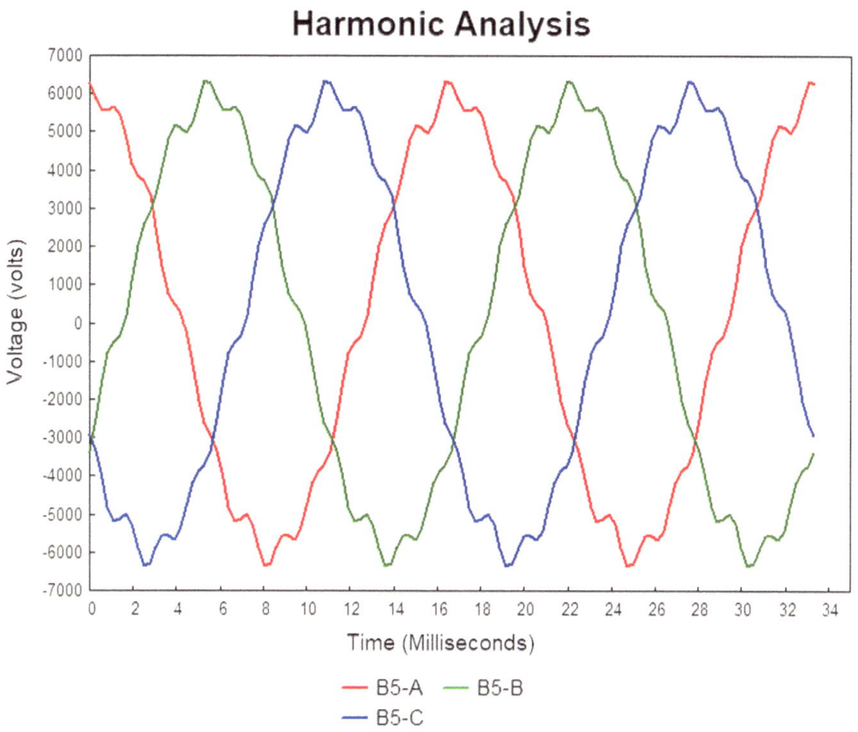

Fig. 10.6 Voltage distortions profile for bus 5

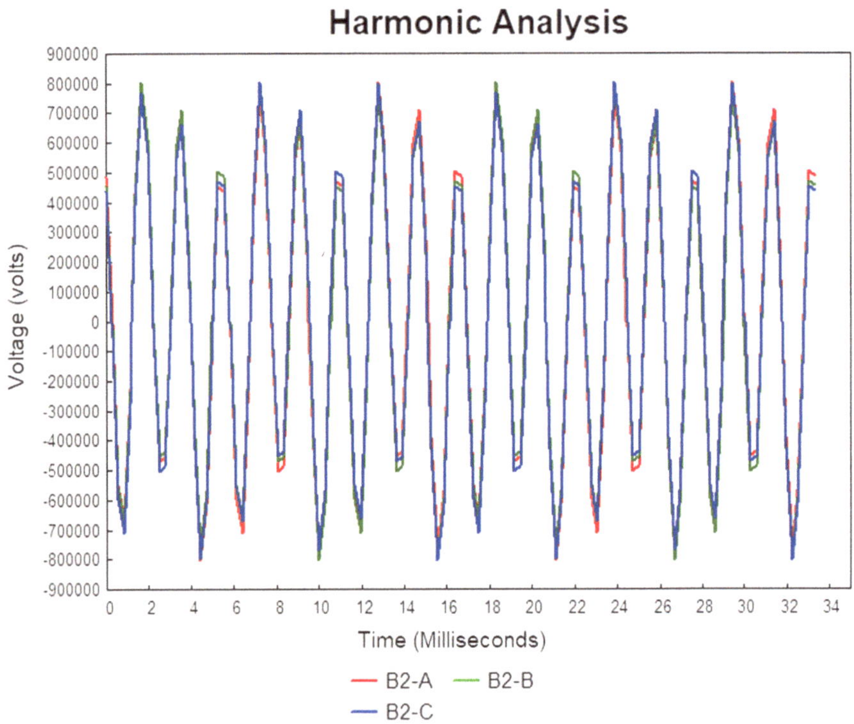

Fig. 10.7 Voltage distortions profile for bus 2

10.13 Power Quality Parameters and Measurement

Power quality has become important issues for power utility companies and electrical contractors due to the use of sophisticated control equipment at the substations and due to nonlinear loads. Equipment such as smart devices, battery chargers, and other variable frequency devices create disturbances in the power system network. There are sophisticated equipment for monitoring and measuring power quality anomalies such as transients, sags, swells, harmonic distortions, voltage flickers, overvoltages, and undervoltages [5, 7].

Sags: A decrease in rms voltage or current to between 0.1 and 0.9 pu at the power frequency for durations between of 0.5 cycles to 1 min. It is often known as voltage dips. Sags or voltage dips occur due to the energization of heavy loads, starting of large induction motor and single-line-to-ground fault. In a single-line-to-ground fault, the magnitude of the current is very high, but the voltage is very low.

Swells: It is the opposite of sag. An increase in rms voltage or current to between 1.1 and 1.8 pu at the power frequency for durations of 0.5 cycles to 1 min. The

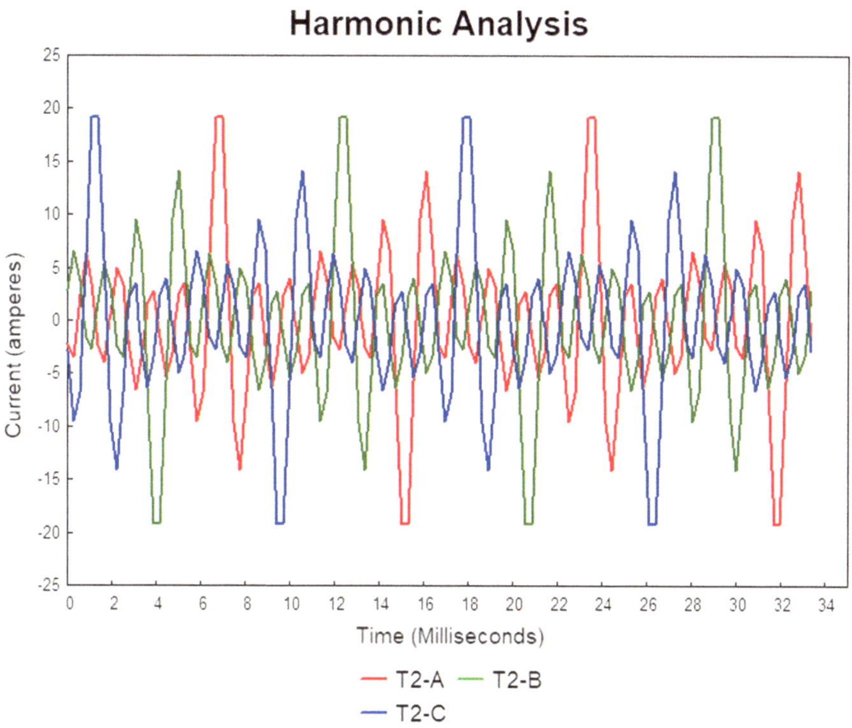

Fig. 10.8 Current distortions profile for transfer 2

main causes of swells are switching off large loads, energizing a capacitor bank, and momentarily overvoltage.

Undervoltage: A reduction in rms voltage between to 0.8 pu to 0.9 pu at the power frequency for more than one minute is known as undervoltage, and the main causes of undervoltage are due to less supply capability.

Overvoltage: An increase in rms voltage between to 1.1 pu to 1.2 pu at the power frequency for more than one minute is known as overvoltage, and the main causes of undervoltage are due to lightning, ferroresonance, high voltage equipment switching, and insulation fault.

Undervoltage: A reduction in rms voltage between to 0.8 pu to 0.9 pu at the power frequency for more than one minute is known as undervoltage, and the main causes of undervoltage are due to draw a high current by an induction motor, high reactive power with low power factor, etc.

PCC: It is abbreviated as the point of common coupling as shown in Fig. 10.13. The common connection point where the power utility and the consumer have common access to measure harmonic indices is known as the point of common coupling.

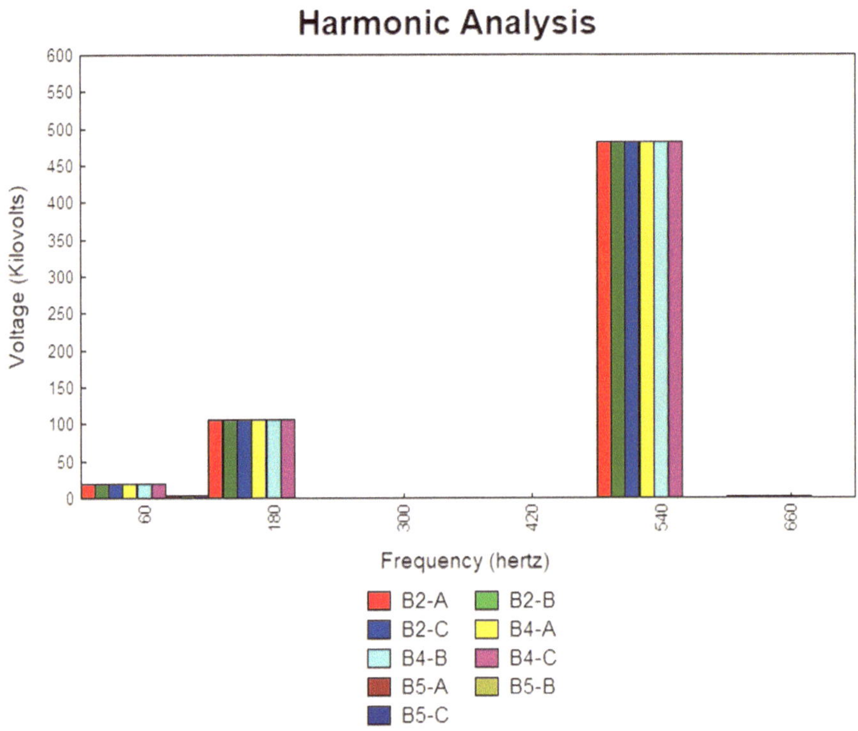

Fig. 10.9 Voltage distortions bar profiles for different buses

Interruptions: A decrease in the voltage level of less than 0.1 pu for up to one-minute duration is known as interruptions. It is categorized as short-term (up to 3 min) and long-term interruptions (more than 3 min). The main causes of interruptions are due to equipment failure or malfunctioning, damage to transmission lines, or poles, etc.

Flicker: The flicker means fluctuation of line voltage. The fluctuation of line voltage occurs due to continuous and rapid changes in load current. The voltage will fluctuate due to changing the load current, which causes changing a light. This voltage fluctuation on a light or lamp is usually perceived as flicker by the human eyes.

Electrical Noise: It is an unwanted high-frequency signal (less than 200 kHz) which is superimposed on a low-frequency voltage or current in a power system. There are many causes for the electrical nose. Some examples are improper grounding, variable speed drive devices, corona, electrical interference etc.

DC offset: DC offset is the presence of DC voltage and current in an AC system. The main causes of dc offset are due to the operation of electronic devices.

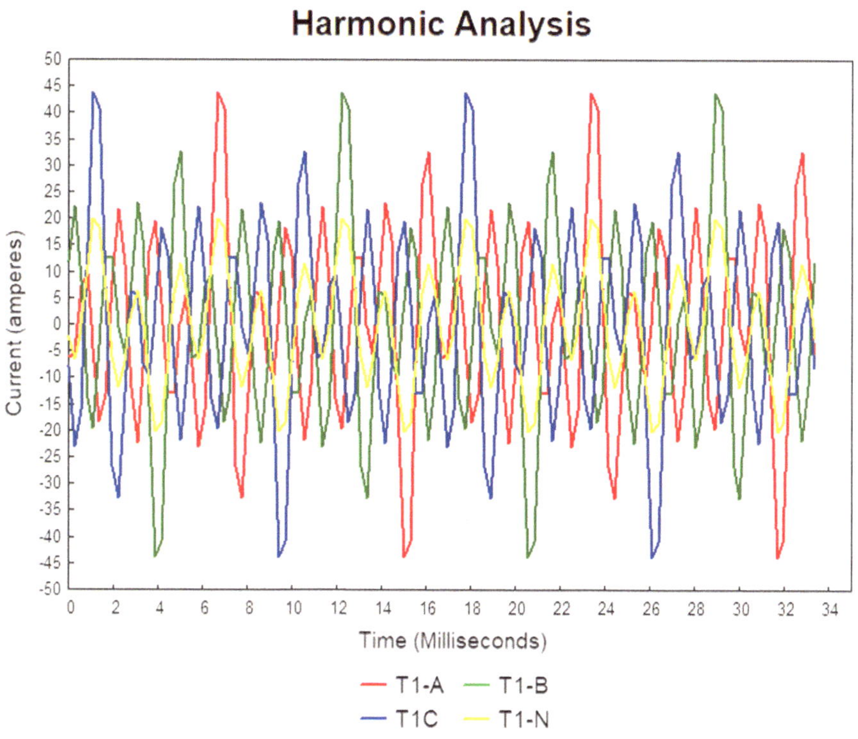

Fig. 10.10 Current distortions profiles for transformer 1

THD: The total harmonic distortions due to voltage and current are already defined in Eqs. (10.109) and (10.110), respectively. However, it can be extended as the ratio of the square root of the individual harmonic components squared to the fundamental component of the current and can be expressed as,

$$THD_I = \frac{\sqrt{I_3^2 + I_5^2 + I_7^2}}{I_1} \tag{10.162}$$

TDD: It is abbreviated as the total demand distortion. It is defined as the ratio of the square root of the individual harmonic components squared to the maximum demand load current (I_L at the fundamental frequency) at PCC, and it can be expressed as,

$$TDD_I = \frac{\sqrt{I_3^2 + I_5^2 + I_7^2}}{I_L} \tag{10.163}$$

According to IEEE Std 519-2014 [6], at the PCC, system owners or operators should limit line-to-neutral (phase) voltage harmonics as daily 99th percentile very

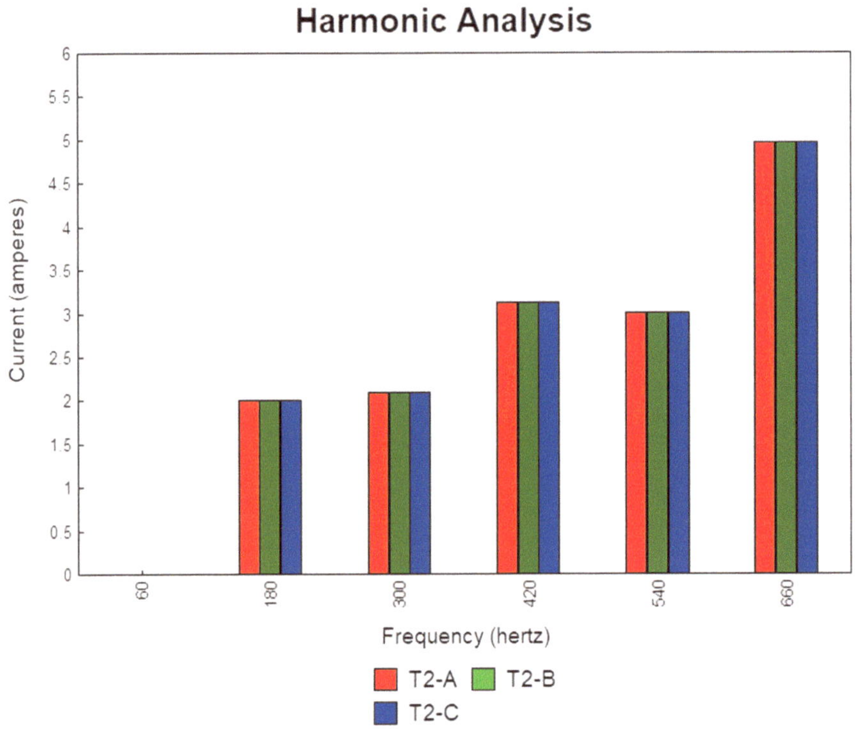

Fig. 10.11 Current distortions bar profile for transformer 2

short time (3 s) and weekly 95th percentile short time (10 min) values should be less than the values given in Table 10.5.

Again, according to IEEE Std 519-2014, the current distortion limits for the users connected to a system where the rated voltage at the PCC is 120 V to 69 kV and is mentioned in Table 10.6. In Table 10.6, the following points are summarized.

- Daily 99th percentile very short time (3 s) harmonic currents should be less than 2.0 times the values given in Table 10.6.
- Weekly 99th percentile short time (10 min) harmonic currents should be less than 1.5 times the values given in Table 10.6.
- Weekly 95th percentile short time (10 min) harmonic currents should be less than the values given in Table 10.6.
- TDD up to 50th order only.
- The maximum demand current value is established at the PCC and should be taken as the sum of the currents corresponding to the maximum demand during each of the twelve previous months divided by 12.

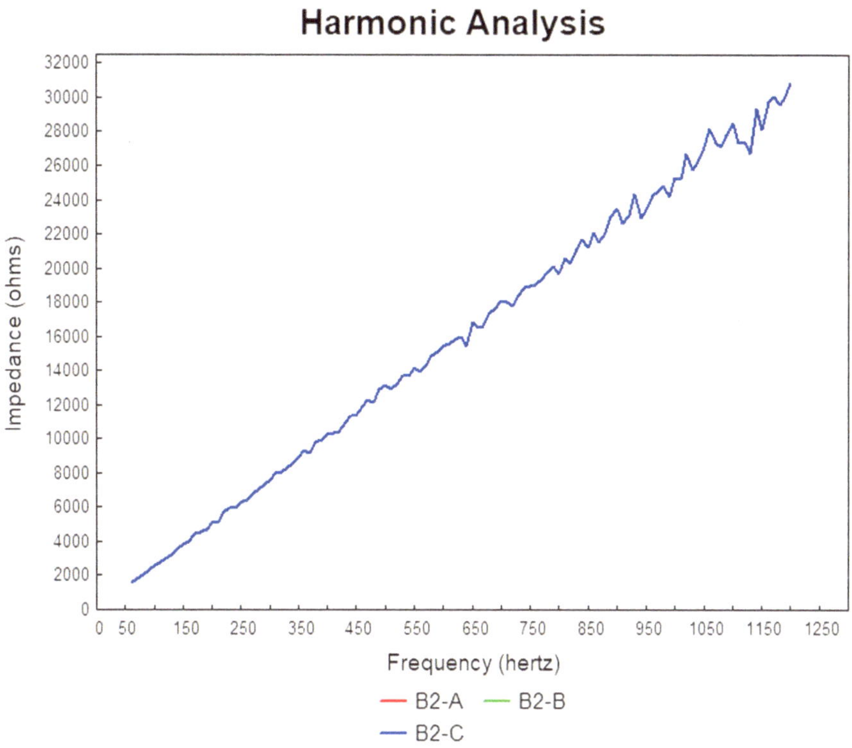

Fig. 10.12 Impedance profile bus 2

Table 10.5 Voltage distortion limits [7]

Bus voltage V_b at PCC	Individual harmonic (%)	THD (%)
$V_b \leq 1\,\text{kV}$	5.0	8.0
$1\,\text{kV} < V_b \leq 69\,\text{kV}$	3.0	5.0
$69\,\text{kV} < V_b \leq 161\,\text{kV}$	1.5	2.5
$161\,\text{kV} < V_b$	1.0	1.5[a]

[a] High voltage equipment can reach up to 2%

Fluke Corporation developed a sophisticated power quality analyzer (F438-2) which is used widely as shown in Fig. 10.14. It is a "Class A" tool that complies with international standards with a measurement accuracy of ±0.1%, 500 kHz bandwidth, and 200 kHz sample rate. It has a wideband measurement capability and can measure harmonics of up to 80 kHz. The tool provides an inverter measurement facility and can measure power and efficiency across two separate circuits.

Table 10.6 Current distortion limits [7]

Maximum harmonic current distortion in percent of I_L						
Individual harmonic order (odd harmonics)[a,b]						
I_{sc}/I_L	$3 \leq h < 11$	$11 \leq h < 17$	$17 \leq h < 23$	$23 \leq h < 35$	$35 \leq h \leq 50$	*TDD*
$< 20^c$	4.0	2.0	1.5	0.6	0.3	5.0
$20 < 50$	7.0	3.5	2.5	1.0	0.5	8.0
$50 < 100$	10.0	4.5	4.0	1.5	0.7	12.0
$100 < 1000$	12.0	5.5	5.0	2.0	1.0	15.0
> 1000	15.0	7.0	6.0	2.5	1.4	20.0

[a] Even harmonics are limited to 25% of the odd harmonic limits above

[b] Current distortions that result in a dc offset, i.e., half-wave rectifier converters are not allowed

[c] All power generation equipment are limited to these values of current distortion, regardless of the actual I_{sc}/I_L

I_{sc} is the maximum current at PCC

I_L is the maximum demand load current (fundamental frequency) at PCC

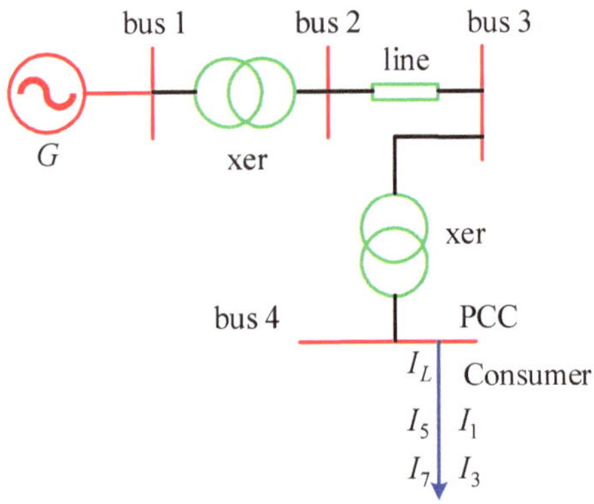

Fig. 10.13 Identification of PCC between utility and consumer

Different types of power line anomalies have been measured by the Fluke Corporation power quality analyzer F438-2, and the measurement results are shown in Fig. 10.15.

Example 10.5

A three-phase, 60 Hz source delivers power to a car manufacturing company. Section 10.1 draws a current of 100A at a fundamental frequency. Section 10.2 draws 100A at a fundamental frequency along with 14A, 11A, 8A for the third, fifth, and seventh harmonic components of current, respectively. Calculate the THD_I and TDD_I.

Solution

The value of the load current at the fundamental frequency is calculated as,

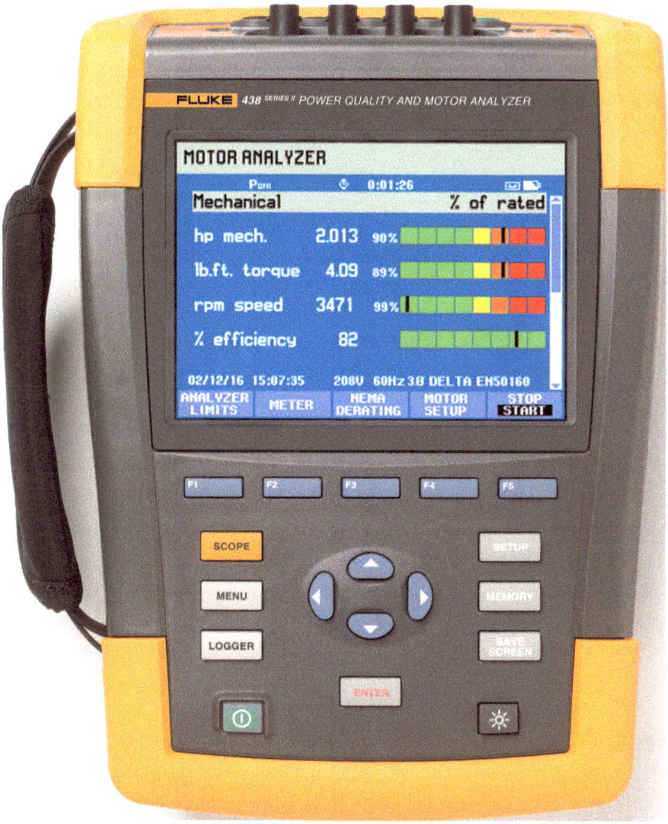

Fig. 10.14 A F438-2 Power Quality Analyzer courtesy by Fluke Corporation

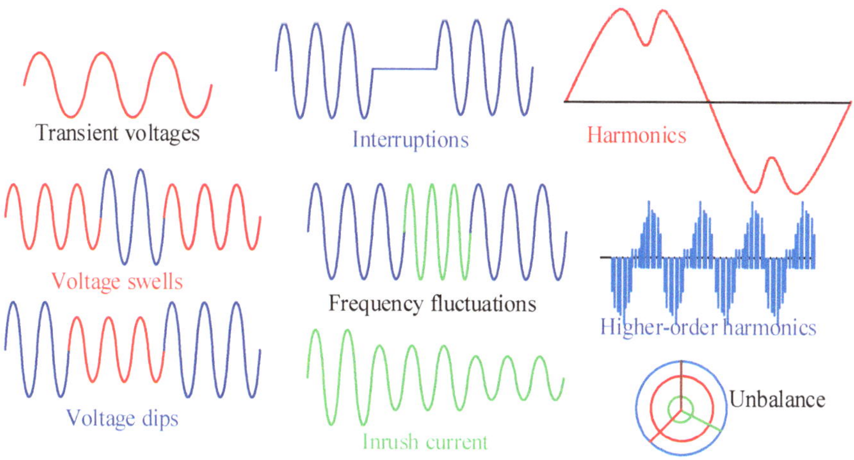

Fig. 10.15 Power line voltage and current variations

$$I_L = 100 + 100 = 200\,\mathrm{A} \tag{10.164}$$

The value of the THD_I can be calculated as,

$$THD_I = \frac{\sqrt{I_3^2 + I_5^2 + I_7^2}}{I_1} = \frac{\sqrt{14^2 + 11^2 + 8^2}}{100} = 0.20 \tag{10.165}$$

The value of the TDD_I can be calculated as,

$$TDD_I = \frac{\sqrt{I_3^2 + I_5^2 + I_7^2}}{I_L} = \frac{\sqrt{14^2 + 11^2 + 8^2}}{200} = 0.10 \tag{10.167}$$

Practice Problem 10.5

A three-phase, 60 Hz source delivers power to an automobile battery manufacturing company. Section 10.1 draws a current of 100A at a fundamental frequency. Section 10.2 draws 100A at a fundamental frequency along with 13A, 11A, 7A, 3 A for the third, fifth, seventh, and ninth harmonic components of current, respectively. Calculate the THD_{I} and TDD_{I}.

References

1. Das JC (2015) Power system harmonics and passive filter design. First Edition, IEEE Press Wiley, pp 1–844, USA
2. Emanuels AE (2010) Power definitions and the physical mechanism of power flow. 1st edn, Wiley, pp 1– 273, UK
3. Chattopadhaya S, Mitra M, Sengupta S (2011) Electric power quality. First, Springer, pp 1–279, New York
4. Arrillaga Jos, Watson Neville R (2003) Power system harmonics. Second Edition, Wiley, pp 1–412
5. Mikkili S, Panda AK (2016) Power quality issues-current harmonics. 1st edn., CRC Press, pp 1–185, USA
6. Berg GJ (1973) Power-system load representation. IEEE Proc 120(3):344–348
7. IEEE Std 519–2014, IEEE Recommended Practices and Requirements for Harmonic Control in Electric Power Systems, Institute of Electrical and Electronics Engineers, Inc. pp. 1–50, USA

Exercise Problems

10.1 A linear load of impedance $Z_L = 3 + 8\,\Omega$ is connected in series with a voltage source $v(t) = \sqrt{2}(110 \sin \omega t)$. Calculate the total instantaneous power, instantaneous active power, instantaneous reactive power, average power, reactive power, apparent power and power factor.

10.2 The fundamental, second, third and fourth harmonic components of the current of a 110 V, 0.95 pf electrical system are found to be 10A, 7A, 4A and 2A, respectively. Calculate the H, P, Q, and distribution factor (DF).

10.3 The distribution factor of a 120 V, 0.90 pf electrical system is found to be 0.31. The fundamental and second harmonic components of the current are 3A and 5A, respectively. Calculate the third harmonic component of the current, P, Q, and H.

10.4 The single-phase current and voltage expressions of an electrical system are given as, $v(t) = 45 + \sqrt{2} \times 110 \sin \omega t + \sqrt{2} \times 25 \sin(2\omega t - 20°) + \sqrt{2} \times 13 \sin(3\omega t - 40°)$ and $i(t) = 9 + \sqrt{2} \times 7 \sin(\omega t - 35°) + \sqrt{2} \times 5 \sin(2\omega t - 70°) + \sqrt{2} \times 3 \sin(3\omega t - 105°)$. Find the average active power due to dc component, average fundamental, harmonic active and reactive powers.

10.5 The single-phase current and voltage expressions of an electrical system are given as,$v(t) = \sqrt{2} \times 75 \sin \omega t + \sqrt{2} \times 35 \sin(3\omega t - 20°) + \sqrt{2} \times 21 \sin(7\omega t - 60°)$ and $i(t) = \sqrt{2} \times 19 \sin(\omega t) + \sqrt{2} \times 15 \sin(3\omega t - 70°) + \sqrt{2} \times 11 \sin(7\omega t - 105°)$. Calculate the rms values of voltage and current, average apparent power, fundamental and harmonic active and reactive powers.

10.6 The single-phase current and voltage expressions of an electrical system are given as, $v(t) = 60 + \sqrt{2} \times 90 \sin \omega t + \sqrt{2} \times 65 \sin(3\omega t - 25°) + \sqrt{2} \times 35 \sin(5\omega t - 50°) + \sqrt{2} \times 15 \sin(7\omega t - 75°)$ $i(t) = 10 + \sqrt{2} \times$ $10 \sin(\omega t - 25°) + \sqrt{2} \times 9 \sin(3\omega t - 50°) + \sqrt{2} \times 7 \sin(5\omega t - 75°) + \sqrt{2} \times 5 \sin(7\omega t - 100°)$ Calculate the S_H, THD_V, THD_I, D_V, D_I and S_N.

10.7 A three-phase, 60 Hz source delivers power to a plastic manufacturing company. The moulding section draws a current of 100A at a fundamental frequency. The finishing section draws 100A at a fundamental frequency along with 9A, 7A, 3A for the third, fifth and seventh harmonic components of current, respectively. Calculate the THD_{I} and TDD_{I}.

Chapter 11
Overhead Line Insulators and Sags

11.1 Introduction

There are several components of overhead lines used for electrical power transmission and distribution systems. Some of these components are an insulator, crossarm, tower, pole, conductor, etc. Insulators of different types are available for power transmission and distribution networks. Some of these insulators are cap-and-pin, long-rod, shackle, and line-post composite, etc. The shackle insulators are normally used at the dead-end or at a sharp turn of the distribution lines where there is a high tensile load. Suspension insulators are used at the transmission lines as vertical, horizontal, and V-string configurations. The line-post insulators are used at the grid substation with the relevant high voltage equipment as well as at the busbar. Sag is formed when the transmission line conductors are supported between the two equal electrical towers. A perfect value of sag is maintained by the conductors to avoid excess stretch or tension. In this chapter, details on insulators and sags will be discussed.

11.2 String Efficiency of Suspension Insulators

The three cap-and-pin porcelain insulators are arranged in a suspension string and hung to the crossarm as shown in Fig. 11.1. The porcelain portion of each insulator is arranged between two metal parts. Therefore, each insulator will form a capacitor known as mutual capacitance. Another capacitor will be formed between the metal parts of the insulator to the tower. This capacitor is known as shunt capacitance. The equivalent circuit of Fig. 11.1 is shown in Fig. 11.2. Let us consider the voltage across the first, second, and third insulators as V_1, V_2, and V_3, respectively. Let us consider the mutual capacitance of each insulator as C and the shunt capacitance C_1 as a fraction m of mutual capacitance. This assumption can be expressed as [1, 2],

Md. A. Salam, *Fundamentals of Electrical Power Systems Analysis*,
https://doi.org/10.1007/978-981-15-3212-2_11

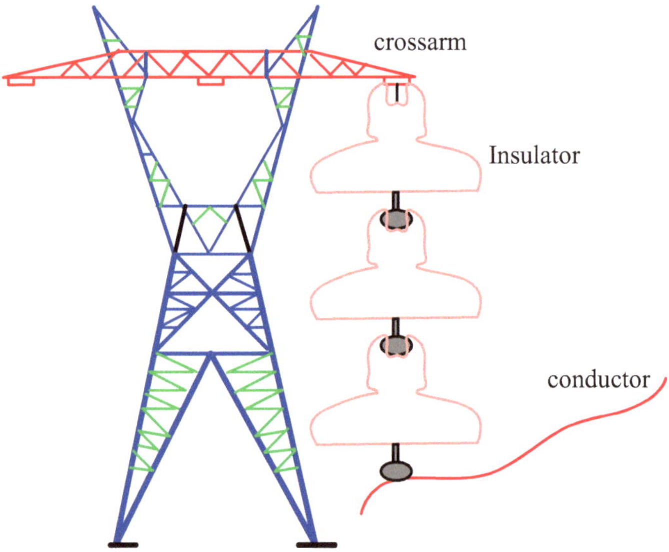

Fig. 11.1 A tower with suspension insulator

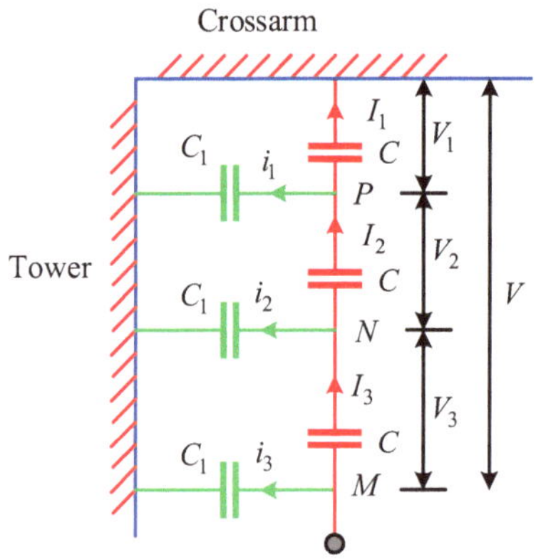

Fig. 11.2 An equivalent circuit of a tower with suspension insulator

$$C_1 = mC \tag{11.1}$$

Applying KCL at node P yields,

$$I_2 = I_1 + i_1 \tag{11.2}$$

$$V_2 \omega C = V_1 \omega C + V_1 \omega C_1 \tag{11.3}$$

Substituting Eq. (11.1) into Eq. (11.3) yields,

$$V_2 \omega C = V_1 \omega C + V_1 \omega m C \tag{11.4}$$

$$V_2 = V_1(1 + m) \tag{11.5}$$

Again, applying KCL at node N yields,

$$I_3 = I_2 + i_2 \tag{11.6}$$

$$V_3 \omega C = V_2 \omega C + (V_1 + V_2)\omega C_1 \tag{11.7}$$

Substituting Eq. (11.1) into Eq. (11.7) yields,

$$V_3 \omega C = V_2 \omega C + (V_1 + V_2)\omega m C \tag{11.8}$$

$$V_3 = V_1 m + V_2(1 + m) \tag{11.9}$$

Substituting Eq. (11.5) into Eq. (11.9) yields,

$$V_3 = V_1 m + V_1(1 + m)^2 \tag{11.10}$$

$$V_3 = V_1(m^2 + 3m + 1) \tag{11.11}$$

The total voltage across the insulators string, i.e., the voltage between the conductor and earth (V) is calculated as,

$$V = V_1 + V_2 + V_3 \tag{11.12}$$

Substituting Eqs. (11.5) and (11.11) into Eq. (11.12) yields,

$$V = V_1 + V_1(1 + m) + V_1(m^2 + 3m + 1) \tag{11.13}$$

$$V = V_1(m^2 + 4m + 3) = V_1(m + 1)(m + 3) \tag{11.14}$$

The string efficiency is defined as the ratio of the voltage across the string to the product of a number of the string and the voltage across the insulator nearest to the conductor. Mathematically, it can be defined as,

$$\eta_s = \frac{V}{3 \times V_3} \times 100 \tag{11.15}$$

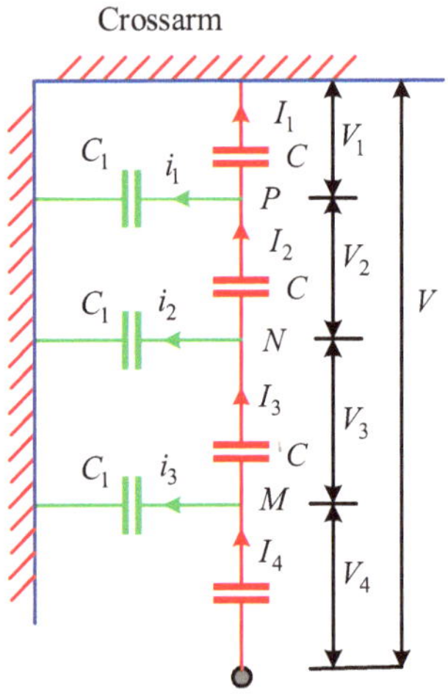

Fig. 11.3 An equivalent circuit of a tower with the four insulator strings

In general, for n number of insulator string, Eq. (11.16) can be written as,

$$\eta_s = \frac{V}{n \times V_n} \times 100 \tag{11.16}$$

From Eq. (11.16), it is seen that the string efficiency will decrease with increasing the number of insulator.

Again, the four insulators are considered as a string as shown in Fig. 11.3.

Applying KCL at node M yields,

$$I_4 = I_3 + i_3 \tag{11.17}$$

$$V_4\omega C = V_3\omega C + (V_1 + V_2 + V_3)\, C_1 \tag{11.18}$$

Substituting Eqs. (11.1), (11.5), and (11.11) into Eq. (11.18) yields,

$$V_4\omega C = V_1(m^2 + 3m + 1)\omega C + [V_1 + V_1(m+1) + V_1(m^2 + 3m + 1)]\, m\omega C \tag{11.19}$$

$$V_4 = m^2 + 3m + 1 + m + m(m+1) + m(m^2 + 3m + 1) \tag{11.20}$$

$$V_4 = m^2 + 3m + 1 + m + m^2 + m + m^3 + 3m^2 + m \tag{11.21}$$

$$V_4 = m^3 + 5m^2 + 6m + 1 \tag{11.22}$$

From Eq. (11.5), (11.11), and (11.22), it is concluded that the voltage distributions across the insulators are not equal.

Example 11.1

The three units of the cap-and-pin insulator are used as a string for the transmission lines tower as shown in Fig. 11.4. The value of the shunt capacitance is 9% of the mutual capacitance. The maximum peak voltage across the unit is found to be 11 kV. Calculate the voltage distribution across the string and string efficiency.

Solution

The value of the m factor is,

$$m = \frac{C_1}{C} = 0.09 \tag{11.23}$$

The phase voltage across the string is calculated as,

$$V_3 = \frac{11}{\sqrt{2}} = 7.78\,\text{kV} \tag{11.24}$$

Applying KCL at node P yields,

$$I_2 = I_1 + i_1 \tag{11.25}$$

$$V_2\omega C = V_1\omega C + V_1\omega(0.09C) \tag{11.26}$$

$$V_2 = 1.09V_1 \tag{11.27}$$

Again, applying KCL at node N yields,

$$I_3 = I_2 + i_2 \tag{11.28}$$

$$V_3\omega C = V_2\omega C + (V_1 + V_2)\omega(0.09C) \tag{11.29}$$

$$V_3 = V_2 + (V_1 + V_2)0.09 \tag{11.30}$$

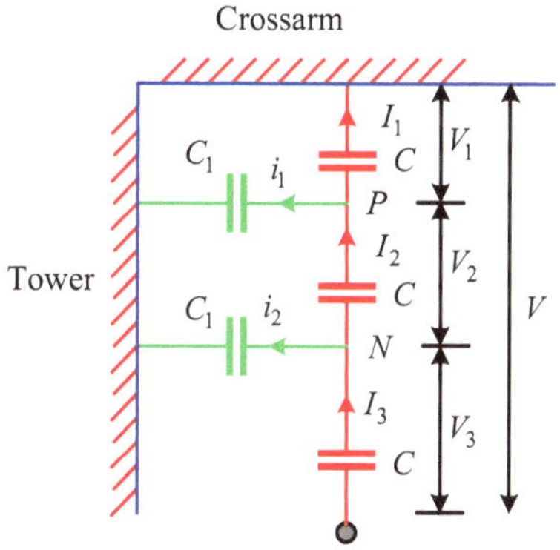

Fig. 11.4 An equivalent circuit for Example 11.1

Substituting Eq. (11.27) into Eq. (11.30) yields,

$$V_3 = 1.09V_1 + (V_1 + 1.09V_1)0.09 \tag{11.31}$$

Substituting Eq. (11.24) into Eq. (11.31) yields,

$$7.78 = 1.09V_1 + (V_1 + 1.09\,V_1)0.09 \tag{11.32}$$

$$V_1 = 6.09\,\text{kV} \tag{11.33}$$

Substituting Eq. (11.31) into Eq. (11.27) yields,

$$V_2 = 1.09 \times 6.09 = 6.64\,\text{kV} \tag{11.34}$$

String efficiency is calculated as,

$$\eta_s = \frac{V}{n \times V_n} \times 100 = \frac{6.09 + 6.64 + 7.78}{3 \times 7.78} \times 100 = 87.87\% \tag{11.35}$$

Practice Problem 11.1

The three units of the cap-and-pin insulator are used as a string for the 33 kV transmission lines tower as shown in Fig. 11.5. The value of the shunt capacitance is 12% of the mutual capacitance. Calculate the voltage distribution across the string and string efficiency.

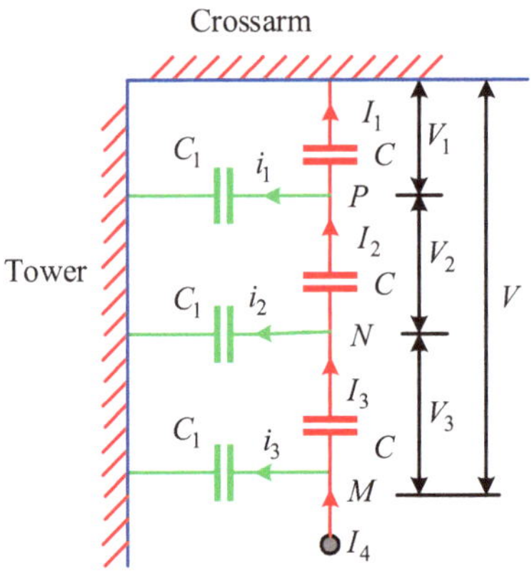

Fig. 11.5 An equivalent circuit for Practice Problem 11.1

11.3 Equalization of Voltage Distributions

Generally, the voltage distributions across the insulators are not uniform. It is due to when leakage current flows from the insulator pin to tower, this leakage current cannot be eliminated. It is necessary to equalize the voltage distribution across the insulators of the string to improve the string efficiency. Proper selection of the ratio of earth capacitance to mutual capacitance, capacitance grading, and guard ring is used to equalize the voltage distribution across the string.

Capacitance grading: In a capacitance grading, the value of the top unit (near the crossarm) capacitor should be minimum and the value of the lower unit (near the conductor) should be maximum [2]. Let the top unit capacitor is C, and other capacitors are C_2, C_3, and C_4 as shown in Fig. 11.6. The voltage across each unit is assumed V.

Applying KCL at node P of Fig. 11.6 yields,

$$I_2 = I_1 + i_1 \tag{11.36}$$

$$V\omega C_2 = V\omega C + m\omega CV \tag{11.37}$$

$$C_2 = (1 + m)C \tag{11.38}$$

Applying KCL at node N of Fig. 11.6 yields,

$$I_3 = I_2 + i_2 \tag{11.39}$$

$$V\omega C_3 = V\omega C_2 + m\omega C(2V) \tag{11.40}$$

$$C_3 = C_2 + 2\,mC \tag{11.41}$$

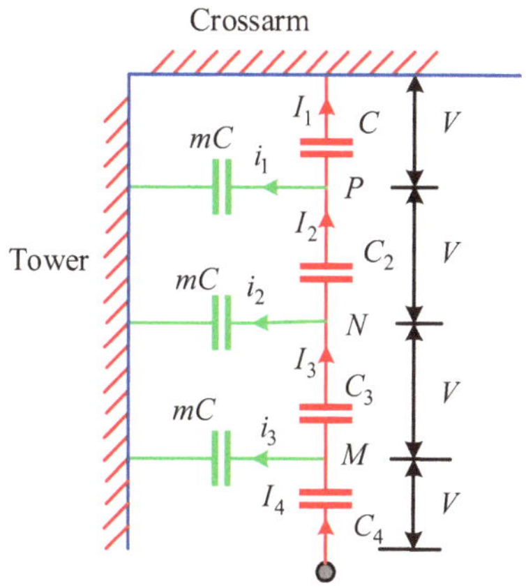

Fig. 11.6 An equivalent circuit for capacitance grading

Substituting Eq. (11.38) into Eq. (11.41) yields,

$$C_3 = (1+m)C + 2\,mC \tag{11.42}$$

$$C_3 = (1+3\,m)\,C \tag{11.43}$$

Again, applying KCL at node M of Fig. 11.6 yields,

$$I_4 = I_3 + i_3 \tag{11.44}$$

$$V\omega C_4 = V\omega C_3 + m\omega C(3V) \tag{11.45}$$

$$C_4 = (1+3m)C + 3mC \tag{11.46}$$

$$C_4 = (1+6m)C \tag{11.47}$$

Guard Ring method: It is a large metal ring that is electrically connected to the conductor and surrounds the bottom unit of the insulator as shown in Fig. 11.7. As a result, this arrangement increases the capacitance between the metal fittings and line. Let the capacitances between the metal fittings and the guard ring are C_1, C_2, and C_3. Also, the self-capacitance is C and the earth capacitance is mC.

The capacitance of each unit is the same. Therefore, the current passes through them will be the same and it can be written as,

$$I_4 = I_3 = I_2 = I_1 \tag{11.48}$$

Applying KCL at node P of Fig. 11.7 yields,

$$I_2 + i_{1x} = I_1 + i_1 \tag{11.49}$$

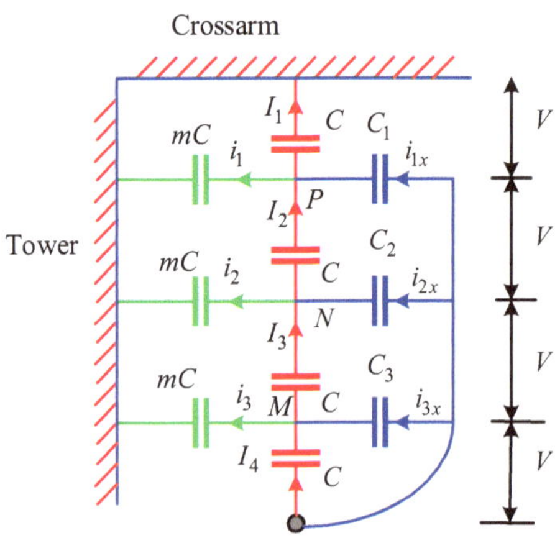

Fig. 11.7 An equivalent circuit with a guard ring

Substituting Eq. (11.48) into Eq. (11.49) yields,

$$i_{1x} = i_1 \tag{11.50}$$

$$\omega C_1(3V) = m\omega CV \tag{11.51}$$

$$C_1 = \frac{1}{3}mC \tag{11.52}$$

Applying KCL at node *N* of Fig. 11.7 yields,

$$I_3 + i_{2x} = I_2 + i_2 \tag{11.53}$$

Substituting Eq. (11.48) into Eq. (11.53) yields,

$$i_{2x} = i_2 \tag{11.54}$$

$$\omega C_2(2V) = m\omega C(2V) \tag{11.55}$$

$$C_2 = mC \tag{11.56}$$

Applying KCL at node *M* of Fig. 11.7 yields,

$$I_4 + i_{3x} = I_3 + i_3 \tag{11.57}$$

Substituting Eq. (11.48) into Eq. (11.57) yields,

$$i_{3x} = i_3 \tag{11.58}$$

$$\omega C_3(V) = m\omega C(3V) \tag{11.59}$$

$$C_3 = 3mC \tag{11.60}$$

Example 11.2
The three insulators are used as a string for high voltage transmission lines. The capacitance between the metal fitting and earth (earth capacitance) is 0.18 *C* and the capacitance between the metal fitting and guard ring is 0.12 *C* as shown in Fig. 11.8. Calculate the string efficiency.

Solution
Applying KCL at node *P* of Fig. 11.7 yields,

$$I_2 + i_{1x} = I_1 + i_1 \tag{11.61}$$

$$\omega CV_2 + \omega(0.12C)(V_2 + V_3) = \omega CV_1 + \omega(0.18C)V_1 \tag{11.62}$$

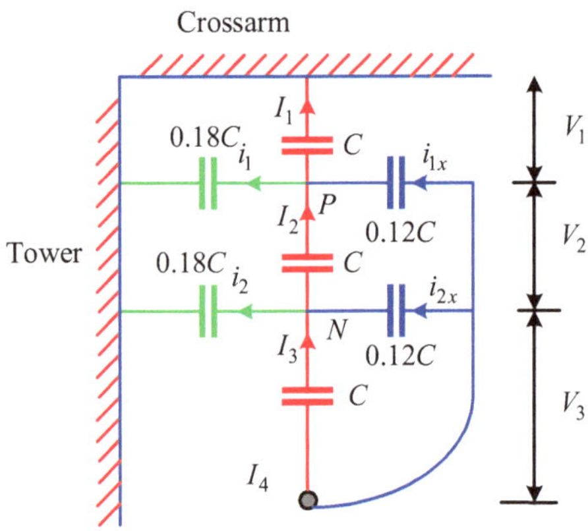

Fig. 11.8 An equivalent circuit for Example 11.2

$$V_2 + 0.12(V_2 + V_3) = V_1 + 0.18V_1 \tag{11.63}$$

$$1.18V_1 - 1.12V_2 - 0.12V_3 = 0 \tag{11.64}$$

Applying KCL at node N of Fig. 11.7 yields,

$$I_3 + i_{2x} = I_2 + i_2 \tag{11.65}$$

$$\omega CV_3 + \omega(0.12C)V_3 = \omega CV_2 + \omega(0.18C)(V_1 + V_2) \tag{11.66}$$

$$V_3 + 0.12V_3 = V_2 + 0.18(V_1 + V_2) \tag{11.67}$$

$$0.18V_1 + 1.18V_2 - 1.12V_3 = 0 \tag{11.68}$$

Dividing Eq. (11.64) by $0.12V_3$ and Eq. (11.68) by $1.12V_3$ yields,

$$9.83\frac{V_1}{V_3} - 9.33\frac{V_2}{V_3} = 1 \tag{11.69}$$

$$0.16\frac{V_1}{V_3} + 1.05\frac{V_2}{V_3} = 1 \tag{11.70}$$

Equating Eqs. (11.69) and (11.70) yields,

$$9.83\frac{V_1}{V_3} - 9.33\frac{V_2}{V_3} = 0.16\frac{V_1}{V_3} + 1.05\frac{V_2}{V_3} \tag{11.71}$$

$$9.67\frac{V_1}{V_3} = 10.38\frac{V_2}{V_3} \tag{11.72}$$

$$V_1 = 1.07V_2 \tag{11.73}$$

Substituting Eq. (11.73) into Eq. (11.70) yields,

$$0.16\frac{1.07V_2}{V_3} + 1.05\frac{V_2}{V_3} = 1 \tag{11.74}$$

$$V_2 = 0.82V_3 \tag{11.75}$$

Substituting Eq. (11.75) into Eq. (11.73) yields,

$$V_1 = 1.07 \times 0.82V_3 = 0.88V_3 \tag{11.76}$$

The total voltage across the string is calculated as,

$$V = V_1 + V_2 + V_3 \tag{11.77}$$

Substituting Eqs. (11.75) and (11.76) into Eq. (11.77) yields,

$$V = 0.88V_2 + 0.82V_3 + V_3 = 2.7V_3 \tag{11.78}$$

The string efficiency is calculated as,

$$\eta_s = \frac{V}{n \times V_n} \times 100 = \frac{2.78V_3}{3 \times V_3} \times 100 = 92.67\% \tag{11.79}$$

Practice Problem 11.2

The three insulators are used as a string for 66 kV high voltage transmission lines. The capacitance between the metal fitting and earth (earth capacitance) is 0.14 C and the capacitance between the metal fitting and the guard ring is 0.10 C as shown in Fig. 11.9. Calculate the voltage distribution.

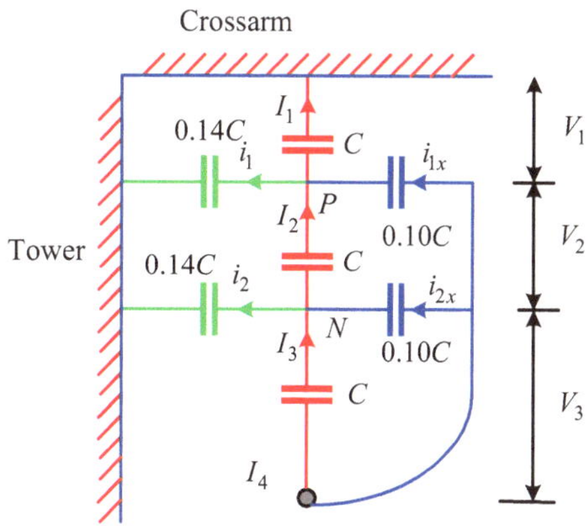

Fig. 11.9 An equivalent circuit for Practice Problem 11.2

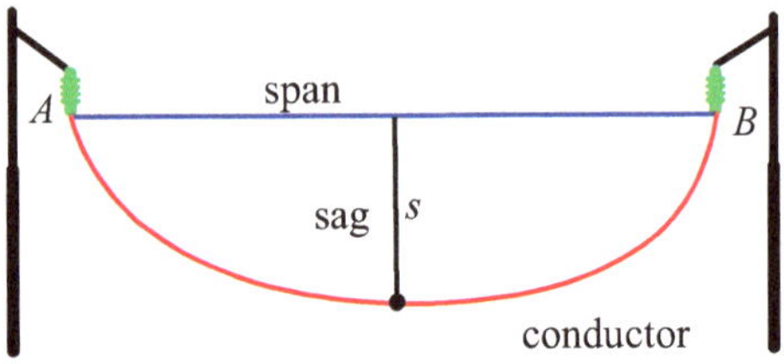

Fig. 11.10 A conductor with a sag

11.4 Transmission Lines Sag

Conductors of transmission lines are generally supported by electric tower or pole. The distance between the two towers is known as span. According to the IEEE standard, conductors should be supported with minimum ground clearance. Therefore, the sag is the distance between the two points of support and the lowest point on the conductor as shown in Fig. 11.10.

Sag reduces excessive tension and increases the life of the conductor. It is also used to cater to ice loading in the winter season. Sag is very small compared to the span of the conductor. The sag looks like a parabolic (span in between 120m and 160 m) when the load of the conductor is distributed horizontally, whereas it looks like a catenary (span is more than 1000 ft or 304.8 m) when the load of the conductor is concentered at the center. The tension at any point of the conductor acts horizontally, and the horizontal component of the tension is constant throughout the length of the conductor. Sag depends on the following factors.

- Sag will increase if the weight of the conductor increases.
- Sag depends on the location of the conductor installation.
- Sag will increase with the increase of the span, and it is directly proportional to the square of the span.
- Conductor expands at a higher temperature, which increases the sag. The conductor shrinks at a lower temperature, which decreases the sag.

11.5 Sag Calculation with Equal Supports

Consider a conductor is suspended by the two equal supports A and B as shown in Fig. 11.11. Let w be the weight of the conductor per unit length, and T is the tension [3, 4].

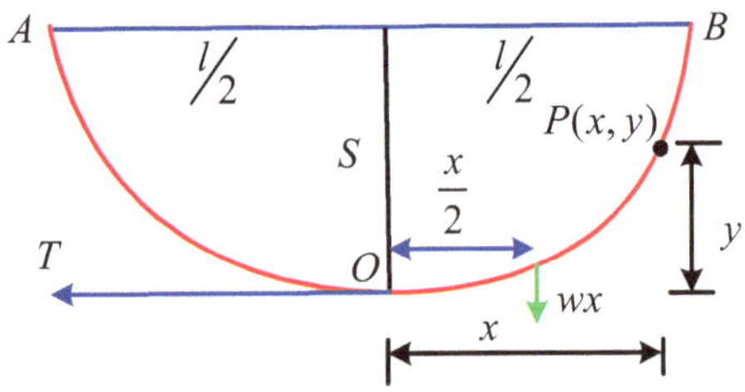

Fig. 11.11 A conductor with equal height supports

Equating two moments for two forces (T and wx) around the point O yields,

$$T \times y = wx \times \frac{x}{2} \tag{11.80}$$

$$y = \frac{wx^2}{2T} \tag{11.81}$$

From Fig. 11.11, it is seen that the sag is maximum from the two supporting points A and B. In this case, at $x = l/2$, $y = S$. Substituting these conditions in Eq. (11.81) yields,

$$S = \frac{wl^2}{8T} \tag{11.82}$$

Alternate approach: Consider, $P(x, y)$ is the equilibrium point at a distance of arc and ds is the differential distance. Tension T acting is at the point P on the conductor, which is hanged by the equal supports as shown in Fig. 11.12. Let w be weight per unit length of the conductor, l is the span, and H is the lowest point.

From Fig. 11.12, the following relations can be written as,

$$T_0 = wc \tag{11.83}$$

$$T_x = T\cos\phi \tag{11.84}$$

$$T_y = T\sin\phi \tag{11.85}$$

$$\tan\phi = \frac{dy}{dx} \tag{11.86}$$

$$ds^2 = dx^2 + dy^2 \tag{11.87}$$

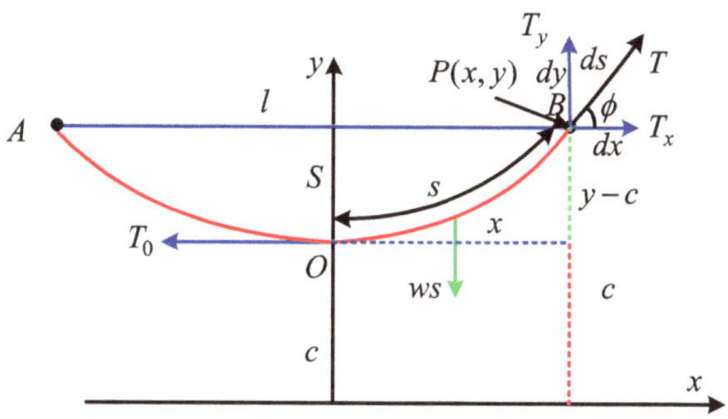

Fig. 11.12 A conductor with equal height supports and resolving the tension

Equating the horizontal and the vertical tensions yields,

$$T_x = T\cos\phi = T_0 = wc \tag{11.88}$$

$$T_y = T\sin\phi = ws \tag{11.89}$$

Dividing Eq. (11.85) by Eq. (11.84) yields,

$$\frac{dy}{dx} = \frac{T_y}{T_x} \tag{11.90}$$

Substituting Eqs. (11.88) and (11.89) into Eq. (11.90) yields,

$$\frac{dy}{dx} = \frac{ws}{wc} = \frac{s}{c} \tag{11.91}$$

Equation (11.87) can be rearranged as,

$$ds^2 = dx^2\left(1 + \frac{dy^2}{dx^2}\right) \tag{11.92}$$

$$\frac{ds^2}{dx^2} = \left(1 + \frac{dy^2}{dx^2}\right) \tag{11.93}$$

$$\frac{ds}{dx} = \sqrt{1 + \frac{dy^2}{dx^2}} \tag{11.94}$$

Substituting Eq. (11.91) into Eq. (11.94) yields,

$$\frac{ds}{dx} = \sqrt{1 + \frac{s^2}{c^2}} = \sqrt{\frac{c^2 + s^2}{c^2}} \tag{11.95}$$

$$\frac{cds}{\sqrt{c^2 + s^2}} = dx \tag{11.96}$$

Consider the following expression,

$$s = c\sinh\phi \tag{11.97}$$

Differentiate Eq. (11.97) with respect to ϕ yields,

$$ds = c\cosh\phi\, d\phi \tag{11.98}$$

Substituting Eqs. (11.97) and (11.98) into Eq. (11.96) yields,

$$\frac{c^2 \cosh \phi d\phi}{c\sqrt{1+\sinh^2 \phi}} = dx \tag{11.99}$$

$$\frac{c^2 \cosh \phi d\phi}{c \cosh \phi} = dx \tag{11.100}$$

$$c d\phi = dx \tag{11.101}$$

Integrating Eq. (11.101) yields,

$$c \int d\phi = \int dx \tag{11.102}$$

$$c\phi = x + m \tag{11.103}$$

Considering the following boundary conditions to find the integrating constant (m).

$$x = 0, \quad s = 0, \quad \phi = 0 \tag{11.104}$$

Substituting Eq. (11.104) into Eq. (11.103) yields,

$$m = 0 \tag{11.105}$$

Substituting Eq. (11.105) into Eq. (11.103) yields,

$$\phi = \frac{x}{c} \tag{11.106}$$

Substituting Eq. (11.106) into Eq. (11.97) yields,

$$\frac{s}{c} = \sinh\left(\frac{x}{c}\right) \tag{11.107}$$

Again, substituting Eq. (11.91) into Eq. (11.107) yields,

$$\frac{dy}{dx} = \sinh\left(\frac{x}{c}\right) \tag{11.108}$$

$$dy = \sinh\left(\frac{x}{c}\right) dx \tag{11.109}$$

Integrating Eq. (11.109) yields,

$$\int dy = \int \sinh\left(\frac{x}{c}\right)dx \tag{11.110}$$

$$y = \frac{\cosh\left(\frac{x}{c}\right)}{\frac{1}{c}} + n \tag{11.111}$$

Considering the following boundary conditions to find the integrating constant (n).

$$x = 0, \quad y = c \tag{11.112}$$

Substituting Eq. (11.112) into Eq. (11.111) yields,

$$c = c + n \tag{11.113}$$

$$n = 0 \tag{11.114}$$

Substituting Eq. (11.114) into Eq. (11.111) yields,

$$y = c\cosh\left(\frac{x}{c}\right) \tag{11.115}$$

Squaring Eqs. (11.88) and (11.89) and adding yields,

$$T^2 = (ws)^2 + (wc)^2 \tag{11.116}$$

$$T^2 = w^2c^2\left\{1 + \left(\frac{s}{c}\right)^2\right\} \tag{11.117}$$

Substituting Eq. (11.107) into Eq. (11.117) yields,

$$T^2 = w^2c^2\left\{1 + \sinh^2\left(\frac{x}{c}\right)\right\} \tag{11.118}$$

$$T^2 = w^2c^2\cosh^2\left(\frac{x}{c}\right) \tag{11.119}$$

$$T = wc\cosh\left(\frac{x}{c}\right) \tag{11.120}$$

$$T = wy \tag{11.121}$$

$$y = c\cosh\left(\frac{x}{c}\right) \tag{11.122}$$

$$y = c\left[1 + \frac{(x/c)^2}{2!} + \frac{(x/c)^4}{4!} + - -\right] \tag{11.123}$$

Neglecting higher terms of Eq. (11.123) yields,

$$y = c + \frac{x^2}{2c} \tag{11.124}$$

$$y - c = \frac{x^2}{2c} \tag{11.125}$$

From Eq. (11.125), the sag (S = y-c) is expressed as,

$$S = \frac{x^2}{2c} \tag{11.126}$$

The sag will be maximum at $x = l/2$ and substituting Eq. (11.83) into Eq. (11.126) yields,

$$S = \frac{l^2}{8c} = \frac{wl^2}{8T_0} = \frac{wl^2}{8T} \tag{11.129}$$

From Eq. (11.107), the length of the conductor is calculated as,

$$\frac{s}{c} = \sinh\left(\frac{x}{c}\right) = \left[\frac{(x/c)}{1!} + \frac{(x/c)^3}{3!} + \frac{(x/c)^5}{5!} + - -\right] \tag{11.130}$$

$$s = c\left[\frac{(x/c)}{1!} + \frac{(x/c)^3}{3!}\right] \tag{11.131}$$

$$s = x + \frac{x^3}{6c^2} \tag{11.132}$$

However, the total length of the conductor (L_{cond}) is 2 s. The following condition can be written as,

$$x = \frac{l}{2}, \quad s = \frac{L_{cond}}{2} \tag{11.133}$$

$$\frac{L_{cond}}{2} = \frac{l}{2} + \frac{l^3}{48c^2} \tag{11.134}$$

$$L_{cond} = l + \frac{l^3}{24c^2} \tag{11.134}$$

From Fig. 11.12, the following relation (approximate) can be written as,

$$s^2 = x^2 + (y - c)^2 \quad (11.135)$$

$$y^2 - s^2 = 2yc - c^2 - x^2 \quad (11.135)$$

Substituting Eq. (11.124) into Eq. (11.135) yields,

$$y^2 - s^2 = 2c\left(c + \frac{x^2}{2c}\right) - c^2 - x^2 \quad (11.136)$$

$$y^2 - s^2 = c^2 \quad (11.137)$$

Example 11.3

A 33 kV transmission line is crossing a river by the two equal supports. The span is 210 m, and the weight of the conductor is found to be 560 kg/km. The ultimate strength and safety factor of the conductor are given as 2500 kg and 2, respectively. Calculate the sag, length of the conductor, and the height of the supporting towers if the ground clearance is 11 m.

Solution

The weight of the conductor is calculated as,

$$w = \frac{560}{1000} = 0.56\,\mathrm{kg/m} \quad (11.138)$$

Working tension of the conductor is calculated as,

$$T = \frac{\text{Ultimate strength}}{\text{safety factor}} = \frac{2500}{2} = 1250\,\mathrm{kg} \quad (11.139)$$

The sag is calculated as,

$$S = \frac{wl^2}{8T} = \frac{0.85 \times 210^2}{8 \times 1250} = 3.75\,\mathrm{m} \quad (11.140)$$

The distance c is calculated as,

$$c = \frac{T}{w} = \frac{1250}{0.85} = 1470.59\,\mathrm{m} \quad (11.141)$$

The sag is calculated as,

$$S = \frac{l^2}{8c} = \frac{210^2}{8 \times 1470.59} = 3.75\,\text{m} \tag{11.142}$$

The length of the conductor is calculated as,

$$L_{cond} = l + \frac{l^3}{24c^2} = 210 + \frac{210^3}{24 \times 1470.59^2} = 210.18\,\text{m} \tag{11.143}$$

The height of the tower is calculated as,

$$H_{tower} = 11 + 3.75 = 15.75\,\text{m} \tag{11.144}$$

Practice Problem 11.3

A 120 kV transmission line is crossing a river by the two equal supports at the height of 50 m. The span of the two towers and the weight of the conductor are found to be 150 m and be 0.72 kg/m. The working tension of the conductor is given 950 kg. Calculate the sag, length of the conductor, and the clearance between the conductor and water at the midpoint if the tow height 20 m.

11.6 Sag Calculation with Unequal Supports

Unequal supports of the conductor are used when the transmission lines are installed in uneven areas such as the mountain areas. Consider a conductor is hung by the two unequal supports A and B as shown in Fig. 11.13. Let S_1 be sag at the point A at a distance x_1 from the lowest point O. Similarly, S_2 be sag at point B at a distance x_2 from the lowest point O. Let W be the weight of the conductor per unit length.

According to Eq. (11.81), the expression of sag for the distance x_1 is written as,

$$S_1 = \frac{wx_1^2}{2T} \tag{11.145}$$

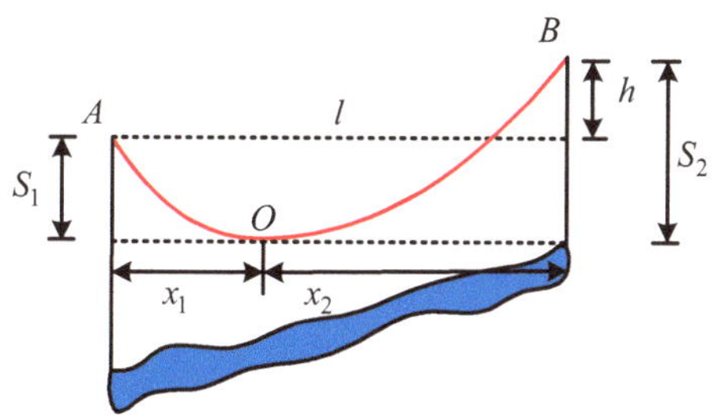

Fig. 11.13 A conductor with unequal height supports

Similarly, the expression of sag for the distance x_2 is written as,

$$S_2 = \frac{wx_2^2}{2T} \tag{11.146}$$

The difference between the two sags is written as,

$$h = S_2 - S_1 \tag{11.147}$$

Substituting Eqs. (11.145) and (11.146) into Eq. (11.147) yields,

$$h = \frac{wx_2^2}{2T} - \frac{wx_1^2}{2T} \tag{11.148}$$

$$h = \frac{w}{2T}\left(x_2^2 - x_1^2\right) = \frac{w}{2T}(x_2 - x_1)(x_2 + x_1) \tag{11.149}$$

The length of span is expressed as,

$$l = x_1 + x_2 \tag{11.150}$$

Substituting Eq. (11.150) into Eq. (11.149) yields,

$$h == \frac{wl}{2T}(x_2 - x_1) \tag{11.151}$$

$$x_2 - x_1 = \frac{2hT}{wl} \tag{11.152}$$

From Eqs. (11.150) and (11.152), the following expressions can be written as,

$$x_1 = \frac{l}{2} - \frac{hT}{wl} \tag{11.153}$$

$$x_2 = \frac{l}{2} + \frac{hT}{wl} \tag{11.154}$$

Example 11.4
The horizontal distance between the two 40 and 80 m unequal towers is found to be 400 m as shown in Fig. 11.14. The weight of the conductor and working tension are given as 1.25 kg/m and 920 kg, respectively. Calculate the clearance between the lowest point of the conductor and the ground level. Also, calculate the sag and clearance from the ground at the midpoint.

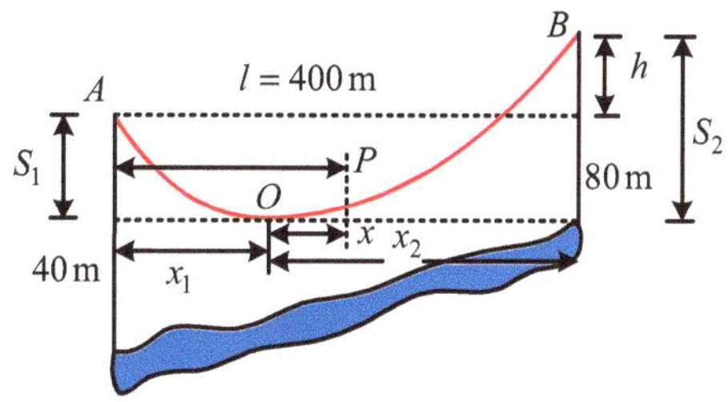

Fig. 11.14 A conductor with unequal height supports for Example 11.4

Solution
The value of the distance from the lowest point to the lower support is calculated as,

$$x_1 = \frac{l}{2} - \frac{hT}{wl} = \frac{400}{2} - \frac{40 \times 920}{1.25 \times 400} = 126.4\,\text{m} \tag{11.155}$$

The sag at the distance x_1 is calculated as,

$$S_1 = \frac{wx_1^2}{2T} = \frac{1.25 \times 126.4^2}{2 \times 920} = 10.85\,\text{m} \tag{11.156}$$

The clearance between the lowest point of the conductor and the ground level from the lower support is calculated as,

$$l_{\text{clearance}} = 40 - 14.24 = 25.76\,\text{m} \tag{11.157}$$

Let P be the midpoint at a distance x from the lowest point O. The value of this distance is calculated as,

$$x = 200 - x_1 = 200 - 126.4 = 73.6\,\text{m} \tag{11.158}$$

The sag at this point is calculated as,

$$S_p = \frac{wx^2}{2T} = \frac{1.25 \times 73.6^2}{2 \times 920} = 3.68\,\text{m} \tag{11.159}$$

The clearance of midpoint above the ground is calculated as,

$$l_{\text{clearance-mid}} = 25.76 + 3.68 = 29.44\,\text{m} \tag{11.160}$$

Practice Problem 11.4
A conductor is hanged by the two unequal supports as shown in Fig. 11.15. The weight of the conductor and working tension are given as 1.02 kg/m and 835 kg, respectively. The length of the span and the difference between the two supports are 350 m and 18 m, respectively. Calculate the sag from the smaller support.

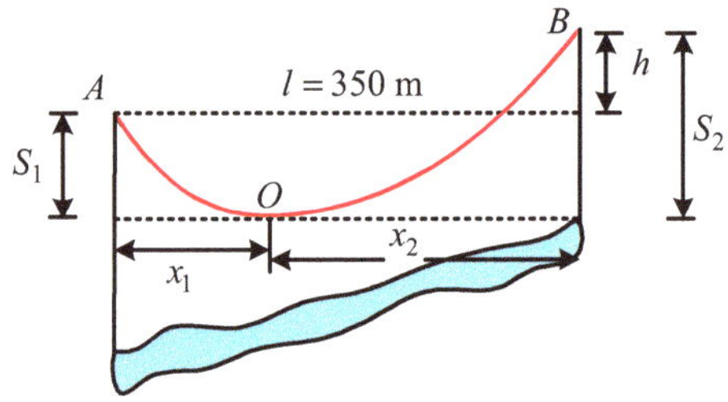

Fig. 11.15 A conductor with unequal height supports for Practice Problem 11.4

11.7 Sag Calculation with the Effect of Ice and Wind

There are some countries such as China, Canada, and many European countries where transmission lines are covered unevenly by ice (or snow) in addition to snowstorms as shown in Fig. 11.16. Icing is considered as one of the most important factors affecting the reliability of power system networks in many countries. Therefore, the combined effect of icing and wind severely disturbs the safe operation of transmission and distribution lines.

Consider a cross section of a transmission line conductor that is uniformly covered by ice as shown in Fig. 11.17. Let d is the diameter of the conductor and t is the thickness of the ice around the conductor. In this condition, this conductor can be compared to the hollow cylinder whose inner and outer diameters are D and $(D + 2t)$, respectively. The volume per unit length of this cylinder is expressed as,

$$V_{pul} = A \times 1 \tag{11.161}$$

Fig. 11.16 Sample transmission lines covered with ice

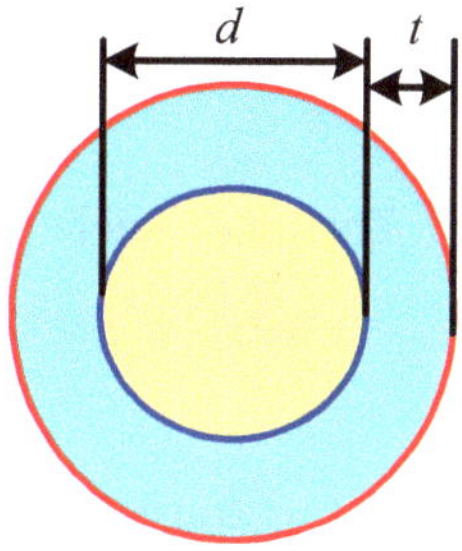

Fig. 11.17 A conductor surrounded by ice

The area of the cylinder is expressed as,

$$A = \frac{\pi}{4}[(d + 2t)^2 - d^2] = \frac{\pi}{4}(4dt + 4t^2) = \pi t(d + t) \quad (11.162)$$

Substituting Eq. (11.162) into Eq. (11.161) yields,

$$V_{pul} = \pi t(d + t) \times 1 \quad (11.163)$$

The density of ice (ρ) is defined as the ratio of mass (m) to the volume per unit length (V_{pul}), and it can be expressed as,

$$\rho = \frac{m}{V_{pul}} \quad (11.164)$$

Substituting Eq. (11.163) into Eq. (11.164) and the weight of the ice per unit length of the conductor is expressed as,

$$w_i = \rho \pi t(d + t) \quad (11.165)$$

The weight of the ice (w_i in kg/m) will act vertically downward with the weight of the conductor (W) and the force per unit length (w_w or f_w in kg/m) due to wind will act horizontally as shown in Fig. 11.18. Let the sag S makes an angle θ with the vertical load and acts in the direction of the total load as shown in Fig. 11.18. This angle is calculated as,

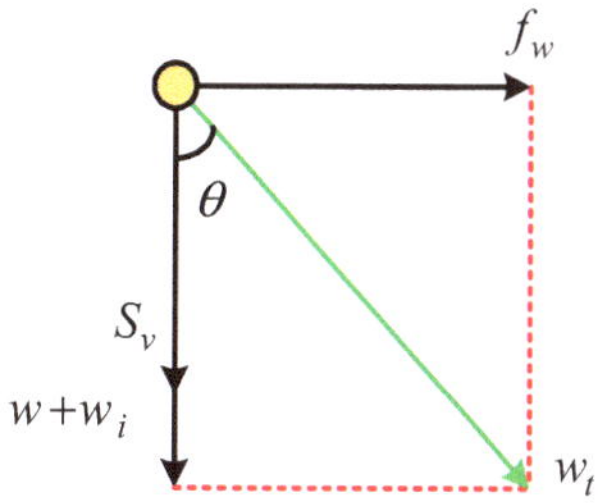

Fig. 11.18 A loading diagram of a conductor surrounded by ice

$$\tan\theta = \frac{f_w}{w + w_i} \quad (11.166)$$

The total weight per unit length is calculated as,

$$w_t = \sqrt{(w + w_i)^2 + f_w^2} \quad (11.167)$$

Let us consider P is the wind pressure per unit projected area of the conductor, the wind load per unit length of the ice-covered conductor is expressed as,

$$w_w = P(d + 2t) \times l \quad (11.168)$$

The sag with the total weight of the conductor is expressed as,

$$S = \frac{w_t l^2}{2T} \quad (11.169)$$

The vertical sag is acting in the direction of the ice and the conductor load is expressed as,

$$S_v = S\cos\theta \quad (11.170)$$

Example 11.5
The horizontal distance between the towers is found to be 130 m. The weight and diameter of the conductor are given as 0.96 kg/m and 1.25 cm, respectively. The conductor is covered with ice of 0.92 cm thickness by a wind pressure of 4.6 gm/cm^2. The ultimate strength and safety factor of the conductor are 1850 kg and 2, respectively. Assume the weight of ice is 0.65 gm/cm^3. Calculate the sag and the vertical sag.

Solution
The value of the working tension is calculated as,

$$T = \frac{1850}{2} = 925\,\text{kg/m} \quad (11.171)$$

The volume of ice per meter (100 cm) length of the conductor is calculated as,

$$V = \pi t(d + t) \times 100 = \pi \times 0.92(1.25 + 0.92) \times 100 = 627.19\,\text{cm}^3 \quad (11.172)$$

The weight of ice per unit meter of length of the conductor is calculated as,

$$w_i = 0.65 \times 627.19 = 0.41\,\text{kg} \quad (11.173)$$

The wind force per meter length of the conductor is calculated as,

$$w_w = P(d + 2t) \times l = 4.6(1.25 + 2 \times 0.92) \times 100 = 1.42\,\text{kg} \quad (11.174)$$

The total weight per meter length of the conductor is calculated as,

$$w_t = \sqrt{(w + w_i)^2 + f_w^2} = \sqrt{(0.96 + 0.41)^2 + 1.42^2} = 1.97\,\text{kg} \quad (11.175)$$

The sag of the conductor is calculated as,

$$S = \frac{w_t l^2}{2T} = \frac{1.97 \times 130^2}{2 \times 925} = 18\,\text{m} \quad (11.176)$$

The angle for the vertical sag is calculated as,

$$\theta = \tan^{-1}\left(\frac{1.42}{0.96 + 0.41}\right) = 46.03^\circ \quad (11.178)$$

The value of the vertical sag is calculated as,

$$S_v = S\cos\theta = 18\cos 46.03^\circ = 12.5\,\text{m} \quad (11.179)$$

Practice Problem 11.5
The span of the transmission line is given as 145 m, and the diameter of the conductor is found to be 1.5 cm. The conductor is covered with ice, and the combined loading of conductor and ice is given as 1.65 km/m. The wind loading is 1.22 kg/m, and the tension of the cable is 1230 kg. Calculate the sag and the vertical sag.

References

1. Wildi T (2014) Electrical machines, drives and power systems, 6th edn, Pearson Education Ltd, USA, pp 1–920
2. Satyanarayana S. Sivanagaraju S (2009) Electric power transmission and distribution, 1st edn. Pearson, Canada, pp 1–632
3. Nagsarkar TK, Sukhija MS (2014) Power system analysis, 2nd edn. Oxford University Press, Oxford, pp 1–726
4. Metha VK, Metha R (2005) Principles of power system, 3rd edn. S. Chad and Co. Ltd, India, pp 1–608

Exercise Problems

11.1 The transmission lines tower uses the three units of the cap-and-pin insulator as a string as shown in Fig. P11.1. The value of the shunt capacitance is 15% of the mutual capacitance. The line voltage across the string is found to be 22 kV. Calculate the voltage distribution across each insulator and string efficiency.

11.2 The four units cap-and-pin insulator is used as a string for the 66 kV transmission lines tower as shown in Fig. P11.2. The value of the shunt capacitance is 16% of the mutual capacitance. Calculate the voltage distribution across each insulator and string efficiency.

11.3 A guard ring is fitted with a string that consists of the three insulators of the transmission lines. The capacitance between the metal fitting and the earth (earth capacitance) is 0.09 C, and the capacitance between the metal fitting and guard ring is 0.11 C as shown in Fig. P11.3. Find voltage the string efficiency.

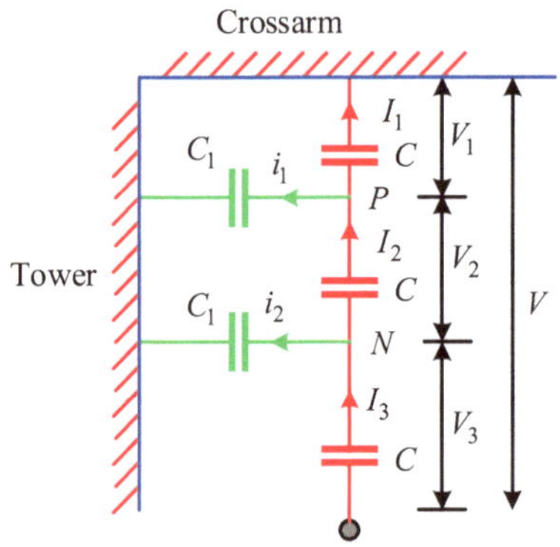

Fig. P11.1 An equivalent circuit for Problem 11.1

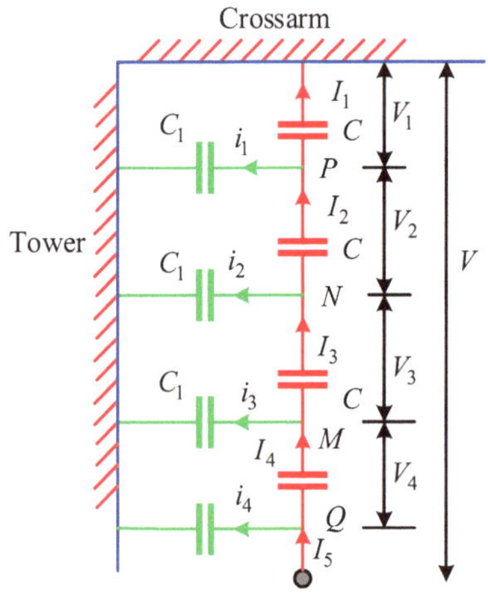

Fig. P11.2 An equivalent circuit for Problem 11.2

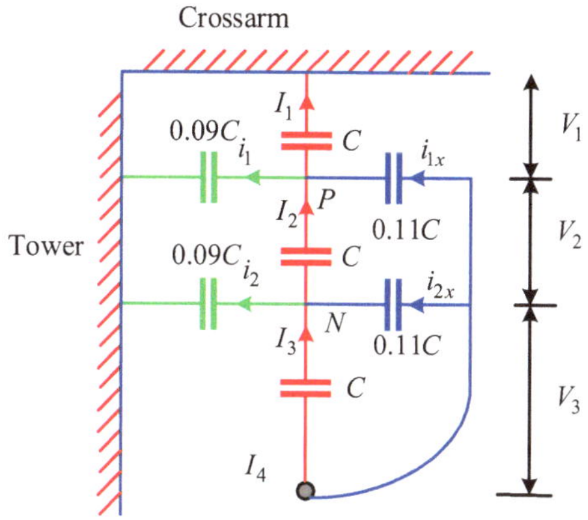

Fig. P11.3 An equivalent circuit for Problem 11.3

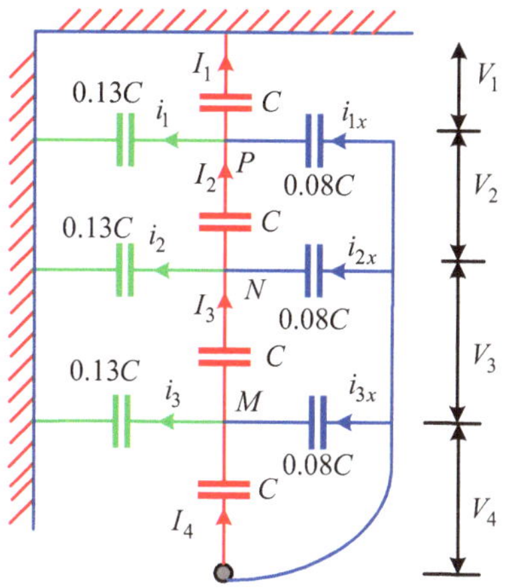

Fig. P11.4 An equivalent circuit for Problem 11.4

11.4 A string consists of four insulators for high voltage transmission networks. The capacitance between the metal fitting and the earth (earth capacitance) is 0.13 *C*, and the capacitance between the metal fitting and guard ring is 0.08 *C* as shown in Fig. P11.4. Find the string efficiency of this arrangement.

11.5 The transmission line conductor is hanged between the two equal supports with a span of 220 m as shown in Fig. P11.5. The sag and weight of the cable are found to be 20 m and 1.67 kg/m. Calculate the length of the conductor and the horizontal tension of the conductor.

11.6 A 30 m conductor suspended the two equal supports with a suitable span as shown in Fig. P11.6. The sag and weight of the cable are found to be 12 m and 30 kg/m. Calculate the length of the span and the maximum tension of the conductor.

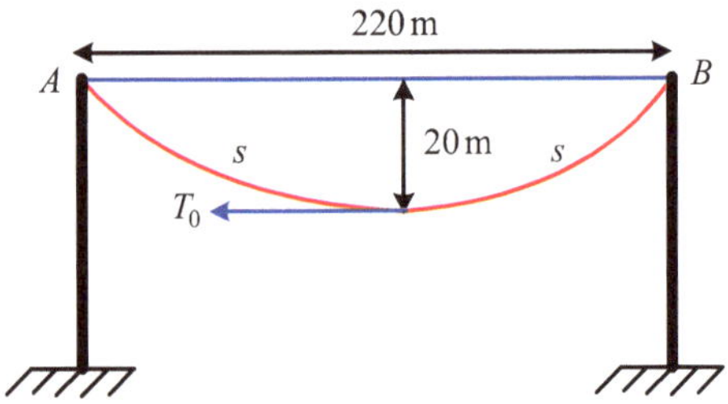

Fig. P11.5 An equivalent circuit for Problem 11.5

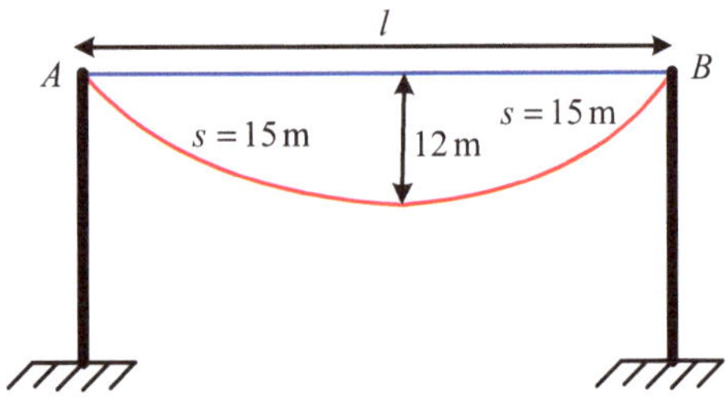

Fig. P11.6 An equivalent circuit for Problem 11.6

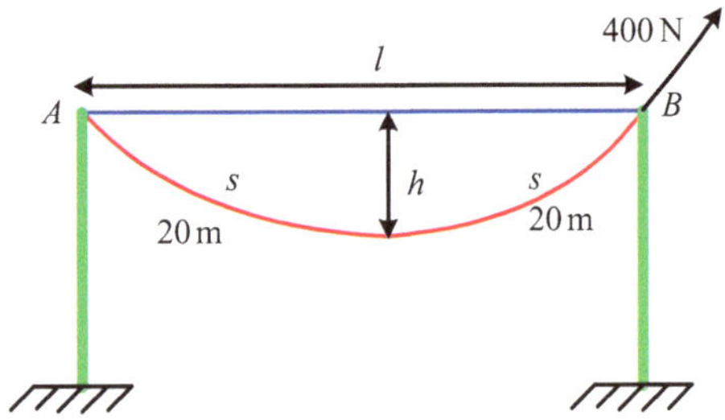

Fig. P11.7 An equivalent circuit for Problem 11.7

11.7 A 40 m conductor is suspended at equal heights with a suitable span as shown in Fig. P11.7. The maximum tension and the mas of the conductor are found to be 400 N and 25 kg. Determine the length of the span and the sag of the conductor.

11.8 A 50 m long conductor is suspended between a fixed support A and a moveable support B with a suitable span as shown in Fig. P11.8. The mass of the conductor is found to be 0.2 kg/m. Calculate the span and sag.

11.9 The span between the two 30 and 50 m unequal towers is found to be 185 m as shown in Fig. P11.9. The weight of the conductor and working tension are found to be 3.29 kg/m and 526 kg, respectively. Find the clearance between the lowest point of the conductor and the ground level. Also, find the sag and clearance from the ground at the midpoint.

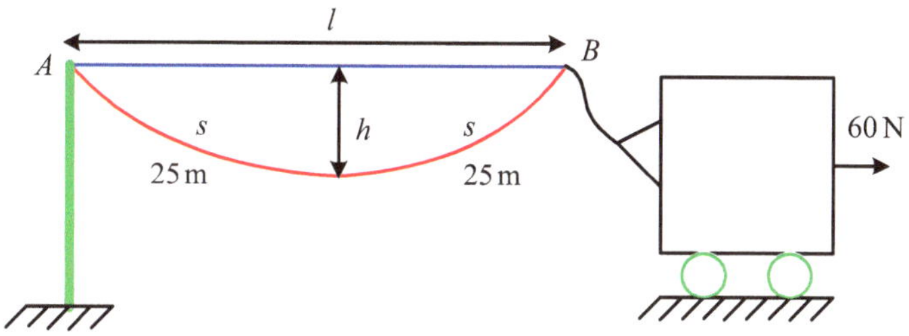

Fig. P11.8 An equivalent circuit for Problem 11.8

Fig. P11.9 A circuit for Problem 11.9

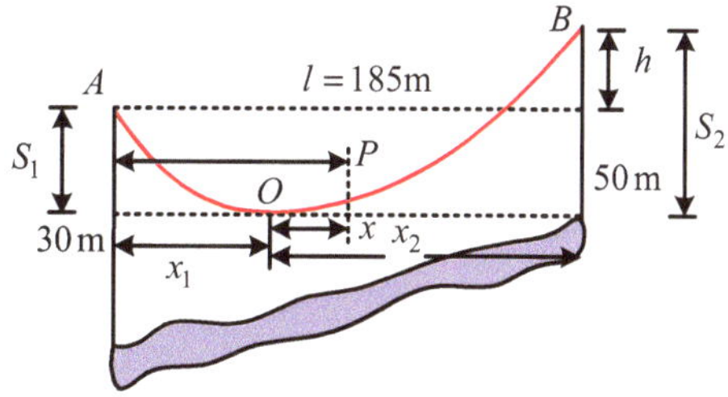

11.10 The span between the towers is found to be 160 m. The weight and diameter of the conductor are found to be 1.97 kg/m and 1.12 cm, respectively. The conductor is covered with an ice of 0.95 cm thickness by a wind pressure of 6.9 gm/cm^2. The ultimate strength and safety factor of the conductor are 2050 kg and 2, respectively. Assume the weight of ice is 0.75 gm/cm^3. Calculate the sag and the vertical sag.

Appendix
Answers to Practice and Exercise Problems

Chapter1
Practice Problems

1.1 $v(t) = 20\sin(\omega t + 35^\circ)$V, $p(t) = 98.48 - 100\cos(2\omega t - 20^\circ)$W
1.2 16.26 W, 6.92 W, 6.56 W
1.3 600 VA, 8.49 Ω, 0.027 H
1.4 15.6 VA, 9.75 W, 12.18 Var
1.5 12 k W + j 3.94 kVar, 4.19 Ω
1.6 $8.89 - j17.79$VA, $2.34 + j14.35$VA, $44.35 - j2.08$VA
1.7 29.25 kW
1.8 $260.68\angle 48.34^\circ$V
1.9 $V_{an} = 200\angle 130^\circ$V, $V_{cn} = 200\angle -230^\circ$V
1.10 $V_{an} = 115.47\angle -30^\circ$V, $V_{bn} = 115.47\angle -150^\circ$V, $V_{cn} = 115.47\angle -270^\circ$V
1.11 $I_{Aa} = 49.08\angle 35^\circ$A, $I_{Bb} = 49.08\angle -85^\circ$A, $I_{Cc} = 49.08\angle 155^\circ$A, 6.25 kW, 6.25 kW, $18.74 - j18.74$kVA

Exercise Problems

1.1 $p(t) = 6.3 - 6.4\cos(20t - 10^\circ)$W
1.2 $p(t) = 39.39 - 40\cos(200t - 30^\circ)$W
1.3 107.33W, 197.12W, 48.91W, 9.07 W
1.4 109.91 W, 141.13 W
1.5 52.53 W, 18.58 W
1.6 1.34 kW, 406.83 W
1.7 483.22 W, 224.4 W
1.8 13.89 Ω, 0.05 mF
1.9 $22.98 - j19.28$VA, 22.98 W, 19.28 Var
1.10 $40.74 + j87.36$kVA, 40.74 W, 87.36 Var

Md. A. Salam, *Fundamentals of Electrical Power Systems Analysis*,
https://doi.org/10.1007/978-981-15-3212-2

1.11 $124.36 + j121.27$VA
1.12 $140.9 + j176.95$VA
1.13 0.028mF
1.14 $V_s = 477.25 \angle 79.59°$V
1.15 $I_s = 152.08 \angle -0.14°AV_s = 620.96 \angle 78.43°$V
1.16 $V_{an} = 254.03 \angle -30°$V, $V_{bn} = 254.03 \angle -150°$V, $V_{cn} = 254.03 \angle 90°$V
1.17 $V_{bn} = 100 \angle -130°$V, $V_{cn} = 100 \angle 110°$V

Chapter 2
Practice Problems

2.1 880 V
2.2 10, 0.09 Wb, 13.63 A, 136.36A
2.3 2, 32 Ω, 3.75 A
2.4 21.69 Ω, 5.4 Ω
2.5 2.5%
2.6 109.54 kVA, 97.33%
2.7 666.67 Ω, 1000 Ω
2.8 7.65 Ω, 5.53 Ω, 0.076 Ω, 0.055 Ω, 4.6%
2.9 600 kVA, 720 kVA
2.10 2.38 kV, 207.85 kV
2.11 0.0045, 0.027
2.12 50, 25

Exercise Problems

2.1 $V_2 = 48$ V
2.2 $V_1 = 1000$ V
2.3 $I_2 = 5$ A
2.4 $I_1 = 2$ A
2.5 $\frac{V_1}{V_2} = \frac{I_1}{I_2} = 0.1$, $V_2 = 10$ kV
2.6 $\frac{V_1}{V_2} = \frac{N_1}{N_2} = 0.25$, $N_1 = 1250$
2.7 $V_2 = 220$ V, $N_2 = 100$
2.8 $I_2 = 0.96$ A
2.9 $I_2 = 4.8$ A, $V_2 = 15$ V
2.10 $I_1 = 3$ A, $V_2 = 115$ V
2.11 $I_1 = 6.8$ A, $I_2 = 68$ A
2.12 $I_1 = 4.5$ A, $I_2 = 21.6$ A
2.13 $\phi_m = 0.02$Wb
2.14 $E_1 = 6660$ V
2.15 $N_1 = 330$, $N_2 = 33$
2.16 $N_1 = 991$, $N_2 = 9.91 = 10$
2.17 $a = 2$, $I_1 = 2.5$ A, $\phi_m = 1.98$mWb

2.18 $a = 4.8$, $N_2 = 104$, $B_m = 1.17\ \text{Wb/m}^2$
2.19 $B_m = 0.79\ \text{Wb/m}^2$, $E_2 = 2192.3\ \text{V}$
2.20 $A = 0.006\ \text{m}^2$
2.21 $f = 55\ \text{Hz}$
2.22 $N_1 = 220$, $N_2 = 22$, $A = 0.075\ \text{m}^2$
2.23 $B_m = 1.8\ \text{Wb/m}^2$
2.24 $I_w = 0.32\ \text{A}$, $I_m = 0.39\ \text{A}$
2.25 $I_w = 0.2\ \text{A}$, $I_m = 0.46\ \text{A}$
2.26 $R_0 = 178.8\Omega$, $X_0 = 115.2\Omega$
2.27 $\phi_m = 1.8\ \text{mWb}$, $I_w = 0.15\ \text{A}$, $I_m = 0.25\ \text{A}$
2.28 $I_w = 0.58\ \text{A}$, $I_m = 0.55\ \text{A}$, $P_{cu} = 0.70\text{W}$, $P_{core} = 255.70\text{W}$
2.29 $I_1 = 17.7\ \text{A}$
2.30 $I_1 = 19.6\ \text{A}$, $\cos\phi_1 = 0.97$
2.31 $Z_{01} = 19.4\Omega$, $Z_{02} = 0.194\Omega$
2.32 $Z_{01} = 25\Omega$, $Z_{02} = 0.11\Omega$, $P_{cu} = 34.9\text{kW}$
2.33 $VR = 2.8\%$
2.34 $VR = 0.7\%$
2.35 $\eta_{full-load} = 96.9\%$, Output kVA = 30.9kVA, $\eta = 97.6\%$
2.36 $\eta_{full-load} = 96.5\%$, Output kVA = 28.3kVA, $\eta = 97.6\%$
2.37 $\eta_{full-load} = 94.6\%$, Output kVA = 18.5kVA, $\eta = 96\%$
2.38 $\eta = 97.6\%$, $\eta = 98.9\%$
2.39 $\eta_{full-load} = 43\%$, $\eta = 99\%$
2.40 $Z_{01} = 7.5\Omega$ $R_{01} = 5.6\Omega$, $X_{01} = 4.9\Omega$, $Z_{02} = 0.06\Omega$ $R_{02} = 0.05\Omega$, $X_{02} = 0.04\Omega$, $VR = 3.8\%$
2.41 $Z_{01} = 12.5\Omega R_{01} = 9.8\Omega$, $X_{01} = 7.8\Omega$, $Z_{02} = 0.5\Omega R_{02} = 0.39\Omega$, $X_{02} = 0.31\Omega$, $VR = 2.05\%$
2.42 $R_0 = 314.2\Omega$, $X_0 = 183.3\Omega$, $Z_{01} = 1.37\Omega$, $R_{01} = 0.64\Omega$, $X_{01} = 1.2\Omega$
2.43 $I_w = 0.66\ \text{A}$, $I_m = 1\text{A}, R_0 = 333.33\Omega$, $X_0 = 220\Omega$
2.44 $\text{kVA}_1 = 599.8\text{kVA}$, $\text{kVA}_2 = 119.9\text{kVA}$
2.45 $N_1 = 53.86 = 54$, $N_2 = 12.43 = 12$
2.46 (a) $V_{L2} = 2.54\text{kV}$, (b) $V_{L2} = 0.85\text{kV}$
2.47 $I_b = 5.45\ \text{A}$, $Z_b = 4036.7\Omega$
2.48 $I_b = 524.86\ \text{A}$, $Z_b = 125.75\Omega$
2.49 $X_{pu} = 0.59\text{pu}$
2.50 $Z_{actual} = 48.4\Omega$
2.51 $X_{pu1} = X_{pu2} = 0.034\text{pu}$
2.52 $V_{line} = 1120.96\ \underline{|78.66^\circ}\ \text{V}$
2.53 $X_{T1} = 0.32\text{pu}$, $X_{T2} = 0.24\text{pu}$, $X_{line} = 0.05\text{pu}$, $Z_{load} = 1.58 + j2.1\text{pu}$, $E_G = 253\ \text{kV}$
2.54 $X_G = 0.18\text{pu}$, $X_{T1} = 0.48\text{pu}$, $X_{line-1} = 0.08\text{pu}$, $X_{T2} = 0.4\text{pu}$, $X_{T3} = 0.32\text{pu}$, $X_{line-2} = 0.4\text{pu}$, $X_{T4} = 0.24\text{pu}$, $Z_{load} = 3.2 + j2.4\text{pu}$

2.55 $X_{G1} = 0.6\text{pu}$, $X_{T1} = 0.17\text{pu}$, $X_{G2} = 0.2\text{pu}$, $X_{line} = 0.056\text{pu}$, $X_{T2} = 0.062\text{pu}$, $X_M = 0.066\text{pu}$, $Z_{load} = 0.64 + j0.39\text{pu}$

2.56 $X_{G1} = 0.21\text{pu}$, $X_{T1} = 0.22\text{pu}$, $X_{line} = 0.17\text{pu}$, $X_{T2} = 0.2\text{pu}$, $X_{G2} = 0.27\text{pu}$

Chapter 3
Practice Problems

3.1 $F_{def} = 0.44$, Av.demand = 3424.66kW, $F_{LF} = 0.17$

3.2 $F_{dif} = 1.24$, Av.demand = 799.09kW, $F_{LF} = 0.32$

3.3 Max.demand = 60kW, $UG_{day} = 840\text{MWh}$, Av.Load = 35MW, $F_{LF} = 0.58$

3.4 Annual charge = B$ 940, Equivalent flat rate = B$ 0.11

3.5 Unit consumed per year = 192720kWh, Annual charge = B$ 24772, Overall cost per kWh = B$ 0.13

Exercise Problems

3.1 Total energy = 315.36×10^3kWh

3.2 $F_{LF} = 0.49$

3.3

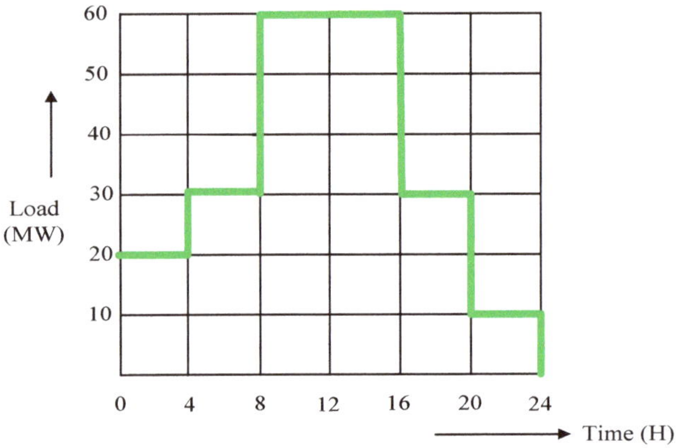

$F_{def} = 0.62$, Av.demand = 10273.97kW, $F_{LF} = 0.26$

3.4 $F_{def} = 0.44$, Av.demand = 6849.32kW, $F_{LF} = 0.34$

3.5 $F_{dif} = 1.06$, Av.demand = 627.85kW, $F_{LF} = 0.21$

3.6 540kW

3.7 Max.demand = 12230.77MW

3.8 Energy = 78.84×10^6kWh, 0.3, 30MW, 10MW

3.9 $F_{CF} = 0.27$

3.10 Max.demand = 80MW, Unit generated per day = 1100×10^3kWh, Av. load = 45833.33kW, $F_{LF} = 0.57$

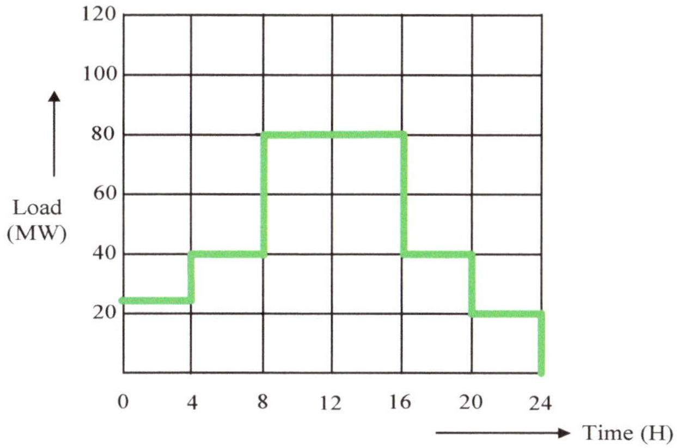

3.11 Max.demand = 60MW, Unit generated per day = 740×10^3kWh, Av. load = 30833.33kW, $F_{LF} = 0.51$

3.12

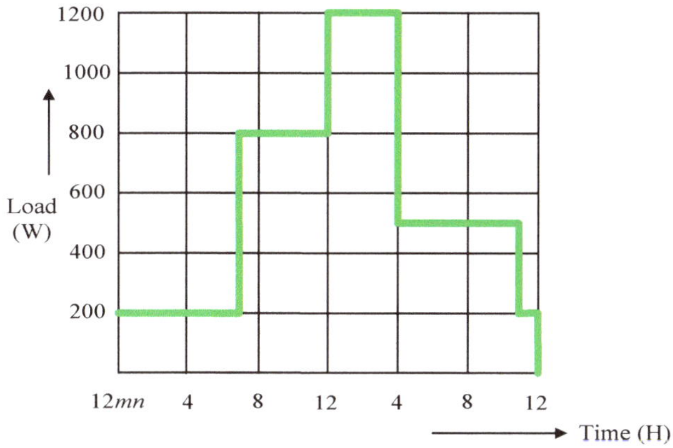

Max.demand = 1200W, $F_{DiF} = 1.33$

3.13 Total annual charges = B\$ 340, Equivalent flat rate = B\$ 0.039

3.14 Total units consumed per year = 245280kWh,
Annual charges = B\$ 18664, Overall cost per kWh = B\$ 0.08

Chapter 4
Practice Problems

4.1 $R_{30} = 28.23\Omega$

4.2 $GMD = 13.44$m, $GMR_A = 0.429$m, $GMR_B = 0.078$m, $L_A = 6.88 \times 10^{-7}$H/m, $L_B = 1.03 \times 10^{-6}$H/m, $L_t = 1.719 \times 10^{-6}$H/m

4.3 $L = 1.25 \times 10^{-6}$H/m, $X_L = 13.42\Omega$

4.4 $L = 1.21 \times 10^{-6}\text{H/km}$, $L = 0.15\text{H}$
4.5 $C = 0.52\mu\ \text{F}$, $I_c = 3.11\text{A}$
4.6 $C = 0.94\mu\ \text{F}$, $I_c = 5.63\text{A}$
4.7 $C = 0.21\mu\ \text{F}$, $X_c = 15157.61\Omega$
4.8 1.03m

Exercise Problems

4.1 $R_{25} = 15.93\Omega$
4.2 $R_{20} = 16.2\Omega$
4.3 $r = 0.704\text{cm}$
4.4 $L = 0.014\text{H}$
4.5 $GMR = 1.70r$
4.6 $GMR = 1.25r$
4.7 $GMR = 1.72r$
4.8 4.29 m, 0.18 m, 0.06 m, $L = 0.00148\text{H/km}$
4.9 4.23m, $L = 0.133 \times 10^{-3}\text{H/km}$
4.10 4.48 m, 0.15 m, $L = 1.63 \times 10^{-3}\text{H/km}$
4.11 $L = 1.28 \times 10^{-3}\text{H/km}$
4.12 4.8 m
4.13 $L = 7.76 \times 10^{-4}\text{H/km}$
4.14 $L = 1.02 \times 10^{-3}\text{H/km}$
4.15 $L = 9.67 \times 10^{-4}\text{H/km}$, 0.23H
4.16 $L = 8.82 \times 10^{-4}\text{H/km}$
4.17 $L = 1.12 \times 10^{-3}\text{H/km}$
4.18 0.09 H, 28.52 Ω
4.19 1.18 cm
4.20 105.96 m
4.21 1.27 μF, 27.63 A
4.22 $C = 1.24 \times 10^{-8}\mu\text{F/km}$, 0.07 A
4.23 $C = 1.11 \times 10^{-11}\text{F/m}$, 2394521 Ω
4.24 $C = 2.25 \times 10^{-9}\text{F}$, 1178897.73 Ω

Chapter 5
Practice Problems

5.1 $V_s = 18096.94\angle 2.53^\circ\text{V}$, $VR = 4.5\%$, $\eta = 98.23\%$
5.2 $V_s = 45865.37\angle 9.19^\circ\text{V}$, $V_{s(L-L)} = 79.44\text{kV}$, $VR = 20.36\%$
5.3 $V_{s(L-L)} = 56.13\text{kV}$, $\cos\phi_s = 0.7$
5.4 $V_{s(L-L)} = 79.79\text{kV}$, $I_s = 210.53\angle -26.98^\circ\text{A}$
5.5 $A = 1$, $C = 0.01\text{S}$, $B = 4\Omega$, $D = 1.04$
5.6 $V_s = 33395.8\angle 12.98^\circ\text{V}$
5.7 $A = 0.65\angle 4.15^\circ$, $C = 0.0011\angle 97^\circ\text{S}$, $B = 14.27\angle 35.25^\circ\Omega$, $D = 0.65\angle 4.15^\circ$
5.8 $\beta = 0.0015\text{rad/km}$, $v = 204124.15\text{km/s}$, $\lambda = 4082.48\text{km}$, $Z_c = 195.96\Omega$
$V_s = 231497.32\angle 5.31^\circ\text{V}$

Exercise Problems

5.1 $V_s = 896.14 \angle 41.87°$V, $\cos\phi_s = 0.5$, $\eta = 50.97\%$
5.2 $V_s = 10.93$kV, $VR = -0.64\%$ $\eta = 99.36\%$
5.3 $V_s = 6.6$kV, $VR = 0.91\%$
5.4 $V_s = 11.04$kV, $\eta = 99.58\%$
5.5 $V_r = 5.98$kV, $I_L = 74.23$A, $\eta = 96.03\%$
5.6 $V_s = 10.66$kV, $\delta = 15.04°$
5.7 $V_s = 11.38$kV, $P_s = 243.33$kW, $S_s = 246.2\text{kW} + j152.59\text{kVAR}$
5.8 $I_L = 408.53 \angle -68.2°$A, $P_r = 2.29$MW, $\phi_s = 48.81°$(lag)
5.9 $A = 0.99 \angle 0.01°$, $C = 0.0003 \angle 90°$S, $B = 5.14 \angle 76.5°\Omega$, $D = 0.99 \angle 0.01°$, $V_s = 66.65$kV $I_s = 198 \angle -28.98°$A, $\cos\phi_s = 0.86$
5.10 $A = D = 0.99 \angle 0.06°$, $C = 0.000216 \angle 90°$S, $B = 54.59 \angle 80.57°\Omega$, $V_{s(L-L)} = 46.92$kV $I_s = 463.7 \angle 26.36°$A, $\cos\phi_s = 0.79$
5.11 $A = D = 0.99 \angle 0.2°$, $C = 0.000348 \angle 90.1°$S, $B = 82.46 \angle 75.96°\Omega$, $V_{s(L-L)} =$ 92.69kV $I_s = 269.65 \angle -15.31°$A, $\cos\phi_s = 0.80$
5.12 $V_{s(L-L)} = 117.63$kV, $I_s = 161.61 \angle -27.57°$A, $\eta = 89.38\%$
5.13 $V_s = 58897.97 \angle 15.77°$V, $VR = 59.34\%$
5.14 $V_s = 64530.23 \angle -0.33°$V, $I_s = 243.05 \angle -23.74°$A
5.15 $V_s = 52752.47 \angle 16.51°$V, $I_s = 79.91 \angle 34.72°$A
5.16 $A = 0.68 \angle 4.68°$, $C = 0.0015 \angle 105.66°$S, $B = 19.55 \angle 40°\Omega$, $D = 0.68 \angle 4.68°$
5.17 $\beta = 0.0012$rad/km, $v = 2.65 \times 10^5$km/s, $\lambda = 5303.72$km, $Z_c = 209.5\Omega$

Chapter 6
Practice Problems

6.1 $I'' = 42939.07$ A, $I' = 23618.85$A, $I = 4204.16$A, $I_{total} = 64408.61$A
6.2 $\text{MVA}_{SC} = 19.04$, $I_{SC} = 999.17$A
6.3 $\text{MVA}_{SC} = 662.25$, $I_{SC} = 29411.63$A
6.4 55 MVA, 29159.05 A
6.5 $I_{a0} = 22.27 \angle 24.36°$A, $I_{a1} = 4.46 \angle -85.51°$A, $I_{a2} = 7.55 \angle -140.68°$A
6.6 $I_a = 14.22 \angle 31.05°$A, $I_b = 4.64 \angle -143.71°$A, $I_c = 2.68 \angle 58.88°$A

6.7

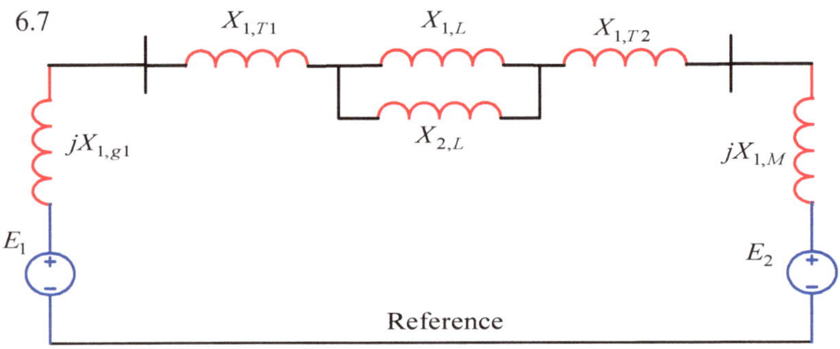

Positive sequence

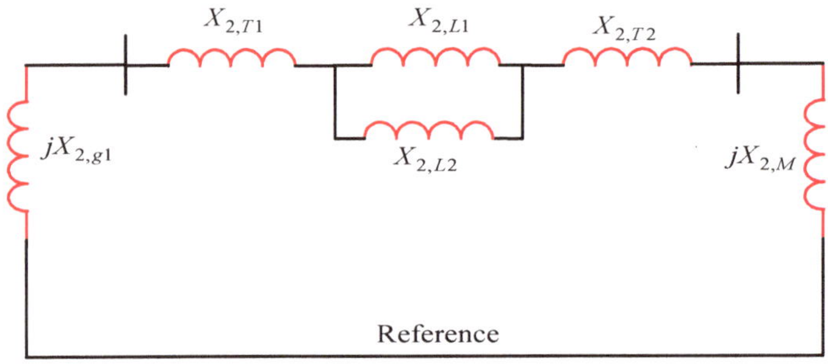

Negative sequence

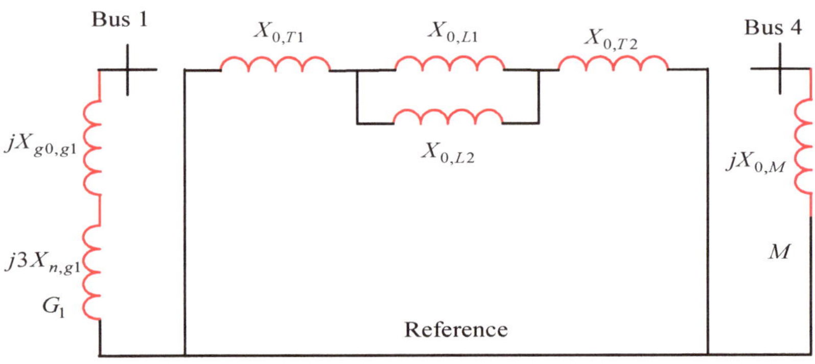

Zero sequence

6.8 $I_a = 4849.75\ \angle -90^\circ\text{A}$

6.9 $I_{a0} = I_{b0} = I_{c0} = 50\text{A},\ I_{a1} = 50\ \angle 120^\circ\text{A},\ I_{b1} = 50\ \angle 360^\circ\text{A},\ I_{c1} = 50\ \angle 240^\circ\text{A},$ $I_{a2} = 50\ \angle 240^\circ\text{A},\ I_{b2} = 50\ \angle 360^\circ\text{A},\ I_{c2} = 50\ \angle 480^\circ\text{A}$

6.10 $V_{ab} = 1.2\ \angle 0^\circ\text{V},\ V_{bc} = 0\ \angle 0^\circ\text{V},\ V_{ca} = 1.2\ \angle 180^\circ\text{V}$

6.11 $I_a = 0\ \angle 0^\circ\text{pu},\ I_b = 5.01\ \angle 148.02^\circ\text{pu},\ I_c = 5.01\ \angle 31.98^\circ\text{pu}$

Exercise Problems

6.1 (i) $-0.5 - j0.86$, (ii) $-4 + j1.732$, (iii) $-2.5 + j2.6$

6.2 $I_{ao} = 14.97\ \angle 35.56^\circ\text{A},\quad I_{a1} = 3.45\ \angle -110.66^\circ\text{A},\quad I_{a2} = 2.31\ \angle -169.11^\circ\text{A},$ $I_{bo} = 14.97\ \angle 35.56^\circ\text{A},\ I_{b1} = 3.45\ \angle 129.34^\circ\text{A},\ I_{b2} = 2.31\ \angle -49.11^\circ\text{A}$

6.3 $V_{ao} = 191.69\ \angle 48.12^\circ\text{V},\ V_{a1} = 47.79\ \angle -87.20^\circ\text{V},\ V_{a2} = 60.41\ \angle -142.51^\circ\text{V}$

6.4 $V_a = 537.27\ \angle 73.87^\circ\text{V},\ V_b = 263.19\ \angle -126.04^\circ\text{V},\ V_c = 218.85\ \angle -41.28^\circ\text{V}$

6.5 $I_{bo} = 169.71\ \angle 45^\circ\text{A},\ I_{b1} = 25.05\ \angle 175.21^\circ\text{A},\ I_{b2} = 16.96\ \angle 27.41^\circ\text{A}$

6.6 $I_a = 11.95\ \angle 15.03^\circ\text{A},\ I_b = 3.93\ \angle 7.63^\circ\text{A},\ I_c = 7.11\ \angle 104.83^\circ\text{A}$

6.7 $I_{bo} = 500\ \angle 0^\circ\text{A},\ I_{b1} = 500\ \angle 240^\circ\text{A},\ I_{b2} = 500\ \angle 120^\circ\text{A}$

6.8 $I_{ao} = 0\ \angle 0^\circ\text{A},\quad I_{a1} = 25.72\ \angle 30^\circ\text{A},\quad I_{a2} = 25.72\ \angle -30^\circ\text{A},\quad I_{bo} = 0\ \angle 0^\circ\text{A},$ $I_{b1} = 25.72\ \angle 270^\circ\text{A},\ I_{b2} = 25.72\ \angle 90^\circ\text{A},\ I_{co} = 0\ \angle 0^\circ\text{A},\ I_{c1} = 25.72\ \angle 150^\circ\text{A},$ $I_{c2} = 25.72\ \angle 210^\circ\text{A}$

6.9 $I_a = 4204.16\text{A}$

6.10 $I_b = 2171.32\text{A},\quad V_{ao} = 0\ \angle 0^\circ\text{pu},\quad V_{a1} = 0.4\ \angle 0^\circ\text{pu},\quad V_{a2} = 0.6\ \angle 0^\circ\text{pu},$ $V_a = 1\ \angle 0^\circ\text{pu},\ V_b = 0.53\ \angle 160.89^\circ\text{pu},\ V_c = 0.53\ \angle -160.89^\circ\text{pu}$

6.11 $I_a = 0\ \angle 0^\circ\text{pu},\ I_b = 3.52\ \angle 143.64^\circ\text{pu},\ I_c = 3.52\ \angle 36.36^\circ\text{pu}$

6.12 $I_{f(s-l-g)} = 25426.689\text{A},\ I_{f(l-l)} = 17819.149\text{A},\ I_{f(d-l-g)} = 24377.308\text{A}$

6.13 $I_{f(s-l-g)} = 4425.082\text{A},\ I_{f(l-l)} = 4291.539\text{A},\ I_{f(d-l-g)} = 4734.115\text{A}$

6.14 $I_{f(s-l-g)} = 4756.745\text{A},\ I_{f(l-l)} = 4860.853\text{A},\ I_{f(d-l-g)} = 5280.761\text{A}$

6.15 $I_{f(s-l-g)} = 2853.425\text{A},\ I_{f(l-l)} = 5588.138\text{A},\ I_{f(d-l-g)} = 5662.692\text{A}$

Chapter 7

Practice Problems

7.1 $X = 10.83\Omega$

7.2
$$Y_{bus} = \begin{bmatrix} 7.97\ \angle 41.25^\circ & -3\ \angle 45^\circ & 0 & -5\ \angle 45^\circ \\ -3\ \angle 45^\circ & 4.98\ \angle 49^\circ & 3\ \angle 55^\circ & 0 \\ 0 & 2\ \angle 55^\circ & 5.81\ \angle 34.89^\circ & 4\ \angle 25^\circ \\ -5\ \angle 45^\circ & 0 & 4\ \angle 25^\circ & 8.86\ \angle 36.12^\circ \end{bmatrix}$$

7.3 $V_1 = 0.0825,\ V_2 = 0.8829,\ V_3 = 1.5904$

7.4

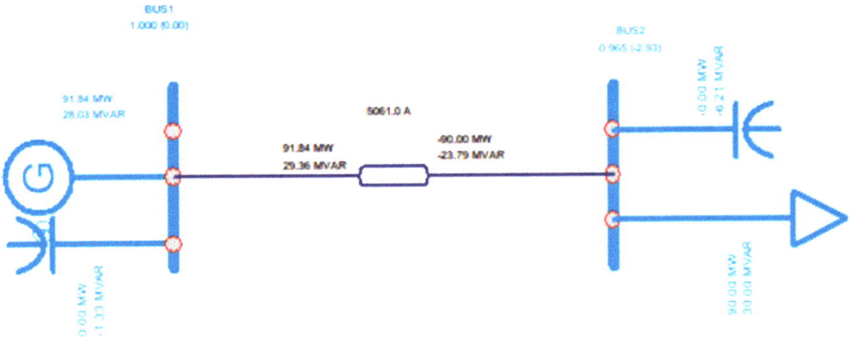

7.5

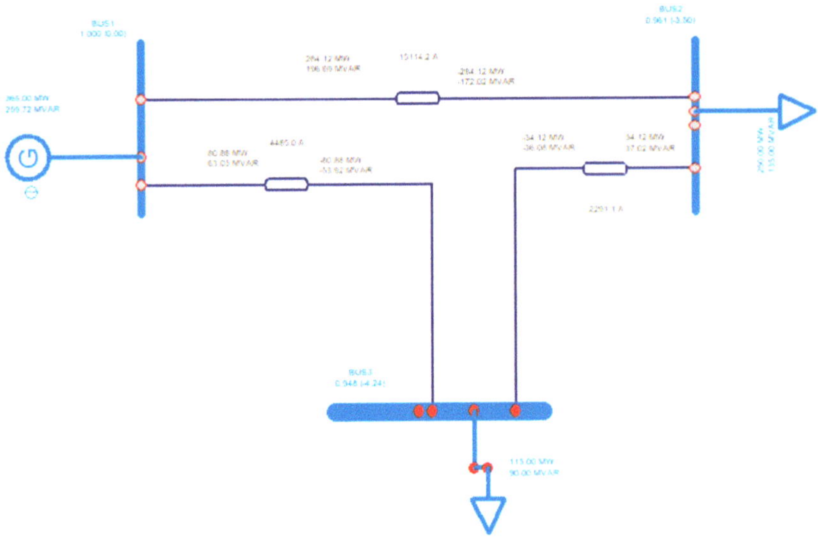

7.6 $x = 1.047$

7.7 Student can solve it

7.8

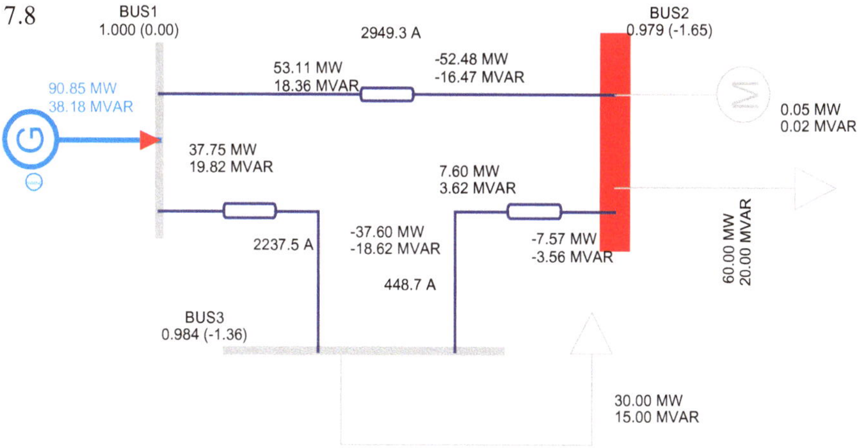

7.9

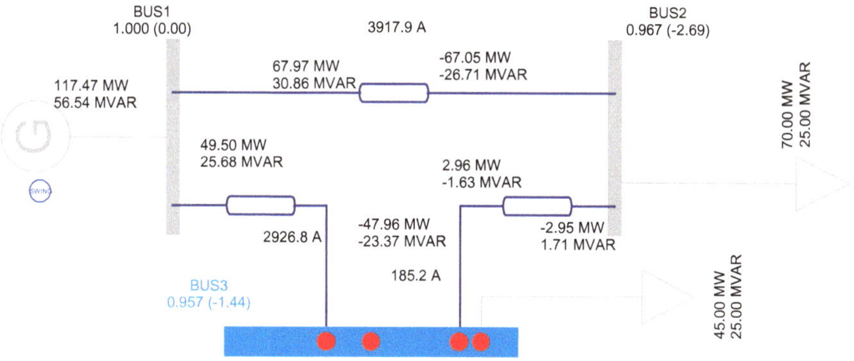

Exercise Problems

7.1 $Q_{av} = 233.33\text{VAR}$

7.2 $X = 17.5\Omega$

7.3 $V_r = 66.33\text{V}$

7.4 $Y_{bus} = \begin{bmatrix} -j6 & j1 \\ -j1 & j1.77 \end{bmatrix}$

7.5 $Y_{bus} = \begin{bmatrix} j5 & -j2 & 0 \\ j2 & -j10 & j5 \\ 0 & j5 & -j5 \end{bmatrix}$

7.6 $Y_{bus} = \begin{bmatrix} 5-j10 & -5+j10 \\ -5+j10 & 5-j10 \end{bmatrix}$

7.7 $Y_{bus} = \begin{bmatrix} 8.33 - j16.67 & -5 + j10 & -3.33 + j6.67 \\ -5 + j10 & 9 - j18 & -4 + j8 \\ -3.33 + j6.67 & -4 + j8 & 7.33 + j14.67 \end{bmatrix}$

7.8 $Y_{bus} = \begin{bmatrix} 10.63 - j20.85 & -5 + j10 & -2.55 + j5.47 & -3.08 + j5.38 \\ -5 + j10 & 9 - j18 & -4 + j8 & 0 \\ -2.55 + j5.47 & -4 + j8 & 9.88 - j20.14 & -3.33 + j6.67 \\ -3.08 + j5.38 & 0 & -3.33 + j6.67 & 6.41 - j12.05 \end{bmatrix}$

7.9 $V_1 = 1.3846\text{V}$, $V_2 = 2.0384\text{V}$

7.10 $V = 0.4384$

7.11

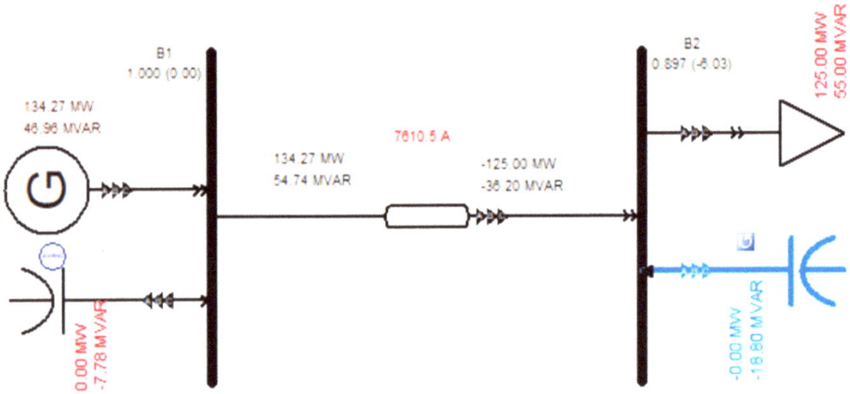

7.12

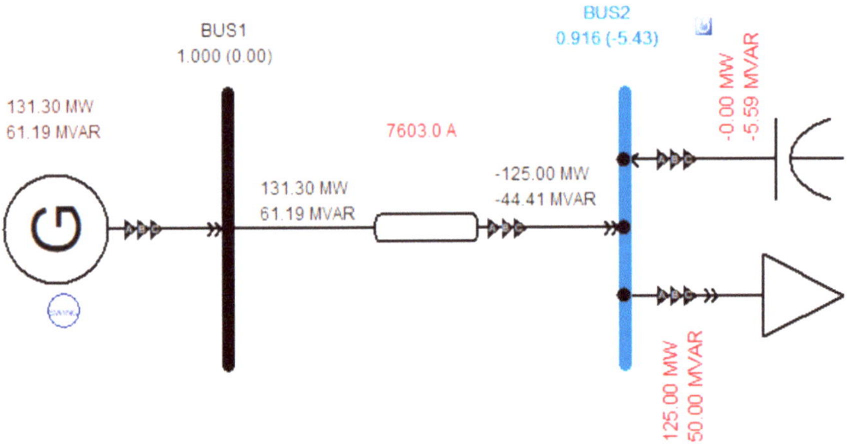

7.13

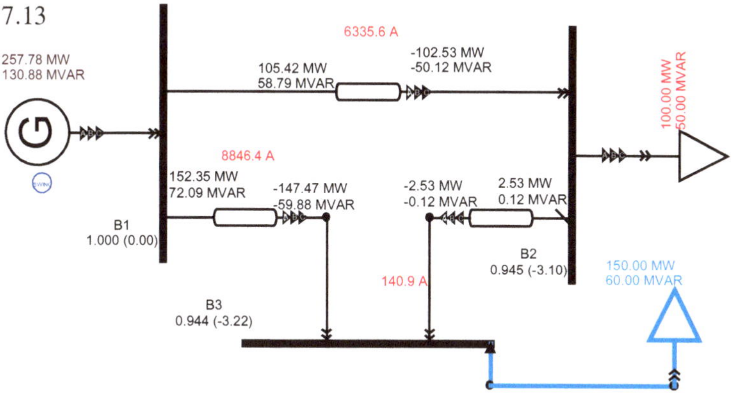

7.14 $V = 1.8414$
7.15 $V = 1.04709$
7.16

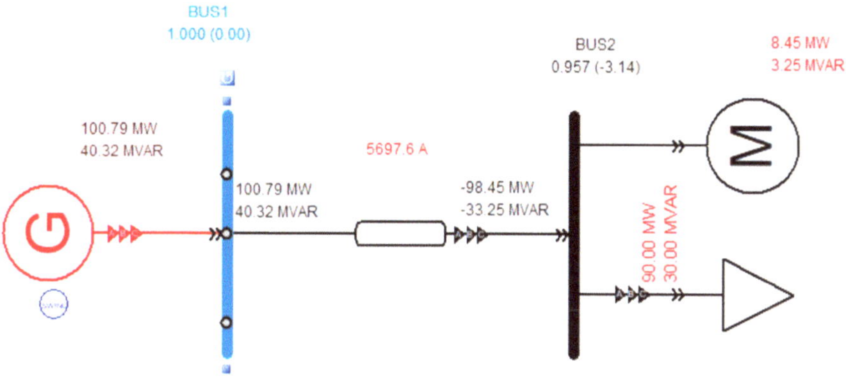

7.17

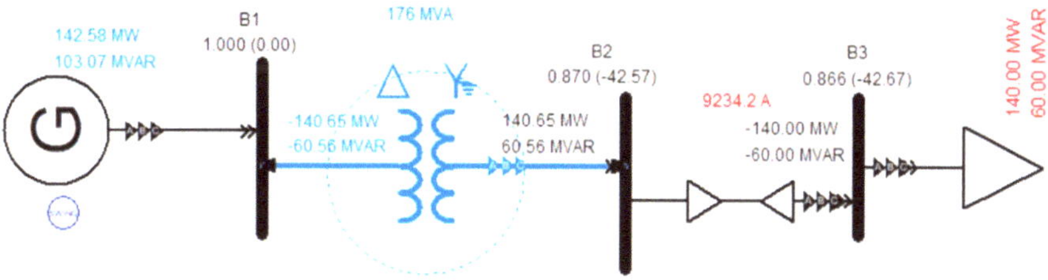

7.18

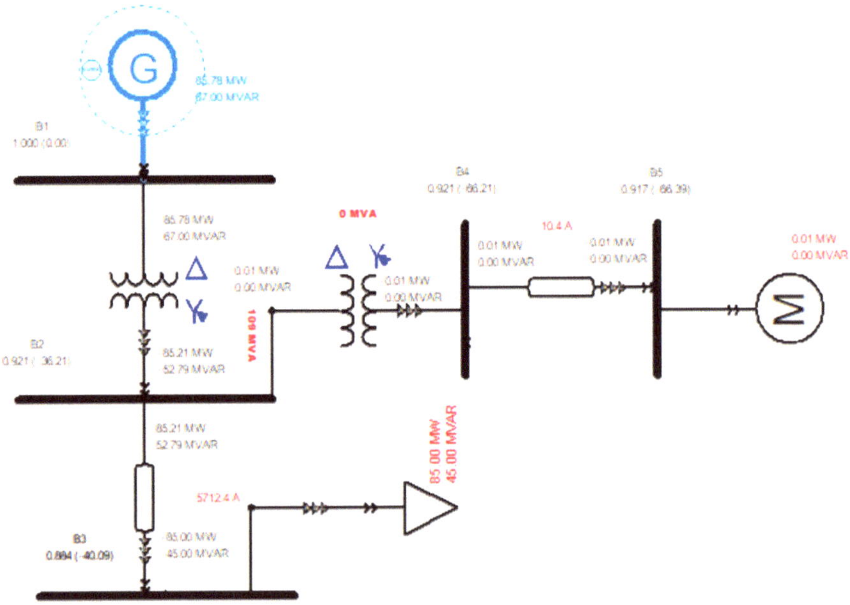

Chapter 8
Practice Problems

8.1 $\rho = 2.61 \times 10^{13}\Omega - \text{m}$
8.2 2.28 cm, 14.59 kV/cm, 0.84 cm
8.3 $r_1 = 2.24\text{cm}$, $r_2 = 3\text{cm}$, $V = 46.01\text{kV}$, $V_{safe} = 32.54\text{kV}$
8.4 $V_{s1} = 32.86\text{kV}$, $V_{s2} = 14.46\text{kV}$
8.5 6184.65 V, 480.42 W, 471.92 W
8.6 $C = 0.64\mu$ F
8.7 $C_n = 4.46\mu$ F, $I_c = 10.68\text{A}$

Exercise Problems

8.1 975.6 MΩ
8.2 53.31 MΩ
8.3 $E_{\max} = 15.90\text{V/cm}$, $E_{\min} = 5.02\text{V/cm}$, $t_{ins} = 1.3\text{cm}$, (iv) $r = 0.70\text{cm}$
8.4 $E_{\max} = 28.09\text{V/cm}$, $E_{\min} = 9.66\text{V/cm}$, $r = 1.18\text{cm}$
8.5 $t_{ins} = 3.025\text{cm}$, $V = 43.62\text{kV}$
8.6 $r = 0.196\text{cm}$, $r_{econom} = 1.1\text{cm}$
8.7 $r_1 = 1.5\text{cm}$, $r_2 = 1.8\text{cm}$, $V = 41.16\text{kV}$ $V_{safe} = 29.10\text{kV}$
8.8 $t_{in} = 1.04\text{cm}$, $t_{out} = 2.76\text{cm}$, $V_{peak} = 71.18\text{kV}$, $V_{safe} = 50.33\text{kV}$
8.9 $V_{s1} = 7.17\text{kV}$, $V_{s2} = 3.41\text{kV}$
8.10 $P_{lc} = 40.33\text{W}$, $P_{dl} = 1368.48\text{W}$, $P_{dh} = 1328.15\text{W}$
8.11 $P_{lc} = 0.92\text{W}$, $C = 0.189\mu\text{F}$, $P_{dh} = 1199.08\text{W}$

8.12 $C = 0.680\mu$ F, $I_c = 2.39$A
8.13 $C = 0.272\mu$ F, $R = 2.22$cm
8.14 $C_{cs} = 0.345\mu$ F, $C_{cc} = 0.145\mu$ F, $C_n = 0.78\mu$ F, $I_c = 0.062$A
8.15 $C_{cs} = 1.45\mu$ F, $C_{cc} = 0.515\mu$ F, $C_n = 3\mu$ F

Chapter 9
Practice Problems

9.1 $P_e = 26.86$MW, $I_L = 3169.82 \angle{-42.15^\circ}$A
9.2 $H = 8$MJ/MVA, $P_e = 2$MW, $\frac{d^2\delta}{dt^2} = 3.92\text{rad/s}^2$
9.3 $S_p = 1.643$, $\omega_n = 5.03\text{rad/s}$, $f_n = 0.8$Hz, $T = 1.25$s
9.4 6.7MJ/MVA, 670MJ, $\frac{d^2\delta}{dt^2} = 339.84\text{rad/s}^2$
9.5 $106.7^\circ = 1.86$rad, 0.46s, $64.53^\circ = 1.12$rad, 0.16s

Exercise Problems

9.1 $\delta = 78.74^\circ$, $I_L = 4460.29 \angle{32.18^\circ}$A, $Q_{line} = 59.68$Mvar, $Q_{3-\phi} = 89.73$Mvar
9.2 138.43 MW, $I_L = 4460.291482.04 \angle{-157.52^\circ}$A
9.2 $X_c = 14.99\Omega$
9.3 640 MJ, 8 MW, 216 elec.deg/s^2
9.4 7 MJ/MVA, 210.52 elec.deg/s^2
9.5 0.66, $I_L = 0.66 \angle{-18.19^\circ}$, $E_g = 0.95 \angle{17.67^\circ}$, 1.96, 7.84 rad/s
9.6 2.53, 8.91 rad/s, 0.24 s
9.7 1833.33 MW
9.8 14.5 MJ/MVA, 1450 MJ, 64.29 elec.deg/s^2
9.9 21.8 MJ/MVA, 2180 MJ, 375 elec.deg/s^2
9.10 9.49°, 121.52°, 0.48 s, 105.11°, 0.18 s
9.11 95.42°, 0.34 s
9.12 Use the software to solve
9.13 Use the software to solve

Chapter 10
Practice Problems

10.1 $p(t) = 2367.25 - 6740.71\cos(2\omega t - 69.44^\circ)$W,
$p_a = 4734.51(1 - \cos 2\omega t)$W, $q_a = -12622.71\sin 2\omega t$Var, $P = 4734.51$W,
$Q = 12622.71$Var, $P_{av} = 4766.4$W
10.2 187 W, 115.89 Var, 4.07 A, 627.72 VA
10.3 800 W, 2598.07 W, 1125.83 W, 650 Var
10.4 2054.79 VA
10.5 0.18, 0.09

Exercise Problems

10.1 $p(t) = 994.85 - 2832.81\cos(2\omega t - 69.44^\circ)$W,
$p_a = 944.85(1 - \cos 2\omega t)$W, $q_a = -1875.51\sin 2\omega t$Var, $P = 944.85$W,
$Q = 1875.51$Var, $P_{av} = 351.73$W, 0.45
10.2 913.73 VA, 1045 W, 343.47 Var, 1.2

10.3 8.23 A, 324 W, 156.92 Var, 1161.6 VA
10.4 405 W, 630.75 W, 96.83 W, 131.1 Var
10.5 85.38 V, 26.59 A, 2270.25 VA, 1425 W, 0 Var, 1925.8 W, 565.51 Var
10.6 937.85 VA, 0.84, 1.24, 753.3 VA, 1120.5 VA, 1643.94 VA
10.7 0.12, 0.06

Chapter 11
Practice Problems

11.1 5.46 kV, 6.11 kV, 7.48 kV, 84.67%
11.2 13.98 kV, 13.10 kV, 11.01 kV
11.3 256.58 m, 17.87 m
11.4 10.65 m, 17.54 m, 14.1 m

Exercise Problems

11.1 3.51 kV, 4.03 kV, 5.16 kV, 82.05%
11.2 6.61 kV, 7.67 kV, 9.98 kV, 13.81 kV, 68.9%
11.3 99%
11.4 82.75%
11.5 129.39 m, 216.09 kg
11.6 18 m, 461.4 kg
11.7 38.34 m, 3.14 m
11.8 4424 m, 9.6 m
11.9 12.31 m, 0.93 m, 13.24 m
11.10 39.83 m, 30.25 m

Index

Md. A. Salam, *Fundamentals of Electrical Power Systems Analysis*,
https://doi.org/10.1007/978-981-15-3212-2

MIX
Papier aus verantwortungsvollen Quellen
Paper from responsible sources
FSC® C105338

If you have any concerns about our products,
you can contact us on
ProductSafety@springernature.com

In case Publisher is established outside the EU,
the EU authorized representative is:
Springer Nature Customer Service Center GmbH
Europaplatz 3, 69115 Heidelberg, Germany

Printed by Libri Plureos GmbH
in Hamburg, Germany